Lecture Notes in Computer Science   **9827**

Commenced Publication in 1973
Founding and Former Series Editors:
Gerhard Goos, Juris Hartmanis, and Jan van Leeuwen

More information about this series at http://www.springer.com/series/7409

Sven Hartmann · Hui Ma (Eds.)

Database and Expert Systems Applications

27th International Conference, DEXA 2016
Porto, Portugal, September 5–8, 2016
Proceedings, Part I

 Springer

Editors
Sven Hartmann
Clausthal University of Technology
Clausthal-Zellerfeld
Germany

Hui Ma
Victoria University of Wellington
Wellington
New Zealand

ISSN 0302-9743   ISSN 1611-3349   (electronic)
Lecture Notes in Computer Science
ISBN 978-3-319-44402-4   ISBN 978-3-319-44403-1   (eBook)
DOI 10.1007/978-3-319-44403-1

Library of Congress Control Number: 2016947400

LNCS Sublibrary: SL3 – Information Systems and Applications, incl. Internet/Web, and HCI

Printed on acid-free paper

This Springer imprint is published by Springer Nature
The registered company is Springer International Publishing AG Switzerland

Preface

This volume contains the papers presented at the 27th International Conference on Database and Expert Systems Applications (DEXA 2016), which was held in Porto, Portugal, during September 5–8, 2016. On behalf of the Program Committee, we commend these papers to you and hope you find them useful.

Database, information, and knowledge systems have always been a core subject of computer science. The ever-increasing need to distribute, exchange, and integrate data, information, and knowledge has added further importance to this subject. Advances in the field will help facilitate new avenues of communication, to proliferate interdisciplinary discovery, and to drive innovation and commercial opportunity.

DEXA is an international conference series which showcases state-of-the-art research activities in database, information, and knowledge systems. The conference and its associated workshops provide a premier annual forum to present original research results and to examine advanced applications in the field. The goal is to bring together developers, scientists, and users to extensively discuss requirements, challenges, and solutions in database, information, and knowledge systems.

DEXA 2016 solicited original contributions dealing with any aspect of database, information, and knowledge systems. Suggested topics included but were not limited to:

- Acquisition, Modeling, Management and Processing of Knowledge
- Authenticity, Privacy, Security, and Trust
- Availability, Reliability and Fault Tolerance
- Big Data Management and Analytics
- Consistency, Integrity, Quality of Data
- Constraint Modeling and Processing
- Cloud Computing and Database-as-a-Service
- Database Federation and Integration, Interoperability, Multi-Databases
- Data and Information Networks
- Data and Information Semantics
- Data Integration, Metadata Management, and Interoperability
- Data Structures and Data Management Algorithms
- Database and Information System Architecture and Performance
- Data Streams, and Sensor Data
- Data Warehousing
- Decision Support Systems and Their Applications
- Dependability, Reliability and Fault Tolerance
- Digital Libraries, and Multimedia Databases
- Distributed, Parallel, P2P, Grid, and Cloud Databases
- Graph Databases
- Incomplete and Uncertain Data
- Information Retrieval

- Information and Database Systems and Their Applications
- Mobile, Pervasive, and Ubiquitous Data
- Modeling, Automation and Optimization of Processes
- NoSQL and NewSQL Databases
- Object, Object-Relational, and Deductive Databases
- Provenance of Data and Information
- Semantic Web and Ontologies
- Social Networks, Social Web, Graph, and Personal Information Management
- Statistical and Scientific Databases
- Temporal, Spatial, and High-Dimensional Databases
- Query Processing and Transaction Management
- User Interfaces to Databases and Information Systems
- Visual Data Analytics, Data Mining, and Knowledge Discovery
- WWW and Databases, Web Services
- Workflow Management and Databases
- XML and Semi-structured Data

Following the call for papers, which yielded 137 submissions, there was a rigorous review process that saw each paper reviewed by three to five international experts. The 39 papers judged best by the Program Committee were accepted for long presentation. A further 29 papers were accepted for short presentation.

As is the tradition of DEXA, all accepted papers are published by Springer. Authors of selected papers presented at the conference were invited to submit extended versions of their papers for publication in the Springer journal *Transactions on Large-Scale Data- and Knowledge-Centered Systems (TLDKS)*.

We wish to thank all authors who submitted papers and all conference participants for the fruitful discussions. We are grateful to Bruno Buchberger and Gottfried Vossen, who accepted to present keynote talks at the conference.

The success of DEXA 2016 is a result of the collegial teamwork from many individuals. We like to thank the members of the Program Committee and external reviewers for their timely expertise in carefully reviewing the submissions. We are grateful to our general chairs, Abdelkader Hameurlain, Fernando Lopes, and Roland R. Wagner, to our publication chair, Vladimir Marik, and to our workshop chairs, A Min Tjoa, Zita Vale, and Roland R. Wagner.

We wish to express our deep appreciation to Gabriela Wagner of the DEXA conference organization office. Without her outstanding work and excellent support, this volume would not have seen the light of day.

Finally, we would like to thank GECAD (Research Group on Intelligent Engineering and Computing for Advanced Innovation and Development) at ISEP (Instituto Superior de Engenharia do Porto) for being our hosts for the wonderful days in Porto.

July 2016   Sven Hartmann
Hui Ma

Organization

General Chairs

Abdelkader Hameurlain   IRIT, Paul Sabatier University Toulouse, France
Fernando Lopes   LNEG - National Research Institute, Portugal
Roland R. Wagner   Johannes Kepler University Linz, Austria

Program Committee Chairs

Hui Ma   Victoria University of Wellington, New Zealand
Sven Hartmann   Clausthal University of Technology, Germany

Publication Chair

Vladimir Marik   Czech Technical University, Czech Republic

Program Committee

Afsarmanesh, Hamideh   University of Amsterdam, The Netherlands
Albertoni, Riccardo   Italian National Council of Research, Italy
Anane, Rachid   Coventry University, UK
Appice, Annalisa   Università degli Studi di Bari, Italy
Atay, Mustafa   Winston-Salem State University, USA
Bakiras, Spiridon   Michigan Technological University, USA
Bao, Zhifeng   National University of Singapore, Singapore
Bellatreche, Ladjel   ENSMA, France
Bennani, Nadia   INSA Lyon, France
Benyoucef, Morad   University of Ottawa, Canada
Berrut, Catherine   Grenoble University, France
Biswas, Debmalya   Swisscom, Switzerland
Bouguettaya, Athman   RMIT, Australia
Boussaid, Omar   University of Lyon, France
Bressan, Stephane   National University of Singapore, Singapore
Camarinha-Matos, Luis M.   Universidade Nova de Lisboa, Portugal
Catania, Barbara   DISI, University of Genoa, Italy
Ceci, Michelangelo   University of Bari, Italy
Chen, Cindy   University of Massachusetts Lowell, USA
Chen, Phoebe   La Trobe University, Australia
Chen, Shu-Ching   Florida International University, USA
Chevalier, Max   IRIT - SIG, Université de Toulouse, France
Choi, Byron   Hong Kong Baptist University, Hong Kong, SAR China

Christiansen, Henning	Roskilde University, Denmark
Chun, Soon Ae	City University of New York, USA
Cuzzocrea, Alfredo	University of Trieste, Italy
Dahl, Deborah	Conversational Technologies, USA
Darmont, Jérôme	Université de Lyon (ERIC Lyon 2), France
de vrieze, cecilia	Bournemouth University, UK, Switzerland
Decker, Hendrik	Ludwig-Maximilians-Universität München, Spain
Deng, Zhi-Hong	Peking University, China
Deufemia, Vincenzo	Università degli Studi di Salerno, Italy
Dibie-Barthélemy, Juliette	AgroParisTech, France
Ding, Ying	Indiana University, USA
Dobbie, Gill	University of Auckland, New Zealand
Dou, Dejing	University of Oregon, USA
du Mouza, Cedric	CNAM, France
Eder, Johann	University of Klagenfurt, Austria
El-Beltagy, Samhaa	Nile University, Egypt
Embury, Suzanne	The University of Manchester, UK
Endres, Markus	University of Augsburg, Germany
Fazzinga, Bettina	ICAR-CNR, Italy
Fegaras, Leonidas	The University of Texas at Arlington, USA
Felea, Victor	Al. I. Cuza University of Iasi, Romania
Ferilli, Stefano	University of Bari, Italy
Ferrarotti, Flavio	Software Competence Center Hagenberg, Austria
Fomichov, Vladimir	National Research University Higher School of Economics, Moscow, Russian Federation
Frasincar, Flavius	Erasmus University Rotterdam, The Netherlands
Freudenthaler, Bernhard	Software Competence Center Hagenberg, Austria
Fukuda, Hiroaki	Shibaura Institute of Technology, Japan
Furnell, Steven	Plymouth University, UK
Garfield, Joy	University of Worcester, UK
Gergatsoulis, Manolis	Ionian University, Greece
Grabot, Bernard	LGP-ENIT, France
Grandi, Fabio	University of Bologna, Italy
Gravino, Carmine	University of Salerno, Italy
Groppe, Sven	Lübeck University, Germany
Grosky, William	University of Michigan, USA
Grzymala-Busse, Jerzy	University of Kansas, USA
Guerra, Francesco	Università degli Studi Di Modena e Reggio Emilia, Italy
Guzzo, Antonella	University of Calabria, Italy
Hameurlain, Abdelkader	Paul Sabatier University, France
Hamidah, Ibrahim	Universiti Putra Malaysia, Malaysia
Hara, Takahiro	Osaka University, Japan
Hartmann, Sven	TU Clausthal, Germany
Hsu, Wynne	National University of Singapore, Singapore
Hua, Yu	Huazhong University of Science and Technology, China
Huang, Jimmy	York University, Canada

Huptych, Michal	Czech Technical University in Prague, Czech Republic
Hwang, San-Yih	National Sun Yat-Sen University, Taiwan
Härder, Theo	TU Kaiserslautern, Germany
Iacob, Ionut Emil	Georgia Southern University, USA
Ilarri, Sergio	University of Zaragoza, Spain
Imine, Abdessamad	Inria Grand Nancy, France
Ishihara, Yasunori	Osaka University, Japan
Jin, Peiquan	University of Science and Technology of China, China
Kao, Anne	Boeing, USA
Karagiannis, Dimitris	University of Vienna, Austria
Katzenbeisser, Stefan	Technische Universität Darmstadt, Germany
Kim, Sang-Wook	Hanyang University, Republic of Korea
Kleiner, Carsten	University of Applied Sciences and Arts Hannover, Germany
Koehler, Henning	Massey University, New Zealand
Kosch, Harald	University of Passau, Germany
Krátký, Michal	Technical University of Ostrava, Czech Republic
Kremen, Petr	Czech Technical University in Prague, Czech Republic
Küng, Josef	University of Linz, Austria
Lammari, Nadira	CNAM, France
Lamperti, Gianfranco	University of Brescia, Italy
Laurent, Anne	LIRMM, University of Montpellier 2, France
Léger, Alain	FT R&D Orange Labs Rennes, France
Lhotska, Lenka	Czech Technical University, Czech Republic
Liang, Wenxin	Dalian University of Technology, China
Ling, Tok Wang	National University of Singapore, Singapore
Link, Sebastian	The University of Auckland, New Zealand
Liu, Chuan-Ming	National Taipei University of Technology, Taiwan
Liu, Hong-Cheu	University of South Australia, Australia
Liu, Rui	HP Enterprise, USA
Lloret Gazo, Jorge	University of Zaragoza, Spain
Loucopoulos, Peri	Harokopio University of Athens, Greece
Lumini, Alessandra	University of Bologna, Italy
Ma, Hui	Victoria University of Wellington, New Zealand
Ma, Qiang	Kyoto University, Japan
Maag, Stephane	TELECOM SudParis, France
Masciari, Elio	ICAR-CNR, Università della Calabria, Italy
May, Norman	SAP SE, Germany
Medjahed, Brahim	University of Michigan - Dearborn, USA
Mishra, Harekrishna	Institute of Rural Management Anand, India
Moench, Lars	University of Hagen, Germany
Mokadem, Riad	IRIT, Paul Sabatier University, France
Moon, Yang-Sae	Kangwon National University, Republic of Korea
Morvan, Franck	IRIT, Paul Sabatier University, France
Munoz-Escoi, Francesc	Universitat Politecnica de Valencia, Spain
Navas-Delgado, Ismael	University of Málaga, Spain

Ng, Wilfred	Hong Kong University of Science and Technology, Hong Kong, SAR China
Ozsoyoglu, Gultekin	Case Western Reserve University, USA
Pallis, George	University of Cyprus, Cyprus
Paprzycki, Marcin	Polish Academy of Sciences, Warsaw Management Academy, Poland
Pastor Lopez, Oscar	Universidad Politecnica de Valencia, Spain
Pivert, Olivier	Ecole Nationale Supérieure des Sciences Appliquées et de Technologie, France
Pizzuti, Clara	ICAR-CNR, Italy
Poncelet, Pascal	LIRMM, France
Pourabbas, Elaheh	National Research Council, Italy
Qin, Jianbin	University of New South Wales, Australia
Rabitti, Fausto	ISTI, CNR Pisa, Italy
Raibulet, Claudia	Università degli Studi di Milano-Bicocca, Italy
Ramos, Isidro	Technical University of Valencia, Spain
Rao, Praveen	University of Missouri-Kansas City, USA
Resende, Rodolfo F.	Federal University of Minas Gerais, Brazil
Roncancio, Claudia	Grenoble University/LIG, France
Ruckhaus, Edna	Universidad Simon Bolivar, Venezuela
Ruffolo, Massimo	ICAR-CNR, Italy
Sacco, Giovanni Maria	University of Turin, Italy
Saltenis, Simonas	Aalborg University, Denmark
Sansone, Carlo	Università di Napoli Federico II, Italy
Sarda, N.L.	I.I.T. Bombay, India
Savonnet, Marinette	University of Burgundy, France
Sawczuk da Silva, Alexandre	Victoria University of Wellington, New Zealand
Scheuermann, Peter	Northwestern University, USA
Schewe, Klaus-Dieter	Software Competence Center Hagenberg, Austria
Schweighofer, Erich	University of Vienna, Austria
Sedes, Florence	IRIT, Paul Sabatier University, Toulouse, France
Selmaoui, Nazha	University of New Caledonia, New Caledonia
Siarry, Patrick	Université Paris 12 (LiSSi), France
Skaf-Molli, Hala	Nantes University, France
Srinivasan, Bala	Monash University, Australia
Sunderraman, Raj	Georgia State University, USA
Taniar, David	Monash University, Australia
Teisseire, Maguelonne	Irstea - TETIS, France
Tessaris, Sergio	Free University of Bozen-Bolzano, Italy
Teste, Olivier	IRIT, University of Toulouse, France
Teufel, Stephanie	University of Fribourg, Switzerland
Teuhola, Jukka	University of Turku, Finland
Thevenin, Jean-Marc	University of Toulouse 1 Capitole, France
Torra, Vicenc	University of Skövde, Sweden
Truta, Traian Marius	Northern Kentucky University, USA

Tzouramanis, Theodoros	University of the Aegean, Greece
Vaira, Lucia	University of Salento, Italy
Vidyasankar, Krishnamurthy	Memorial University of Newfoundland, Canada
Vieira, Marco	University of Coimbra, Portugal
Wang, Guangtao	NTU, Singapore
Wang, Junhu	Griffith University, Australia
Wang, Qing	The Australian National University, Australia
Wang, Wendy Hui	Stevens Institute of Technology, USA
Wijsen, Jef	Université de Mons, Belgium
Wu, Huayu	Institute for Infocomm Research, A*STAR, Singapore
Yang, Ming Hour	Chung Yuan Christian University, Taiwan
Yang, Xiaochun	Northeastern University, China
Yin, Hongzhi	The University of Queensland, Australia
Yokota, Haruo	Tokyo Institute of Technology, Japan
Zhao, Yanchang	RDataMining.com, Australia
Zhu, Qiang	The University of Michigan, USA
Zhu, Yan	Southwest Jiaotong University, China

External Reviewers

Liliana Ibanescu	UMR MIA-Paris, INRA, France
Paola Podestà	Italian National Council of Research, Italy
Luke Lake	Department of Human Services, Australia
Roberto Corizzo	University of Bari, Italy
Pasqua Fabiana Lanotte	University of Bari, Italy
Corrado Loglisci	University of Bari, Italy
Gianvito Pio	University of Bari, Italy
Weiqing Wang	The University of Queensland, Australia
Stephen Carden	Georgia Southern University, USA
Arpita Chatterjee	Georgia Southern University, USA
Tharanga Wickramarachchi	Georgia Southern University, USA
Hastimal Jangid	University of Missouri-Kansas City, USA
Loredana Caruccio	University of Salerno, Italy
Giuseppe Polese	University of Salerno, Italy
Valentina Indelli Pisano	University of Salerno, Italy
Virginie Thion	University of Rennes 1/IRISA, France
Grégory Smits	University of Rennes 1/IRISA, France
Hélène Jaudoin	University of Rennes 1/IRISA, France
Yves Denneulin	Grenoble INP, France
Ermelinda Oro	ICAR-CNR, Italy
Harekrishna Misra	Institute of Rural Management Anand, India
Vijay Ingalalli	LIRMM, France

Gang Qian	University of Central Oklahoma, USA
Lubomir Stanchev	California Polytechnic State University, USA
Xianying (Steven) Liu	IBM Almaden Research Center, USA
Alok Watve	Broadway Technology, USA
Xin Shuai	Thomson Reuters, USA
María del Carmen Rodríguez-Hernández	University of Zaragoza, Spain
Óscar Urra	University of Zaragoza, Spain
Samira Pouyanfar	Florida International University, USA
Hsin-Yu Ha	Florida International University, USA
Miroslav Blaško	Czech Technical University in Prague, Czech Republic
Bogdan Kostov	Czech Technical University in Prague, Czech Republic
Yosuke Watanabe	Nagoya University, Japan
Atsushi Keyaki	Tokyo Institute of Technology, Japan
Miika Hannula	The University of Auckland, New Zealand
Dominik Bork	University of Vienna, Austria
Michael Walch	University of Vienna, Austria
Nikolaos Tantouris	University of Vienna, Austria
Jingjie Ni	Hewlett-Packard Enterprise Company, USA
Prajwol Sangat	Monash University, Australia
Xiaotian Hao	HKUST, Hong Kong, SAR China
Ji Cheng	HKUST, Hong Kong, SAR China
Yiling Dai	Kyoto University, Japan
Arnaud Castelltort	University of Montpellier, France
Sabin Kafle	University of Oregon, USA
Shih-Wen George Ke	Chung Yuan Christian University, Taiwan
Yi-Hung Wu	Chung Yuan Christian University, Taiwan
Jorge Martinez-Gil	Software Competence Center Hagenberg, Austria
Loredana Tec	Software Competence Center Hagenberg, Austria
Senen Gonzalez	University of Chile, Chile
Nicolas Travers	CNAM, France
Fayçal Hamdi	CNAM, France
Camelia Constantin	University of Pierre et Marie Curie - Paris 6, France
Daichi Amagata	Osaka University, Japan
Masumi Shirakawa	Osaka University, Japan
Eleftherios Kalogeros	Ionian University, Greece
Stéphane Jean	LIAS/ISAE-ENSMA, France
Selma Khouri	LIAS/ISAE-ENSMA, France
Soumia Benkrid	ESI, Algiers, Algeria
Andrea Esuli	ISTI-CNR, Italy
Giuseppe Amato	ISTI-CNR, Italy
Imen Megdiche	IRIT, France
Fotini Michailidou	University of the Aegean, Greece
Christos Kalyvas	University of the Aegean, Greece

Eirini Molla	University of the Aegean, Greece
Sajib Mistry	RMIT University, Australia
Tooba Aamir	RMIT University, Australia
Azadeh Ghari Neiat	RMIT University, Australia
Rahma Jlassi	RMIT University, Australia

Keynotes

From Natural Language to Automated Reasoning

Bruno Buchberger

We outline the possible interaction between knowledge mining, natural language processing, sentiment analysis, data base systems, ontology technology, algorithm synthesis, and automated reasoning for enhancing the sophistication of web-based knowledge processing.

We focus, in particular, on the transition from parsed natural language texts to formal texts in the frame of logical systems and the potential impact of automating this transition on methods for finding hidden knowledge in big (or small) data and the automated composition of algorithms (cooperation plans for networks of application software).

Simple cooperation apps like IFTTT and the new version of SIRI demonstrate the power of (automatically) combining clusters of existing applications under the control of expressions of desires in natural language.

In the Theorema Working Group of the speaker quite powerful algorithm synthesis methods have been developed that can generate algorithms for relatively difficult mathematical problems. These methods are based on automated reasoning and start from formal problem specifications in the frame of predicate logic. We ask ourselves how the deep reasoning used in mathematical algorithm synthesis could be combined with recent advances in natural language processing for reaching a new level of intelligence in the communication between humans and the web for every-day and business applications.

The talk is expository and tries to draw a big picture of how we could and should proceed in this area but will also explain some technical details and demonstrate some surprising results in the formal reasoning aspect of the overall approach.

The Price of Data

Gottfried Vossen[1,2]

[1] ERCIS, University of Münster, Münster, Germany
vossen@wi.uni-muenster.de
[2] The University of Waikato Management School, Hamilton, New Zealand
vossen@waikato.ac.nz

Abstract. As data is becoming a commodity similar to electricity, as individuals become more and more transparent thanks to the comprehensive data traces they leave, and as data gets increasingly connected across company boundaries, the question arises of whether a price tag should be attached to data and, if so, what it should say. In this talk, the price of data is studied from a variety of angles and applications areas, including telecommunication, social networks, advertising, and automation; the issues discussed include aspects such as fair pricing, data quality, data ownership, and ethics. Special attention is paid to data market-places, where nowadays everybody can trade data, although the currency in which buyers are requested to pay may no longer be what they expect.

The term "Big Data" will always be remembered as *the* big buzzword of 2013 and, somewhat surprisingly, of several years thereafter. According to Bernard Marr[1], "the basic idea behind the phrase 'Big Data' is that everything we do is increasingly leaving a digital trace (or data), which we (and others) can use and analyze. Big Data therefore refers to that data being collected and our ability to make use of it." In earlier times, it was not unusual to leave analog traces, like purchase receipts from the grocery store, and neither was the idea to somehow monetize these traces. The owner of the grocery store would know his regular customers, and would try to keep old ones and attract new ones by offering them discount coupons or other incentives. With digital traces, business along such lines has exploded, become possible at a world-wide scale, and has reached nuances of everyday life that nobody would ever have thought of. So it is time to ask whether that data comes with a price tag and, if so, what it says.

This talk looks at the price of data from a variety of angles and application areas for which pricing is relevant. In telecommunication, for example, prices for making phone calls as well as for data (e.g., surfing the Web) have come down enormously over the last 20 years, due to increasingly cheaper technology as well as more and more competition. Search engines have made it popular to make money through advertising, where participants bid on keywords that may occur in search queries, and social networks generate revenue from letting companies have access to their user profiles and all the data that these contain. So what is the value of a user profile?

---

[1] http://www.datasciencecentral.com/profile/BernardMarr.

Data marketplaces [2, 4, 5, 9], on the other hand, are an emerging species of digital platform that revisits traditional marketplaces and their mechanisms. In a data marketplace, producers of data provide query answers to consumers in exchange for payment. In general, a data marketplace integrates public Web data with other data sources, and it allows for data extraction, data transformation and data loading, and it comprises meta data repositories describing data and algorithms. In addition, it consists of technology for 'uploading' and optimizing operators with user-defined-functionality, as well as trading and billing components. In return, the 'vendor' of this functionality receives a monetary contribution from a buyer. Essentially, everybody can trade data nowadays, and the roles of sellers and buyers may be swapped over time and be exchangeable. For a seller, the interesting issue is the question of how valuable some data may be for a customer (or what the competition is charging for the same or similar data); if that could be figured out, the seller could adapt the price he is asking accordingly.

From a more technical perspective, the pricing problem can be tackled from the point of view of data quality, and here it is possible to establish a notion of *fair pricing*. [6, 8] cast this problem into a universal-relation setting and study the impact of quantifiable data quality; they follow [1] who argue that relational *views* can be interpreted as versions of the 'information good' data and hence study the issue of pricing for competing data sources that provide essentially the same data but in different quality.

Fair pricing has been addressed in depth by [7], by demonstrating how the quality of relational data products can be adapted to match a buyer's willingness to pay by employing a *Name Your Own Price* (NYOP) model. Under that model, data providers can discriminate customers so that they realize the maximum price a customer is willing to pay, and data customers receive a product that is tailored to their own data quality needs and budgets. To balance customer preferences and vendor interests, a model is developed which translates fair pricing into a Multiple-Choice Knapsack optimization problem, thereby making it amenable to an algorithmic solution. The concept of trading data quality for a discount was previously suggested in [10, 11] and applied to both relational as well as XML data.

A final aspect to be mentioned in this context is that of data used in automation. Following [3], automation has become pervasive in recent years and has lead to the danger that people lose their specific abilities when supported or even replaced by machines, robots, or generally automated devices. Carr explains this, for example, with auto-pilots in airplanes: Often pilots are so reliant on an auto-pilot that they do not want to accept the fact the a decision the device has just made is wrong, and he gives examples where this has ended in disaster more than once. Hence the danger is that we overestimate the truth in data, that we trust it too much, so that, as a consequence, the quest for its price becomes obsolete.

References

[1] Balazinska, M., et al.: A discussion on pricing relational data. In: Tannen, V., et al. (eds) In Search of Elegance in the Theory and Practice of Computation. LNCS, vol. 8000, pp. 167–173. Springer, Heidelberg (2013)

[2] Balazinska, M., et al.: Data markets in the cloud: an opportunity for the database community. In: PVLDB 4.12, pp. 1482–1485 (2011)

[3] Carr, N.: The Glass Cage — Automation and Us. W.W. Norton & Company (2014)

[4] Muschalle, A., et al.: Pricing approaches for data markets. In: Proceedings of 6th BIRTE Workshop 2012. Istanbul, Turkey, pp. 129–144

[5] Schomm, F., et al.: Marketplaces for data: an initial survey. In: SIGMOD Record 42.1, pp. 15–26 (2013). http://doi.acm.org/10.1145/2481528.2481532

[6] Stahl, F., et al.: Fair knapsack pricing for data marketplaces. In: Proceedings of 20th East-European Conference on Advances in Databases and Information Systems (ADBIS). LNCS. Springer (2016)

[7] Stahl, F.: High-quality web information provisioning and quality-based data pricing. PhD thesis. University of Münster (2015)

[8] Stahl, F., et al.: Data quality scores for pricing on data marketplaces. In: Proceedings 8th ACIIDS Conference. Da Nang, Vietnam, pp. 214–225 (2016)

[9] Stahl, F., et al.: Data marketplaces: an emerging species. In: Haav, H., et al. (eds.) Databases and Information Systems VIII - Selected Papers from the Eleventh International Baltic Conference, DB&IS 2014, 8–11 June 2014, Tallinn, Estonia. Frontiers in Artificial Intelligence and Applications, vol. 270, pp. 145–158. IOS Press (2014). http://dx.doi.org/10.3233/978-1-61499-458-9-145

[10] Tang, R., et al.: Get a sample for a discount. In: Decker, H., et al. (eds.) Database and Expert Systems Applications. LNCS, vol. 8644, pp. 20–34. Springer International Publishing, Switzerland (2014)

[11] Tang, R., et al.: What you pay for is what you get. In: Decker, H., et al. (eds.) Database and Expert Systems Applications. LNCS, vol. 8056, pp. 395–409. Springer, Berlin (2013)

Contents – Part I

Temporal, Spatial, and High Dimensional Databases

Target-Oriented Keyword Search over Temporal Databases   3
    Xianyan Jia, Wynne Hsu, and Mong Li Lee

General Purpose Index-Based Method for Efficient MaxRS Query   20
    Xiaoling Zhou, Wei Wang, and Jianliang Xu

An Efficient Method for Identifying MaxRS Location in Mobile Ad Hoc
Networks   37
    Yuki Nakayama, Daichi Amagata, and Takahiro Hara

Data Mining

Discovering Periodic-Frequent Patterns in Transactional Databases
Using All-Confidence and Periodic-All-Confidence   55
    J.N. Venkatesh, R. Uday Kiran, P. Krishna Reddy,
    and Masaru Kitsuregawa

More Efficient Algorithms for Mining High-Utility Itemsets with Multiple
Minimum Utility Thresholds.   71
    Wensheng Gan, Jerry Chun-Wei Lin, Philippe Fournier-Viger,
    and Han-Chieh Chao

Mining Minimal High-Utility Itemsets   88
    Philippe Fournier-Viger, Jerry Chun-Wei Lin, Cheng-Wei Wu,
    Vincent S. Tseng, and Usef Faghihi

Authenticity, Privacy, Security, and Trust

Automated k-Anonymization and l-Diversity for Shared Data Privacy   105
    Anne V.D.M. Kayem, C.T. Vester, and Christoph Meinel

Context-Based Risk-Adaptive Security Model and Conflict Management   121
    Mahsa Teimourikia, Guido Marilli, and Mariagrazia Fugini

Modeling Information Diffusion via Reputation Estimation.   136
    Bao-Thien Hoang, Kamel Chelghoum, and Imed Kacem

Data Clustering

Mining Arbitrary Shaped Clusters and Outputting a High Quality
Dendrogram... 153
   Hao Huang, Song Wang, Shuangke Wu, Yunjun Gao, Wei Lu,
   Qinming He, and Shi Ying

Hierarchically Clustered LSH for Hierarchical Outliers Detection 169
   Konstantinos Georgoulas and Yannis Kotidis

Incorporating Clustering into Set Similarity Join Algorithms: The *SjClust*
Framework.. 185
   Leonardo Andrade Ribeiro, Alfredo Cuzzocrea,
   Karen Aline Alves Bezerra, and Ben Hur Bahia do Nascimento

Distributed and Big Data Processing

"Overloaded!" — A Model-Based Approach to Database Stress Testing 207
   Jorge Augusto Meira, Eduardo Cunha de Almeida, Dongsun Kim,
   Edson Ramiro Lucas Filho, and Yves Le Traon

A Cost Model for DBaaS Storage................................ 223
   Djillali Boukhelef, Jalil Boukhobza, and Kamel Boukhalfa

A Query Processing Framework for Array-Based Computations 240
   Leonidas Fegaras

Decision Support Systems, and Learning

Creative Expert System: Result of Inference and Machine Learning
Integration 257
   Bartlomiej Sniezynski, Grzegorz Legien, Dorota Wilk-Kołodziejczyk,
   Stanislawa Kluska-Nawarecka, Edward Nawarecki,
   and Krzysztof Jaśkowiec

A Reverse Nearest Neighbor Based Active Semi-supervised Learning
Method for Multivariate Time Series Classification 272
   Yifei Li, Guoliang He, Xuewen Xia, and Yuanxiang Li

Leveraging Structural Hierarchy for Scalable Network Comparison 287
   Rakhi Saxena, Sharanjit Kaur, Debasis Dash, and Vasudha Bhatnagar

Data Streams

Incremental Stream Processing of Nested-Relational Queries 305
   Leonidas Fegaras

Incremental Continuous Query Processing over Streams and Relations with
Isolation Guarantees   321
  *Salman Ahmed Shaikh, Dong Chao, Kazuya Nishimura,
  and Hiroyuki Kitagawa*

An Improved Method of Keyword Search over Relational Data Streams
by Aggressive Candidate Network Consolidation   336
  Savong Bou, Toshiyuki Amagasa, and Hiroyuki Kitagawa

Data Integration, and Interoperability

Evolutionary Database Design: Enhancing Data Abstraction Through
Database Modularization to Achieve Graceful Schema Evolution   355
  *Gustavo Bartz Guedes, Gisele Busichia Baioco,
  and Regina Lúcia de Oliveira Moraes*

Summary Generation for Temporal Extractions   370
  *Yafang Wang, Zhaochun Ren, Martin Theobald, Maximilian Dylla,
  and Gerard de Melo*

SuMGra: Querying Multigraphs via Efficient Indexing   387
  Vijay Ingalalli, Dino Ienco, and Pascal Poncelet

Semantic Web, and Data Semantics

Re-constructing Hidden Semantic Data Models by Querying SPARQL
Endpoints   405
  *María Jesús García-Godoy, Esteban López-Camacho,
  Ismael Navas-Delgado, and José F. Aldana-Montes*

A New Formal Approach to Semantic Parsing of Instructions and to File
Manager Design   416
  Alexander A. Razorenov and Vladimir A. Fomichov

Ontology-Based Deep Restricted Boltzmann Machine   431
  Hao Wang, Dejing Dou, and Daniel Lowd

Author Index   447

Contents – Part II

Social Networks, and Network Analysis

A Preference-Driven Database Approach to Reciprocal User
Recommendations in Online Social Networks.   3
    Florian Wenzel and Werner Kießling

Community Detection in Multi-relational Bibliographic Networks.   11
    Soumaya Guesmi, Chiraz Trabelsi, and Chiraz Latiri

Quality Prediction in Collaborative Platforms: A Generic Approach
by Heterogeneous Graphs.   19
    Baptiste de La Robertie, Yoann Pitarch, and Olivier Teste

Analyzing Relationships of Listed Companies with Stock Prices and News
Articles   27
    Satoshi Baba and Qiang Ma

Linked Data

Approximate Semantic Matching over Linked Data Streams.   37
    Yongrui Qin, Lina Yao, and Quan Z. Sheng

A Mapping-Based Method to Query MongoDB Documents with SPARQL. . .   52
    Franck Michel, Catherine Faron-Zucker, and Johan Montagnat

Incremental Maintenance of Materialized SPARQL-Based Linkset Views. . . .   68
    Elisa S. Menendez, Marco A. Casanova, Vânia M.P. Vidal,
    Bernardo P. Nunes, Giseli Rabello Lopes, and Luiz A.P. Paes Leme

Data Analysis

Aggregate Reverse Rank Queries   87
    Yuyang Dong, Hanxiong Chen, Kazutaka Furuse,
    and Hiroyuki Kitagawa

Abstract-Concrete Relationship Analysis of News Events Based on a 5W
Representation Model 102
    Shintaro Horie, Keisuke Kiritoshi, and Qiang Ma

Detecting Maximum Inclusion Dependencies without Candidate Generation . . . 118
    Nuhad Shaabani and Christoph Meinel

NoSQL, NewSQL

Footprint Reduction and Uniqueness Enforcement with Hash Indices
in SAP HANA   137
  Martin Faust, Martin Boissier, Marvin Keller, David Schwalb,
  Holger Bischoff, Katrin Eisenreich, Franz Färber, and Hasso Plattner

Benchmarking Replication in Cassandra and MongoDB NoSQL Datastores . . .   152
  Gerard Haughian, Rasha Osman, and William J. Knottenbelt

τJSchema: A Framework for Managing Temporal JSON-Based NoSQL
Databases.   167
  Safa Brahmia, Zouhaier Brahmia, Fabio Grandi, and Rafik Bouaziz

Multimedia Data

Enhancing Similarity Search Throughput by Dynamic Query Reordering   185
  Filip Nalepa, Michal Batko, and Pavel Zezula

Creating a Music Recommendation and Streaming Application for Android . . .   201
  Elliot Jenkins and Yanyan Yang

A Score Fusion Method Using a Mixture Copula   216
  Takuya Komatsuda, Atsushi Keyaki, and Jun Miyazaki

Personal Information Management

Axiomatic Term-Based Personalized Query Expansion Using Bookmarking
System   235
  Philippe Mulhem, Nawal Ould Amer, and Mathias Géry

A Relevance-Focused Search Application for Personalised Ranking Model. . .   244
  Al Sharji Safiya, Martin Beer, and Elizabeth Uruchurtu

Aggregated Search over Personal Process Description Graph   254
  Jing Ouyang Hsu, Hye-young Paik, Liming Zhan, and Anne H.H. Ngu

Inferring Lurkers' Gender by Their Interest Tags   263
  Peisong Zhu, Tieyun Qian, Zhenni You, and Xuhui Li

Semantic Web and Ontologies

Data Access Based on Faceted Queries over Ontologies.   275
  Tadeusz Pankowski and Grażyna Brzykcy

Incremental and Directed Rule-Based Inference on RDFS   287
Jules Chevalier, Julien Subercaze, Christophe Gravier,
and Frédérique Laforest

Top-*k* Matching Queries for Filter-Based Profile Matching in Knowledge
Bases.   295
Alejandra Lorena Paoletti, Jorge Martinez-Gil,
and Klaus-Dieter Schewe

FETA: Federated QuEry TrAcking for Linked Data   303
Georges Nassopoulos, Patricia Serrano-Alvarado, Pascal Molli,
and Emmanuel Desmontils

Database and Information System Architectures

Dynamic Power-Aware Disk Storage Management in Database Servers   315
Peyman Behzadnia, Wei Yuan, Bo Zeng, Yi-Cheng Tu,
and Xiaorui Wang

FR-Index: A Multi-dimensional Indexing Framework for Switch-Centric
Data Centers.   326
Yatao Zhang, Jialiang Cao, Xiaofeng Gao, and Guihai Chen

Unsupervised Learning for Detecting Refactoring Opportunities
in Service-Oriented Applications.   335
Guillermo Rodríguez, Álvaro Soria, Alfredo Teyseyre, Luis Berdun,
and Marcelo Campo

A Survey on Visual Query Systems in the Web Era   343
Jorge Lloret-Gazo

Query Answering and Optimization

Query Similarity for Approximate Query Answering   355
Verena Kantere

Generalized Maximal Consistent Answers in P2P Deductive Databases   368
Luciano Caroprese and Ester Zumpano

Computing Range Skyline Query on *Uncertain Dimension*.   377
Nurul Husna Mohd Saad, Hamidah Ibrahim, Fatimah Sidi,
Razali Yaakob, and Ali Amer Alwan

Aging Locality Awareness in Cost Estimation for Database Query
Optimization.   389
Chihiro Kato, Yuto Hayamizu, Kazuo Goda, and Masaru Kitsuregawa

XXVIII   Contents – Part II

Information Retrieval, and Keyword Search

Constructing Data Graphs for Keyword Search 399
  Konstantin Golenberg and Yehoshua Sagiv

Generating Pseudo Search History Data in the Absence of Real Search
History 410
  Ashraf Bah and Ben Carterette

Variable-Chromosome-Length Genetic Algorithm for Time Series
Discretization 418
  Muhammad Marwan Muhammad Fuad

Approximate Temporal Aggregation with Nearby Coalescing 426
  Kai Cheng

Data Modelling, and Uncertainty

A Data Model for Determining Weather's Impact on Travel Time 437
  Ove Andersen and Kristian Torp

Simplify the Design of XML Schemas by Type Dependencies 445
  Jia Liu and Husheng Liao

An Efficient Initialization Method for Probabilistic Relational Databases 454
  Hong Zhu, Caicai Zhang, and Zhongsheng Cao

Author Index 463

Temporal, Spatial, and High Dimensional Databases

Target-Oriented Keyword Search over Temporal Databases

Xianyan Jia$^{(\boxtimes)}$, Wynne Hsu, and Mong Li Lee

School of Computing, National University of Singapore, Singapore, Singapore
{jiaxiany,whsu,leeml}@comp.nus.edu.sg

Abstract. Keyword search in relational databases has gained popularity due to its ease of use. However, existing methods do not handle keyword search in temporal databases. In this paper, we extend keyword queries to allow temporal information to be associated with keywords, as well as support temporal relationships between two keywords. We design a target-oriented search over an augmented data graph to efficiently evaluate such temporal keyword queries. Experiments on 3 datasets demonstrate the efficiency of the proposed approach to answer complex temporal keyword queries.

1 Introduction

Temporal data has become prevalent in many applications such as finance, business, bank, and health care. While SQL:2011 provides the efficient querying of data on their temporal characteristics, it requires users to write complicated SQL queries [13]. Keyword queries provide a simple and user-friendly query interface to access relational databases [1,8,12]. However, existing relational keyword search techniques assume that keywords are not associated to time constraints and there is no relationship among keywords in the queries.

Figure 1 shows a relational database with two snapshot relations (`Patient` and `Doctor`) and two temporal relations (`Visit` and `Symptom`). The `Visit` relation records the date at which a patient sees a doctor, while the `Symptom` relation gives the start and end dates where a patient experiences various symptoms. For example, the first two tuples (id s_1 and s_2) in the `Symptom` relation depict that a patient p_1 complained of fever and headache in the same consultation visit. These two different symptoms occurred over different periods of time. On the other hand, tuples with id s_5 and s_7 show that the same patient p_2 visited the doctor on different occasions for his cough.

If a user wants to find *patients who have fever on 1 January 2015* in this database, s/he can issue a keyword query such as {Patient, fever, 01/01/2015}. However, this query will return additional answers such as patient p_2 who is born on 1 January 2015 but has fever on 9 March 2015. In order to retrieve answers that match the user's intention, we need to associate the time information to the appropriate keywords. Here, we use square brackets to indicate this association. Hence, the query {Patient, fever[01/01/2015]} refers to the patients who

© Springer International Publishing Switzerland 2016
S. Hartmann and H. Ma (Eds.): DEXA 2016, Part I, LNCS 9827, pp. 3–19, 2016.
DOI: 10.1007/978-3-319-44403-1_1

Patient

pid	YOB	Gender	Name	Ethnicity
p_1	02/03/1982	F	Anna	Indian
p_2	01/01/2015	M	Andy	Chinese
p_3	09/01/1986	F	John	Eurasian

Doctor

did	Name	Gender
d_1	Anna	F
d_2	Ben	M

Visit

vid	pid	did	date
v_1	p_1	d_1	04/01/2015
v_2	p_1	d_1	10/02/2015
v_3	p_2	d_1	11/03/2015
v_4	p_2	d_2	04/04/2015
v_5	p_3	d_2	17/07/2015

Symptom

sid	vid	Name	start	end
s_1	v_1	fever	01/01/2015	03/01/2015
s_2	v_1	headache	02/01/2015	03/01/2015
s_3	v_2	cough	03/02/2015	10/02/2015
s_4	v_2	headache	04/02/2015	06/02/2015
s_5	v_3	cough	08/03/2015	09/03/2015
s_6	v_3	fever	09/03/2015	10/03/2015
s_7	v_4	cough	31/03/2015	02/04/2015
s_8	v_4	headache	02/04/2015	04/04/2015
s_9	v_5	cough	10/07/2015	15/07/2015
s_{10}	v_5	fever	10/07/2015	16/07/2015

Fig. 1. Example clinic database

have fever on 1 January 2015 while the query {Patient[01/01/2015], fever} refers to the patients who are born on 1 January 2015 and have fever at some point in time.

We further extend the time information to support queries with intervals. For example, the query {Patient, fever [01/01/2015 – 01/31/2015]} will return patient p_1 who has fever in the month of January 2015. Besides associating a keyword with time information, we also support queries with temporal relationships between keywords. The work in [2] identified 13 temporal relationships between two time intervals including OVERLAP, BEFORE which form the set of reserved words in our temporal keyword queries. For example, query {Patient, fever BEFORE cough} will return patient p_1 who has fever before cough.

A naive approach to handle temporal keyword queries is to use the traditional inverted list to retrieve all tuples containing the keywords and then filter them based on time constraints. However, this leads to many wasted computations to obtain candidate answers which eventually do not satisfy the time constraints. In this work, we address the problem of keyword search in temporal databases by providing support for complex queries with time-associated keywords and pre-defined temporal relationships between keywords. We design a target-oriented search algorithm to evaluate such queries. We augment selected nodes in the data graph with time boundaries to enable time-aware pruning during the search process. We also incorporate overlapping interval partitioning into the keyword inverted lists to filter nodes that do not satisfy the time constraints. Experiment results show that the proposed approach is efficient and effective in pruning invalid answers early. To the best of our knowledge, this is the first attempt to support keyword search over temporal databases.

The rest of the paper is organized as follows. Section 2 summarizes the related works. Section 3 gives the preliminaries. Section 4 presents our proposed solution *ATQ*. Section 5 contains performance study and we conclude in Sect. 6.

2   Related Work

Relational keyword search can be classified into schema graph approach [1,11, 15,18,19] and data graph approach [5,8,9,12,14]. The schema graph approach models the database schema as a directed graph where each node is a relation and edges are key-foreign key reference between two relations. The work in [11] proposes a breadth-first traversal on the schema graph to generate a set of candidate networks and limit the maximum number of joins allowed. [10,15] focus on finding top-k answers and use AND/OR-semantics. [19] returns answers that are semantically meaningful by identifying the query context and interpreting search target.

In the data graph approach, the database is modeled as a graph where nodes represent tuples and edges represents key-foreign key. [5] uses backward expansion search algorithm to find Steiner trees that contain all the keywords. [12] improves the search efficiency of [5] with bidirectional search technique. [8] employs a dynamic programming technique to identify the top-k minimal Steiner trees. [9] uses a bi-level index to quickly compute the shortest distances. All these works do not handle queries with time intervals and they regard dates as standard keywords for matching.

While there are some works [6,16] that provide support for simple temporal keyword queries on XML, their evaluation is based on the hierarchical structure of XML trees and is not applicable to the general graph model of relational data. Further, complex temporal queries involving relationship between two keywords and associating time constraints to different keywords are not considered.

3   Preliminaries

Temporal databases are known to support two time dimensions: the transaction time and the valid time [17]. Here, we focus on the *valid time* where the attribute value holds.

We model a temporal database D as a data graph, $G = (V, E)$, where V is the set of nodes and E is the set of edges. A node corresponds to a tuple in D, and an edge $(u, v) \in E$, represents a key-foreign key constraint between two tuples. Each node is labeled with tuple id. Figure 2(a) shows the data graph of our example database in Fig. 1, while Fig. 2(b) shows the corresponding schema graph where each node is a relation and an edge denotes the key-foreign key constraint between two relations.

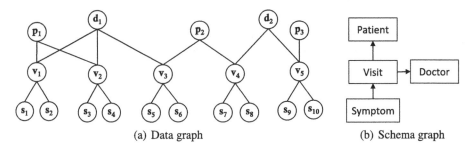

Fig. 2. Data graph and schema graph of clinic database in Fig. 1

We represent a temporal keyword query as $\{head : body\}$ where

1. *head* is a set of keywords indicating the search target. The search target is the user's search intention when issuing a query. Here, we give users the option to explicitly indicate his search target in the head of the query. If the user does not specify any search target, we would use existing methods to identify them [3,4,19], and rewrite the query into the above temporal keyword query format where *head* is the search targets identified.
2. *body* is a set of keywords indicating the query condition. Some of these keywords may be constrained by time intervals, and the user may specify temporal relationships among the keywords.

Table 1 gives the syntax of temporal keyword query in Backus-Naur Form (BNF). Based on the grammar, we can formulate a variety of temporal keywords queries as shown in Table 2. Queries C_1 to C_4 are similar to standard keyword queries, except that the search target is explicitly specified at the head of the query to facilitate the efficient retrieval of relevant answers. Queries C_5 to C_9 involve time information and temporal relationships between keywords which are not handled by existing keyword queries.

An *answer* to a temporal keyword query Q over a data graph G is a subgraph which contains nodes that match all the keywords in Q. Figure 3 shows the possible answers to the query C_2 which finds patients who have fever and cough.

Table 1. Syntax of temporal keyword query in BNF

\<query\>	::=	\{\<head\> : \<body\>\}
\<head\>	::=	ϵ | \<search_list\>
\<search_list\>	::=	\<relation\> | \<value\> | \<relation\>, \<search_list\> | \<value\>, \<search_list\>
\<body\>	::=	\<cond\> | \<cond\>, \<body\>
\<cond\>	::=	\<term\> | \<term\>\<temporal_relation\>\<term\>
\<term\>	::=	\<keyword\> | \<time_associated_keyword\>
\<keyword\>	::=	\<relation\> | \<value\>
\<time_associated_keyword\>	::=	keyword [\<time\>] | keyword [\<time\>, \<time\>]
\<temporal_relation\>	::=	BEFORE | AFTER | EQUAL | MEET | MET_BY | START | STARTED_BY | OVERLAP | OVERLAPPED_BY | CONTAIN | DURING | FINISH | FINISHED_BY

Table 2. Temporal keyword queries for clinic database

Query	Meaning
C_1 {Patient : fever }	Find patients who have fever
C_2 {Patient : fever, cough}	Find patients who have fever and cough
C_3 {Patient, male : fever, cough}	Find male patients who have fever and cough
C_4 {Doctor, Patient : fever, cough }	Find doctors and patients pairs with fever and cough
C_5 {Patient : fever BEFORE cough }	Find patients who have fever before cough
C_6 {Patient : fever[1/1/2015, 31/1/2015], cough[1/1/2015, 31/1/2015] }	Find patients who have fever and cough in January 2015
C_7 {Patient : fever[1/1/2015, 31/1/2015] BEFORE cough[1/1/2015, 31/1/2015] }	Find patients who have fever before cough in January 2015
C_8 {Doctor, Patient : Visit[1/1/2015, 31/1/2015] }	Find doctors and patients pairs with consultation visits in January 2015
C_9 {Doctor, Patient : Visit[1/1/2015, 31/1/2015], fever[1/1/2015, 31/1/2015]}	Find doctors and patients pairs with consultation visits for fever in January 2015

Nodes that match the keywords in the query body are highlighted and patients p_1, p_2, and p_3 are retrieved.

Note that the placement of a keyword in the query head or query body may lead to different answers. For example, Fig. 4 shows the possible answers to the query {Patient: male, fever, cough} which include male patients who have fever and cough (Fig. 4(a) and (b)) as well as female patients who have seen male doctors for fever and cough (Fig. 4(c)). However, if the keyword "male" is in the head of the query as in query C_3, the answers will consist of only Fig. 4(a) and (b). This allows user to clearly indicate his search intention.

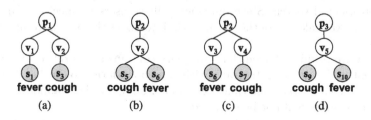

Fig. 3. Candidate answers for query C_2 = {Patient: fever, cough}

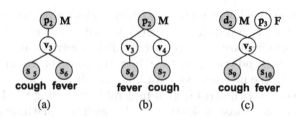

Fig. 4. Possible answers for query {Patient: male, fever, cough}

We parse a temporal keyword query into the following 3 sets:

a. K_{head} is a set of $<k,t>$ pairs where k is a keyword that occurs in the query head and t is the time information associated with k.
b. K_{body} is a set of $<k,t>$ pairs where k is a keyword that occurs in query body and t is the time information associated with k.
c. TR is a set of (p_1, tr, p_2) where $p_1 \in K_{body}$ and $p_2 \in K_{body}$ and tr is the temporal relationship between p_1 and p_2.

Consider query C_5. We have $K_{head} = \{< \texttt{Patient}, \_ >\}$, $K_{body} = \{< \texttt{fever}, \_ >,$ $< \texttt{cough}, \_ >\}$ and $TR = \{(< \texttt{fever}, \_ >, \texttt{BEFORE}, < \texttt{cough}, \_ >)\}$. For query C_6, we have $K_{head} = \{< \texttt{Patient}, \_ >\}$, $K_{body} = \{< \texttt{fever}, [\texttt{1/1/2015}, \texttt{31/1/2015}] >,$ $< \texttt{cough}, [\texttt{1/1/2015}, \texttt{31/1/2015}] >\}$ and $TR = \emptyset$. These information will be utilized in the proposed target-oriented search algorithm described in the next section.

4   Proposed Solution

We design a target-oriented search algorithm to answer keyword queries over a temporal database modelled as a data graph. Existing data graph keyword search techniques such as BANKS [5] and Bidirectional [12] regard time constraints as keywords to be matched and will return answers that may not satisfy users' search intention. A naive approach to process temporal keyword queries is to extend these methods by first ignoring the time constraints to retrieve all the possible matches and then using the time constraints to filter out invalid answers. This is computationally inefficient.

The proposed algorithm, called ATQ, utilizes the following two strategies to prune the search space:

1. Target-oriented search. Since our query allows users to specify their search intention, we make use of the schema graph to direct the search to the relevant nodes.
2. Time-aware pruning. Given that our query contains temporal constraints, we augment nodes in the data graph with time boundaries to quickly determine if a subtree can satisfy the time constraints. Subtrees that cannot satisfy the time constraints will not be explored.

4.1   Target-Oriented Search

The ATQ algorithm begins by finding matching nodes for the keywords in K_{head} and K_{body}. Since our keywords may be associated with time information, it is not efficient to use the standard keyword inverted list to retrieve all the tuples that contain the keyword, and then filter them based on time constraints. Instead, we group the tuples in the inverted list according to their relations, and index these tuples with the state-of-the-art Overlap Interval Partitioning [7]. This allows us to quickly retrieve only those tuples that overlap with the time interval associated with the keyword.

Having found these matching nodes, we construct answers to the query by connecting them. The work in [5] uses Dijkstra's algorithm to find the connecting paths between all pairs of matching nodes. This leads to overwhelming number of answers, many of which are complex and do not satisfy the user's search intention. The Occam's razor principle states that the simplest answer is always favored and this translates to the shortest path that connects the matching nodes. Here, we utilize the schema graph to find the shortest path between the relations corresponding to the matching nodes.

Figure 2(b) shows the schema graph of the clinic database in Fig. 1. Each node is a relation and an edge denotes the key-foreign key constraint between two relations. For example, in query $C_5 = \{$Patient: fever BEFORE cough$\}$, the keyword *Patient* in K_{head} corresponds to the Patient relation, while the keywords fever and cough in K_{body} correspond to the Symptom relation. Based on the schema graph, the shortest path between these relations is via the Visit relation. As such, when we traverse the data graph to construct query answers, we do not need to visit nodes that correspond to the Doctor relation as they are not part of the shortest path.

With this, our target-oriented search comprises of two phases. The first phase aims to construct a partial answer by starting from a node that matches a keyword in K_{body} to find a connected component involving nodes that match all the keywords in K_{head}. The second phase completes the search process by finding nodes that match the remaining keywords in K_{body} as well as satisfy the temporal constraints, if any.

Consider the query C_5 and the data graph in Fig. 2(a). We start with s_1, a matching node for the keyword fever, and visit the node v_1, followed by p_1. Note that we do not need to visit d_1 as it corresponds to the Doctor relation which does not lie on the shortest path from Symptom to Patient (see Fig. 2(b)). At this point, we have found a partial answer, that is, patient p_1 with fever. Next, we complete the search by checking if p_1 has a cough which occurs after fever. We traverse the data graph from p_1 to the $Visit$ nodes v_1 and v_2. The node v_1 does not have any neighbour nodes that match the keyword cough, whereas v_2 has the matching node s_3. Comparing the time intervals of s_1 and s_3, since they satisfy the temporal relationship BEFORE, we return this subtree $(s_1 - v_1 - p_1 - v_2 - s_3)$ as an answer to the query.

4.2 Time-Aware Pruning

In general, a node may have a large number of neighbours. Here, we want to use the temporal constraints in a query to prune subtrees that will not contribute to the query answer. We allow nodes in the data graph to be augmented with time boundaries. In selecting which relations whose nodes need to be augmented with time boundaries, we focus on relations which have a key-foreign key constraint. Given two such relations R_1 and R_2 where R_2 contains the foreign key, we estimate the pruning power obtained by augmenting the nodes of R_2 as $|R_2|/|R_1|$. For our example clinic application, suppose the Patient relation has 100 tuples and the Visit relation has 5000 tuples, then it will be useful to augment Visit nodes with time boundaries to direct the search since each patient will have an average of 50 visits.

Let u be a node in the data graph, S_u be the set of nodes in the subtree rooted at u, and $S_u[R]$ be the set of nodes in S_u that belong to the relation R. Suppose $min(S_u[R])$ and $max(S_u[R])$ are the earliest and latest time of the nodes in $S_u[R]$. Then we associate u with the triplet $< R, min(S_u[R]), max(S_u[R]) >$ to indicate the time boundary of a subset of nodes for R. We use this information to eliminate subtrees whose time boundaries are outside the query's time constraints.

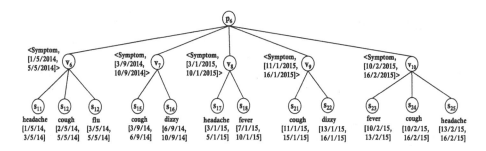

Fig. 5. Augmented data graph with time boundary

Figure 5 shows a data graph where the Visit nodes of a new patient p_6 are augmented with the time boundaries of the Symptom nodes. For Visit node v_8, it has two Symptom nodes s_{17} and s_{18} spanning the periods $[3/1/2015, 5/1/2015]$ and $[7/1/2015, 10/1/2015]$ respectively. Thus, the time boundary covered by node v_8 is $[3/1/2015, 10/1/2015]$. A partial answer for the query $C_5 = \{$Patient : fever BEFORE cough$\}$ over this data graph is $s_{18} - v_8 - p_6$, indicating that patient p_6 has fever from $7/1/2015$ to $10/1/2015$.

Recall that the BEFORE relation in Allen's Algebra [2] requires that the start time of the second interval must be greater than the end time of the first interval. Hence, when we try to check if p_6's fever is BEFORE cough, we do not need to check all p_6's Visit nodes. Instead, only cough that occurs after $10/1/2015$ up to the current date ($currentDate$) can contribute to the query answer. Our time-aware pruning strategy determines a valid range $[10/1/2015, currentDate]$ and checks if this range overlaps with the time boundaries of p_6's Visit nodes. In this example, we only need to traverse v_9 and v_{10} since their time boundaries overlap with the valid range.

On the other hand, suppose cough is associated with a time interval as in query $C_7 = \{$Patient : fever[1/1/2015, 31/1/2015] BEFORE cough[1/1/2015, 31/1/2015]$\}$. Then the valid range for cough should be $[10/1/2015, 31/1/2015]$. In this case, only the time boundary of v_9 overlaps with this valid range.

Table 3 shows the valid ranges corresponding to all possible temporal relationships when we are given the interval of a partial answer $I_1 = [s_1, e_1]$ and the interval $I_2 = [s_2, e_2]$ of a time-associated keyword. A dash entry ($'-'$) indicates that there is no valid range, and the partial answer can be pruned in this case.

Table 3. Computation of valid range

	BEFORE	MEET	OVERLAP	FINISHED BY	CONTAINS	STARTS	EQUALS
(interval)	$[s_2,e_2]$	-	-	-	-	-	-
(interval)	$[s_2,e_2]$	$[e_1,e_2]$	-	-	-	-	-
(interval)	$[e_1,e_2]$	$[e_1,e_2]$	$[s_2,e_1]$	$[s_2,e_1]$	$[s_2,e_1]$	-	-
(interval)	-	$[e_1,e_2]$	$[s_2,e_2]$	$[s_2,e_1]$	$[s_2,e_1]$	-	-
(interval)	-	-	$[s_2,e_2]$	-	$[s_2,e_2]$	-	-
(interval)	$[e_1,e_2]$	$[e_1,e_2]$	$[s_2,e_1]$	$[s_2,e_1]$	$[s_2,e_1]$	$[s_1,e_2]$	$[s_1,e_1]$
(interval)	-	$[e_1,e_2]$	$[s_2,e_2]$	$[s_2,e_1]$	$[s_2,e_2]$	$[s_1,e_2]$	$[s_1,e_1]$
(interval)	-	-	$[s_2,e_2]$	-	$[s_2,e_2]$	$[s_1,e_2]$	-
(interval)	$[e_1,e_2]$	$[e_1,e_2]$	$[s_1,e_1]$	$[s_2,e_1]$	$[s_1,e_1]$	$[s_1,e_2]$	$[s_1,e_1]$
(interval)	-	$[e_1,e_2]$	$[s_1,e_2]$	$[s_2,e_1]$	$[s_1,e_1]$	$[s_1,e_2]$	$[s_1,e_1]$
(interval)	-	-	$[s_1,e_2]$	-	$[s_1,e_2]$	$[s_1,e_2]$	-
(interval)	-	-	-	-	-	-	-
(interval)	-	-	-	-	-	-	-

4.3 Algorithms

We incorporate the target oriented search strategy and time-aware pruning strategy into our ATQ (**A**nswering **T**emporal **Q**uery) algorithm. Details are given in Algorithm 1.

We first parse an input query into three sets: K_{head}, K_{body} and TR (Line 1). For each tuple $< k, t >$ in the set K_{head}, we retrieve the set of relations corresponding to the nodes that match k (Lines 2–3). For each tuple $< k, t >$ in the set K_{body}, we retrieve the set of nodes that match k and satisfy its associated time constraint t (Lines 4–5). We select the set $V_{k_{min}}$ that has the least number of matched nodes for a keyword in K_{body} to start the search (Line 6). For example, in query C_4, the nodes that match the keyword **fever** are $\{s_1, s_6, s_{10}\}$, and the nodes that match **cough** are $\{s_3, s_5, s_7, s_9\}$. We start the search with the smaller set as it enables us to narrow the search space quickly.

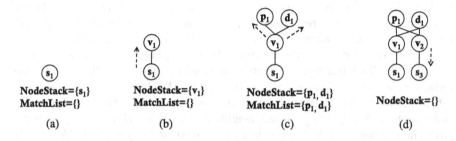

Fig. 6. Construction of a partial answer tree for query C_4

For each node $v \in V_{k_{min}}$, we search from v along the shortest path based on the schema graph to connect nodes that can match the keywords in K_{head} (Lines 7–25). We maintain two stacks: *NodeStack* keeps the traversed nodes in G, and *Partial* stores the subtrees of partial answers built during the search process. We also maintain a *MatchList* to keep track of the keywords in K_{head} that we have found so far. In our example, suppose we start with node s_1. We first add it to *NodeStack*, and a partial tree is created with s_1 as shown in Fig. 6(a). Since s_1's relation does not match any keyword in K_{head}, we get its relevant neighbor v_1 in the shortest path $\{Symptom - Visit - Patient\}$, add v_1 to *NodeStack* and connect v_1 to the partial answer tree (see Fig. 6(b)).

When a node v matches some keyword in K_{head}, we add v to *MatchList* (Lines 14–15). If not, we call function *getRelevantNeighbours()* to find the set of nodes to traverse next (Lines 26–39). From Fig. 6(b), we see that v_1 does not match any keyword in K_{head}. Hence, we obtain v_1's relevant neighbor p_1. Since p_1's relation matches *Patient*, we add p_1 to the *MatchList* and connect p_1's node to the partial answer tree. At this point, *MatchList* has not satisfied K_{head} as we still need to match *Doctor*. Hence, the algorithm continues with the next relevant neighbor of v_1. This time, d_1 is found and is added to the *MatchList*. The partial answer tree obtained is shown in Fig. 6(c).

When *MatchList* satisfies K_{head}, we check if the partial answer $tree_v$ satisfies K_{body} (Lines 16–17). If so, $tree_v$ is an answer to the query and we add it into the result set *Results* (Lines 18). Otherwise, we get the set of lowest common ancestors (LCA) for the nodes in *MatchList* (Lines 20). In our example, since p_1's relation matches the keyword Patient in K_{head} and d_1's relation matches the keyword Doctor in K_{head}, we add p_1 and d_1 to *MatchList*. Although *MatchList* satisfies K_{head}, the partial answer tree does not satisfy K_{body}. As such, we obtain the LCA of the nodes in *MatchList*, that is, $\{v_1, v_2\}$ in this case.

For each node in the LCA set, we call Algorithm *reverseSearch* to find nodes that match the remaining keywords in K_{body} (Lines 21–23). This algorithm returns a tree that is an answer to the query and is added to the result set (Line 24). Algorithm reverseSearch (see Algorithm 2) takes as input a partial answer tree and tries to construct the complete answer by finding nodes that match the remaining keywords in K_{body}. It also uses a stack *NodeStack* to keep track of the nodes to be processed and calls function *getRelevantNeighbours()* to find the set of nodes to traverse next (Lines 5–6). For each node u to be traversed, if u matches a keyword in K_{body}, we check that u satisfies the time constraints and connect u to the answer tree (Lines 7–10). When *tree* matches all the keywords in K_{body}, we have an answer (Lines 11–12). If u does not match a keyword in K_{body}, we perform time-aware pruning by calling the function *hasOverlap()* (Lines 14–17). This function computes the valid range and checks if this range overlaps with the time boundary of node u (Lines 19–26).

Continuing with our example in Fig. 6, we try to match the remaining keyword cough in K_{body}. The relevant neighbor of v_1 is s_2. However, s_2 does not match the keyword cough. We proceed to the next node v_2 in the LCA set. The relevant neighbors of v_2 are $\{s_3, s_4\}$. Since s_3 matches the keyword *cough*, s_3 is

---

Algorithm 1. ATQ Algorithm

---

input : query Q, data graph G, schema graph H
output: Result set *Results*

1   Parse query Q to get K_{head}, K_{body}, and *TR*
2   **foreach** *tuple* $\langle k, t \rangle$ *in* K_{head} **do**
3     $R_k \leftarrow$ the set of relations corresponding to the nodes that match k

4   **foreach** *tuple* $\langle k, t \rangle$ *in* K_{body} **do**
5     $V_k \leftarrow$ the set of nodes in G that match k and satisfy the time constraint t

6   Let k_{min} be the keyword in K_{body} with the least number of matched nodes,
7   **foreach** $v \in V_{k_{min}}$ **do**
8   Initialize *NodeStack, Partial* to empty stacks;
9     $tree_v \leftarrow$ create a tree with root v
10   push(v, *NodeStack*); push($tree_v$, *Partial*)
11   MatchList$\leftarrow \emptyset$
12   **while** *NodeStack is not empty* **do**
13      $u \leftarrow$ pop(*NodeStack*); $tree_v \leftarrow$ pop(*Partial*)
14   **if** *u's relation matches some keyword in* K_{head} **then**
15   add u to MatchList
16   **if** *MatchList satisfy* K_{head} **then**
17   **if** $tree_v$ *satisfy* K_{body} **then**
18   add $tree_v$ to *Results*
19   **else**
20         $W \leftarrow$ getLCA(MatchList)
21   **foreach** $w \in W$ **do**
22   let $tree_v'$ be a copy of $tree_v$
23          $tree \leftarrow$ reverseSearch($tree_v'$, w, K_{body} , *TR*)
24   add $tree$ to *Results*
25   MatchList$\leftarrow \emptyset$

26     $R = \bigcup_{k \in K_{head}} R_k$
27     $N =$ getRelevantNeighbours(u, R, H)
28   **foreach** *node n in N* **do**
29   let $tree_v'$ be a copy of $tree_v$
30   connect n to $tree_v'$
31   push(n, *NodeStack*)
32   push($tree_v'$, *Partial*)

33 **Function** *getRelevantNeighbours(u, R, H)*
34    $N \leftarrow \emptyset$
35   Let N_u be the set of nodes that are one hop away from u
36   **foreach** v *in* N_u **do**
37   **if** *relation(v) is on the shortest path from relation(u) to some relation in R in the schema graph H* **then**
38      $N \leftarrow N \bigcup \{v\}$
39   return N

Algorithm 2. reverseSearch $(tree, v, K_{body}, TR$)

---

    input : partial answer $tree$, LCA node v, K_{body}, temporal relationship TR
    output: result $tree$
1 Initialize $NodeStack$ to an empty stack
2 push(v, $NodeStack$)
3 **while** $NodeStack$ is not empty **do**
4     $u \leftarrow$ pop($NodeStack$)
5   Let R be the set of relations that correspond to the remaining keywords in K_{body} that has not been matched in $tree$
6     $N =$ getRelevantNeighbours(u, R, H)
7   **foreach** node u in N **do**
8   **if** u matches keyword in K_{body} **then**
9   **if** u satisfies the time constraints **then**
10   connect u to $tree$
11   **if** $tree$ matches all the keywords in K_{body} **then**
12   return $tree$

13   **else**
14   Let I be the interval constrained by $tree$
15   **if** hasOverlap(I, u, K_{body}, TR) **then**
16   connect u to $tree$;
17   push(u, $NodeStack$)

18 return $\emptyset$

19 **Function** hasOverlap(I, u, K_{body}, TR)
20   **foreach** $\langle k, t \rangle \in K_{body}$ **do**
21   Let $TR_k \subset TR$ be the set of temporal relationships involving k
22   **foreach** $tr \in TR_k$ **do**
23             $range \leftarrow$ getValidRange(I, tr, t)
24   **if** $range$ overlap $Boundary[u]$ **then**
25   return true

26   return false

---

connected to the partial tree as shown in Fig. 6(d). We return this tree as an answer to query C_4 since it contains all the keywords in K_{body}.

5   Performance Study

We evaluate the performance of ATQ and compare it with BANKS [5] and Bidirectional [12]. All the algorithms are implemented in Java and experiments are carried out on a 1.4 GHz Intel Core i5 CPU with 4 GB RAM. Each experiment is repeated 10 times and we report the average results. We use the following datasets in our experiments.

1. *Clinic dataset*[1]. It contains information about patient consultations with doctors. We use 565 records from the real world dataset as seeds whereby we generate 50 visits per day from 2006 to 2016, and randomly choose a patient and a doctor for each generated visit. For each visit, we randomly assign up to 5 symptoms. The start date of each symptom varies between 1 to 14 days before the visit date. The end date of each symptom is set to be the visit date.
2. *Employee dataset*[2]. This dataset contains the job histories of employees, as well as the departments where the employees have worked in from 1985 to 2003.
3. *ACMDL dataset*[3]. This publication dataset is contains information about authors, proceedings, editors and publishers from 1969 to 2011.

Table 4 shows the schema of these datasets and the number of tuples in each relation. We design two sets of queries for each dataset. The first set does not involve any time constraint, while the second set contains keywords associated with time information and temporal relationships. Queries for the Clinic dataset is shown in Table 2, while queries for the Employee and ACMDL are listed in Tables 5 and 6 respectively.

Table 4. Dataset schemas and number of tuples for each relation

Clinic	# of tuples
Doctor(did, dname, gender)	149
Patient(pid, pname, gender, birthday, ethnicity, postalCode)	1,033
Visit(vid, date, pid, did)	182,600
Symptom(sid, sname, startDate, endDate, vid)	430,470
Employee	**# of tuples**
Department(dept_no, dept_name)	9
Employees(emp_no, fname, lname, gender, hire_date)	300,024
Dept_emp(deid, emp_no, dept_no, from_date, to_date)	331,603
Title(tid, title, emp_no, from_date, to_date)	443,308
ACMDL	**# of tuples**
Publisher(publisherid, code, name)	40
Proceeding(procid, title, date, area, publisherid)	4,176
Editor(editorid, fname, lname)	20,008
Edit(editorid, procid)	20,712
Paper(paperid, procid, date, ptitle)	248,185
Author(authorid, fname, lname)	257,694
Write(authorid, paperid)	550,000

[1] This dataset is not available due to patient confidentiality.
[2] https://dev.mysql.com/doc/employee/en/.
[3] http://dl.acm.org/.

Table 5. Temporal keywords queries for `Employee` dataset

Query	Intended meaning
E_1 {Employee: Engineer}	Find employees who are engineers.
E_2 {Employee: Engineer, Manager}	Find employees who have been engineer and manager before.
E_3 {Employee, Female: Engineer, Manager}	Find female employees who have been engineer and manager before.
E_4 {Employee, Department: Engineer}	Find employees who are engineers and their departments.
E_5 {Employee: Engineer BEFORE Manager}	Find employees who are engineers before coming managers.
E_6 {Employee: Manager[1/1/1990, 1/1/2000], Engineer[1/1/1990, 1/1/2000]}	Find employees who have been engineer and manager from 1990 to 2000
E_7 {Employee: Manager[1/1/1990, 1/1/2000] BEFORE Engineer[1/1/1990, 1/1/2000] }	Find employees who are engineers before becoming managers from 1990 to 2000
E_8 {Employee, Department: Engineer[1/1/1990,1/1/2000]}	Find employees and departments where these employees are engineers from 1990 to 2000
E_9 {Employee, Department: Manager[1/1/1990,1/1/2000], Engineer[1/1/1990,1/1/2000]}	Find employees who have been engineer and manager from 1990 to 2000 and their departments

Table 6. Temporal keywords queries for `ACMDL` dataset

Query	Intended meaning
A_1 {Author: Integration}	Find authors who has published papers on "Integration"
A_2 {Author: Integration, Cleaning}	Find authors who has published papers on "Integration" and "Cleaning"
A_3 {Proceeding, SIGMOD: Integration}	Find papers published in the "SIGMOD" proceeding that are on "Integration"
A_4 {Publisher, Proceeding: Data, Integration}	Find publishers and proceedings pair where the proceedings contain papers on "Data Integration"
A_5 {Author: Media BEFORE AI}	Find authors who have published papers in "Media" proceedings prior to publishing papers in "AI" proceedings
A_6 {Author: Media[01/01/2000, 01/01/2008], AI[01/01/2000, 01/01/2008]}	Find authors who have published papers in both "Media" and "AI" proceedings from 2000 to 2008
A_7 {Author: Media[01/01/2000, 01/01/2008] BEFORE AI[01/01/2000,01/01/2008]}	Find authors who have published papers in "Media" proceedings before publishing papers in "AI" proceedings from 2000 to 2008
A_8 {Proceeding, Publisher: Integration[1/1/2000, 1/1/2008]}	Find the publishers and proceedings that have included papers on "Integration" from 2000 to 2008
A_9 {Proceeding, Publisher: Integration[1/1/2000, 1/1/2008], Data[1/1/2000, 1/1/2008]}	Find the publishers and proceedings that have included papers on "Data Integration" from 2000 to 2008

5.1   Experiments on Queries Without Time Constraints

We first evaluate the performance of our approach using queries that do not involve time information. These queries correspond to C_1 to C_4 in Table 2, E_1 to E_4 in Table 5, and A_1 to A_4 in Table 6. We compare the runtime of ATQ with $BANKS$ [5] and $Bidirectional$ [12]. Since both $BANKS$ and $Bidirectional$ do not handle keywords that match relation names, we modify these algorithms to consider all the nodes of the queried relation as matching nodes. For fair comparison, we report the time taken by these methods to return the first 20 answers.

Figure 7 shows the results for the 3 datasets. We observe that ATQ outperforms $Bidirectional$ and $BANKS$ for all the queries, with $BANKS$ being the slowest. This indicates the advantage of our target-oriented search strategy. For the $Clinic$ dataset, we see that the runtimes of ATQ for queries C_2 and C_3 are lower than C_1 although these queries have more keywords than C_1. This is because ATQ will make use of the keyword with the least number of matching nodes to generate a small set of partial answers. This reduces the time needed to check if these partial answers are valid during the $reverseSearch$ process to obtain the complete answers. On the other hand, the runtime of ATQ for query C_4 increases compared to C_2 and C_3. This is because C_4 has an additional search target relation in the head of the query, leading to a larger number of matching nodes, thus the time needed to find the partial answers is longer. We observe similar trends for the queries on the $Employee$ and $ACMDL$ datasets.

5.2   Experiments on Queries with Time Constraints

Next, we evaluate the performance of our approach to process keyword queries that involve time. These queries correspond to C_5 to C_9 in Table 2, E_5 to E_9 in Table 5, and A_5 to A_9 in Table 6. We extend existing methods $BANKS$ and $Bidirectional$ to handle temporal keyword queries by ignoring the time intervals and temporal relationships in these queries and processing the keywords to obtain candidate answers. Answers that do not satisfy the time constraints are filtered by a post-processing step.

At the same time, we implemented ATQ^-, a variant of the ATQ algorithm which does not utilize the augmented data graph (time boundaries in the nodes) and the overlapping time interval in the inverted lists for the keywords. Instead, ATQ^- also has a post-processing step to filter invalid answers.

Figure 8 shows the results for the 3 datasets. We observe that both ATQ and ATQ^- outperform BANKS and Bidirectional for all the queries by a large margin. Further, we see that time-aware pruning strategy enables ATQ to be faster than ATQ^-. In particular, for query C_7, we observe that ATQ is very much faster than ATQ^-. This is because the combination of time interval constraints and temporal relationships leads to a narrow valid range that allows more invalid partial answers can be pruned.

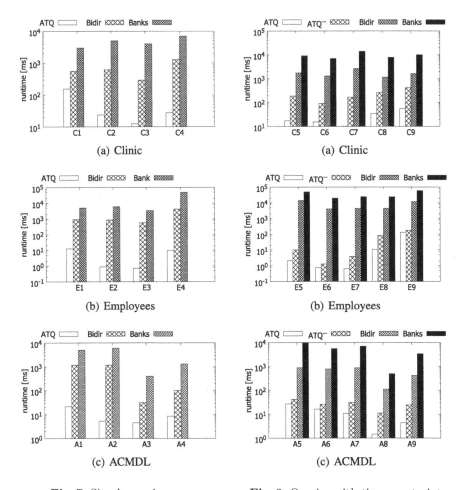

Fig. 7. Simple queries   Fig. 8. Queries with time constraints

6   Conclusion

In this paper, we have examined how keyword queries can be expressed and supported over temporal databases. We introduced a new representation for users to specify their search target, associate keywords with time constraints and indicate temporal relationships between keywords. This enables flexible querying of complex temporal relationships in the databases. We have designed an efficient *ATQ* algorithm that incorporates a target-oriented search process and time-aware pruning to retrieve answers to these queries. Experimental results on 3 datasets showed that the proposed approach outperforms current state-of-the-art keyword search methods. Future work includes a time-aware ranking scheme and extending temporal keyword queries to handle uncertainty.

References

1. Agrawal, S., Chaudhuri, S., Das, G.: DBXPlorer: a system for keyword-based search over relational databases. In: IEEE ICDE (2002)
2. Allen, J.: Maintaining knowledge about temporal intervals. Commun. ACM **26**, 832–843 (1983)
3. Bergamaschi, S., Domnori, E., Guerra, F., Trillo Lado, R., Velegrakis, Y.: Keyword search over relational databases: a metadata approach. In: ACM SIGMOD (2011)
4. Bergamaschi, S., Guerra, F., Interlandi, M., Trillo-Lado, R., Velegrakis, Y.: Quest: a keyword search system for relational data based on semantic and machine learning techniques. VLDB J. **6**, 1222–1225 (2013)
5. Bhalotia, G., Hulgeri, A., Nakhe, C., Chakrabarti, S., Sudarshan, S.: Keyword searching and browsing in databases using banks. In: IEEE ICDE (2002)
6. Bin-Thalab, R., El-Tazi, N., El-Sharkawi, M.: TMIX: temporal model for indexing XML documents. In: International Conference on Computer Systems and Applications (2013)
7. Dignös, A., Böhlen, M., Gamper, J.: Overlap interval partition join. In: ACM SIGMOD (2014)
8. Ding, B., Yu, J.X., Wang, S., Qin, L., Zhang, X., Lin, X.: Finding top-k min-cost connected trees in database. In: IEEE ICDE (2007)
9. He, H., Wang, H., Yang, J., Yu, P.S.: BLINKS: ranked keyword searches on graphs. In: ACM SIGMOD (2007)
10. Hristidis, V., Gravano, L., Papakonstantinou, Y.: Efficient IR-style keyword search over relational databases. In: VLDB (2003)
11. Hristidis, V., Papakonstantinou, Y.: Discover: keyword search in relational databases. In: VLDB (2002)
12. Kacholia, V., Pandit, S., Chakrabarti, S., Sudarshan, S., Desai, R., Karambelkar, H.: Bidirectional expansion for keyword search on graph databases. In: VLDB (2005)
13. Kulkarni, K., Michels, J.: Temporal features in SQL:2011. SIGMOD Rec. **41**, 34–43 (2012)
14. Li, G., Ooi, B., Feng, J., Wang, J., Zhou, L.: EASE: efficient and adaptive keyword search on unstructured, semi-structured and structured data. In: ACM SIGMOD (2008)
15. Luo, Y., Lin, X., Wang, W., Zhou, X.: SPARK: top-k keyword query in relational databases. In: ACM SIGMOD (2007)
16. Manica, E., Dorneles, C., Galante, R.: Supporting temporal queries on XML keyword search engines. J. Inf. Data Manag. **1**(3), 471 (2010)
17. Özsoyoğlu, G., Snodgrass, R.: Temporal and real-time databases: a survey. IEEE Trans. Knowl. Data Eng. **7**(4), 513–532 (1995)
18. Sandeep, T., Guy, M.: SQAK: doing more with keywords. In: ACM SIGMOD (2008)
19. Zeng, Z., Bao, Z., Le, T., Lee, M., Ling, T.: ExpressQ: identifying keyword context and search target in relational keyword queries. In: CIKM (2014)

General Purpose Index-Based Method for Efficient MaxRS Query

Xiaoling Zhou[1(✉)], Wei Wang[1], and Jianliang Xu[2]

[1] University of New South Wales, Sydney, Australia
{xiaolingz, weiw}@cse.unsw.edu.au
[2] Hong Kong Baptist University, Hong Kong, China
xujl@comp.hkbu.edu.hk

Abstract. The Maximizing Range Sum problem is widely applied in facility locating, spatial data mining, and clustering problems. The current most efficient method solves it in time $O(n \log n)$ for a particular given rectangle size. This is inefficient in cases where the queries are frequently called with different parameters. Thus, in this paper, we propose an index-based method that solves the maxRS query in time $O(\log n)$ for any given query. Besides, our method can be used to solve the k-enclosing problem in time $O(1)$ for any given k value if indexes are sorted according to the optimizing criteria, or $O((n - k)^2 k + n \log n)$ without using any index, which is comparative to the current most efficient work.

Keywords: Maximizing range sum · Index construction · Query processing

1 Introduction

In this paper, we study the *Maximizing Range Sum* (maxRS) problem [1–5], which is also known as the *Maximum-enclosing Rectangle Problem* in computational geometry. Given a set of n points P, each with a positive weight $w(p)$, the maxRS problem finds the placement of an axis-parallel rectangle r of given size $\alpha \times \beta$ that maximizes the weight sum of points covered by r. The maxRS problem is well-motivated in recent work [3–5], and is generally applied in facility location problems [14] for finding the best facility location with maximum number of potential clients, spatial data mining for extracting interesting locations from log data [19], and point enclosing problems.

However, existing work targeted at finding efficient algorithms to answer the query given a particular rectangle size, and they need query time superlinear in n. Currently, the best method solves the exact maxRS problem in time $O(n \log n)$ [2] based on the plane-sweep algorithm [1], and the best algorithm for the $1 - \epsilon$ approximate version of the problem works in time $O(n \log \frac{1}{\epsilon} + n \log \log n)$ via grid sampling. This is undesirable in cases where the maxRS queries are asked frequently with different parameters. For example, in computational geometry, maxRS serves as a subroutine and is called many times with different rectangle

© Springer International Publishing Switzerland 2016
S. Hartmann and H. Ma (Eds.): DEXA 2016, Part I, LNCS 9827, pp. 20–36, 2016.
DOI: 10.1007/978-3-319-44403-1_2

Table 1. Comparison of algorithms in terms of space-time tradeoffs

Algorithm	Query	Space	Comment
[2]	$O(n \log n)$	$\Theta(n)$	Exact
[4]	$O(n \log \log n)$	$O(n)$	$(1 - \epsilon)$ approximate with $\geq 1 - \frac{1}{n}$ probability
This paper	$O(\log n)$	$\Theta(n \cdot \lambda) = O(n^3)$	Exact
This paper	$O(\log n)$	$\Theta(\log n \cdot \lambda) = O(n^2 \log n)$	$(1 - \epsilon)$ approximate guaranteed

Notes: (1) λ is the maximum size of any k-line, and (2) assume ϵ is a constant.

sizes in point enclosing problems [16]; and in real life applications, a web service that answers queries like finding the location in a city with most number of tourist attractions within a given reachable area, will be enquired by tons of users with different parameters.

Therefore, in this paper, we propose new solutions to this problem that answer queries much more efficient than previous methods, by making use of a special precomputed index. The idea of using precomputed index to accelerate query processing is a common one in databases, and our study expands the spectrum of space-time tradeoff to the maxRS problem. In addition, our method immediately gives comparable or superior results to other related problems compared with existing solutions.

The main idea of our method is to try to precompute and store as few as possible the maxRS results for a limited number of rectangle sizes, and yet still be able to answer any arbitrary query in an efficient manner. Based on the index, we answer the maxRS query in $O(\log n)$ time, which compares favorably with existing bond $O(n \log n)$. Besides, our method solves the k-enclosing rectangle problem in constant time if the index is sorted according to the optimizing criteria, or $O((n - k)^2 k + n \log n)$ without using index, which is comparative to the best time bound achieved so far [11] for general k values. We solve the k-enclosing problem for all possible k values in time $O(n^3 \log n)$, while the direct adaption of existing method [11] takes time $O(n^4)$. On top of the above, our index can be used to answer maximum point enclosing problems with $O(\log n)$ query time. Details are presented in Sect. 6.

We highlight our main contributions in the following:

- We design an index structure that supports very fast maxRS queries. The index is based on novel concepts of changing points and k-lines. The method provides the possibility of efficient batch query, and space-time tradeoff for the maxRS problem (Sect. 3).
- We design a query processing technique achieves time $O(\log n)$ by nontrivially adapting the idea of Fractional Cascading based on a tree structure (Sect. 4).
- We present applications of our method on other problems, and the superiority of our method compared with existing works (Sect. 6).
- We perform experiments on synthetic as well as real datasets to show the feasibility and efficiency of our methods compared with the state-of-the-art methods (Sect. 7).

2   Related Work

The MaxRS Problem. Nandy et al. [2] proposed an $O(n \log n)$ time algorithm to solve the maxRS problem using the plane-sweeping technique [1] with interval trees. Choi et al. [3] proposed an external memory solution following the distribution sweep paradigm [7], and the work was further extended in [5] by providing solutions to the AllMaxRS problem, which retrieves all the locations of rectangles achieving the maximum total covered weight. Recognized the need for further speedup, Tao et al. [4] studied the approximate MaxRS problem, and obtained a $(1 - \epsilon)$-approximate answer with high confidence in time $O(n \log \frac{1}{\epsilon} + n \log \log n)$ via grid sampling. Instead of finding a rectangle with maximum cover weight, Das et al. [6] retrieved the highest density axis-parallel rectangle r where density was defined as the ratio between the covered weight of r and its area.

All the above works solved the exact or approximate maxRS problem for a fixed query rectangle size only, required at least query time superlinear in n, and they did not consider using an index. Our work is the first that builds a special index to speed up the query processing time to $O(\log n)$, hence provides a different solution exploiting the space-time tradeoff and is beneficial to frequent queries.

k-Enclosing Problem. Driven by the wide application in pattern recognition [13], facility locating [14], and VLSI chip design [18], the problem of finding the axis-aligned rectangle, square, or circle that encloses k of n points in $\mathcal{R}^2$ and is optimal on some particular criteria (area, perimeter, or diameter etc.) has been long studied in a large number of works [8–12,15,16]. Aggarwal et al. [8] obtained the smallest k-enclosing rectangle or square in time $O(nk^2 \log n)$ and space $O(nk)$ based on higher-order Voronoi diagrams in 1991. The complexity was improved several times by subsequent works [9–11]. The work in [10] improved the time complexity to $O(nk^2 + n \log n)$ using space $O(n)$ based on the idea of testing sets of the $O(k)$ nearest neighbors to each point. Segal et al. [11] solved the problem in time $O(n + k(n - k)^2)$ using $O(n)$ space, which performed better than previous methods on large k values, and they also proposed solution on d (> 2) dimensions that takes time $O(dn + dk(n-k)^{2(d-1)})$ and space $O(dn)$.

Datta et al. [9] solved the smallest k-enclosing square problem in $O(n \log n + n \log^2 k)$ time and $O(n)$ space. The bound was then improved to $O(n + (n - k) \log^2(n - k))$ using $O(n)$ space by Mahapatra et al. [16], which searched the target square by means of prune and search technique and adopted the maxRS problem as a subroutine to guide the search. Das et al. [17] addressed the generalized version of above problem, where the desired rectangle may be of arbitrary orientation, and their method runs in $O(n^2 \log n + kn(n-k)(n-k+\log k))$ time using $O(n)$ space.

3   The New Index

We formally define the problem.

Definition 1. *Let P be a set of n points[1] in 2D Euclidean space $\mathcal{R}^2$, and each point $p \in P$ carries a positive value $w(p)$ as its weight. Given non-negative values α and β, the goal of the maximizing range sum (maxRS) problem is to place a $\alpha \times \beta$ rectangle r in $\mathcal{R}^2$ to maximize the covered weight of r, defined as: covered-weight$(r) = \sum_{p \in P \cap r} w(p)$.*

Note that the rectangles can be placed at any position. We denote $maxRS(\alpha, \beta)$ as the maximum total weight that can be achieved given query (α, β).

In this paper, we deal with points with equal weights, a.k.a., the *max-enclosing problem*. Our method can be easily extended to handle general weighted cases (See Sect. 6).

Notations. We introduce some notations used in the rest of the paper. Let $P = \{p_1, p_2, \ldots, p_n\}$ denotes the given n data points. $p_i.x$ (resp. $p_i.y$) denotes the x-coordinate (resp. y-coordinate) of point p_i. We can sort P into P^x by the x-coordinates of the points, and obtain P^y analogously. For a pair of different points p_i and p_j ($i \neq j$), we can obtain an x-interval $a = |p_i.x - p_j.x|$. We have $O(n^2)$ distinct x-intervals, and they are collectively denoted as list A. We obtain list B of all distinct y-intervals analogously. Given an x-interval a_i, and the two points p_l and p_r ($l, r \in [1, n]$) in P that form this x-interval (w.l.o.g., assume $p_l.x < p_r.x$), we denote the list of points $p_j \in P$ such that $p_l.x \leq p_j.x \leq p_r.x$ sorted according to y-coordinate as S_{a_i}. Note that p_l and p_r are included in S_{a_i}. More notations will be introduced at their first use in the rest of the paper.

3.1   A Naïve Solution

A naïve idea is to index the maxRS values for *all possible* queries. While there are infinite number of possible queries, Lemma 1 divides them into $O(n^4)$ *equivalent classes*, and this immediately leads to a naïve index and query processing method.

Lemma 1. *Given a query (α, β), let a^* be the largest value in A such that $a^* \leq \alpha$, and define b^* analogously. Then $maxRS(\alpha, \beta) = maxRS(a^*, b^*)$.*

The index is essentially an $O(n^2) \times O(n^2)$ matrix, which stores the precomputed maxRS values for every $(a_i, b_j) \in A \times B$. An example dataset and its naïve index matrix is shown in Fig. 1(a), (b). We call the matrix cell (a^*, b^*) as the *target cell* of the query. The values of a^* and b^* can be obtained using binary search in A and B, respectively. Therefore, the index size is $O(n^4)$, and the query time is $O(\log n)$.

3.2   Index the Changing Points into k-Lines

The naïve index contains many redundant values. Our important observation is that maxRS values in the matrix follow certain pattern and can be exploited to further reduce the index size.

---

[1] W.l.o.g., we assume that no two points have the same x (or y) coordinate.

Definition 2 (Changing Point). *A changing point (CP) in the matrix is a cell $M[i,j]$ such that (1) $M[i-1,j] = M[i,j-1] = M[i-1,j-1]$ if any of them exists, and (2) $M[i,j] > M[i-1,j]$.*

We highlight all the CPs in the example dataset in Fig. 1(b).

We observe that all the CPs with same maxRS value k collectively form a *skyline*, which we call k-line. There are n number of k-lines, and they divide the maxRS matrix into n separate regions, as demonstrated as shaded regions of different colors in Fig. 1(c). Our new index just stores these n k-lines (each is organized as a list of CPs sorted on the x-axis). This gives us an index of size $O(n \cdot \lambda)$, where λ is the maximum size of any k-line.

Lemma 2. λ *is at most* $O(n^2)$.

The proof is based on the observation that each CP in L_k has distinct x(or y) intervals, and the size of A and B are bounded by $O(n^2)$. Thus, the index size is upper bounded by $O(n^3)$. Note that in real practice, λ is demonstrated to be consistently linear in n in our empirical evaluation, which leads to cn^2 index space, where c is a constant. Hence, the index size is significantly smaller than the naïve index.

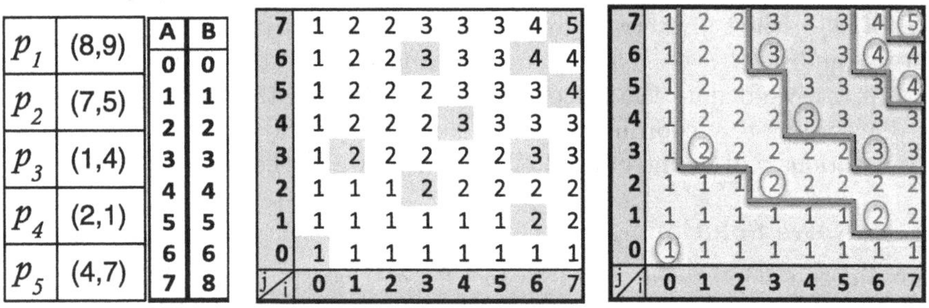

		A	B
p_1	(8,9)	0	0
		1	1
p_2	(7,5)	2	2
p_3	(1,4)	3	3
		4	4
p_4	(2,1)	5	5
		6	6
p_5	(4,7)	7	8

(a) Dataset

MaxRS Matrix (b):

y\x	0	1	2	3	4	5	6	7
7	1	2	2	3	3	3	4	5
6	1	2	2	3	3	3	4	4
5	1	2	2	2	3	3	3	4
4	1	2	2	2	3	3	3	3
3	1	2	2	2	2	2	3	3
2	1	1	1	2	2	2	2	2
1	1	1	1	1	1	1	2	2
0	1	1	1	1	1	1	1	1

(b) MaxRS Matrix

Changing Points (c):

y\x	0	1	2	3	4	5	6	7
7	1	2	2	3	3	3	4	(5)
6	1	2	2	(3)	3	3	(4)	4
5	1	2	2	2	3	3	3	(4)
4	1	2	2	2	(3)	3	3	3
3	1	(2)	2	2	2	2	(3)	3
2	1	1	1	(2)	2	2	2	2
1	1	1	1	1	1	1	(2)	2
0	(1)	1	1	1	1	1	1	1

(c) Changing Points

Fig. 1. Index (Color figure online)

This new index poses challenges on query processing, as the cell of (a^*, b^*) for the query may not be a CP, hence is not stored in the index; this renders the previous $O(\log n)$ query algorithm inapplicable. Nevertheless, we devise a novel $O(\log n)$ query processing method and introduce it in Sect. 4. The detailed construction of index is presented in Sect. 5.

4   Query Processing

Next, we introduce our query processing algorithm for the new index based on k-lines. We start with an algorithm with $O(\log^2 n)$ complexity and then further improve it to $O(\log n)$ by adapting the Fractional Cascading (FC) technique.

4.1   Dominance Relationship

We first introduce a few notations and useful Lemmas.

Given a query Q with search parameters (α, β), we can map it into a two dimensional point (α, β) (called *query point*). Note that every changing point is also a valid query point, and the same mapping applies. A k-line can be mapped into a polyline with axis-parallel segments by connecting two adjacent changing points, q_1 and q_2 ($q_1.x < q_2.x$), via inserting another point $(q_2.x, q_1.y)$ in between (refer to blue lines in Fig. 1(c)). In the rest of this section, we will abuse the notation of k-line and L_k to denote the polyline it maps to. Given a k-line (L_k), it divides the first quadrant into two disjoint regions, the one on the lower-left side of it (called $LOW(L_k)$) and the other on the upper-right side (called $HIGH(L_k)$). Points on L_k is included in $HIGH(L_k)$.

We can define a **dominance** relationship between any two query/matrix points as follows.

Definition 3 ((Point) Dominance). *Let q_1 and q_2 be two points. q_1 dominates q_2, denoted as $q_1 \prec q_2$, if and only if $q_1.x \leq q_2.x \wedge q_1.y \leq q_2.y$ and $q_1 \neq q_2$.*

Lemma 3. *If $q_1 \prec q_2$, then $maxRS(q_1) \leq maxRS(q_2)$.*

We can generalize the dominance to be between a k-line and a point Q.

Definition 4. ((Line-Point) Dominance). *Given a k-line and a point q, q has to be either in $LOW(L_k)$, or in $HIGH(L_k)$. We say q dominates L_k, denoted as $q \prec L_k$ for the former case, and $L_k \prec q$ for the latter case.*

Unlike the point dominance where it is possible that two points do not dominate each other, a k-line and a point always fall into one of the two dominance orders. We also have the following Lemma.

Lemma 4. *A query point $Q \in HIGH(L_k)$ if and only if there is a changing point q in L_k such that q dominates Q.*

4.2   The $O(\log^2 n)$ Query Processing Method

Given a query (α, β), we only need to find its target cell with parameters (a^*, b^*). Note that after mapping queries and matrix cells to points, the region defined by the axes and two adjacent k-lines has the same maxRS values. Therefore, we only need to determine which region the target cell falls into. This can be performed by a standard *outer* binary search on the n disjoint regions formed by the k-lines. Technically, this requires us to determine the largest k-line that dominates the target point (a^*, b^*). This step can be performed by another *inner* binary search among the CPs of a k-line, thanks to Lemmas 4 and 5. The inner binary search takes time $O(\log \lambda) = O(\log n)$, and hence the total query cost is $O(\log^2 n)$.

In the interest of space, in the following, we focus on the inner binary search step to determine the line-point dominance.

Algorithm 1. DominanceCheck(L_i, Q, *low*, *high*)

    Input   : L_i is a i-th sky-line; Q is the query point.
    Output: 0 if Q is found in L_i; 1 if L_i dominates Q; -1 if L_i does not dominate Q.
1 **while** *high* $\geq$ *low* **do**
2      $mid \leftarrow \lfloor (low + high)/2 \rfloor$; $q \leftarrow L_i[mid]$;
3   **if** $q = Q$ **then return** *0* ;
4   **else if** $q \prec Q$ **then return** *1* ;
5   **else if** $Q \prec q$ **then return** *-1* ;
6   **else**
7   **if** $q.x \geq Q.x$ and $q.y \leq Q.y$ **then**
8 | $high \leftarrow mid - 1$;
9   **else**   /* must be $q.x \leq Q.x$ and $q.y \geq Q.y$ */
10          $\lfloor$ $low \leftarrow mid + 1$;

11 **return** *-1*   /* No dominating point found $\Rightarrow Q \prec L_i$ */

The pseudocode of sub-routine to determine the dominance relationship between the current L_i and the query point Q is given in Algorithm 1. Initially, *low* and *high*, the index values into L_i, are set to 1 and $|L_i|$, respectively. At each binary search iteration, we take the middle point q in the current search scope, and check the dominance relationship between q and Q. There are four cases:

1. $q = Q$ (Line 3): the query point is a CP in L_i, hence we return 0. The outer binary search will terminate immediately.
2. q dominates Q (Line 4), hence L_i dominates Q, and we return 1. The outer binary search will continue to search among L_js, where $j \in [i + 1, n]$.
3. Q dominates q (Line 5), hence Q dominates L_i, and we return -1. The outer binary search will continue to search among L_js, where $j \in [1, i - 1]$.
4. q and Q do not dominate each other (Lines 7–10). We need to investigate this further by considering another changing point in L_i. This is handled by moving q towards Q for continuing the binary search. Intuitively, this abandons the half part of L_i that no point inside there could dominate/is dominated by Q.

Finally, if we exit the **while** loop, it means no changing points in L_i can dominate Q and vice versa, we know that Q dominates L_i in this case due to Lemma 4.

    The correctness of the algorithm is provided by the following Lemma.

Lemma 5. *Algorithm 1 correctly determines the line-point dominance relationship between L_i and Q.*

An Example of Algorithm 1. Using the same example as Fig. 1, we assume the algorithm input is $Q(2, 5)$ and L_3. We have $L_3 = \{(3, 6), (4, 4), (6, 3)\}$. The search starts with the middle entry $q(4, 4)$ in L_3. We cannot decide the dominance relationship between Q and q since $q.x > Q.x \wedge q.y < Q.y$, but we know the right half of L_3 must not contain point dominates or is dominated by Q, as all the points q' on right of q have $q'.x > q.x > Q.x$ and $q'.y < q.y < Q.y$. Therefore, we search the left half of L_3 and find Q dominates entry $(3,6)$, so -1 is returned.

4.3   The $O(\log n)$ Query Processing Algorithm

In this section, we further improve the previous query processing algorithm to achieve $O(\log n)$ complexity by adapting our problem to use the Fractional Cascading (FC) technique. FC can reduce binary searches on multiple sorted lists into one by judiciously sampling elements from other lists to one single list. A typical case is to find the smallest number in all the lists no smaller than a query number.

FC cannot be directly applied due to two major technical challenges. One is that our inner search is based on dominance checking, which is essentially 2D, while traditional FC essentially works in 1D lists as the lists need to be sorted. We overcome this by reducing our problem to 1D thanks to the following Lemma.

Lemma 6. *Given Q and L_k, $Q \in HIGH(L_k)$ (i.e. $maxRS(Q) \geq k$) if and only if q dominates Q, where q is the CP in L_k with the largest x-coordinate yet $q.x \leq Q.x$.*

We call such q the **anchor point** of Q in L_k. Then, for each L_k, the goal is to find the anchor point of Q from L_k, so that we can resolve the dominance relationship between Q and L_k.

The second challenge is that by applying FC directly, the total query cost will be $O(\log n + n)$, even worse than $O(\log^2 n)$. We overcome this by observing that we only need to check $\log n$ (rather than n) k-lines (i.e., lists). In addition, though the set of $\log n$ lists inspected will be different for different Q, the visiting order between any two lists is fixed due to the binary search procedure (e.g., the middle list is always inspected before one of the two quadrant lists). Therefore, we can construct a binary tree that reflects the binary search process on the k lists, and apply FC technique bottom up for every parent-child list pairs. We prove later that the total space is still linear in the total size of the lists.

We show the data structure created for the example dataset (Fig. 1) in Fig. 2 and explain it below:

1. Construct a binary tree following the order of binary search on n values. Each node with key value i (inside the node) is logically associated with list L_i.
2. Each node actually stores a list U_i, constructed as a result of applying FC recursively in the subtree rooted at it. For a leaf node v, $U_i = L_i$. For an internal node representing L_i, $U_i = L_i \cup \text{even}(U_{\text{left}}) \cup \text{even}(U_{\text{right}})$, where U_{left} and U_{right} are the U list for the left and right child node of v, respectively.

Additionally, we store three pointers $[t_1, t_2, t_3]$ for each entry e in U_i (See the blue values shown in Fig. 2(a) above each element e), where t_1, t_2, t_3 point to the position of the *anchor point* of e in L_i, U_{left}, and U_{right}, respectively.

L_1	(0,0)
L_2	(1,3)(3,2)(6,1)
L_3	(3,6)(4,4)(6,3)
L_4	(6,6)(7,5)
L_5	(7,8)

U_1	(0,0)
U_2	[1,1,∅] [2,1,1] [2,1,2] [3,1,3] (1,3) (3,2) (4,4) (6,1)
U_3	(3,6) (4,4) (6,3)
U_4	[∅,2,∅][1,4,∅][1,4,∅][2,4,1] (3,2) (6,1) (6,6) (7,5)
U_5	(7,8)

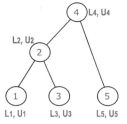

(a) Auxiliary Lists   (b) Query Binary Tree

Fig. 2. Query processing based on FC

Query Processing. We answer a query Q by top-down traversing the tree nodes and processing their associated U_i lists. We first search the anchor point of Q in U_{root} using standard binary search. Then according to the dominance relationship between Q and L_{root}, we search U_{left} or U_{right} following the pointers in U_{root}, until we determine $maxRS(Q)$ or reach a leaf. In other words, the use of Algorithm 1 is now replaced with a constant cost operation (of point-wise dominance checking), thanks to FC.

An Example. Let Q be $(2,5)$ and we use the data structure in Fig. 2. Anchor value of Q is 2, and binary search in the root node (i.e., U_4) results in $\emptyset$, indicating no point in U_4 dominates Q, hence $Q \in LOW(L_4)$. We need to continue to search query's anchor point in the left child node (i.e., U_2); this can be done by following the pointer stored in $U_4[1]$ (the last entry we searched in the parent) to go to $U_2[2]$ directly. Since we only sampled even entry of U_2 in U_4, even though $U_2[2]$ is not the anchor point of Q, we still need to check if the previous entry $U_2[1]$ is. The answer is true, and as $U_2[1]$ dominates Q, we get $Q \in HIGH(L_2)$ based on Lemma 6. The next step is to search Q's anchor point in U_3 by following pointer to U_3 attached with $U_2[1]$, and we get $\emptyset$. This indicates no dominate point of Q in U_3, so $Q \in LOW(L_3)$. Now, we can decide $maxRS(Q) = 2$, and the search stops.

Analysis of Space Complexity. We have n L_k lists. Denote the length of each list L_k as l_k, and the total CP lists size is $S = \sum_{k=1}^{n} l_k$. Let leaf node has level 1, and the level of the root is h. Denote S_i (resp. W_i) as the total size of L (resp. U) lists associated in tree level i. Then we have $\sum_{i=1}^{h} S_i = \sum_{k=1}^{n} l_k = S$, and $W_i = S_i + \frac{W_{i-1}}{2}$. Thus, the total size of all U lists is $\sum_{i=1}^{h} W_i = S_1 + (\frac{1}{2}S_1 + S_2) + \cdots + (\frac{1}{2^{h-1}}S_1 + \frac{1}{2^{h-2}}S_2 + \cdots + S_h)) < 2(S_1 + S_2 + \cdots + S_h) = 2S$. Therefore, the total space of U_i and L_i lists is linear in total number of CPs.

Analysis of Time Complexity. The size of U_{root} list is $W_h = \frac{1}{2^{h-1}}S_1 + \frac{1}{2^{h-2}}S_2 + \cdots + S_h < S$. The binary search in U_{root} has cost $O(\log S) = O(\log n)$. We inspect at most $\log n$ non-root nodes in the tree, and each with constant cost $O(1)$. Therefore, the total search cost is $O(\log n + \log n) = O(\log n)$.

5   Index Construction

In this section, we introduce the index construction and complexity analysis.

If we directly apply existing method in [2] to compute the index, for each combination of (a_i, b_i), we initiate a query taking $O(n \log n)$ to find its maxRS value, and finally obtain all the CPs in time $O(n^5 \log n)$. Whereas in the next subsection, we present an algorithm that constructs the index in time $O(n^3 \log n)$.

5.1   Main Idea

Given a CP in cell $M[i, j]$, it means there exists a rectangle r of size $a_i \times b_j$ that contains $M[i, j]$ number of points from P. We refer such a rectangle r as the rectangle **represented** by this CP. Then we have the following lemma:

Lemma 7. *The rectangle represented by each changing point must touch points from P on all its four edges.*[2]

Lemma 7 narrows down our search space into $O(n^4)$ rectangles to find all the CPs. An implied property of CP is that among all the rectangles containing k points from P, the rectangle represented by a CP in L_k is the one that cannot be dominated[3] by any other rectangles. This can be clearly observed from the matrix in Fig. 1(c). Thus, instead of generating all the $O(n^4)$ rectangles whose four edges touch points in P, we find L_k ($k \in [2 \dots n]$)[4] in the following way: for each possible rectangle width a_i, we find the rectangle r_{min} with minimum height (i.e. min b_i) such that the pair of points (p_l, p_r) which forms interval a_i lies on the left and right vertical edges of r_{min} respectively, and r_{min} contains exactly k points. Then r_{min} is a candidate CP in L_k. After all the a_i are processed, we get $O(n^2)$ r_{min} in L_k, and eliminate the ones that are dominated by others, we get all the CPs in L_k.

In this idea, we generate $O(n^3)$ candidate rectangles. The total construction cost is closely related to the cost of finding r_{min} for each a_i and k value. We introduce the structure to find r_{min} next.

Y-coordinate Distance Arrays. Given a_i, recall that points in S_{a_i} are sorted according to y-coordinates, we assume the size of S_{a_i} is $|S_{a_i}|$, and denote $S_{a_i} = \{p_1, p_2, \dots, p_{|S_{a_i}|}\}$. We construct $|S_{a_i}| - 1$ number of y-coordinate distance arrays over points in S_{a_i} as the following:

- $D_2 : \{|p_1.y - p_2.y|, |p_2.y - p_3.y|, |p_3.y - p_4.y|, \cdots, |p_{|S_{a_i}|-1}.y - p_{|S_{a_i}|}.y|\}$
- $D_3 : \{|p_1.y - p_3.y|, |p_2.y - p_4.y|, \cdots, |p_{|S_{a_i}|-2}.y - p_{|S_{a_i}|}.y|\}$
- $\dots$
- $D_{|S_{a_i}|} : \{|p_1.y - p_{|S_{a_i}|}.y|\}$

---

[2] The proof is obvious and omitted due to space limitation.

[3] $R1$ is dominated by $R2$ if both the width and height of $R1$ are larger than or equal to that of $R2$, the dominance relationship is formally defined in Definition 3.

[4] L_1 contains the only point $(0, 0)$.

Assume the two points forming a_i are p_l and p_r ($l < r$) in S_{a_i}, then for each value $k \in [2, |S_{a_i}|]$, we find the height of the minimum rectangle that contains k points by calling $RMQ_{D_k}(r - k + 1, l)$ on D_k. The index of D_k starts from 1. Now the construction cost depends on the cost of answering Range Minimum Queries on D_k.

Update Distance Arrays. While techniques for answering RMQ on 1D array have been extensively studied in current literatures, and the most efficient ones achieve O(1) query time with linear auxiliary index space, they work mainly on static underlying arrays. If adapted to our dynamic distance arrays D_k, which change along with a_i, the cost of updating the auxiliary index structure for RMQ outweighs the cost saved by using it.

Another challenge is that after processing interval a_i, and move to the next closest one, say a_{i+1}, such that S_{a_i} and $S_{a_{i+1}}$ differ by only 1 point p_j, then for each D_k, the number of distance values affected by p_j is $O(k)$. For example, if p_3 is removed from D_3, the entries being affected are $|p_1.y - p_3.y|$, $|p_2.y - p_4.y|$, and $|p_3.y - p_5.y|$. Thus, the total number of entries need to be updated in all D_k ($k \in [2 \dots |S_{a_i}|]$) is $O(n^2)$ for each a_i, and totally $O(n^4)$ for all a_i ($i \in [1 \dots |A|]$).

To overcome the above challenges, we devise a tree-based structure that supports RMQ in time $O(\log n)$ as well as updating of each D_k in $O(\log n)$ according to the following observation.

Lemma 8. *When a point p is removed from list S_{a_i}, the $O(k)$ entries that have to be updated in D_k are continuous, and the updated results are identical to a continuous part of entries in D_{k+1}.*

Left-Complete Binary Tree. We maintain a left-complete binary tree T_k over each list D_k in the sense that the left subtree of each internal node is complete. The leaves of T_k store the elements of D_k, and internal nodes correspond to ranges of consecutive elements of the list. Each internal node v stores a pointer to a leaf μ in the subtree rooted at v with minimum distance value. At any time, the size of the data structure is linear in the number of elements present in the list. The key value of each node is set as the following: (1) Leaf node's key is the index of the element in D_k it stores. (2) The ith internal node in level l has key value $2^l * (i - 1) + 2^{(l-1)} + 0.5$ (l starts from 0, i starts from 1, and leaf has level 0).

When a point is removed from S_{a_i}, and $O(k)$ continuous entries in the range $D_k[l \dots r]$ need to be updated, we firstly search l and r in T_k and T_{k+1} simultaneously. Denote the path of searching l (resp. r) in T_k as FL_{T_k} (resp. FR_{T_k}), and the forest contained between FL_{T_k} and FR_{T_k} as F_k. Updating $D_k[l \dots r]$ is achieved by moving F_{k+1} to the corresponding position of F_k. The above tree structure and node key settings guarantee all the key values appear in $FL_{T_{k+1}}$ and $FR_{T_{k+1}}$ also appear in FL_{T_k} and FR_{T_k}. Thus the moving can be done by simply adjusting pointers of nodes in FL_{T_k}, FR_{T_k}, and root nodes in forest F_{k+1}. After the moving step, a bottom-up traversing of paths FL_{T_k} and FR_{T_k} is performed to update the min pointers in each node. Therefore, for each

a_i, although we need to update $O(n^2)$ distance values, the total update cost is $O(n \log n)$. The RMQ can be solved in $O(\log n)$ time as usual.

5.2   The Complete Algorithm

Algorithm 2 shows our index construction method.

1. We initialize n number of empty sorted lists (Line 1). Each list L_k stores the currently generated candidate rectangles containing k points. When a new rectangle R needs to be inserted into L_k(Lines 13–14), we check if R is dominated by existing entries in L_k. If not, we insert R into L_k, and remove all elements in L_k that are dominated by R. Here, a rectangle R is represented by a pair of value (a, b), where a refers to width and b refers to height.
2. At the beginning, we build the $n-1$ y-coordinate distance arrays for all points in P, and the auxiliary trees for fast RMQ and range update (Lines 4–7).
3. For each x-interval a_i, and for each $k \in [2 \dots |S_{a_i}|]$, we query $b_{min} = RMQ_{T_k}(r - k + 1, l)^5$ on tree T_k to find the minimum height of rectangle containing k points, l and r are the indexes of the two points in S_{a_i} that form the x-interval a_i. Then we add (a_i, b_{min}) as a candidate rectangle to L_k (Lines 10–14). For each a_i, after all the possible k values are processed, the auxiliary trees are updated as described in Sect. 5.1 (Lines 15–16).

Total Cost Analysis. There are totally $O(n^3)$ rectangles generated, each takes $O(1)$, and checking whether it is dominated by existing candidates takes $O(\log n)$, so in total it is $O(n^3 \log n)$ time, and $O(n^3)$ space to store rectangles. The first construction of y-coordinate distance arrays and auxiliary trees costs $O(n^2)$ in time and $O(n^2)$ in space. For each change of x-interval a_i, $O(n^2)$ distance values are updated from all trees, which costs only $O(n \log n)$ in time as analyzed before. There are $O(n^2)$ x-intervals, so total tree update cost is $O(n^3 \log n)$.

Therefore, our index can be constructed in $O(n^3 \log n)$ time, and $O(n^3)$ space.

6   Other Problems

In this section, we present the applications and advantages of our method for solving other classic computational geometry problems.

General Weighted maxRS Problem and Approximate Version. If points in P have different weights, the only difference with analysis in previous sections is that the number of different maxRS values cannot be bounded by $O(n)$. In this case, the number of CPs is upper bounded by $O(n^4)$. The construction can be done in $O(n^4)$ by simply finding all the rectangles whose four edges touch

---

[5]   In the pseudocode of Algorithm 2 Line 12, we make the optimization by calling $RMQ_{T_k}(1, |T_k|)$ rather than $RMQ_{T_k}(r - k + 1, l)$. We omit the proof here.

Algorithm 2. ConstructIndex(list P)

Input : data point list P
Output: The list of changing points L_k for each $k \in [2, n]$

1 $L_k \leftarrow \emptyset, \forall k \in [2, n]$;
2 $P^y \leftarrow P$ sorted in decreasing order of y-coordinates ;
3 $P^x \leftarrow P$ sorted in increasing order of x-coordinates ;
4 **for** $k \leftarrow 2$ **to** n **do**
5       $D_k \leftarrow \{ |p_i^y.y - p_{i+k-1}^y.y| \mid i \in [0, n-k+1] \}$;   /* build dist arrays */;
6       $T_k \leftarrow$ the binary tree constructed from D_k ;
7       $T_k' \leftarrow$ a copy of T_k ;

8 **for** $i \leftarrow 0$ **to** $n - 2$ **do**
9   **for** $j \leftarrow n - 1$ **downto** $i + 1$ **do**
10          $a_i \leftarrow p_j^x.x - p_i^x.x$; /* Process each of the $O(n^2)$ x-intervals */;
11   **for** $k \leftarrow 2$ **to** $j - i + 1$ **do**
12              $(b_{min}, p_c^y, p_d^y) \leftarrow$ RMQ$(1, |T_k|)$;
13   **if** (p_c^y, p_d^y) *covers* (p_i^x, p_j^x) **then**
14   L_k.addresult(a_i, b_{min});   /* add result and remove dominance */;

15   **for** $k \leftarrow 2$ **to** $j - i + 1$ **do**
16   updatetree(T_k, p_j^x); /* update entries affacted by p_j^x in T_k */;

17   **for** $k \leftarrow 2$ **to** $n - i + 1$ **do**
18   updatetree(T_k', p_i^x) ;
19          $T_k \leftarrow T_k'$;   /* restore T_k */;

20 **return** $\{ L_k \mid k \in [2, n] \}$;

---

points and finally get all the CPs. Despite the increase of indexing cost, the query time remains the same as $O(\log n)$.

To answer $1 - \epsilon$ approximate maxRS queries with 100 % guarantee, the only change is to include only the i-lines in the index, where $i = \{ 1, c, c^2, \dots \}$, where $c = \frac{1}{1-\epsilon}$. When c is a constant, this reduces the index size down to $O(n^2 \log n)$ with the same $O(\log n)$ query time complexity (See Table 1).

k-Enclosing Problem. K-enclosing rectangle problem [8–11], is the problem of finding the minimum axis-parallel rectangle that contains exact or at least k points from P, in terms of measurement like rectangle area or perimeter.

The most efficient existing method [11] solves this problem in $O(n + k(n - k)^2) = O(n^3)$ when $k = \frac{n}{2}$. Using our index, we solve the problem in $O(1)$ if each L_k is sorted according the query measurement, or $O(n)$ by linear scan L_k if L_k is not sorted. Both cases are way better than existing time bounds achieved. Even without index, our method can be modified to answer k-enclosing problem in time $O((n - k)^2 k + n \log n) = O(n^3)$ when $k = \frac{n}{2}$, which meets the bound of the best existing method.

Furthermore, our method addresses k-enclosing problem for all possible values of k in time $O(n^3 \log n)$ including both indexing and query time. While existing method takes $O(n^4)$ in total in order to give answer for all possible values of k.

Maximum Point Enclosing Problem. Maximum point enclosing problem can be seen as the inverse problem of k-enclosing problem. It aims at finding the maximum number of points a rectangle can enclose, and the rectangle measurement (area or perimeter) is no larger than given query value Q.

Existing method [2] can be adapted to solve the above problem in $O(n^3 \log n)$ time by trying for each possible width a_i ($O(n^2)$ number of possibilities), together with the largest height b_i computed so that the rectangle size is within given query parameter, finding the maxRS value of (a_i, b_i) in time $O(n \log n)$, and the global maximum maxRS value is the final answer.

Whereas our method solves the problem in $O(\log^2 n)$ by a nested binary search procedure[6] similar as the one presented in Sect. 4 if L_k lists are sorted on query measurement, or $O(n \log n)$ if L_k lists are not sorted. Even if indexing time is considered, we take $O(n^3 \log n)$ total time, but be able to answer any given query parameter efficiently, while existing method sloves each query using $O(n^3 \log n)$ time.

7 Experiment

In this section, we perform empirical experiments to confirm our theoretical analysis of algorithms' performance and demonstrate the substantial query performance improvement by our proposed method in practice.

Experiment Setting. We perform experiments on both synthetic and real datasets. The synthetic dataset contains 10,000 points generated with uniform distribution. The ranges of x and y coordinates are both set to $[0, 10^5]$. The real dataset is drawn from the publicly available NorthEast (NE) dataset[7], which contains 20,000 postal addresses in New York, Philadelphia and Boston.

We compare our method with the plane-sweep algorithm [2] (denoted as **PS**), as it is the most efficient one for the exact maxRS problem; our index-based algorithm is denoted as **Index**. Both methods are implemented in C++, and experiments are conducted on a PC with Intel Core i7 CPU 2.7 GHz with 8 GB of memory.

Varying Query Rectangle Size. We test the performance of both methods on various query rectangle sizes. The five groups of rectangle size ranges are $[0^2, 10^2]$, $[10^2, 100^2]$, $[100^2, 1000^2]$, $[1000^2, 10000^2]$, and $[10000^2, 100000^2]$.

---

[6] Based on the property that $\forall q \in L_k$, there exists $q' \in L_{k-1}$ such that q' dominates q.
[7] www.rtreeportal.org.

We produce 1,000 queries with sizes generated uniformly from each group. We show the *total query time* for each group in Fig. 3.

It can be seen that our method is around 5 orders of magnitude faster than existing method PS. This demonstrates the substantial advantage of our index-based method to efficiently support batch query workloads. When query range increases, the running time of PS remains steadily due to the nature of plane-sweeping, while our method takes slightly more time when query size is in middle range. This is because the size of k-lines with k around $n/2$ tends to be larger than other values — as can be seen in Fig. 1(c).

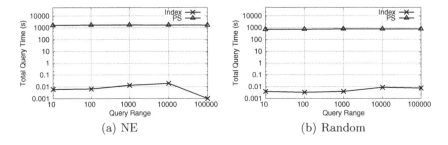

(a) NE   (b) Random

Fig. 3. Vary query range

Vary Dataset Size. We start with a synthetic dataset with 2000 points, and increase its size till 5x, and measure the ratio between PS's total query time over that of Index's. The result is plotted in Fig. 4(a). Clearly, our method has better scalability than PS as the time ratio raises up quickly when the data size increases. This is expected as our query time increases only logarithmically with n, while PS's increases superlinearly.

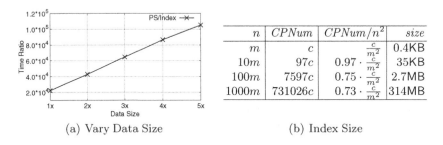

n	$CPNum$	$CPNum/n^2$	$size$
m	c	$\frac{c}{m^2}$	0.4KB
$10m$	$97c$	$0.97 \cdot \frac{c}{m^2}$	35KB
$100m$	$7597c$	$0.75 \cdot \frac{c}{m^2}$	2.7MB
$1000m$	$731026c$	$0.73 \cdot \frac{c}{m^2}$	314MB

(a) Vary Data Size   (b) Index Size

Fig. 4. Other experiments

Index Size. We start with an initial dataset size of $m = 10$ which results in total number of changing points $c = 36$. Then we increase the dataset till 1000 fold and show the increase of CPs in Fig. 4(b). As mentioned in Sect. 3.2, although our current bound on our index size is $O(n^3)$, we conjecture that it could be $O(n^2)$. Hence, we also show in the third column the ratio of index size over square of data size. The result strongly suggests that the conjecture may be true. We show the actual index size in the last column.

8   Conclusion

In this paper, we study the maximizing range sum problem. Existing methods for both exact and approximate query processing require $\Omega(n \log \log n)$ time. We propose a novel method based on indexing changing points, which results in an index of size $\Theta(n\lambda) = O(n^3)$. The index enables us to devise a non-trivial query processing algorithm with $O(\log n)$ complexity. Our method provides new space-time tradeoff for the maxRS and related problems. Experiments on real and synthetic datasets verify the efficiency of our method.

Acknowledgements. This work was partially done when X. Zhou and W. Wang visited Hong Kong Baptist University. W. Wang was supported by ARC DP 130103401 and 130103405. J. Xu was supported by HK-RGC grants 12201615 and HKBU12202414.

References

1. Imai, H., Asano, T.: Finding the connected components and a maximum clique of an intersection graph of rectangles in the plane. J. Algorithms **4**(4), 310–323 (1983)
2. Nandy, S.C., Bhattacharya, B.B.: A unified algorithm for finding maximum and minimum object enclosing rectangles and cuboids. Math. Appl. **29**(8), 45–61 (1995)
3. Choi, D.W., Chung, C.W., Tao, Y.: A scalable algorithm for maximizing range sum in spatial databases. Proc. VLDB Endow. **5**(11), 1088–1099 (2012)
4. Tao, Y., Hu, X., Choi, D.W., Chung, C.W.: Approximate MaxRS in spatial data-bases. PVLDB **6**(13), 1546–1557 (2013)
5. Choi, D.W., Chung, C.W., Tao, Y.: Maximizing range sum in external memory. ACM Trans. Database Syst. **39**(3), 21:1–21:44 (2014)
6. Das, A.S., Gupta, P., Srinathan, K., Kothapalli, K.: Finding maximum density axes parallel regions for weighted point sets. In: CCCG (2011)
7. Goodrich, M.T., Tsay, J.-J., Vengroff, D.E., Vitter, J.S.: External-memory com-putational geometry (preliminary version). In: FOCS, pp. 714–723 (1993)
8. Aggarwal, A., Imai, H., Katoh, N., Suri, S.: Finding k points with minimum diam-eter and related problems. J. Algorithms **12**, 38–56 (1991)
9. Datta, A., Lenhof, H.E., Schwarz, C., Smid, M.: Static and dynamic algorithms for k-point clustering problems. J. Algorithms **19**, 474–503 (1995)
10. Eppstein, D., Erickson, J.: Iterated nearest neighbors and finding minimal poly-topes. Discrete Comput. Geom. **11**, 321–350 (1994)
11. Segal, M., Kedem, K.: Enclosing k points in the smallest axis parallel rectangle. Inf. Process. Lett. **65**, 95–99 (1998)
12. Matouek, J.: On geometric optimization with few violated constraints. Discrete Comput. Geom. **14**, 365–384 (1995)
13. Hartigan, J.A.: Clustering Algorithms. Wiley, New York (1975)
14. Abellanas, M., Hurtado, F., Icking, C., Klein, R., Langetepe, E., Ma, L., Palop, B., Sacristán, V.: Smallest color-spanning objects. In: Meyer auf der Heide, F. (ed.) ESA 2001. LNCS, vol. 2161, pp. 278–289. Springer, Heidelberg (2001)
15. Efrat, A., Sharir, M., Ziv, A.: Computing the smallest k-enclosing circle and related problems. Comput. Geom. Theory App. **4**(3), 119–136 (1994)

16. Mahapatra, P.R.S., Karmakar, A., Das, S., Goswami, P.P.: k-enclosing axis-parallel square. In: Murgante, B., Gervasi, O., Iglesias, A., Taniar, D., Apduhan, B.O. (eds.) ICCSA 2011, Part III. LNCS, vol. 6784, pp. 84–93. Springer, Heidelberg (2011)
17. Das, S., Goswami, P.P., Nandy, S.C.: Smallest k-point enclosing rectangle and square of arbitrary orientation. Inf. Process. Lett. **94**, 259–266 (2005)
18. Mukherjee, M., Chakraborty, K.: A polynomial time optimization algorithm for a rectilinear partitioning problem with applications in VLSI design automation. Inf. Process. Lett. **83**, 41–48 (2002)
19. Tiwari, S., Kaushik, H.: Extracting region of interest (ROI) details using LBS infrastructure and web databases. In: MDM 2012, pp. 376–379 (2012)

An Efficient Method for Identifying MaxRS Location in Mobile Ad Hoc Networks

Yuki Nakayama$^{(\boxtimes)}$, Daichi Amagata, and Takahiro Hara

Department of Multimedia Engineering, Graduate School of Information Science and Technology, Osaka University, Yamadaoka 1-5, Suita-shi, Osaka, Japan
`nakayama.yuki@ist.osaka-u.ac.jp`

Abstract. In this paper, we address the problem of MaxRS (Maximizing Range Sum) query processing in mobile ad hoc networks (MANETs). We assume a MANET consisting of nodes which hold data items with scores and location information. A query originating node issues a MaxRS query with a size of a rectangle. Then, it retrieves the location of the rectangle which maximizes the sum of the scores of all data items covered by the rectangle (MaxRS location). We can employ MaxRS queries to enhance rescue operations in disaster sites, which is a typical example of MANETs. In this example, we can find a dense location where many victims exist in a disaster site. In a naive approach to processing MaxRS queries, each node forwards its entire dataset to the query originating node, and this node locally computes the result. This approach is inefficient because the network bandwidth is limited in MANETs. We therefore propose a communication-efficient method which retrieves the MaxRS location in two phases. Phase 1 plays the important role of pruning data items not necessary to identify the MaxRS location, and enables the retrieval of only necessary data items in phase 2. Simulation experiments demonstrate the efficiency of our approach.

Keywords: Mobile ad hoc networks · MaxRS query · Location data

1 Introduction

Due to advances in wireless technology enabling peer-to-peer communication, such as IEEE802.11, Bluetooth, and Wi-Fi Direct, mobile ad hoc networks (MANETs) have been receiving much research attention [9,11]. MANETs need no fixed communication infrastructures, hence we can employ a MANET in an environment where we cannot utilize the Internet, e.g., disaster sites [2].

In such a disaster site, we assume that rescue workers perform rescue operations by constructing a MANET consisting of wireless nodes. During the rescue operations, they store victim information on their holding nodes. An important requirement of the above application is to enhance the rescue operations; for example, it would be useful to find the location where help is needed most. MaxRS queries [6] achieve this. A MaxRS query requires a size of a rectangle, and it retrieves the location of a rectangle which maximizes the sum of the scores

© Springer International Publishing Switzerland 2016
S. Hartmann and H. Ma (Eds.): DEXA 2016, Part I, LNCS 9827, pp. 37–51, 2016.
DOI: 10.1007/978-3-319-44403-1_3

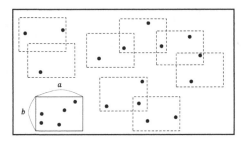

Fig. 1. An example of a MaxRS query

of all data items covered by the rectangle. In the above scenario, the locations and damaged scores of victims correspond to the locations and the scores of data items, respectively. We can intuitively see that a MaxRS query can find a location where many injured victims exist. We describe a concrete example below.

Example 1.1. In Fig. 1, the exterior rectangle denotes a network region, the dashed and solid rectangles denote rectangles whose size is specified by the query originator $(a \times b)$, and the black points denote the locations of data items with scores. For simplicity, let us assume that the scores of all data items are 1. In this example, the solid rectangle covers the maximum number of data items among all rectangles; hence, it is one of the answers to this query.

Consider another application of MaxRS queries in MANETs.

Example 1.2. We consider a mobile sensor network, which is a kind of MANET. Let us assume that each sensor node records its location every 1 s. If we execute a MaxRS query on the set of location data items held by sensor nodes, we can find the location where sensor nodes exist densely. As mobile sensor network applications require to collect the sensor readings on various locations, detecting such dense-node locations is valuable.

Although, as illustrated above, there are practical applications, MaxRS query processing has not been addressed in MANETs. Motivated by this, we tackle the problem of MaxRS query processing in MANETs. This problem has several challenges. A simple method for processing MaxRS queries is to collect all data items in a given network. This is because local aggregation is infeasible to obtain the exact result. However, this method is inefficient, since MANETs require communication-efficiency [2,9], to avoid packet losses and communication delay. Therefore, in this paper, we propose a communication-efficient method for MaxRS query processing in MANETs. Our proposed method processes MaxRS queries in two phases. In phase 1, a query originating node limits a search region which could be the answer to the MaxRS query. Then, in phase 2, it retrieves the data items whose locations are on the region and calculates the MaxRS location. By reducing unnecessary data forwarding, our proposed method avoids packet losses and communication delay, resulting in efficient retrieval of the MaxRS location.

Contribution. We summarize our contributions as follows.

- We address the problem of MaxRS query processing in MANETs. To the best of our knowledge, this is the first work to address this problem.
- We propose an efficient method for processing MaxRS queries in MANETs. Phase 1 limits the search region and prunes data items not necessary to identify the MaxRS location. Phase 2 retrieves the data items that have not been pruned in phase 1.
- Through simulation experiments, we show that our method reduces traffic while keeping accuracy, and is more efficient than the naive method.

Organization. In Sect. 2, we define the problem and describe the naive method. Section 3 describes our proposed method, and we show our experimental results in Sect. 4. We review related works in Sect. 5. Finally, Sect. 6 concludes the paper.

2   Preliminaries

We first introduce our assuming environment, and then formally define MaxRS queries in Sect. 2.1. We describe a naive method for MaxRS query processing in Sect. 2.2.

2.1   Problem Definition

Network Model. A MANET consists of n nodes which communicate with each other by the identical communication range. Each node is aware of the network region where all the nodes can move, and has its own dataset O_i. Each data item $o \in O_i$ has 2-dimensional location information $(o.x, o.y)$, and a score $(o.score)$ calculated by a certain scoring function, with $o.score > 0$. Note that we assume no data replications, and we argue that the location information on o $(o.x, o.y)$ is independent of the location of the node holding o.

DEFINITION 1 (MAXRS PROBLEM) [6]. *Given an infinite set of points in the network region P, $O = \cup O_i$, and a rectangle of a given size, this problem is to find the location $p \in P$ that maximizes $\sum_{o \in O_{R(p)}} o.score$. $R(p)$ is a rectangle whose center and size are p and the given size, respectively. $O_{R(p)}$ is the set of data items covered by $R(p)$.*

We can see that solving the above problem is impractical, since P is an infinite set. As [10] argues, however, this problem can be solved by transforming another problem, which is defined below.

DEFINITION 2 (RECTANGLE INTERSECTION PROBLEM (RI PROBLEM)) [10]. *Given a network region and a set of rectangles R, this problem is to find the region which intersects with the maximum number of rectangles in the network region.*

Literature [10] shows that the result of a MaxRS problem is obtained from the region returned by the RI problem, if the scores of given data items are 1. We illustrate this in Fig. 2.

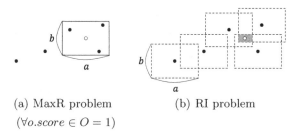

(a) MaxR problem   (b) RI problem

$(\forall o.score \in O = 1)$

Fig. 2. An example of MaxRS problem transformation

Example 2.1. In Fig. 2(a), the black points denote the locations of data items, and the white point is an answer to the MaxRS problem (the size of the rectangle is $a \times b$). Then, we consider the RI problem where the center locations of the given rectangles are the black points, and their sizes are $a \times b$ (Fig. 2(b)). The answer to the RI problem in Fig. 2(b) is the shaded region. Figure 2(b) shows that the white point, an answer of the MaxRS problem, exists in the shaded region.

By solving the RI problem, we can obtain an answer to the MaxRS problem. Algorithms for the RI problem have been proposed in [8,10], and their time complexity is $O(n \log n)$, where n is the number of rectangles. We here remove the assumption that the scores of all data items are 1, and introduce the following definition.

DEFINITION 3 (REGION SCORE). *Given O and a rectangle of a given size, we obtain a set of rectangles, R, whose centers and sizes are locations of the data items and the given size, respectively. Let the score of a rectangle in R be o.score when the center of the rectangle is the location of the data item o. Then, the score of a region is the sum of the scores of the rectangles on the region.*

We can find the region with the largest score, utilizing the algorithms of [8,10], by considering the region score [6,7]. We are now ready to define MaxRS queries below.

DEFINITION 4 (MAXRS QUERIES). *Given a dataset O and a rectangle of a given size, a MaxRS query retrieves the region with the largest score in the network region.*

In this paper, we address the problem of processing MaxRS queries as defined above. Note that we do not use node locations but *data locations* for MaxRS query processing. In addition, we do not deal with rotation of a rectangle [4], and focus on the usual case, as well as [1,6,7,12].

2.2 Naive Method

In this section, we describe a naive method for MaxRS query processing. In a nutshell, this method retrieves all data items in a given network, and then calculates an answer to the MaxRS query. The details of the naive method are described below.

First, a query originating node N_{org} floods a data request message ($DReq$) over the entire network. A $DReq$ consists of source node ID, parent node ID, and hop count from the query originating node. Each node N_i, which receives a $DReq$ for the first time, sets the source node of the $DReq$ as its parent node. Then, N_i broadcasts a $DReq$ with an incremented hop count, its ID as the source node ID, and its parent's ID as the parent node ID. An ACK message (ACK_{req}) is also sent to its parent node. An ACK_{req} consists of source node ID and destination node ID. In addition, N_i sets a waiting time WT to receive data items from its children nodes, as follows:

$$WT = Delay_{max} \cdot \frac{HOP_{max} - HOP_{count}}{HOP_{max}}$$

Note that $Delay_{max}$ is the maximum time for waiting, HOP_{max} is the maximum hop count in the network, and HOP_{count} is the hop count from the query originating node. Each node N_j, which receives a $DReq$ whose parent node ID is its ID or receives an ACK_{req}, sets the source node of the $DReq$ or the ACK_{req} as its child node. We can see that a tree topology is constructed in this $DReq$ flooding procedure. We explain the data reply procedure below.

If a node N_i has no children nodes, N_i sends a data reply message ($DRep$) with its own data items O_i to its parent node. A $DRep$ consists of source node ID, destination node ID, and data items. Each node N_j, which has received $DReps$ from all its children nodes or whose WT has expired, sends to its parent node a $DRep$ with its own data items O_j and the data items received from its children nodes. Each node which receives a $DRep$ sends an ACK message (ACK_{rep}) to the source node of the $DRep$. If nodes receive an ACK_{rep}, they consider that their parent nodes have received their forwarded data items, and thus no longer send the data items. If not, they re-send the data items to their parent nodes. In this way, N_{org} retrieves all the data items in the network. It then calculates an answer to the MaxRS query, based on the retrieved data items and the specified rectangle, using the algorithm proposed in [8].

This method retrieves all data items in a given network, and thus is inefficient when the number of data items in the network is large, resulting in large traffic, packet losses, and communication delay.

3 Proposed Method

3.1 Overview

First, we describe the overview of the proposed method. To reduce traffic, it is desirable to retrieve the data items on a limited region which includes the exact MaxRS location. To this end, we employ two-phase query processing. In phase 1, N_{org} divides the network region into specified cells, and collects cell scores from all the nodes. A cell score is the sum of the scores of the data items on the cell. Based on the collected cell scores, N_{org} finds a region which could be an answer to the query. In phase 2, N_{org} retrieves the data items on this region.

N_{org} then calculates an answer to the query based on the retrieved data items. This approach obtains a correct answer without retrieving all data items in the network region. We describe the detail of the proposed method in the following sections.

3.2 Phase 1: Finding a Region Which Could Be a Query Answer

In phase 1, N_{org} finds a region which could be an answer to the MaxRS query. To find the region, N_{org} divides the network region into rectangular cells of the same size; the width and length of the cells are $C.width = \frac{a}{k}$ and $C.length = \frac{b}{l}$ ($k, l \in \mathbb{N}$), respectively (a and b are respectively the width and length of the rectangle specified by N_{org}). Then, N_{org} attaches the width and length of the cells to a query, and floods the query over the entire network to share the division of the network region with all nodes. While flooding the query, a tree topology is constructed, with N_{org} as the root; that is, each node sets its parent node, its children nodes, and its WT as in the naive method.

Query receiving nodes, as well as N_{org}, divide the network region into cells specified by the width and length in the query. Figure 3 shows an example of such a division of the network region. In this figure, the network region is divided into 25 cells ($c_{00} - c_{44}$). Then, the query receiving nodes prepare an array which can store all cell scores, and add the scores of their own data items to the corresponding cell scores (i.e., array elements). Specifically, they add the score of each data item to the score of the cell on which the data item is. Note that the scores of cells containing no data items are 0.

We describe a reply procedure in phase 1 below. Each leaf node attaches the array to a reply message (Rep_{phase1}) and sends it to its parent node. A Rep_{phase1} consists of source node ID, destination node ID, and an array. Each node N_i which receives a Rep_{phase1} sends an ACK message (ACK_{rep1}) to the source node of the Rep_{phase1}. An ACK_{rep1} is processed in the same way as an ACK_{rep} of the naive method. Additionally, N_i merges the array in the Rep_{phase1} with its prepared array. Recall that the division of the network region is shared with all nodes; thus, all nodes know which element of the array has stored each cell score. When nodes have received Rep_{phase1}s from all of their children nodes or their WT has expired, they send a Rep_{phase1} with their array to their parent nodes. We explain this procedure below.

Example 3.1. Figure 4 shows an example of the reply procedure in phase 1. The network region is divided as shown in Fig. 3, each Rep_{phase1} thus includes an array which can store the scores of 25 cells. In Fig. 4, the table on the left shows the data items held by each node, and the balloons show the components of the Rep_{phase1} array. Although, for simplicity, Fig. 4 represents the locations of data items using cell IDs, these locations are in fact represented by 2-dimensional values such as latitude and longitude. Leaf node N_1 holds the data items D_1, D_2, and D_3, and they are on c_{00}, c_{00}, and c_{01}, respectively. N_1 thus conducts the following assignments to an array: $c_{00} = 10 + 10 = 20$, $c_{01} = 10$, and the other cell scores are 0. N_1 then sends a Rep_{phase1} including the array to its parent node

C.width   C.length

c_{40}	c_{41}	c_{42}	c_{43}	c_{44}
c_{30}	c_{31}	c_{32}	c_{33}	c_{34}
c_{20}	c_{21}	c_{22}	c_{23}	c_{24}
c_{10}	c_{11}	c_{12}	c_{13}	c_{14}
c_{00}	c_{01}	c_{02}	c_{03}	c_{04}

Fig. 3. An example of a network region division

Node	ID	Score	Location
N_1	D_1	10	c_{00}
	D_2	10	c_{00}
	D_3	10	c_{01}
N_2	D_4	10	c_{01}
	D_5	10	c_{11}
	D_6	10	c_{01}
N_3	D_7	30	c_{11}
	D_8	30	c_{11}
	D_9	10	c_{12}
N_4	D_{10}	50	c_{13}
	D_{11}	20	c_{20}
	D_{12}	30	c_{20}
	D_{13}	70	c_{21}

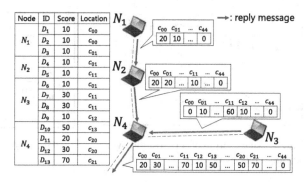

Fig. 4. An example of a reply procedure in phase 1

N_2. Leaf node N_3 sends a Rep_{phase1} to its parent node N_4, similarly to N_1. When N_4 receives the Rep_{phase1} from N_3, N_4 merges the cell scores of N_3 with its own cell scores. When N_2 receives the Rep_{phase1} from N_1, N_2 merges the cell scores of N_1 with its own cell scores. That is, N_2 sets $c_{00} = 20$, $c_{01} = 10 + 10 = 20$, and $c_{11} = 10$. Then, N_2 sends a Rep_{phase1} with the updated array to its parent node N_4. When N_4 receives the Rep_{phase1} from N_2, N_4 merges the cell scores of N_2 with its cell scores: N_4 sets $c_{00} = 20$, $c_{01} = 30$, $c_{11} = 70$, $c_{12} = 10$, $c_{13} = 50$, $c_{20} = 50$, $c_{21} = 70$. N_4 then sends a Rep_{phase1} with the cell scores to its parent node since N_4 has received a Rep_{phase1} from all its children nodes.

In phase 1, N_{org} retrieves all cell scores in the entire network. N_{org} next determines the set of cells which cover the locations of the data items necessary to find an exact answer to the MaxRS query, which we call *data request cell set*. We first define the relevant notations, and then explain how the data request cell set is determined.

Let c_{ij} represent an arbitrary cell, and $c_{i(j+1)}$ and $c_{(i+1)j}$ represent the cell to the right of c_{ij} and the cell above c_{ij}, respectively. We define the region of C as the union of the regions of all the cells in C, where C is a set of arbitrary cells. In addition, we define $score(C) = \sum_{\forall c_{ij} \in C} c_{ij}.score$, where $c_{ij}.score$ is the cell score of c_{ij}. Also, $C_{ij}^{kl} = \cup_{j \leq x < j+k} \cup_{i \leq y < i+l} c_{yx}$, hence $score(C_{ij}^{kl}) = \sum_{j \leq x < j+k} \sum_{i \leq y < i+l} c_{yx}.score$. Finally, C_{all} represents the set of all cells in the network region, and let τ^* be the score of a MaxRS location.

We are now ready to explain how the data request cell set is determined. Since the size of the specified rectangle is equal to the size of the region of C_{ij}^{kl} due to $a = C.width \times k$ and $b = C.length \times l$, we can see $\tau^* \geq \max_{c_{ij} \in C_{all}} \{score(C_{ij}^{kl})\}$. This means that $\max_{c_{ij} \in C_{all}} \{score(C_{ij}^{kl})\} (= \tau_{lb})$ is the lower bound score of a MaxRS location. We next consider where the rectangle answering to the MaxRS query could exist. The rectangle could overlap the cells in $C_{ij}^{(k+1)(l+1)}$ as shown in Fig. 5. In this case, the score of the rectangle could be $score(C_{ij}^{(k+1)(l+1)})$ (the region of $C_{ij}^{(k+1)(l+1)}$ is the shaded region in Fig. 5). If $score(C_{ij}^{(k+1)(l+1)}) \geq \tau_{lb}$, the data items on $C_{ij}^{(k+1)(l+1)}$ are necessary to calculate an exact answer to the

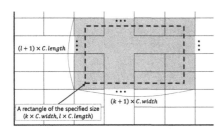

Fig. 5. An example of a rectangle over-lapping cells in $C_{ij}^{(k+1)(l+1)}$

Fig. 6. An example of cell scores

MaxRS query. On the other hand, the data items on c_{yx}, where all of $C_{ij}^{(k+1)(l+1)}$ that includes c_{yx} satisfy $score(C_{ij}^{(k+1)(l+1)}) < \tau_{lb}$, are not necessary, since an answer to the MaxRS query is not on c_{yx}.

N_{org} therefore calculates the $score(C_{ij}^{kl})$ of each cell $c_{ij} \in C_{all}$, and obtains τ_{lb}. Then, N_{org} calculates the $score(C_{ij}^{(k+1)(l+1)})$ of each cell $c_{ij} \in C_{all}$. If $score(C_{ij}^{(k+1)(l+1)}) \geq \tau_{lb}$, N_{org} sets the cells in $C_{ij}^{(k+1)(l+1)}$ as elements of the data request cell set. In this manner, the data request cell set is determined. Below, we illustrate an example of this procedure in the case where $k = l = 1$.

Example 3.2. Figure 6 shows an example of the cell scores retrieved by N_{org}. The values enclosed with a rectangle in a cell denote the scores of the respective cells, e.g., the scores of c_{00} and c_{01} are 100 and 80, respectively. τ_{lb} is $\max_{c_{ij} \in C_{all}}\{score(C_{ij}^{11})\} = 300$ at c_{11}. N_{org} calculates the $score(C_{ij}^{22})$ of each cell $c_{ij} \in C_{all}$. For example, $score(C_{03}^{22})$ is 80 (the sum of the scores of c_{03}, c_{04}, c_{13}, and c_{14}). After this calculation, N_{org} can see that the cell sets C_{ij}^{22}, where $score(C_{ij}^{22}) \geq \tau_{lb}$, are $\{c_{00}, c_{01}, c_{10}, c_{11}\}$, $\{c_{01}, c_{02}, c_{11}, c_{12}\}$, $\{c_{10}, c_{11}, c_{20}, c_{21}\}$, and $\{c_{11}, c_{12}, c_{21}, c_{22}\}$. Therefore, the data request cell set is $\{c_{00}, c_{01}, c_{02}, c_{10}, c_{11}, c_{12}, c_{20}, c_{21}, c_{22}\}$, and the region of the cell set is the shaded region in Fig. 6.

3.3   Phase 2: Retrieving Data Items in a Limited Region

In phase 2, N_{org} attaches the cell IDs in the data request cell set to a data request message ($DReq$), and floods the $DReq$ over the entire network in the same manner as in phase 1. Let S_{DReq} be the region covered by the data request cell set. In the $DReq$ flooding procedure, all nodes obtain S_{DReq}.

If each leaf node has data items on S_{DReq}, it sends a data reply message ($DRep$) with these data items to its parent node. If not, it sends a $DRep$ with no data items to its parent node. Each node which receives a $DRep$ sends an ACK message to the source node of the $DRep$. This ACK message is processed in the same manner as in the naive method's reply procedure. Each node N_j, which has received data items from all its children nodes or whose WT has expired, sends to its parent node a $DRep$ with the received data items and its own data items on S_{DReq}. Besides, if N_j does not receive the query in phase 1, N_j does not know the information on the division of the network region, and therefore sends

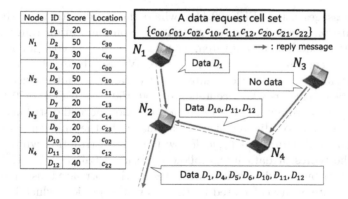

Node	ID	Score	Location
N_1	D_1	20	c_{20}
	D_2	50	c_{30}
	D_3	30	c_{40}
N_2	D_4	70	c_{00}
	D_5	50	c_{10}
	D_6	20	c_{11}
N_3	D_7	20	c_{13}
	D_8	20	c_{14}
	D_9	20	c_{23}
N_4	D_{10}	20	c_{02}
	D_{11}	30	c_{12}
	D_{12}	40	c_{22}

Fig. 7. An example of a reply procedure in phase 2

all its data items to its parent node. If the parent node has received the query in phase 1, it knows S_{DReq}, hence it can know the data items not on S_{DReq}. These data items are no longer sent. We illustrate an example of this reply procedure below.

Example 3.3. Figure 7 shows an example of the reply procedure in phase 2. Leaf nodes N_1 and N_3 send a $DRep$ to their parent nodes. N_1 holds a data item on S_{DReq}, D_1, and thus N_1 sends D_1 to its parent node N_2. On the other hand, N_3 sends a $DRep$ with no data items to its parent node N_4, since N_3 does not have data items on S_{DReq}. When N_4 receives a $DRep$ from its child node N_3, N_4 sends its data items on S_{DReq}, D_{10}, D_{11}, D_{12}, to its parent node N_2. When N_2 receives $DReps$ from all its children nodes (N_1 and N_4), N_2 sends the received data items D_1, D_{10}, D_{11}, and D_{12}, and its own data items D_4, D_5, and D_6 to its parent node.

In this way, N_{org} retrieves data items on S_{DReq}. N_{org} then calculates an answer to the MaxRS query, based on the retrieved data items.

4 Experiments

In this section, we show the experimental results of our method. We used a network simulator, Qualnet 6.1[1]. The algorithm proposed in [8] was implemented in C++, and all experiments were conducted on a PC with 3.6 GHz Intel Core i7 CPU and 16 GB RAM.

4.1 Simulation Environment

We specify the network region 600 [m] × 600 [m] and deploy 80 nodes in the region. Nodes move according to the random waypoint model [3] with the movement speed chosen from [0, 1.0] [m/s] and the pause time set at 30 [s]. Nodes

---

[1] http://www.scalable-networks.com.

transmit messages using IEEE802.11b, where the bandwidth is 11 Mbps, and the transmission power is selected so that the communication range is about 100 [m]. Each node holds d data items, and each data item consists of data ID, 2-dimensional coordinate (x, y), and a score. We use cluster distribution and uniform distribution for the coordinates of the data items. In the cluster distribution, we draw 4 clusters whose centers are (150, 150), (150, 450), (450 150), and (450, 450). The coordinates of data items (x, y) follow normal distribution whose mean is the center coordinates of a given cluster. In the uniform distribution, the coordinates of data items (x, y) follow uniform distribution whose range is [0, 600]. The scores of data items follow uniform distribution of [1, 1000], or correlated distribution with their coordinates. In the correlated distribution, the scores of data items are calculated by the following equation, which is based on the probability density function of pareto distribution:

$$o.score = \frac{MaxScore}{2} \cdot \frac{pq^p}{w^{(p+1)}} + 20z, \text{ where } w = \frac{2Distance}{Distance_{max}} + 1$$

$MaxScore$ is the maximum value of each cluster, which is determined by uniform distribution whose range is [200,1000]. $Distance$ is the distance between a location of a data item and the center of the cluster to which the data item belongs. $Distance_{max}$ is the maximum value of $Distance$. Besides, z follows normal distribution, while p and q are parameters of pareto distribution and we set $p = 2$ and $q = 1$. Note that the scores of data items are discrete integer values. Generated data items are randomly distributed to each node, thus the locations of the data items held by the node are not related to the location of the node. In our experiment, a query originating node specifies a square with sides of L [m] as the rectangle in a MaxRS query.

We compared our method with the naive method introduced in Sect. 2.2. We set $HOP_{max} = 9$ in all methods. We also set the $Delay_{max}$ of each method based on accuracy in a preliminary experiment. The naive method (NAIVE) set $Delay_{max} = 28$ [s] and the proposed method (TWOPHASE) set $Delay_{max} = 8$ [s] in phase 1 and $Delay_{max} = 20$ [s] in phase 2. Moreover, TWOPHASE(HALF), whose $Delay_{max}$s are half of TWOPHASE, set $Delay_{max} = 4$ [s] in phase 1 and $Delay_{max} = 10$ [s] in phase 2.

In the above simulation environment, we generated 50 queries, with randomly chosen query originators, and measured the average of the following criteria.

- Traffic: is the total volume of messages sent during a query processing. Table 1 shows the size of messages in the respective methods. Here, D is the number of data items in a given message, $AllCell$ is the number of cells in the network, and $ReqCell$ is the number of cells in the data request cell set.
- Accuracy: is the ratio of the number of retrieved data items on an answer rectangle for a given MaxRS query to the total number of all data items on the rectangle.
- Search time: is the time required for the query originating node to obtain an answer to its MaxRS query.

Table 1. Size of messages

Method	Type	Size [B]
Naive method	Data request message	16
	Data reply message	$24 + 16 \times D$
Proposed method	Query message	24
	Reply message in phase 1	$16 + 4 \times AllCell$
	Data request message	$24 + 4 \times ReqCell$
	Data reply message	$24 + 16 \times D$
	Ack message to reply message in phase 1	12
Both methods	Ack message to query or data request message	12
	Ack message to data reply message	16

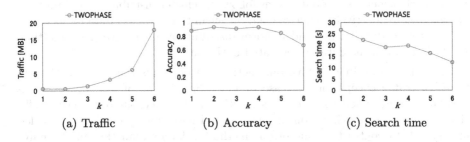

(a) Traffic   (b) Accuracy   (c) Search time

Fig. 8. Impact of k

4.2 Results

Impact of k. First, we investigated the impact of parameters k and l, which are related to the size of cells (Sect. 3.2), and Fig. 8 shows the result. For simplicity, we set $k = l$ since the number of the potential combination of these values is infinite. In this investigation, we used both cluster distribution and correlated distribution, and set $d = 80$ and $L = 60$.

In Fig. 8(a), we can see that traffic increases as k becomes large. When k is large, the number of cells is large due to $C.width = \frac{a}{k}$ and $C.length = \frac{b}{l}$, and this in turn results in large sized reply messages in phase 1. As we see in Fig. 8(b), accuracy is high when k is between 1 and 5. However, accuracy decreases at $k = 6$. This is because the size of reply messages in phase 1 is so large, and thus more packet losses occur, and the probability of finding the correct region which could be a MaxRS location decreases. In Fig. 8(c), search time is reduced as k becomes large. When k is large, the number of forwarded data items is small because the limited region is small. This results in fewer packet losses, thus more nodes forward data items before their WTs expire.

Based on the highest accuracy in this result, we set $k = l = 2$ in the following investigations.

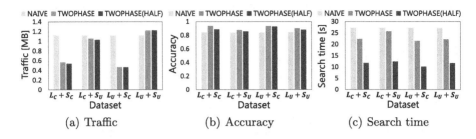

(a) Traffic   (b) Accuracy   (c) Search time

Fig. 9. Impact of a dataset.

Impact of Datasets. We investigated the impact of datasets, and Fig. 9 shows the result. In this investigation, we set $d = 80$ and $L = 60$. In Fig. 9, L_C and L_U mean that locations of data items follow the cluster distribution and uniform distribution, respectively, while S_C and S_U mean that scores of data items follow the correlated distribution and uniform distribution, respectively. In addition, $L_C + S_C$ denotes the dataset generated by L_C and S_C.

As can be seen in Fig. 9(a), our method outperforms the naive method in terms of traffic except for the case of $L_U + S_U$. Traffic of our method is in particular small when the score distribution is S_C, because our method efficiently limits the search region in phase 1, and fewer data items are forwarded in phase 2. In contrast, traffic of our method is larger than that of the naive method in the case of $L_U + S_U$. If the score distribution is uniform, phase 1 of our method is not helpful to limit the search region. In Fig. 9(b), we can see that accuracy of our method is higher than that of the naive method. These results show that our method retrieves only necessary data items, demonstrating communication-efficiency. In Fig. 9(c), location distribution and score distribution have no significant effect on search time in the naive method. On the other hand, search time of our method is smallest in $L_U + S_C$. The number of necessary data items is also smallest in this case, and thus fewer packet losses occur.

Impact of the Number of Data Items. We investigated the impact of the number of data items held by each node, and Fig. 10 shows the result. In this investigation, we used $L_C + S_C$ and $L = 60$.

Figures 10(a) and (b) show that traffic increases and accuracy decreases as d becomes large in all methods. When d is large, the number of forwarded data items is large, which leads to more packet losses and lower accuracy. When d is large, traffic of TWOPHASE is smaller than that of the naive method, and yet TWOPHASE keeps high accuracy. This shows that our approach, which involves retrieval of data items on a limited region, is efficient for finding the MaxRS location. Accuracy of TWOPHASE(HALF) is lower than that of TWOPHASE because more packet losses occur due to the shorter $Delay_{max}$. As Fig. 10(c) shows, in all methods, search time increases as d becomes large, and finally reaches $Delay_{max}$ in the naive method or the sum of $Delay_{max}$ in phase 1 and $Delay_{max}$ in phase 2 in our method, respectively. This is because, when d is

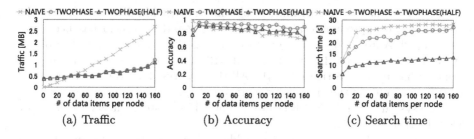

Fig. 10. Impact of d

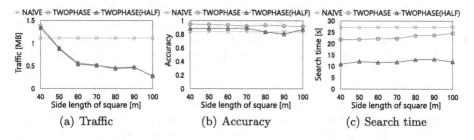

Fig. 11. Impact of L

large, the number of packet losses increases, and thus the probability that nodes cannot receive data items from their children nodes increases. As a consequence, more nodes wait until their WTs expire.

Impact of the Side Length of the Square. We investigated the impact of the side length of the square, and Fig. 11 shows the result. In this investigation, we used $L_C + S_C$ and $d = 80$.

In Fig. 11(a), traffic of the naive method is constant regardless of L, whereas traffic of our method is affected by L. This is because the naive method retrieves all the data items in the network, whereas in our method, the message size in phase 1 becomes large when L is small, and this results in increased traffic. In Figs. 11(b) and (c), we can see that both accuracy and search time of the naive method are, like traffic, unaffected by L, while our method outperforms the naive method in terms of accuracy in most cases.

5 Related Work

Although several efforts have been conducted on query processing in MANETs, all have assumed queries with totally different semantics from MaxRS queries, such as top-k [2], kNN [9], and reverse kNN [11]. Therefore, we focus on works dealing with MaxRS queries in different environments.

Numerous works have tackled the MaxRS problem in centralized database systems [1,4,6–8,10,12]. The works in [8,10] have proposed in-memory algorithms based on the *plane-sweep* algorithm, while the works in [6,7] have proposed external-memory algorithms, to obtain an exact MaxRS location. On the other hand, an approximate algorithm for the MaxRS problem has been proposed in [12]. A literature [4] has proposed an algorithm for the MaxRS problem with rotation of the specified rectangle. MaxRS monitoring in data stream environments has been investigated in [1]. All these works, however, have proposed algorithms to reduce the time complexities and I/O complexities in centralized database systems. Since MANETs are decentralized systems, it is hard to apply these algorithms simply.

Several works have addressed the problem of region retrieval (as in a MaxRS query) in wireless sensor networks [5,13]. The work in [13] has proposed an energy-efficient method to process max regional aggregate (MaxRA) queries, which retrieve the region that maximizes the aggregate value (e.g., sum and average) of sensor readings in a region of a given size; a MaxRS query is thus a kind of MaxRA query. In addition, an approximation approach which selects certain regions and sensors has been proposed. It is hard to apply the approach since the approach assumes that nodes' locations are equal to the data locations held by the nodes (it is general in wireless sensor networks). Region-based queries have also been proposed in [5], where the query retrieves the region satisfying the condition that the aggregate value of sensor readings in the region of a given size falls within a given interval. By clustering sensor nodes, this work achieves energy-efficient query processing.

6   Conclusion

In this paper, we addressed the problem of identifying a MaxRS location in MANETs. To efficiently find a MaxRS location, we employ two phase query processing. Phase 1 retrieves not data items but the aggregate scores of data items in order to limit the search region which could contain a correct MaxRS location. Phase 2 then retrieves only the necessary data items on this search region. Our approach achieves fewer packet losses and high accuracy due to low traffic, which is demonstrated by our experimental results.

When the number of data items is large, however, our method retrieves a large number of data items; and in addition, the performance of our method is affected by the distribution of scores of data items. As a part of future work, we plan to design a method which can further reduce traffic and is less affected by the distribution of scores of data items.

Acknowledgement. This research is partially supported by the Grant-in-Aid for Scientific Research (A) (26240013) of the Ministry of Education, Culture, Sports, Science and Technology, Japan, and JST, Strategic International Collaborative Research Program, SICORP.

References

1. Amagata, D., Hara, T.: Monitoring MaxRS in spatial data streams. In: EDBT, pp. 329–340 (2016)
2. Amagata, D., Sasaki, Y., Hara, T., Nishio, S.: CTR: an efficient cluster-based top-k query routing in MANETs. In: MoMM, pp. 225–234 (2014)
3. Camp, T., Boleng, J., Davies, V.: A survey of mobility models for ad hoc network research. Wirel. Commun. Mob. Comput. **2**(5), 483–502 (2002)
4. Chen, Z., Liu, Y., Wong, R.C.W., Xiong, J., Cheng, X., Chen, P.: Rotating MaxRS queries. Inf. Sci. **305**, 110–129 (2015)
5. Choi, D.W., Chung, C.W.: REQUEST+: a framework for efficient processing of region-based queries in sensor networks. Inf. Sci. **248**, 151–167 (2013)
6. Choi, D.W., Chung, C.W., Tao, Y.: A scalable algorithm for maximizing range sum in spatial databases. PVLDB **5**(11), 1088–1099 (2012)
7. Choi, D.W., Chung, C.W., Tao, Y.: Maximizing range sum in external memory. ACM TODS **39**(3), 21:1–21:44 (2014)
8. Imai, H., Asano, T.: Finding the connected components and a maximum clique of an intersection graph of rectangles in the plane. J. Algorithms **4**(4), 310–323 (1983)
9. Komai, Y., Sasaki, Y., Hara, T., Nishio, S.: kNN query processing methods in mobile ad hoc networks. IEEE TMC **13**(5), 1090–1103 (2014)
10. Nandy, S.C., Bhattacharya, B.B.: A unified algorithm for finding maximum and minimum object enclosing rectangles and cuboids. Comput. Math. Appl. **29**(8), 45–61 (1995)
11. Nghiem, T.P., Maulana, K., Nguyen, K., Green, D., Waluyo, A.B., Taniar, D.: Peer-to-peer bichromatic reverse nearest neighbours in mobile ad-hoc networks. J. Parallel Distrib. Comput. **74**(11), 3128–3140 (2014)
12. Tao, Y., Hu, X., Choi, D.W., Chung, C.W.: Approximate MaxRS in spatial databases. PVLDB **6**(13), 1546–1557 (2013)
13. Zhuang, Y., Chen, L.: Max regional aggregate over sensor networks. In: ICDE, pp. 1295–1298 (2009)

Data Mining

Discovering Periodic-Frequent Patterns in Transactional Databases Using All-Confidence and Periodic-All-Confidence

J.N. Venkatesh[1]([✉]), R. Uday Kiran[2], P. Krishna Reddy[1],
and Masaru Kitsuregawa[2,3]

[1] Kohli Center on Intelligent Systems (KCIS),
International Institute of Information Technology Hyderabad, Hyderabad, India
jn.venkatesh@research.iiit.ac.in,pkreddy@iiit.ac.in
[2] Institute of Industrial Science, The University of Tokyo, Tokyo, Japan
{uday_rage,kitsure}@tkl.iis.u-tokyo.ac.jp
[3] National Institute of Informatics, Tokyo, Japan

Abstract. Periodic-frequent pattern mining involves finding all frequent patterns that have occurred at regular intervals in a transactional database. The basic model considers a pattern as periodic-frequent, if it satisfies the user-specified minimum support ($minSup$) and maximum periodicity ($maxPer$) constraints. The usage of a single $minSup$ and $maxPer$ for an entire database leads to the *rare-item problem*. When confronted with this problem in real-world applications, researchers have tried to address it using the item-specific $minSup$ and $maxPer$ constraints. It was observed that this extended model still generates a significant number of uninteresting patterns, and moreover, suffers from the issue of specifying item-specific $minSup$ and $maxPer$ constraints. This paper proposes a novel model to address the rare-item problem in periodic-frequent pattern mining. The proposed model considers a pattern as interesting if its *support* and *periodicity* are close to that of its individual items. The *all-confidence* is used as an interestingness measure to filter out uninteresting patterns in *support* dimension. In addition, a new interestingness measure, called *periodic-all-confidence*, is being proposed to filter out uninteresting patterns in *periodicity* dimension. We have proposed a model by combining both measures and proposed a pattern-growth approach to resolve the rare-item problem and extract interesting periodic-frequent patterns. Experimental results show that the proposed model is efficient.

Keywords: Data mining · Rare-item problem · Periodic patterns

1 Introduction

Periodic-frequent pattern mining is an important model in data mining. It involves finding all frequent patterns that are occurring at regular intervals in a transactional database. The periodic-frequent patterns provide useful

© Springer International Publishing Switzerland 2016
S. Hartmann and H. Ma (Eds.): DEXA 2016, Part I, LNCS 9827, pp. 55–70, 2016.
DOI: 10.1007/978-3-319-44403-1_4

Table 1. Running example: a transactional database

tid	Items	tid	Items
1	a, b	7	a, b, c, e
2	a, b, d	8	c, d
3	c, d, g	9	c, d
4	c, e, f	10	a, b, e, f
5	a, b	11	c, d, g
6	h	12	a, e, f

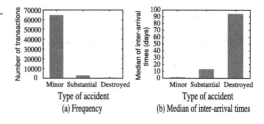

Fig. 1. Statistics on different damage types in FAA data set.

information in many real-world applications. Examples include finding regularities in body sensor networks [14], intrusion detection in computer networks [8], and finding minor events in twitter data [7].

The basic model of periodic-frequent patterns [13] is as follows. Let I be the set of items, and $X \subseteq I$ be a **pattern** (or an itemset). A pattern containing β number of items is called a β-**pattern**. A **transaction**, $t_k = (tid, Y)$ is a tuple, where $tid \in \mathbb{R}$ represents the transaction-id (or timestamp) at which the pattern Y has occurred. A **transactional database** TDB over I is a set of transactions, $TDB = \{t_1, \cdots, t_m\}$, $m = |TDB|$, where $|TDB|$ can be defined as the number of transactions in TDB. For a transaction $t_k = (tid, Y)$, $k \geq 1$, such that $X \subseteq Y$, it is said that X occurs in t_k and such transaction-id is denoted as tid^X. Let $TID^X = \{tid_j^X, \cdots, tid_k^X\}$, $j, k \in [1, m]$ and $j \leq k$, be an **ordered set of transaction-ids** where X has occurred in TDB. In this paper, we call this list of transaction-ids of X as **tid-list** of X. The number of transactions containing X in TDB is defined as the **support** of X and denoted as $sup(X)$. That is, $sup(X) = |TID^X|$. Let tid_q^X and tid_r^X, $j \leq q < r \leq k$, be the two consecutive transaction-ids in TID^X. The time difference (or an inter-arrival time) between tid_r^X and tid_q^X is defined as a **period** of X, say p_a^X. That is, $p_a^X = tid_r^X - tid_q^X$. Let $P^X = (p_1^X, p_2^X, \cdots, p_r^X)$ be the set of all *periods* for a pattern X. The **periodicity** of X denoted as $per(X) = max(p_1^X, p_2^X, \cdots, p_r^X)$. The pattern X is a **frequent pattern** if $sup(X) \geq minSup$, where $minSup$ refers to the user-specified *minimum support* constraint. The frequent pattern X is said to be **periodic-frequent** if $per(X) \leq maxPer$, where $maxPer$ refers to the user-specified *maximum periodicity* constraint. The **problem definition** of periodic-frequent pattern mining involves discovering all patterns in TDB that satisfy the user-specified $minSup$ and $maxPer$ constraints.

Example 1. Table 1 shows the transactional database with the set of items $I = \{a, b, c, d, e, f, g, h\}$. The set of items '$a$' and '$b$,' i.e., '$ab$' is a pattern. This pattern contains only two items. Therefore, this is a 2-pattern. In the first transaction, $t_1 = \{1 : ab\}$, 1 denotes the transaction-id at which the pattern 'ab' has appeared in the database. In the entire database, this pattern appears at the transaction-ids of $1, 2, 5, 7$ and 10. Therefore, $TID^{ab} = \{1, 2, 5, 7, 10\}$. The *support* of '$ab$,' i.e., $sup(ab) = |TID^{ab}| = |1, 2, 5, 7, 10| = 5$. If the user-specified $minSup = 5$, then 'ab' is a frequent pattern as $sup(ab) \geq minSup$. The periods for this pattern are: $p_1^{ab} = 1 \ (= 1-tid_{ini}), p_2^{ab} = 1 \ (= 2-1), p_3^{ab} = 3 \ (= 5-2), p_4^{ab} = 2 \ (= 7-5), p_5^{ab} = 3 \ (10-7)$ and $p_6^{ab} = 2 \ (= tid_{fin}-10)$, where $tid_{ini} = 0$ represents the transaction-id of initial transaction and $tid_{fin} = 12$ represents the transaction-id of final transaction in the database. Therefore, $P^{ab} = (1,1,3,2,3,2)$. The *periodicity* of 'ab,' i.e., $per(ab) = max(1,1,3,2,3,2) = 3$. If the user-defined $maxPer = 3$, then the frequent pattern 'ab' is a periodic-frequent pattern because $per(ab) \leq maxPer$.

The *support* and *periodicity* are two dimensions to determine the interest-ingness of a periodic-frequent pattern. The constraints, $minSup$ and $maxPer$, determine the interestingness of a pattern with respect to these two dimensions. Since only a single $minSup$ and $maxPer$ is used for the whole database, the model works efficiently in the databases in which all the items have uniform *support* and similar *periodicity*. However, this is often not the case as items are non-uniformly distributed in many real-world databases. That is, some items appear very frequently in the data, while others rarely appear. Moreover, rare items typically have longer *periods* (or inter-arrival times) as compared with the *periods* of the frequent items.

Example 2. Consider a case of accident database in which reports related to the completely *destroyed* vehicles do not appear as frequently as the reports related to the *minor* damages to a vehicle. As a result, former type of accidents generally have less frequency and longer inter-arrival times as compared against the latter type of accidents. The same can be observed from Figs. 1(a) and (b), which respectively show the *frequency* and *median* of inter-arrival times of three different damage types reported in the Federal Aviation Administration (FAA) database [1]. For a domain expert, any information pertaining to *destroyed* aircrafts may be found useful as it includes materialistic and/or human loss.

Henceforth, finding periodic-frequent patterns with a single $minSup$ and $maxPer$ leads to the following problems:

- If the $minSup$ is set too high and/or the $maxPer$ is set too short, we will miss the interesting periodic-frequent patterns involving rare items.
- In order to find the interesting periodic-frequent patterns involving rare items, we have to set a low $minSup$ and a long $maxPer$. However, this may result in combinatorial explosion producing too many patterns, because frequent items can combine with one another in all possible ways and many of them will be meaningless.

This dilemma is known as the *rare-item problem* [15] (refer Example 3). In this paper, we make an effort to address this problem in periodic-frequent pattern mining.

Example 3. Consider the rare items 'e' and 'f' in Table 1. If we set a high $minSup$ and a short $maxPer$, say $minSup = 5$ and $maxPer = 3$, we will miss the periodic-frequent patterns containing these rare items. In order to discover the periodic-frequent patterns containing these rare items, we have to set a low *minSup* and a long *maxPer*, say *minSup* = 2 and *maxPer* = 6. All periodic-frequent patterns discovered at these threshold values are shown in the column titled **I** in Table 2. It can be observed from this table that setting a low *minSup* and a long *maxPer* has not only resulted in finding '*ef*' as a periodic-frequent pattern, but also resulted in generating the uninteresting patterns '*ce*' and '*cd*' as periodic-frequent patterns. The pattern '*ce*' is uninteresting (with respect to *support* dimension), because the rare item '*e*' is randomly occurring with a frequent item '*c*' in very few transactions. The pattern '*cd*' is uninteresting (with respect to *periodicity* dimension), because it contains the frequent items '*c*' and '*d*' appearing together at very long inter-arrival times (or *periodicity*).

Table 2. Periodic-frequent patterns discovered from Table 1. The terms *Pat, s, allConf, p* and *perAllConf* refer to *pattern, support, all-confidence, periodicity* and *periodic-all-confidence*, respectively. The columns titled **I, II** and **III** represent the periodic-frequent patterns generated using basic model, extending *all-confidence* to the basic model and the proposed model, respectively.

Pat	s	allConf	p	perAllConf	I	II	III	Pat	s	allConf	p	perAllConf	I	II	III
a	6	1	3	1	✓	✓	✓	f	3	1	6	1	✓	✓	✓
b	5	1	3	1	✓	✓	✓	ab	5	0.833	3	1	✓	✓	✓
c	6	1	3	1	✓	✓	✓	ef	3	0.75	6	1.5	✓	✓	✓
d	5	1	5	1	✓	✓	✓	ce	2	0.4	5	1.67	✓	×	×
e	4	1	4	1	✓	✓	✓	cd	4	0.8	5	1.67	✓	✓	×

In this paper, we propose a novel model to extract the interesting periodic-frequent patterns by addressing the *rare-item problem*. We consider a pattern as interesting if it satisfies the following two conditions: (*i*) if the *support* of a pattern is close to the *support* of its individual items, and (*ii*) if the *periodicity* of a pattern is close to the *periodicity* of its individual items. For this, we employ two measures. To filter out patterns based on *support* and resolve rare-item problem in *support* dimension, we employ *all-confidence* [9]. Similarly, to filter out patterns based on *periodicity* and resolve rare-item problem in *periodicity* dimension, we propose a new measure called *periodic-all-confidence*. These two measures facilitate us to achieve the objective of generating interesting periodic-frequent patterns involving rare items without causing the generation of too many uninteresting patterns. A pattern-growth algorithm, Extended Periodic-Frequent pattern-growth (EPF-growth), has also been proposed to extract all interesting periodic-frequent patterns. Experimental results demonstrate that the proposed model can discover useful information and is efficient as compared to the existing approaches.

The rest of the paper is organized as follows. Section 2 describes the related work of finding periodic-frequent patterns in a transactional database. Section 3 introduces the extended model of periodic-frequent patterns. Section 4 describes the EPF-growth algorithm. Section 5 reports on the experimental results. Finally, Sect. 6 concludes the paper with future research directions.

2   Related Work

The problem of finding periodic patterns has been widely studied in time series data [3,17]. These studies consider time series data as a symbolic sequence, and therefore, do not take into account the actual temporal information of the items within the data. Ozden et al. [10] have enhanced the transactional database by a time attribute that describes the time when a transaction has appeared and investigated the periodic behavior of the patterns to discover cyclic association rules. In this study, a database needs to be fragmented into non-overlapping subsets with respect to time. Henceforth, the drawback of this study is that patterns (or association rules) that span multiple windows cannot be discovered.

Tanbeer et al. [13] have proposed a simplified periodic-frequent model, which does not require data fragmentation. This model implicitly assumes all items occur uniformly in the data, and henceforth, suffer from *rare-item problem*.

Kiran et al. [6] have tried to address the *rare-item problem* by finding periodic-frequent patterns using multiple *minSup* and *maxPer* values. In that model, every item in the database is specified with the *minimum item support* (*minIS*) and the *maximum item periodicity* (*maxIP*). Next, the *minSup* and *maxPer* of a pattern X are specified as follows:

$$minSup(X) = min(minIS(i_j)|\forall i_j \in X) \tag{1}$$
$$maxPer(X) = max(maxIP(i_j)|\forall i_j \in X) \tag{2}$$

where, $minIS(i_j)$ and $maxIP(i_j)$ represent the *minimum item support* and *maximum item periodicity* of an item $i_j \in X$. A pattern-growth algorithm, MCPF-growth has been discussed to find the patterns. The periodic-frequent patterns discovered by that model do not satisfy the *anti-monotonic property*. That is, all non-empty subsets of a periodic-frequent pattern may not be periodic-frequent patterns. Henceforth, MCPF-growth is computationally expensive or impracticable in very large real-world databases.

Surana et al. [11] have proposed an alternative model to address the rare-item problem. In that model, the *minSup* and *maxPer* of a pattern are specified as:

$$minSup(X) = max(minIS(i_j)|\forall i_j \in X) \tag{3}$$
$$maxPer(X) = min(maxIP(i_j)|\forall i_j \in X) \tag{4}$$

A pattern-growth algorithm, MaxCPF-growth has been discussed to find the patterns. The periodic-frequent patterns discovered by that model satisfy the *anti-monotonic property*. Therefore, MaxCPF-growth is computationally inexpensive than MCPF-growth [6], and practicable in very large real-world databases.

The limitations of these two approaches and the proposed model are discussed in next section.

3 Extended Model of Periodic-Frequent Patterns

3.1 Limitations of Existing Approaches

An open problem that is common to above two studies [6,11] is the methodology to specify items' $minIS$ and $maxIP$ values. Kiran et al. [6] have described the following methodology to address this problem:

$$minIS(i_j) \;=\; max(\gamma \times S(i_j),\; LS)$$
$$and \tag{5}$$
$$maxIP(i_j) \;=\; max(\beta \times S(i_j) + Per_{max},\; Per_{min})$$

where $i \in I$ and $S(i)$ is the *support* of the item i. In Eq. 5, LS is the user-specified lowest *minimum item support* allowed and $\gamma \in [0,1]$ is a parameter that controls how the $minIS$ values for items should be related to their *supports*. In Eq. 5, Per_{max} and Per_{min} are the user-specified maximum and minimum *periodicities* such that $Per_{max} \geq Per_{min}$ and $\beta \in [-1,0]$ is a user-specified constant.

Although Eq. 5 facilitates every item to have different $minIS$ and $maxIP$ values, it suffers from the following limitations: (i) This methodology requires several input parameters from the user. (ii) Equation 5 determines the $maxIP$ of an item by taking into account only its *support*. As a result, this equation implicitly assumes that all items having the same *support* will also have similar *periodicities* in a transactional database. However, this is seldom the case as items with similar *support* can have different *periodicities*. We have observed that employing this methodology to specify items' $maxIP$ values in the transactional databases, where items can have similar *support* but different *periodicities* can still lead to the *rare-item problem*.

Example 4. Consider a hypothetical transactional database containing 100 transactions. Let 'x' and 'y' be two items in the database having the same *support* (say, $sup(x) = sup(y) = 40$), but different *periodicities* (say, $per(x) = 11$ and $per(y) = 30$). Since Eq. 5 determines the $maxIP$ values by taking into account only the *support* of the items, both 'x' and 'y' will be assigned a common $maxIP$ value although their actual *periodicity* is different from one another. This can result either in missing interesting patterns or generating too many patterns. For instance, if we set $\beta = -0.5$, $Per_{min} = 10$ and $Per_{max} = 50$, then $maxIP(x) = maxIP(y) = 20$. In this case, we miss the periodic-frequent patterns containing 'y' because $per(y) \nleq maxIP(y)$. In order to find the periodic-frequent patterns containing both 'x' and 'y' items, we have to set a high β value. When β is set at -0.375, we derive $maxIP(x) = maxIP(y) = 35$. In this case, we find periodic-frequent patterns containing 'y' because $per(y) \leq maxIP(y)$. However, we may also witness too many patterns containing the item 'x' because its $maxIP$ value is three times higher than its *periodicity*.

We now describe the proposed model that do not suffer from this problem.

3.2 Proposed Model

To address the *rare-item problem*, we need an approach that extracts interesting periodic-frequent patterns involving both frequent and rare items yet filtering out uninteresting patterns. After conducting the initial investigation on the nature of interesting patterns found in various databases, we have made a key observation that most of the interesting periodic-frequent patterns discovered in a database have their *support* and *periodicity* close to that of its individual items. The following example illustrates our observation.

Example 5. In a supermarket, cheap and perishable goods (e.g., bread and butter) are purchased more frequently and periodically than the costly and durable goods (e.g., bed and pillow). Among all the possible combinations of the above four items, we normally consider {bread, butter} and {bed, pillow} as interesting patterns, because only these two patterns generally have *support* and *periodicity* close to the *support* and *periodicity* of its individual items. All other uninteresting patterns, {bread, bed}, {bread, pillow}, {butter, bed} and {butter, pillow}, generally have *support* and *periodicity* relatively far away from the *support* and *periodicity* of its individual items as compared against the above two patterns.

Henceforth, in this paper we consider a pattern as interesting if its *support* and *periodicity* are close to the *support* and *periodicity* of its individual items. In this context, we need two measures to determine the interestingness of a pattern with respect to both *support* and *periodicity* dimensions.

In the literature, researchers have discussed several measures to address the *rare-item problem* in *support* dimension [12,16]. Each measure has a selection bias that justifies the significance of a knowledge pattern. As a result, there exists no universally acceptable best measure to judge the interestingness of a pattern for any given database. In this paper, we use *all-confidence* to address the *rare-item problem* in frequency dimension. The reasons for choosing this measure over other measures are as follows: (*i*) The *all-confidence* assesses the interestingness of a pattern by determining how close is its *support* with respect to the *support* of all of its items in a database. (*ii*) The *all-confidence* satisfies the *anti-monotonic property* [9]. This property plays a key role in the practicable ability of our model. (*iii*) The *all-confidence* satisfies the *null-invariance property* [5]. This property facilitate us to discover genuine interesting patterns without being influenced by the item co-absence in the database.

Continuing with the basic model of periodic-frequent patterns (discussed in Sect. 1), the proposed model is as follows.

Definition 1. (*All-confidence of X*) *The all-confidence of X, denoted as* $allConf(X)$*, is the ratio of support of X to the maximal support of an item* $i_j \in X$*. That is,* $allConf(X) = \frac{sup(X)}{max(sup(i_j)|\forall i_j \in X)}$*.*

For a pattern X, $allConf(X) \in (0,1]$. As per the *all-confidence* measure, a pattern is interesting in *support* dimension if its *support* is close to the *support* of all of its items. The parameter *minAllConf* indicates the user-specified minimum *all-confidence* threshold value. Based on *minSup* and *minAllConf* thresholds, all the interesting patterns involving rare items in *support* dimension are extracted.

The usage of *all-confidence* alone is insufficient to completely address the *rare-item problem* in periodic-frequent pattern mining. The reason is this measure does not take into account the *periodicity* dimension of a pattern.

Example 6. The column titled **II** in Table 2 shows the periodic-frequent patterns discovered when *all-confidence* is used along with *support* and *periodicity* measures. *The minSup, minAllConf and maxPer values used to find these patterns are 2, 0.6 and 6, respectively.* It can be observed from the discovered periodic-frequent patterns that though *all-confidence* is able to prune the uninteresting pattern 'ce,' it has failed to prune another uninteresting pattern 'cd' from the list of periodic-frequent patterns generated by the basic model. Henceforth, the *rare-item problem* has to be addressed with respect to both *support* and *periodicity* dimensions.

As there exists no measure in the literature that determines the interestingness of a pattern with respect to the *periodicities* of all of its items, we propose a new measure, ***periodic-all-confidence***, to extract interesting patterns in *periodicity* dimension involving rare items, which is defined as follows.

Definition 2. (***Periodic-all-confidence of X***) *The periodic-all-confidence of* X*, denoted as* $perAllConf(X)$*, is the ratio of periodicity of* X *to the minimal periodicity of an item* $i_j \in X$*. That is,* $perAllConf(X) = \frac{per(X)}{min(per(i_j)|\forall i_j \in X)}$*.*

For a pattern X, $perConf(X) \in [1, \infty)$. As per the *periodic-all-confidence* measure, a pattern is interesting in *periodicity* dimension, if the *periodicity* of a pattern is close to the *periodicity* of all of its items. The parameter *maxPerAllConf* indicates the maximum *periodic-all-confidence* threshold set by the user. Based on *maxPer* and *maxPerAllConf* thresholds, the interesting patterns involving rare items in *periodicity* dimension are extracted.

Henceforth, the periodic-frequent pattern is defined as follows.

Definition 3. (***Periodic-frequent pattern X***) *The pattern* X *is said to be periodic-frequent if* $sup(X) \geq minSup$*,* $allConf(X) \geq minAllConf$*,* $per(X) \leq maxPer$ *and* $perAllConf(X) \leq maxPerAllConf$*. The terms minSup, minAllConf, maxPer and maxPerAllConf, respectively represent the user-specified minimum support, minimum all-confidence, maximum periodicity and maximum periodic-all-confidence.*

Example 7. If the user-specified $minSup = 2$, $minAllConf = 0.6$, $maxPer = 6$ and $maxPerAllConf = 1.5$, then the pattern 'ab' is said to be a periodic-frequent pattern, because $sup(ab) \geq minSup$, $allConf(ab) \geq minAllConf$, $per(ab) \leq maxPer$ and $perAllConf(ab) \leq maxPerAllConf$.

Example 8. The column titled **III** in Table 2 shows the complete set of periodic-frequent patterns discovered from Table 1. It can be observed that the proposed model has not only discovered the periodic-frequent patterns containing rare items but also pruned the uninteresting patterns 'cd' and 'ce.' *This clearly demonstrates that the proposed model discovers periodic-frequent patterns containing rare items without generating too many uninteresting patterns.*

Property 1. If $X \subset Y$, then $TID^X \supseteq TID^Y$. Therefore, $sup(X) \geq sup(Y)$ and $allConf(X) \geq allConf(Y)$.

Property 2. If $X \subset Y$, then $per(X) \leq per(Y)$. Therefore, $perAllConf(X) \leq perAllConf(Y)$ as $\frac{per(X)}{min(per(i_j)\forall i_j \in X)} \leq \frac{per(Y)}{min(per(i_j)\forall i_j \in Y)}$.

The discovered periodic-frequent patterns satisfy the *anti-monotonic property*. The correctness is straightforward to prove from Properties 1 and 2.

Definition 4. ***Problem Definition:*** *Given the database (TDB) and the user-specified minimum support (minSup), minimum all-confidence (minAllConf), maximum periodicity (maxPer) and maximum periodic-all-confidence (maxPerAllConf), the problem of finding periodic-frequent patterns involve discovering all patterns that satisfy the minSup, minAllConf, maxPer and maxPerAllConf thresholds. Please note, the support and periodicity of a pattern can also be expressed in percentage of |TDB|.*

4   Proposed Algorithm

Tanbeer et al. [13] have proposed Periodic-Frequent pattern-growth (PF-growth) to discover periodic-frequent patterns using *support* and *periodicity* measures. Unfortunately, this algorithm cannot be directly used for finding the periodic-frequent patterns with our model. The reason is PF-growth does not determine the interestingness of a pattern using *all-confidence* and *periodic-all-confidence* measures. In this paper, we extend PF-growth to determine the interestingness of a pattern using these two measures. We call the proposed algorithm as Extended Periodic-Frequent pattern-growth (EPF-growth). The proposed algorithm involves two steps: (*i*) construction of Extended Periodic-Frequent pattern-tree (EPF-tree), (*ii*) recursively mining EPF-tree to discover periodic-frequent patterns. Before we describe these two steps, we explain the structure of EPF-tree.

4.1   Structure of EPF-Tree

The structure of EPF-tree consists of a prefix-tree and a EPF-list. The EPF-list consists of three fields: item name (*i*), *support* (*s*) and *periodicity* (*p*). The structure of prefix-tree in EPF-tree is similar to that of the prefix-tree in FP-tree [4]. However, to obtain both *support* and *periodicity* of the patterns, the nodes in EPF-tree explicitly maintain the occurrence information for each transaction by maintaining an occurrence transaction-id list, called *tid-list*, only at the last node of every transaction. Complete details on prefix-tree are available in [13].

4.2   Construction of EPF-Tree

Since the periodic-frequent patterns generated by the proposed model satisfy the *anti-monotonic property*, periodic-frequent items (or 1-patterns) play a key

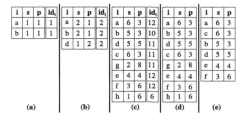 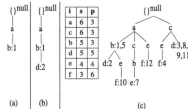

Fig. 2. Construction of EPF-list for Table 1. (a) After scanning the first transaction (b) After scanning the second transaction (c) After scanning the last transaction (d) Updated EPF-list (e) Final EPF-list with sorted list of periodic-frequent items

Fig. 3. Construction of EPF-tree for Table 1. (a) After scanning first transaction (b) After scanning second transaction (c) After scanning every transaction

role in efficient discovery of higher order periodic-frequent patterns. Periodic-frequent items are discovered by populating the EPF-list (lines 1 to 13 in Algorithm 1). Figures 2(a), (b), (c), (d) and (e) show the steps involved in finding periodic-frequent items from EPF-list. The user-specified $minSup$, $minAllConf$, $maxPer$ and $maxPerAllConf$ values are 2, 0.6, 6 and 1.5, respectively.

Algorithm 1. Construction of EPF-tree(TDB: Transactional database, $minSup$: minimum support, $minAllConf$: minimum all-confidence, $maxPer$: maximum periodicity, $maxPerAllConf$: maximum periodic-all-confidence)

1: Let id_l be a temporary array that records the tid of the last appearance of each item in the TDB. Let $t = \{tid_{cur}, X\}$ denote the current transaction with tid_{cur} and X representing the transaction-id of the current transaction and pattern, respectively.
2: **for** each transaction $t \in TDB$ **do**
3:     **if** an item i occurs for the first time **then**
4:         Insert i into the EPF-list with $sup^i = 1$, $per^i = tid_{cur}$ and $id_l^i = 1$.
5:     **else**
6:         $sup^i = sup^i + 1$.
7:         **if** $(tid_{cur} - id_l^i) > per^i$ **then**
8:             $per^i = tid_{cur} - id_l^i$.
9: **for** each item i in EPF-list **do**
10:     **if** $(|TDB| - id_l^i) > per^i$ **then**
11:         $per^i = |TDB| - id_l^i$.
12: Remove items from the EPF-list that do not satisfy $minSup$ and $maxPer$.
13: Sort the remaining items in EPF-list in descending order of their $support$. Let this sorted list of items be EPF.
14: Create a root node in EPF-tree, T, and label it "$null$."
15: **for** each transaction $tr \in TDB$ **do**
16:     Sort the items in t in EPF order. Let this list of sorted periodic-frequent items in t be $[p|P]$, where p is the first item and P is the remaining list. Call $insert\_tree([p|P], tid_{cur}, T)$, which is the same as in [4].

After finding periodic-frequent items, prefix-tree is constructed by performing another scan on the database (lines 14 to 16 in Algorithm 1). The construction of prefix-tree in EPF-tree is similar to the construction of prefix-tree in FP-tree [4]. However, it has to be noted that leaf nodes in EPF-tree maintain the transaction-ids of the transactions. Figures 3(a), (b) and (c) show the construction of EPF-tree after scanning first, second and every transaction in the transactional database, respectively. In EPF-tree, an item header table is built so that each item points to its occurrences in the tree via a chain of node-links, to facilitate tree traversal. For simplicity, we do not show these node-links in trees, however, they are maintained as in FP-tree.

4.3   Mining EPF-Tree

Algorithm 2 describes the procedure for mining periodic-frequent patterns from EPF-tree. The EPF-tree is mined by calling EPF-growth as (EPF-tree, *null*). This algorithm resembles FP-growth. However, the key difference is that once the pattern-growth is achieved for a suffix 1-pattern (or item), it is completely pruned from the EPF-tree by pushing its tid-list to respective parent nodes.

Table 3 summarizes the working of this algorithm. First, we consider item 'f,' which is the bottom-most item in the EPF-list, as a suffix pattern. This item appears in three branches of the EPF-tree (refer Fig. 3(c)). The paths formed by these branches are $\{cef : 4\}$, $\{abef : 10\}$ and $\{aef : 12\}$ (format of these branches is $\{nodes : time\text{-}stamps\}$). Therefore, considering 'f' as a suffix item, its corresponding three prefix paths are $\{ce : 4\}$, $\{abe : 10\}$ and $\{ae : 12\}$, which form its conditional pattern base (refer Fig. 4(a)). Its conditional EPF-tree contains only a single path, $\langle e : 4, 10, 12 \rangle$; '$a$,' '$b$' and '$c$' are not included because their *all-confidence* and *periodic-all-confidence* do not satisfy the *minAllConf* and *maxPerAllConf* respectively. Figure 4(b) shows the conditional EPF-tree of 'f.' The single path generates the pattern $\{ef : 3, 0.75, 6, 1.5\}$ (format is $\{pattern:$ *support, all-confidence, periodicity, periodic-all-confidence*$\}$). The same process of creating prefix-tree and its corresponding conditional tree is repeated for further extensions of 'ef.' Next, 'f' is pruned from the original EPF-tree and its *tid*-lists are pushed to its parent nodes, as shown in Fig. 4(c). All the above processes are once again repeated until EPF-list $= \emptyset$.

5   Experimental Results

In this section, we show that the proposed model discovers interesting patterns pertaining to both frequent and rare items by pruning uninteresting patterns. We also evaluate the proposed model against the existing models of periodic-frequent patterns [6,11,13].

The algorithms, *PF-growth*, *MCPF-growth*, *MaxCPF-growth* and *EPF-growth* are written in C++ and run with Fedora 22 on a 2.66 GHz machine with 8 GB of memory. We have conducted experiments using both synthetic (**T10I4D100K**) and real-world (**Retail** and **FAA-accidents**) databases. The T10I4D100K data-base is generated using the IBM data generator [2]. This

---

Algorithm 2. EPF-growth($Tree$, α)

1: **for** each a_i in the header of $Tree$ **do**
2:    Generate pattern $\beta = a_i \cup \alpha$. Construct an array TID^β, which represents the
      set of transaction-ids at which β has appeared in TDB. Next, compute from
      TID^β, $sup(\beta)$, $allConf(\beta)$, $per(\beta)$ and $perAllConf(\beta)$ and compare them with
      $minSup$, $minAllConf$, $maxPer$ and $maxPerAllConf$, respectively.
3:    **if** $sup(\beta) \geq minSup$, $allConf(\beta) \geq minAllConf$, $per(\beta) \leq maxPer$ and
      $perAllConf(\beta) \leq maxPerAllConf$ **then**
4:       Output β as a periodic-frequent pattern as $\{\beta$: sup, allConf, per, perAllConf$\}$.
5:       Traverse $Tree$ using the node-links of β, and construct β's conditional pattern
         base and β's conditional EPF-tree $Tree_\beta$.
6:       **if** $Tree_\beta \neq \emptyset$ **then**
7:          call EPF-growth($Tree_\beta$, β);
8:    Remove a_i from the $Tree$ and push a_i's tid-list to its parent nodes.

---

database contains 878 items with 100,000 transactions. The **Retail** database contains the market basket data from a Belgian retail store. This database contains 16,471 items with 88,162 transactions. The **FAA-accidents** database is constructed from the accidents data recorded by FAA from 1-January-1978 to 31-December-2014. This database contains 9,290 items with 98,864 transactions.

5.1 Patterns Generated by the Proposed Model

Figure 5 shows the number of patterns generated at different $minAllConf$ and $maxPerAllConf$ values. The $minSup$ and $maxPer$ are set at 0.01 % and 40 %. The following observations can be drawn: (i) The increase in $minAllConf$ results in decrease of periodic-frequent patterns. The reason is that as $minAllConf$ increases, the *support* threshold value of a pattern increases. (ii) The increase in $maxPerAllConf$ results in increase of patterns. The reason is that as $maxPerAllConf$ increases, the *periodicity* threshold value of a pattern increases.

Table 4 shows some of the interesting patterns discovered in FAA database. The $minSup$, $minAllConf$, $maxPer$ and $maxPerConf$ values used are 0.01 %, 0.01, 40 % and 9, respectively. *It can be observed from their support values that*

Table 3. Mining EPF-tree by creating conditional (sub -) pattern bases

Item	sup	per	Cond. Pattern Base	Cond. EPF-tree	Per. Freq. Patterns
f	3	6	$\{ce:4\}$, $\{abe:10\}$, $\{ae:12\}$	$\langle e:4,10,12\rangle$	$\{ef: 3, 0.75, 6, 1.5\}$
e	4	4	$\{c:4\}$, $\{abc:7\}$, $\{ab:10\}$, $\{a:12\}$	−	−
d	5	5	$\{ab:2\}$, $\{c:3,8,9,11\}$	−	−
b	5	3	$\{a:1,2,5,10\}$, $\{ac:7\}$	$\langle a:1,2,5,7,10\rangle$	$\{ab: 5, 0.833, 3, 1\}$
c	6	3	$\{a:7\}$	−	−

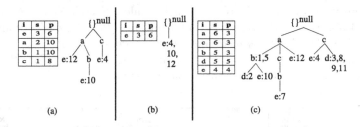

Fig. 4. Mining of EPF-tree for Table 1. (a) Prefix-tree of suffix item '*f*,' i.e., PT_f (b) Conditional tree of suffix item '*f*,' i.e., CT_f (c) EPF-tree after pruning item '*f*.'

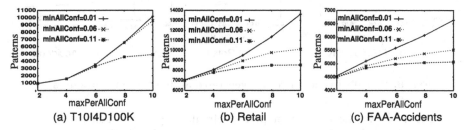

Fig. 5. Periodic-frequent patterns discovered in various databases

our model has discovered interesting patterns involving both frequent and rare items effectively. Please note that the *periodicity* (*per*) is expressed in days.

Table 4. Some of the interesting patterns discovered in FAA-accidents database

S. No.	Patterns	sup	allConf	per	perAllConf
1	{Ultralight Vehicles, Destroyed}	18	0.12	4904	8.01
2	{Starting engines, Destroyed}	13	0.06	4756	7.77
3	{General Operating Rules, Commercial Pilot, Minor}	10,399	0.15	32	6.4

5.2   Comparison of Proposed Model Against the Existing Models

For *MCPF-growth* and *MaxCPF-growth*, we use Eq. 5 to specify items' *minIS* and *maxIP* values. Setting the α and β values in this equation has been a non-trivial task as the patterns discovered by these algorithms can be different from the patterns discovered by *EPF-growth*. After conducting several experiments, we have empirically set the following values for *MCPF-growth* and *MaxCPF-growth* algorithms, such that both algorithms discover almost all periodic-frequent patterns discovered by *EPF-growth*.

Figure 6 shows the number of periodic-frequent patterns generated at different $minSup$ values (Y-$axis$ is plotted on logscale). For EPF-$growth$, we have fixed $minAllConf = 0.01$, $maxPer = 40\%$ and $maxPerAllConf = 9$ and vary $minSup$ values. For $MCPF$-$growth$ and $MaxCPF$-$growth$, we have set $\gamma = 0.01$, $LS = minSup$, $\beta = -0.4$, $Per_{max} = 40\%$ and $Per_{min} = 10\%$. For PF-$growth$, we have set $maxPer = 40\%$ and vary $minSup$ values.

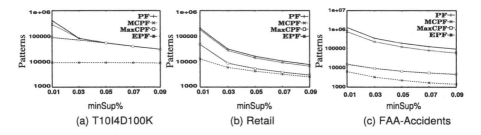

Fig. 6. Periodic-frequent patterns generated at different minSup values

Figure 7 shows the number of periodic-frequent patterns generated at different $maxPer$ values (Y-$axis$ is plotted on logscale). For EPF-$growth$, we have fixed $minSup = 0.01\%$, $minAllConf = 0.01$ and $maxPerAllConf = 9$ and vary $maxPer$ values. For $MCPF$-$growth$ and $MaxCPF$-$growth$, we have set $\gamma = 0.01$, $LS = 0.01\%$, $\beta = -0.4$, $Per_{max} = maxPer$ and $Per_{min} = 10\%$. For PF-$growth$, we have set $minSup = 0.01\%$ and vary $maxPer$ values.

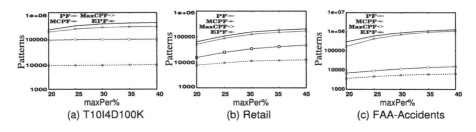

Fig. 7. Periodic-frequent patterns generated at different maxPer values

From Figs. 6 and 7, it can be observed that the proposed model has generated lesser number of periodic-frequent patterns than the other models, because the existing models have suffered from the *rare-item problem*.

Figure 8 shows the runtime taken by various models at different $maxPer$ values (Y-$axis$ is plotted on logscale). It can be observed that, in all the databases the proposed model takes lesser runtime to find periodic-frequent patterns than PF-$growth$ and $MCPF$-$growth$. But the proposed model takes slightly more runtime than $MaxCPF$-$growth$. So the proposed model is not adding any significant overhead in mining periodic frequent patterns. Similar observations can be drawn when $minSup$ is varied. Due to space restrictions, we are not reporting it.

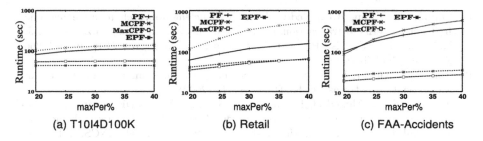

(a) T10I4D100K   (b) Retail   (c) FAA-Accidents

Fig. 8. Runtime requirements of various models at different maxPer values

6   Conclusions and Future Work

This paper introduces a model to address the rare item problem in both *support* and *periodicity* dimensions. A new interestingness measure, *periodic-all-confidence*, is proposed to address the problem in *periodicity* dimension. An efficient pattern-growth algorithm has been proposed to discover all periodic-frequent patterns in a database. Experimental results demonstrate that the proposed model is efficient. As a part of future work, we would like to study the change in periodic behavior of rare items due to noise.

References

1. Faa accidents dataset. http://www.asias.faa.gov/pls/apex/f?p=100:1:0::NO
2. Agrawal, R., Srikant, R.: Quest Synthetic Data Generator. IBM Almaden Research Center
3. Han, J., Gong, W., Yin, Y.: Mining segment-wise periodic patterns in time-related databases. In: KDD, pp. 214–218 (1998)
4. Han, J., Pei, J., Yin, Y., Mao, R.: Mining frequent patterns without candidate generation: a frequent-pattern tree approach. DMKD **8**(1), 53–87 (2004)
5. Kim, S., Barsky, M., Han, J.: Efficient mining of top correlated patterns based on null-invariant measures. In: Gunopulos, D., Hofmann, T., Malerba, D., Vazirgiannis, M. (eds.) ECML PKDD 2011, Part II. LNCS, vol. 6912, pp. 177–192. Springer, Heidelberg (2011)
6. Uday Kiran, R., Krishna Reddy, P.: Towards efficient mining of periodic-frequent patterns in transactional databases. In: Bringas, P.G., Hameurlain, A., Quirchmayr, G. (eds.) DEXA 2010, Part II. LNCS, vol. 6262, pp. 194–208. Springer, Heidelberg (2010)
7. Kiran, R.U., Shang, H., Toyoda, M., Kitsuregawa, M.: Discovering recurring patterns in time series. In: EDBT, pp. 97–108 (2015)
8. Ma, S., Hellerstein, J.: Mining partially periodic event patterns with unknown periods. In: ICDE, pp. 205–214 (2001)
9. Omiecinski, E.R.: Alternative interest measures for mining associations in databases. IEEE Trans. Knowl. Data Eng. **15**(1), 57–69 (2003)
10. Özden, B., Ramaswamy, S., Silberschatz, A.: Cyclic association rules. In: ICDE, pp. 412–421 (1998)

11. Surana, A., Kiran, R.U., Reddy, P.K.: An efficient approach to mine periodic-frequent patterns in transactional databases. In: Cao, L., Huang, J.Z., Bailey, J., Koh, Y.S., Luo, J. (eds.) PAKDD Workshops 2011. LNCS, vol. 7104, pp. 254–266. Springer, Heidelberg (2012)
12. Tan, P.N., Kumar, V., Srivastava, J.: Selecting the right interestingness measure for association patterns. In: KDD, pp. 32–41 (2002)
13. Tanbeer, S.K., Ahmed, C.F., Jeong, B.-S., Lee, Y.-K.: Discovering periodic-frequent patterns in transactional databases. In: Theeramunkong, T., Kijsirikul, B., Cercone, N., Ho, T.-B. (eds.) PAKDD 2009. LNCS, vol. 5476, pp. 242–253. Springer, Heidelberg (2009)
14. Tanbeer, S.K., Hassan, M.M., Alrubaian, M., Jeong, B.-S.: Mining regularities in body sensor network data. In: Di Fatta, G., Fortino, G., Li, W., Pathan, M., Stahl, F., Guerrieri, A. (eds.) IDCS 2015. LNCS, vol. 9258, pp. 88–99. Springer, Heidelberg (2015)
15. Weiss, G.M.: Mining with rarity: a unifying framework. SIGKDD Explor. $6(1)$, 7–19 (2004)
16. Wu, T., Chen, Y., Han, J.: Re-examination of interestingnessmeasures in pattern mining: a unified framework. DMKD $21(3)$, 371–397 (2010)
17. Yang, R., Wang, W., Yu, P.: Infominer+: mining partial periodic patterns with gap penalties. In: ICDM, pp. 725–728 (2002)

More Efficient Algorithms for Mining High-Utility Itemsets with Multiple Minimum Utility Thresholds

Wensheng Gan[1], Jerry Chun-Wei Lin[1(✉)], Philippe Fournier-Viger[2], and Han-Chieh Chao[1,3]

[1] School of Computer Science and Technology,
Harbin Institute of Technology Shenzhen Graduate School, Shenzhen, China
`wsgan001@gmail.com`, `jerrylin@ieee.org`, `hcc@ndhu.edu.tw`
[2] School of Natural Sciences and Humanities,
Harbin Institute of Technology Shenzhen Graduate School, Shenzhen, China
`philfv@hitsz.edu.cn`
[3] Department of Computer Science and Information Engineering,
National Dong Hwa University, Hualien, Taiwan

Abstract. Mining high-utility itemsets (HUIs) is a popular data mining task, which consists of discovering sets of items that yield a high profit in a transaction database. Although HUI mining has numerous applications, a key limitation is that a single minimum utility threshold (*minutil*) is used to assess the utility of all items. This simplifying assumption is unrealistic since in real-life all items do not have the same unit profit, and thus do not have an equal chance of generating a high profit. As a result, if the *minutil* threshold is set high, patterns containing items having a low unit profit are often missed, while if *minutil* is set low, the number of patterns becomes unmanageable. To address this issue, this paper presents an efficient tree-based algorithm named HIMU for mining HUIs using multiple minimum utility thresholds. A novel tree structure called multiple item utility Set-enumeration (MIU)-tree and the *global and conditional downward closure (GDC and CDC) properties* of HUIs in the MIU-tree are proposed. Moreover, a vertical compact utility-list structure is adopted to store the information required for discovering HUIs without performing additional database scans and generating candidates. An extensive experimental study on real-world and synthetic datasets show that this greatly improves the efficiency of the algorithm in terms of runtime and scalability.

Keywords: High-utility itemsets · Rare item problem · Set-enumeration tree · *SDC property* · Pruning strategies

1 Introduction

Knowledge Discovery in Database (KDD) is the process of finding meaningful, unexpected, and useful information in large amounts of data [2,3]. Two fundamental tasks in KDD are frequent itemset mining (FIM) and

© Springer International Publishing Switzerland 2016
S. Hartmann and H. Ma (Eds.): DEXA 2016, Part I, LNCS 9827, pp. 71–87, 2016.
DOI: 10.1007/978-3-319-44403-1_5

association rule mining (ARM) [2,3], which have numerous applications, in many domains. In contrast with traditional FIM and ARM, high-utility itemset mining (HUIM) [5,6,10,11,13,14,16] considers that items may have different unit profits and that purchase quantities may be non binary, to measure how "useful" an item or itemset is. The "utility" of an itemset in HUIM represents its importance to users in real-life applications (e.g., weight, cost, risk, or unit profit). The goal of HUIM is to identify itemsets in transactions that may be frequent or rare, but yield a high profit. HUIM is a key data analysis task, and has been widely utilized to discover valuable knowledge in several domains. Many approaches have been developed to mine high-utility itemsets such as Two-Phase [11], IHUP [5], UP-growth [13], UP-growth+ [14], HUI-Miner [10], and FHM [8], and so on.

However, an important limitation of previous studies is that they rely on a single minimum utility threshold to discover the complete set of HUIs. Using a single threshold value to assess the utility of all items in a database is inadequate since each item is different and thus items should not all be treated the same. Traditional high-utility itemset mining algorithms only let the user specify one *minutil* threshold to assess the utility of all patterns. Using a single threshold implies that all patterns in the database should have an equal chance of having a utility higher than the *minutil* threshold. But this assumption is unrealistic in practical applications [9] since each item generally has a distinct nature, frequency, or importance, and thus different items may tend to exhibit a lower or a higher utility. Hence, using a single fixed threshold, it is difficult to fairly measure the utility of items or itemsets. For example, in a retail store, it may be desirable to view the itemset {*diamond*} as a HUI if it brings more than 5,000\$/week, but to view the itemset {*bread, milk*} as a HUI if its profit is greater than 100\$/week. Using traditional HUI mining algorithms, if the *minutil* threshold is set high, useful patterns having a low utility are missed, and if it is set low, the number of HUIs becomes unmanageable. Thus, assessing the utility of items using a single threshold is inadequate as it does not take the inherent nature of each item (i.e., utility, item importance) into account. It is a non-trivial task and an important challenge to design efficient algorithms that solve this issue.

Mining association rules and frequent itemsets using multiple minimum support thresholds has been extensively studied [7,12,15], but the proposed approaches cannot be directly used in HUIM since HUIM considers non binary purchase quantities, and the unit profits of items. Up to now, few works have addressed the problem of mining HUIs with multiple minimum utility thresholds. To the best of our knowledge, HUI-MMU and the improved HUI-MMU$_{TID}$ algorithms [9] are the only algorithms designed to address this issue. However, a drawback of these algorithms it that they use a level-wise candidate generation-and-test approach to mine HUIs, which may perform poorly on databases containing long transactions or when minimum utility thresholds are set low. In this paper, to improve the efficiency of HUIM with multiple thresholds, an efficient tree-based algorithm named mining **H**igh-utility **I**temsets with **M**ultiple minimum **U**tility thresholds (abbreviated as **HIMU**) is developed. The contributions of this work are fourfold:

- A fast algorithm named HIMU is proposed to reveal useful and meaningful High-utility Itemsets by considering Multiple minimum Utility thresholds. The user can assign a minimum utility threshold to each item based on its real-life utility. This is more flexible and realistic than using a single *minutil* threshold.
- In contrast with previous Apriori-based algorithms, the proposed HIMU algorithm avoids repeatedly scanning the database and generating candidates, thanks to a novel sorted Set-enumeration tree structure named MIU-tree, and the use of a compact utility-list structure that allows obtaining information about an itemset by combining utility-lists of its prefix itemsets.
- Moreover, two novel *global and conditional sorted downward closure (GDC and CDC) properties* guarantee the global and partial anti-monotonicity for mining HUIs in the MIU-tree. Thus, HIMU can easily discover HUIs while pruning a huge number of unpromising itemsets, and only two database scans are performed by HIMU, which is more efficient than previous algorithms.
- Extensive experiments on two real-world datasets show that the proposed algorithms efficiently discover HUIs and outperform the state-of-the-art HUI-MMU and HUI-MMU$_{TID}$ algorithms. In addition, the improved algorithm outperforms the baseline algorithm, in terms of runtime and scalability.

2 Related Works

High-utility itemset mining (HUIM) considers the internal transaction utilities (purchase quantities) and external utilities (unit profits) of items to discover the profitable itemsets in quantitative databases. HUIM was introduced by Chan et al. [6]. Yao et al. then defined a strict unified framework for HUIM [16]. Since the *downward closure property* of ARM does not hold in HUIM, Liu et al. designed the TWU model [11] and a transaction-weighted downward closure (TWDC) property, to greatly reduce the number of unpromising candidates when mining HUIs using a level-wise approach. Several tree-based approaches for HUIM such as IHUP [5], UP-growth [13] and UP-growth+ [14] have been proposed. These pattern-growth approaches, however, generate and keep a huge number of candidates in memory to then obtain the actual HUIs. To address the above limitations of traditional HUIM, the HUI-Miner algorithm was proposed to directly mine HUIs while avoiding performing multiple database scans and generating candidates based on a designed utility-list structure [10]. The FHM algorithm was further proposed to enhance the performance of HUI-Miner using co-occurrences of pair of items [8].

Besides traditional HUIM, several variations of HUIM have been developed. The development of algorithms for HUIM is an active research topic, but most of them consider a single *minutil* threshold. In the field of FIM, several algorithms have been designed to address the "rare item problem such as MSApriori [7], CFP-growth [15], and CFP-growth++ [12]. The key idea of these works is to extract frequent patterns involving rare items using the "multiple minimum supports framework" [7,12,15]. This framework allows the user to specify multiple minimum support thresholds to take into account the nature of each item in

Table 1. An example database

TID	Transaction
T_1	a:1, c:2, d:3
T_2	a:2, d:1, e:2
T_3	b:3, c:5
T_4	a:1, c:3, d:1, e:2
T_5	b:1, d:3, e:2
T_6	b:2, d:2
T_7	b:3, c:2, d:1, e:1
T_8	a:2, c:3
T_9	c:2, d:2, e:1
T_{10}	a:2, c:2, d:1

Table 2. Derived HUIs

Itemset	MIU	Utility	Itemset	MIU	Utility
(b)	65	108	(de)	50	96
(d)	50	126	(acd)	50	76
(ad)	50	90	(bde)	50	93
(bc)	53	79	(cde)	50	55
(bd)	50	126	(bcde)	50	50
(cd)	50	83			

terms of frequency in the database. However, these approaches cannot be directly used in HUIM since HUIM requires to consider the purchase quantities and unit profits of items. There is only one paper that has considered the constraint of multiple minimum utility thresholds for mining HUIs [9].

3   Preliminaries and Problem Statement

Let $I = \{i_1, i_2, \ldots, i_m\}$ be a finite set of m distinct items appearing in a transactional database $D = \{T_1, T_2, \ldots, T_n\}$, where each transaction $T_q \in D$ is a subset of I, and has a unique identifier called its *TID*. A unit profit $pr(i_j)$ is assigned to each item $i_j \in I$, which represents its importance (e.g. profit, interest, risk). Unit profits are stored in a profit-table $ptable = \{pr(i_1), pr(i_2), \ldots, pr(i_m)\}$. An itemset $X \subseteq I$ with k distinct items $\{i_1, i_2, \ldots, i_k\}$ is of length k and is referred to as a k-itemset. An itemset X is said to be contained in a transaction T_q if $X \subseteq T_q$. For an itemset X, let the notation *TIDs(X)* denotes the *TIDs* of transactions in D containing X. For example, Table 1 shows a transactional database containing 10 transactions, and will be used as running example. Assume that the profit-table is defined as in the $ptable = \{pr(a) : 6, pr(b) : 12, pr(c) : 1, pr(d) : 9, pr(e) : 3\}$.

Definition 1. The minimum utility threshold of an item i_j in a database D is denoted as $mu(i_j)$. A structure called *MMU-table* indicates the minimum utility thresholds of each item in D, and is defined as:

$$MMU\text{-}table = \{mu(i_1), mu(i_2), \ldots, mu(i_m)\}. \tag{1}$$

Assume that the minimum utility thresholds of items in the running example are defined as: $MMU\text{-}table = \{mu(a), mu(b), mu(c), mu(d), mu(e)\} = \{56, 65, 53, 50, 70\}$. To avoid the "rare item problem", we consider the smallest utility threshold among items in an itemset as its minimum utility threshold, as defined below.

Definition 2. The minimum utility threshold of a k-itemset $X = \{i_1, i_2, \ldots, i_k\}$ in D is denoted as $MIU(X)$, and defined as the smallest mu value for items in X, that is:

$$MIU(X) = min\{mu(i_j)|i_j \in X, 1 \leq j \leq k\}. \tag{2}$$

For example, $MIU(a) = min\{mu(a)\} = 56$, $MIU(ac) = min\{\mathrm{mu(a)}, \mathrm{mu(c)}\} = min\{56, 53\} = 53$, and $MIU(ace) = min\{mu(a), mu(c), mu(e)\} = 53$.

Definition 3. The utility of an item i_j in a transaction T_q is defined as:

$$u(i_j, T_q) = q(i_j, T_q) \times pr(i_j). \tag{3}$$

Definition 4. The utility of an itemset X in a transaction T_q is defined as:

$$u(X, T_q) = \sum_{i_j \in X \wedge X \subseteq T_q} u(i_j, T_q). \tag{4}$$

Definition 5. The utility of an itemset X in a database D is defined as:

$$u(X) = \sum_{X \subseteq T_q \wedge T_q \in D} u(X, T_q). \tag{5}$$

Definition 6. The transaction utility of a transaction T_q is defined as:

$$tu(T_q) = \sum_{i_j \in T_q} u(i_j, T_q). \tag{6}$$

Definition 7. The transaction-weighted utility of an itemset X is denoted as $TWU(X)$, and defined as:

$$TWU(X) = \sum_{X \subseteq T_q \wedge T_q \in D} tu(T_q). \tag{7}$$

Definition 8. An itemset $X \subseteq I$ is a high transaction-weighted utilization itemset (HTWUI) if its TWU value is no less than the minimum utility threshold [14]. To adapt this definition, we assume that this threshold is $MIU(X)$.

$$HTWUI \leftarrow \{X|TWU(X) \geq MIU(X)\}. \tag{8}$$

Definition 9. An itemset X in a database D is a high-utility itemset (HUI) if and only if its utility is no less than its minimum utility threshold:

$$HUI \leftarrow \{X|u(X) \geq MIU(X)\}. \tag{9}$$

For the running example, the complete set of HUIs when considering multiple minimum utility thresholds is shown in Table 2.

Definition 10. Given a transactional database D and a *MMU-table*, which defines the minimum utility thresholds of each item in D. The problem of mining high-utility itemsets in D with multiple minimum utility thresholds (HUIM-MMU) is to find each itemset X having a utility no less than its threshold $MIU(X)$.

4   Proposed HIMU Algorithm for Mining HUIs

4.1   Search Space of HIMU and the Proposed MIU-Tree

Definition 11 (Total Order $\prec$ on Items). The proposed MIU-tree structure relies on a total order $\prec$ on items. Assume that this order is the ascending order of minimum utility thresholds of items.

Definition 12 (Set-Enumeration Tree with Multiple Minimum Item Utilities, MIU-Tree). The designed MIU-tree structure is a sorted set-enumeration tree where the total order $\prec$ on items is the ascending order of minimum utility thresholds of items.

Definition 13. The extensions (descendant nodes) of an itemset (tree node) X can be obtained by appending an item y to X such that y is greater than all items already in X according to the total order $\prec$.

For example, the proposed MIU-tree used by the HIMU algorithm for the running example is shown in Fig. 1 (left). Based on the designed MIU-tree, the following lemmas can be obtained.

Lemma 1. *The complete search space of the proposed HIMU algorithm for the HUIM-MMU framework can be represented by a MIU-tree where items are sorted according to the ascending order of the mu values on items.*

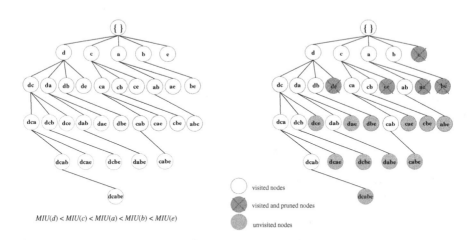

Fig. 1. The MIU-tree representation of the search space.

The traditional TWDC property of the TWU model does not hold in the proposed HUIM-MMU framework. For example, consider items (b), (c), (d) and (e) $(MIU(b) : 65, MIU(c) : 53, MIU(d) : 50$ and $MIU(e) : 70)$. The TWU of an itemset (bce) is calculated as $TWU(bce) = 50$, which is less than the minimum utility values of its subsets $MIU(b)$, $MIU(c)$ and $MIU(e)$. Hence, (bce) is not a HTWUI, and thus the itemset (bcde) and its supersets would be discarded according to the TWDC property. But it can be observed that $TWU(bcde) = 50$, which is equal to $MIU(bcde) = 50$. But as shown in Table 2, it can be seen that itemset (bcde) is actually a HUI. It is thus incorrect to discard the supersets of (bce) based on the TWDC property since (bcde) would not be generated. Therefore, if a k-itemset X^k is a HTWUI (i.e., $TWU(X^k) \geq MIU(X^k)$), we cannot ensure that any subset X^{k-1} of X^k is also a HTWUI (because $MIU(X^{k-1}) \geq MIU(X^k)$). Thus, using this property to prune the search space may fail to discover the complete set of HUIs. To address this limitation, the *Sorted Downward Closure (SDC) property* was proposed in [9].

Theorem 1 (Sorted Downward Closure Property, SDC Property). *Assume that items in itemsets are sorted by ascending order of mu values. Given any itemset $X^k = \{i_1, i_2, \ldots, i_k\}$ of length k, and another itemset $X^{k-1} = \{i_1, i_2, \ldots, i_{k-1}\}$ such that $X^{k-1} \subseteq X^k$. If X^k is a HTWUI then X^{k-1} is also a HTWUI [9].*

Proof. Since $X^{k-1} \subseteq X^k$, the following relationships hold:

(1) By Definition 2, we have that $MIU(X^{k-1}) = min\{mu(i_1), mu(i_2), \ldots, mu(i_{k-1})\}$, and $MIU(X^k) = min\{mu(i_1), mu(i_2), \ldots, mu(i_k)\}$. Since $\{i_1, i_2, \ldots, i_k\}$ is sorted according to the total order $\prec$, $MIU(X^k) = MIU(X^{k-1}) = mu(i_1)$.

(2) Thus, $TWU(X^k) = \sum_{X^k \subseteq T_q \wedge T_q \in D} tu(T_q) \leq \sum_{X^{k-1} \subseteq T_q \wedge T_q \in D} tu(T_q) = TWU(X^{k-1})$. Therefore, if X^k is a HTWUI (i.e., $TWU(X^k) \geq mu(i_1)$), any subset X^{k-1} of X^k is also a HTWUI.

Although the *sorted downward closure (SDC) property* guarantees the anti-monotonicity for HTWUIs, some HUIs would still be missed if items that are HTWUIs are determined using their $MIU(X)$ values. To address this problem, the concept of least minimum utility value (LMU) was developed to guarantee deriving all HUIs when using multiple minimum utility thresholds [9].

Definition 14 (Least Minimum Utility Value, LMU). *The least minimum utility value (LMU) is defined as the smallest value in the MMU-table, that is:*

$$LMU = min\{mu(i_1), mu(i_2), \ldots, mu(i_m)\}, \qquad (10)$$

where m is the total number of items in the database.

For example, the LMU of the given example is calculated as: $min\{mu(a), mu(b), mu(c), mu(d), mu(e)\} = min\{56, 65, 53, 50, 70\} = 50$.

4.2 Proposed Conditional Downward Closure (CDC) and Global Downward Closure (GDC) Properties

Lemma 2. *The MIU value of a node/pattern in the MIU-tree is equal to that of any of its child nodes (extension nodes).*

Proof. Assume that X^{k-1} is a node representing an itemset X in the MIU-tree, and that X^k is any of its child nodes (extensions). By definition, we have that $MIU(X^{k-1}) = min\{mu(i_1), mu(i_2), \ldots, mu(i_{k-1})\}$, and $MIU(X^k) = min\{mu(i_1), mu(i_2), \ldots, mu(i_k)\}$. Since $\{i_1, i_2, \ldots, i_k\}$ is sorted by ascending order of mu values, it can be proven that: $MIU(X^k) = MIU(X^{k-1}) = mu(i_1)$. Thus, the MIU value of a node in the MIU-tree is always equal to the MIU of any of its child nodes.

Lemma 3. *The support of a node in the MIU-tree is no less than the support of any of its child nodes (extension nodes).*

Proof. Since the Set-enumeration MIU-tree is a prefix tree, the relationship of the support of X^k and X^{k-1} can be proven to be $sup(X^k) \leq sup(X^{k-1})$.

Theorem 2 (HUIs $\subseteq$ HTWUIs). *Assume that 1-itemsets having a TWU lower than LMU are discarded and that the total order $\prec$ is applied. We have that HUIs $\subseteq$ HTWUIs, which indicates that if an itemset is not a HTWUI, then it is not a HUI. Moreover, none of its extensions are HTWUIs or HUIs.*

Proof. Let X^k be an itemset such that X^{k-1} is a subset of X^k.

(1) We have that $TWU(X^1) \leq LMU$ and $MIU(X^k) \geq LMU$.
(2) Since items are sorted by ascending order of mu values, $TWU(X^{k-1}) \geq TWU(X^k)$ and $MIU(X^{k-1}) = MIU(X^k) = min\{mu(i_1), mu(i_2), \ldots, mu(i_m)\} = mu(i_1)$.
(3) $u(X) = \sum_{X \subseteq T_q \wedge T_q \in D} u(X, T_q) \leq \sum_{X \subseteq T_q \wedge T_q \in D} tu(T_q) = TWU(X)$.

Thus, if X^{k-1} is not a HTWUI and $TWU(X^{k-1}) < mu(i_1)$, none of its supersets are HUIs.

Lemma 4. *The TWU of any node in the Set-enumeration MIU-tree is no less than the sum of all the actual utilities of any one of its descendant nodes, but not the MIU of its descendant nodes.*

Proof. Let X^{k-1} be a node in the MIU-tree, and X^k be a children (extension) of X^{k-1}. According to Theorem 1 and Lemma 1, we can get $TWU(X^{k-1}) \geq TWU(X^k)$ and the relationship between MIU values. Thus, the lemma holds.

Theorem 3 (Global Downward Closure Property, GDC Property). *In the designed MIU-tree, if the TWU of a tree node X is less than the LMU, X is not a HUI, and all its supersets (not only its child nodes, but all nodes containing X) are also not considered as HUIs.*

Proof. According to Lemma 2 and Theorem 2, this theorem holds.

This theorem ensures that by discarding itemsets with a TWU less than LMU, and their extensions, no HUIs are missed. Thus, the designed *global downward closure (GDC) property* and the LMU guarantee the **completeness** and **correctness** of the proposed HIMU algorithm, when pruning the search space.

In the past, a structure named utility-list was proposed to keep information from transactions in memory to directly mine HUIs [10]. The utility-list structure is efficient and is thus adopted in the proposed HIMU algorithm to store the required information about itemsets, as shown in Fig. 2. The reader can refer to [8,10] for details about the utility-list structure, and the *iu* and *ru* values stored in utility-lists.

(d)				(c)				(a)				(b)				(e)		
tid	*iu*	*ru*		*tid*	*iu*	*ru*		*tid*	*iu*	*ru*		*tid*	*iu*	*ru*		*tid*	*iu*	*ru*
1	9	8		1	2	6		1	6	0		3	12	0		2	6	0
2	9	18		3	5	12		2	12	6		5	36	6		4	6	0
4	27	15		4	3	12		4	6	6		6	24	0		5	6	0
5	18	42		7	2	39		8	12	0		7	36	3		7	3	0
6	18	24		8	3	12		10	12	0						9	3	0
7	9	41		9	2	3												
9	18	5		10	2	12												
10	9	14																

Fig. 2. Constructed utility-lists of 1-itemsets in the running example.

Definition 15. For an itemset X, $X.IU$ and $X.RU$ are respectively the sum of *iu* values and the sum of *ru* values in the utility-list of X, that is:

$$X.IU = \sum_{X \subseteq T_q \wedge T_q \in D} X.iu(T_q); X.RU = \sum_{X \subseteq T_q \wedge T_q \in D} X.ru(T_q).$$

Strategy 1. *When traversing the MIU-tree using a depth-first search, if the TWU of a node X based on its utility-list is less than the LMU, then none of the supersets of node X (note that here supersets contains not only descendant nodes of X, but also other nodes having X as subset) are HUIs.*

Theorem 4 (Conditional Downward Closure Property, CDC Property). *For any node X in the MIU-tree, the sum of $X.IU$ and $X.RU$ in the utility-list of X is no less than the utility of any one of its descendant nodes (extensions). Thus this sum is anti-monotonic and allows pruning itemsets in the MIU-tree.*

Proof. Let X^{k-1} be a (k-1)-itemset, and X^k be a (k)-itemset that is an extension of X^{k-1}. Assume that X^k is a children of X^{k-1} in the MIU-tree, meaning that X^{k-1} is a prefix of X^k. Let the set of items in X^k but not in X^{k-1} be denoted as $(X^k - X^{k-1}) = (X^k \backslash X^{k-1})$, and the set of all the items appearing after X^k in transaction T is denoted as T/X^k. For any transaction $X^k \subseteq T_q$:

$\because X^{k-1} \subset X^k \subseteq T_q \Rightarrow (X^k \backslash X^{k-1}) \subseteq (T_q \backslash X^{k-1}).$

$\therefore$ In each T_q, $X^k.iu = X^{k-1}.iu + (X^k \backslash X^{k-1}).iu = X^{k-1}.iu + \sum_{z \in (X^k \backslash X^{k-1})} z.iu$

$\therefore X^k.iu \leq X^{k-1}.iu + \sum_{z \in (T_q/X^{k-1})} z.iu = X^{k-1}.iu + X^{k-1}.ru$

$\therefore$ In each T_q, $X^k.iu \leq X^{k-1}.iu + X^{k-1}.ru$

$\because X^{k-1} \subset X^k \Rightarrow X^k.tids \subseteq X^{k-1}.tids$

$\therefore$ in D, $X^k.IU = \sum_{T_q \in X^k.tids} X^k.iu \leq \sum_{T_q \in X^k.tids} (X^{k-1}.iu + X^{k-1}.ru)$

$$\leq \sum_{T_q \in X^{k-1}.tids}(X^{k-1}.iu + X^{k-1}.ru) = X^{k-1}.IU + X^{k-1}.RU$$
$$\therefore \text{ in } D, X^k.IU \leq X^{k-1}.IU + X^{k-1}.RU$$

Thus, the sum of the utilities of X^k in D is no greater than $(X^{k-1}.IU + X^{k-1}.RU)$ of X^{k-1} in D.

Strategy 2. *When traversing the MIU-tree using a depth-first search, if the sum of X.IU and X.RU in the utility-list of an itemset X is less than MIU(X), then none of the descendant nodes (extensions) of node X is a HUI since the actual utilities of these extensions will be less than MIU(X).*

In the running example, assume that the node (e) has $TWU(e) < LMU$. Then the visited nodes, pruned nodes, and the skipped nodes are respectively shown in Fig. 1 (right) when applying the **Strategy** 1. And the **Strategy** 2 is used as a conditional strategy to prune all extensions of an unpromising node early.

4.3 Estimated Utility Co-occurrence Pruning Strategy

In this section, we extend the Estimated Utility Co-occurrence Pruning (EUCP) strategy [8], in the proposed algorithm, to provide an additional way of pruning unpromising itemsets early with multiple minimum utility thresholds.

Theorem 5. *Without loss of generality, assume that items in itemsets are sorted by ascending order of mu values. If an itemset X contains a 2-itemset X that is not a HTWUI, then any k-itemset X^k ($k \geq 3$) that is a (transitive) extension of X is not a HTWUI or HUI.*

Proof. Let X be a 2-itemset and X^k be a k-itemset ($k \geq 3$) that is a (transitive) extension of X. According to the GDC property and because $TWU(X^k) \leq TWU(X^{k-1})$, if a 2-itemset is not a HTWUI, then any k-itemset ($k \geq 3$), which is an extension of X is not a HTWUI or HUI.

As mentioned above, not all supersets of a non HTWUI should be pruned but only those having a MIU value greater than the MIU value of this non HTWUI (w.r.t. the extensions of this non HTWUI having higher MIU values). Thus, using the proposed GDC property with the designed total order $\prec$, the **completeness** and **correctness** of the enhanced algorithm named HIMU$_{EUCP}$ is preserved by extending the EUCP strategy. Note that the TWU values of all 2-itemsets are stored in a structure called estimated utility co-occurrence structure $(EUCS)$ [8].

Strategy 3 (EUCP Strategy). *When traversing the MIU-tree using a depth-first search, if the TWU value of a 2-itemset X is less than the MIU value of X according to the EUCS, then X is not a HTWUI; and any k-itemset which is an extension of X will not be a HTWUI or HUI, and they can be pruned directly.*

4.4 Procedure of the HIMU Algorithm and the Enhanced Algorithm

Note that the proposed enhanced HIMU$_{EUCP}$ algorithm is similar to the baseline HIMU algorithm. The difference is that the *EUCS* needs to be constructed initially during the second database scan. Moreover, the mining procedure for deriving HUIs is modified to verify pruning Strategy 3 for each generated itemset. Due to the page limitation, only the details of the HIMU$_{EUCP}$ algorithm are provided.

Input: D; *ptable*; *MMU-table* = $\{mu(i_1), mu(i_2), \ldots, mu(i_m)\}$.
Output: The set of complete high-utility itemsets (HUIs).
1 $i.UL \leftarrow \emptyset, D.UL \leftarrow \emptyset, EUCS \leftarrow \emptyset$;
2 calculate the *LMU* in the *MMU-table*;
3 scan D to calculate the $TWU(i)$ value of each item $i \in I$;
4 find $I^* \leftarrow \{i \in I | TWU(i) \geq LMU\}$, w.r.t. $HTWUI^1$;
5 sort I^* according to the designed total order $\prec$ (ascending order of mu values);
6 scan D to construct the utility-list $i.UL$ of each item $i \in I^*$ and build the *EUCS*;
7 **call HUI-Search**(ϕ, I^*, MMU-*table*, *EUCS*);
8 **return** *HUIs*;

Algorithm 1. The HIMU$_{EUCP}$ algorithm

Input: X, *extensionsOfX*, *MMU-table*, *EUCS*.
Output: The complete set of HUIs.
1 **for** *each itemset* $X_a \in$ *extensionsOfX* **do**
2   obtain the $X_a.IU$ and $X_a.RU$ values from the built $X_a.UL$;
3   **if** $X_a.IU \geq MIU(X_a)$ **then**
4          $HUIs \leftarrow HUIs \cup X_a$;
5   **if** $(X_a.IU + X_a.RU \geq MIU(X_a))$ **then**
6   *extensionsOfX$_a$* $\leftarrow \emptyset$;
7   **for** *each itemset* $X_b \in$ *extensionsOfX such that* X_b *after* X_a **do**
8   **if** $\exists TWU(a, b) \in EUCS \wedge TWU(a, b) \geq MIU(X_a)$ **then**
9                  $X_{ab} \leftarrow X_a \cup X_b$;
10                 $X_{ab}.UL \leftarrow construct(X, X_a, X_b)$;
11   **if** $X_{ab}.UL \neq \emptyset$ **then**
12   *extensionsOfX$_a$* $\leftarrow$ *extensionsOfX$_a$* $\cup X_{ab}.UL$;
13   **call HUI-Search**(X_a, *extensionsOfX$_a$*, $MIU(X_a)$, *EUCS*);
14 **return** *HUIs*

Algorithm 2. The HUI-Search Procedure

As shown in Algorithm 1, the HIMU$_{EUCP}$ algorithm first sets $i.UL$, $D.UL$ and *EUCS* to the empty set (Line 1), and calculates the *LMU* in the MMU-table (Line 2). Then, it scans the database to calculate the $TWU(i)$ value of each item $i \in I$ (Line 3), and then find the potential 1-itemsets which may be

HUIs such that $TWU(i) \geq LMU(I^* \subseteq HTWUI^1)$ (Line 4). After sorting I^* by $\prec$ (ascending order of mu values), the algorithm scans D again to construct the utility-list of each item $i \in I^*$ and build the $EUCS$ (Lines 5 to 6). It is important to notice that only the designed order $\prec$ can guarantee the completeness of HIMU, as previously explained. The utility-list of each item $i \in I^*$ is recursively processed by the depth-first search HUI-Search procedure (Line 7). This latter procedure (cf. Algorithm 2), checks if each 1-extension X_a of an itemset X is a HUI (Lines 2 to 4). Two conditions are then checked to determine whether its child nodes should be considered by the depth-first search (Lines 5 to 12). If an itemset is regarded as a potential HUI, the $Construct(X, X_a, X_b)$ procedure (see [10] for details) is applied to construct the utility-lists of all 1-extensions of X_a (w.r.t. $extensionsOfX_a$) (Lines 9 to 12). Notice that each extension X_{ab} is a 1-extension of itemset X_a, and is added to the set $extensionsOfX_a$ for the later depth-first search (Line 13). The HUI-Search procedure then is recursively called to mine HUIs (Line 13).

5   Experimental Evaluation

The performance of the proposed HIMU and HIMU$_{\text{EUCP}}$ algorithms was evaluated on two real-world datasets, foodmart [4] and mushroom [1]. The foodmart dataset contains customer transactions from an anonymous chain store, and is provided with Microsoft SQL Server. It contains 21,556 transactions and 1,559 distinct items. The mushroom dataset is dense. It has 8,124 transactions and 120 distinct items, and an average transaction length of 23 items. The foodmart dataset contains real utility values, while a simulation model [13] was developed to generate the quantities and profit values of items in transactions for the mushroom dataset, by choosing random values respectively in the [1, 5] and [1, 1000] intervals.

The performance of the designed algorithms was also compared with the state-of-the-art HUI-MMU and HUI-MMU$_{\text{TID}}$ algorithms [9]. To perform a fair comparison, all algorithms were implemented in Java and executed on a computer having an Intel Core2 Duo 2.8 GHz processor and 4GB of main memory, running the 64 bit Microsoft Windows 7 operating system. Moreover, a method to automatically set the mu value of each item was adopted, described in the HUI-MMU algorithm [9]: $mu(i_j) = max[\beta \times pr(i_j), GLMU]$, where β is a constant used to set the mu values of items as a function of their unit profit values. To ensure randomness and diversity in the experiments, β was set in the [1, 100] interval for the foodmart dataset, and in the [1000, 10000] for mushroom. The parameter $GLMU$ is a user-specified global least minimum utility value, and $pr(i_j)$ is the external utility of an item $pr(i_j)$. Note that if β is set to zero, then a single minimum utility value $GLMU$ will be used for all items, and this will be equivalent to traditional HUIM.

5.1   Execution Time

In the conducted experiments, the parameter β was randomly set to a fixed number of items. Figure 3 shows the runtime of the algorithms under various $GLMU$ with a fixed β within an interval, and under various β with a fixed $GLMU$ for different datasets. In Fig. 3, it can be seen that the HIMU and the improved $HIMU_{EUCP}$ algorithms perform well compared to the HUI-MMU, $HUI\text{-}MMU_{TID}$ algorithms under various $GLMU$ with a fixed β, and under various β with a fixed $GLMU$. Moreover, the two MIU-tree-based algorithms are generally up to almost one or two orders of magnitude faster than the level-wise HUI-MMU and $HUI\text{-}MMU_{TID}$ algorithms. $HIMU_{EUCP}$ is faster than the HIMU algorithm on mushroom but not on foodmart, by adopting the EUCP strategy, which is used to avoid join operations for forming the utility-lists of unpromising itemsets. This indicates that the generate-and-test approach has worse performance than the proposed depth-first search approach that utilizes the vertical utility-list structure and additional pruning strategies. The gap between the previous approaches and the proposed MIU-tree-based algorithms becomes large when $GLMU$ and β are decreased. As shown in Fig. 3(a) and (c), $HIMU_{EUCP}$ performs slightly worse than HIMU. The reason is that for the very sparse foodmart dataset, with an average transaction length of 4.4, many unpromising candidates can be directly pruned by the redefined HTWUI and SDC properties, and it is unnecessary to apply the EUCP strategy for pruning unpromising itemsets, and thus construct the $EUCS$. Furthermore, when β is increased, the HIMU and $HIMU_{EUCP}$ algorithms take less time to find the HUIs. The reason is that when β is set to large values, the actual minimum utility threshold of each item is also set to larger

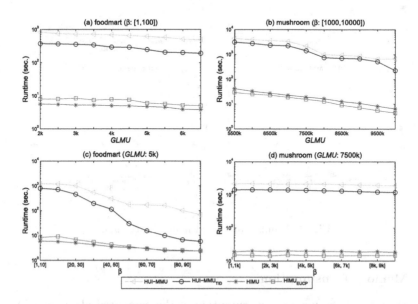

Fig. 3. Runtime performance.

values based on the presented equation. Hence, fewer HUIs and HTWUIs are pruned by the pruning conditions, and the execution time becomes smaller. In summary, the two proposed algorithms considerably outperform the state-of-the-art HUI-MMU and HUI-MMU$_{\text{TID}}$ algorithms.

5.2 Effect of Pruning Strategies

We also evaluated the effectiveness of the EUCP strategy for pruning unpromising itemsets. The number of itemsets (nodes) visited by the HIMU and HIMU$_{\text{EUCP}}$ algorithms are named N_2 and N_3, respectively. Moreover, the number of itemsets generated by combining pairs for determining HTWUIs in HUI-MMU and HUI-MMU$_{\text{TID}}$ is denoted as N_1. Results are shown in Fig. 4. It can be observed that relationship $N_3 \leq N_2$ holds for all datasets no matter how $GLMU$ is set, for a fixed β or under various β with a fixed $GLMU$. Especially, the node reduction obtained by adopting the EUCP strategy in the enhanced algorithm is huge, as shown in $N_3 \leq N_2$. Besides, N_1 is less than N_3 and N_2 for foodmart, but larger on mushroom. It indicates that the search space (in terms of visited nodes in the Set-enumeration MIU-tree) of HIMU may be huge if the effective EUCP pruning strategy is not applied for pruning the search space.

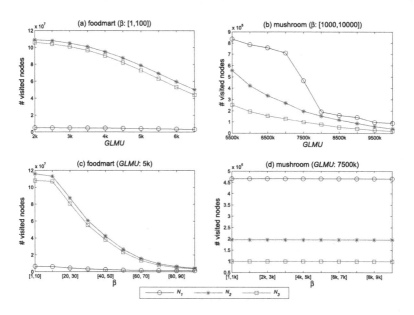

Fig. 4. Number of visited nodes (patterns).

5.3 Memory Consumption

We also assessed the memory consumption of the compared algorithms. Memory measurements were done using the standard Java API. Note that the peak memory consumption of each algorithm was recorded for all datasets. Results are shown in Fig. 5. In this figure, it can be clearly seen that the proposed HIMU algorithms require less memory than the state-of-the-art HUI-MMU and HUI-MMU$_{TID}$ algorithms for various parameters on the two datasets, and by up to 3,000 times on mushroom. Moreover, the HIMU and HIMU$_{EUCP}$ algorithms require nearly constant memory under various parameter values for the two datasets. The memory usage of the level-wise algorithms dramatically increases when $GLMU$ or β are decreased, while the memory usage of the proposed algorithms remain stable. The reason is the same as above. This result is reasonable since the two MIU-tree-based algorithms can quickly traverse the MIU-tree without generating candidates and easily prune unpromising itemsets using the sum of utilities and remaining utilities. Furthermore, the utility-list structure is adopted as a vertical compact structure to store information about itemsets. Thus, less memory is consumed.

5.4   Scalability Analysis

Figure 6 compares the scalability of the algorithms on synthetic data T10I4N4KD$|X|K$ where the transaction count (K) was varied from 100K to 500K, $GLMU = 1,000,000$ and β was varied from 1000 to 10000. It can be seen that the designed algorithms scale well with respect to dataset size and that HIMU$_{EUCP}$ scales better than HIMU. When the dataset size is increased, HIMU$_{EUCP}$ becomes increasingly faster than the other algorithms thanks to the EUCP strategy. HIMU$_{EUCP}$ consumes more memory than HIMU but less than the two level-wise algorithms because it uses the additional $EUCS$ to store TWU values of all 2-itemsets (see Fig. 6(b)). From Fig. 6(c), it can also be seen that

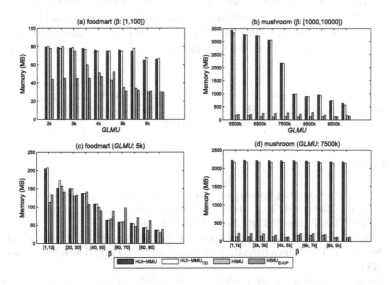

Fig. 5. Memory performance.

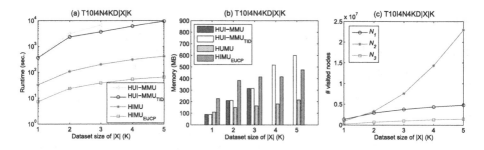

Fig. 6. Scalability of the compared algorithms.

the number of nodes N_3 remains much smaller than N_2. Thus, by utilizing the EUCP strategy, the actual search space of the HIMU$_{\text{EUCP}}$ algorithm is reduced compared to the baseline HIMU algorithm, and it can be concluded that the improved algorithm is acceptable and efficient.

6   Conclusion

In this paper, a novel algorithm named HIMU was presented to discover high-utility itemsets with multiple minimum utility thresholds. A compact Multiple Itemset Utility Set-enumeration tree (MIU-tree) was designed for mining HUIs without candidate generation. Besides, the *global and conditional downward closure (GDC and CDC) properties* were proposed to guarantee the global and partial anti-monotonicity for mining HUIs. Pruning conditions are also incorporated in the proposed algorithms to reduce the search space, and an efficient compact utility-list structure is used to obtain information about any itemset from its prefix itemsets in the designed MIU-tree. An experimental evaluation against the state-of-the-art HUI-MMU and HUI-MMU$_{\text{TID}}$ algorithms on two real-world datasets shows that the two proposed algorithms are highly efficient and scalable.

Acknowledgment. This research was partially supported by the National Natural Science Foundation of China (NSFC) under Grant No. 61503092, and by the Tencent Project under grant CCF-TencentRAGR20140114.

References

1. Frequent itemset mining dataset repository. http://fimi.ua.ac.be/data/
2. Agrawal, R., Imielinski, T., Swami, A.: Database mining: a performance perspective. IEEE Trans. Knowl. Data Eng. **5**(6), 914–925 (1993)
3. Agrawal, R., Srikant, R.: Fast algorithms for mining association rules in large databases. In: The International Conference on Very Large Data Bases, pp. 487–499 (1994)
4. Microsoft. Example database foodmart of Microsoft analysis services. http://www.Almaden.ibm.com/cs/quest/syndata.html

5. Ahmed, C.F., Tanbeer, S.K., Jeong, B.S., Le, Y.K.: Efficient tree structures for high utility pattern mining in incremental databases. IEEE Trans. Knowl. Data Eng. **21**(12), 1708–1721 (2009)
6. Chan, R., Yang, Q., Shen, Y.D.: Mining high utility itemsets. In: The International Conference on Data Mining, pp. 19–26 (2003)
7. Liu, B., Hsu, W., Ma, Y.: Mining association rules with multiple minimum supports. In: ACM SIGKDD International Conference on Knowledge Discovery and Data Mining, pp. 337–341 (1999)
8. Fournier-Viger, P., Wu, C.-W., Zida, S., Tseng, V.S.: FHM: faster high-utility itemset mining using estimated utility co-occurrence pruning. In: Andreasen, T., Christiansen, H., Cubero, J.-C., Raś, Z.W. (eds.) ISMIS 2014. LNCS, vol. 8502, pp. 83–92. Springer, Heidelberg (2014)
9. Lin, J.C.W., Gan, W., Fournier-Viger, P., Hong, T.P.: Mining high-utility itemsets with multiple minimum utility thresholds. In: ACM International Conference on Computer Science & Software Engineering, pp. 9–17 (2015)
10. Liu, M., Qu, J.: Mining high utility itemsets without candidate generation. In: ACM International Conference on Information and Knowledge Management, pp. 55–64 (2012)
11. Liu, Y., Liao, W., Choudhary, A.K.: A two-phase algorithm for fast discovery of high utility itemsets. In: Ho, T.-B., Cheung, D., Liu, H. (eds.) PAKDD 2005. LNCS (LNAI), vol. 3518, pp. 689–695. Springer, Heidelberg (2005)
12. Kiran, R.U., Reddy, P.K.: Novel techniques to reduce search space in multiple minimum supports-based frequent pattern mining algorithms. In: ACM International Conference on Extending Database Technology, pp. 11–20 (2011)
13. Tseng, V.S., Wu, C.W., Shie, B.E., Yu, P.S.: UP-growth: an efficient algorithm for high utility itemset mining. In: ACM SIGKDD International Conference on Knowledge Discovery and Data Mining, pp. 253–262 (2010)
14. Tseng, V.S., Shie, B.E., Wu, C.W., Yu, P.S.: Efficient algorithms for mining high utility itemsets from transactional databases. IEEE Trans. Knowl. Data Eng. **25**(8), 1772–1786 (2013)
15. Hu, Y.H., Chen, Y.L.: Mining association rules with multiple minimum supports: a new mining algorithm and a support tuning mechanism. Decis. Support Syst. **42**(1), 1–24 (2006)
16. Yao, H., Hamilton, J., Butz, C.J.: A foundational approach to mining itemset utilities from databases. In: SIAM International Conference on Data Mining, pp. 211–225 (2004)

Mining Minimal High-Utility Itemsets

Philippe Fournier-Viger[1](✉), Jerry Chun-Wei Lin[2], Cheng-Wei Wu[3],
Vincent S. Tseng[3], and Usef Faghihi[4]

[1] School of Natural Sciences and Humanities,
Harbin Institute of Technology Shenzhen Graduate School, Shenzhen, China
philfv@hitsz.edu.cn
[2] School of Computer Science and Technology, Shenzhen Graduate School,
Harbin Institute of Technology Shenzhen Graduate School, Shenzhen, China
jerrylin@ieee.org
[3] Department of Computer Science, National Chiao Tung University,
Hsinchu, Taiwan
silvemoonfox@gmail.com, vtseng@cs.nctu.edu.tw
[4] Department of Computer Science and Mathematics,
University of Indianapolis, Indianapolis, USA
faghihiu@indy.edu

Abstract. Mining high-utility itemsets (HUIs) is a key data mining task. It consists of discovering groups of items that yield a high profit in transaction databases. A major drawback of traditional high-utility itemset mining algorithms is that they can return a large number of HUIs. Analyzing a large result set can be very time-consuming for users. To address this issue, concise representations of high-utility itemsets have been proposed such as closed HUIs, maximal HUIs and generators of HUIs. In this paper, we explore a novel representation called the *minimal high utility itemsets* (MinHUIs), defined as the smallest sets of items that generate a high profit, study its properties, and design an efficient algorithm named MinFHM to discover it. An extensive experimental study with real-life datasets shows that mining MinHUIs can be much faster than mining other concise representations or all HUIs, and that it can greatly reduce the size of the result set presented to the user.

Keywords: Utility mining · High-utility itemsets · Minimal itemsets

1 Introduction

High-utility itemset mining (HUIM) is an emerging data mining task, which consists of discovering sets of items that have a high utility (yield a high profit) in customer transaction databases [2,5,8–13,15]. HUIM can be viewed as a generalization of *Frequent Itemset Mining* (FIM) [1,3,4,17], where weights are assigned to each item to represent their importance (e.g. unit profit), and purchase quantities of items in transactions are not restricted to binary values. HUIM has applications in many domains such as customer purchase behavior analysis, website click stream analysis, and biomedicine [2,12,15]. HUIM is widely considered

© Springer International Publishing Switzerland 2016
S. Hartmann and H. Ma (Eds.): DEXA 2016, Part I, LNCS 9827, pp. 88–101, 2016.
DOI: 10.1007/978-3-319-44403-1_6

as more difficult than FIM because the utility measure used in HUIM is neither anti-monotonic nor monotonic, unlike the support measure used in FIM [1]. In other words, the utility of an itemset can be lower, equal or higher than the utility of any of its supersets. Hence, techniques for pruning the search space in FIM cannot be directly applied in HUIM. Although HUIM has attracted lots of attention in recent years, an important limitation of traditional high-utility itemset mining algorithms [2,5,8–13,15] is that they can generate a very large amount of HUIs. This can make HUI mining algorithms run out of storage space or even fail to terminate. Moreover, it is very time-consuming for a user to analyze a very large set of HUIs [6,16]. To address this issue, it was proposed to mine concise representations of all HUIs rather than the whole set of HUIs. Three main representations have been proposed in previous work. *Maximal HUIs* are the HUIs that are not included in other HUIs. For a retailer, it answers the question of finding the largest sets of items that yield a high profit. *Closed HUIs* [16] are the HUIs that are not included in another HUIs having the same support. For a retailer, it answers the question of finding the largest groups of items yielding a high profit, which are common to groups of customers. Finally, *Generators of HUIs* [6] answer the question of finding the smallest sets of items common to groups of customers having bought a same high-utility itemset.

In this paper, we investigate a novel representation of HUIs named the *minimal high-utility itemsets* (MinHUIs), defined as the smallest HUIs (HUIs that are not included in another HUI). This representation addresses the problem that HUIM algorithms often find very long HUIs containing many items. But these HUIs often represent rare cases, as in real-life, few customers exactly buy the same large set of items. For marketing purpose, a retailer may be more interested in finding the smallest sets of items that generate a high profit, since it is easier to co-promote a small set of items targeted at many customers rather than a large set of items targeted at few customers. The proposed representation is the opposite of maximal HUIs. It aims at discovering the smallest sets of items that generate a high profit in a database rather than the largest ones. Because this representation has been unexplored, it remains an important challenge to explore the properties of this representation and define an efficient algorithm for mining this representation. In this paper, we address this challenge. We propose a novel algorithm named *MinFHM* to discover this representation efficiently. MinFHM extends FHM, a state-of-the-art algorithm for HUI mining by using a novel pruning property, and several optimizations to mine MinHUIs efficiently. We compare the performance of MinFHM with FHM on several real-life datasets. Results show that mining minimal HUIs is almost two orders of magnitude faster than mining all HUIs, or other concise representations of HUIs and that it can greatly reduce the result set presented to the user. The rest of this paper is organized as follows. Sections 2, 3, 4, 5 and 6 respectively present related work, minimal high-utility itemsets, the MinFHM algorithm, the experimental evaluation and the conclusion.

2   Related Work

The problem of HUIM is defined as follows [5,12,13,15]. Consider a set of items (symbols) denoted as I. A *transaction database* is a set of transactions $D = \{T_0, T_1, ..., T_n\}$ such that for each transaction T_c, $T_c \subseteq I$ and T_c has a unique identifier c called its Tid. Each item $i \in I$ is associated with a positive number $p(i)$, called its external utility, representing its importance (e.g. unit profit). Moreover, for each transaction T_c such that $i \in T_c$, a positive number $q(i, T_c)$ is called the internal utility of i, which represents the purchase quantity of i in transaction T_c. For example, Table 1 shows a transaction database containing five transactions $(T_0, T_1...T_4)$, which will be used as running example. Transaction T_3 indicates that items a, c, and e appear in this transaction with an internal utility of respectively 2, 6, and 2. Table 2 indicates that the external utilities of these items are respectively 5, 1, and 3.

Table 1. A transaction database

TID	Transaction
T_0	$(a,1),(b,5),(c,1),(d,3),(e,1)$
T_1	$(b,4),(c,3),(d,3),(e,1)$
T_2	$(a,1),(c,1),(d,1)$
T_3	$(a,2),(c,6),(e,2)$
T_4	$(b,2),(c,2),(e,1)$

Table 2. External utility values

Item	a	b	c	d	e
Unit profit	5	2	1	2	3

The utility of an item i in a transaction T_c is denoted as $u(i, T_c)$ and defined as $p(i) \times q(i, T_c)$. The utility of an itemset X (a group of items $X \subseteq I$) in a transaction T_c is denoted as $u(X, T_c)$ and defined as $u(X, T_c) = \sum_{i \in X} u(i, T_c)$. The utility of an itemset X (in all transactions of a transaction database) is denoted as $u(X)$ and defined as $u(X) = \sum_{T_c \in g(X)} u(X, T_c)$, where $g(X)$ is the set of transactions containing X. The *problem of high-utility itemset mining* is to discover all high-utility itemsets. An itemset X is a *high-utility itemset* if its utility $u(X)$ is no less than a user-specified minimum utility threshold *minutil* given by the user. For instance, the utility of the itemset $\{a, c\}$ is $u(\{a, c\}) = u(a) + u(c) = u(a, T_0) + u(a, T_2) + u(a, T_3) + u(c, T_0) + u(c, T_2) + u(c, T_3) = 5 + 5 + 10 + 1 + 1 + 6 = 28$. If *minutil* $= 25$, the set of HUIs is $\{a, c\} : 28$, $\{a, c, e\} : 31$, $\{a, b, c, d, e\} : 25$, $\{b, c\} : 28$, $\{b, c, d\} : 34$, $\{b, c, d, e\} : 40$, $\{b, c, e\} : 37$, $\{b, d\} : 30$, $\{b, d, e\} : 36$, $\{b, e\} : 31$ and $\{c, e\} : 27$, where each HUI is annotated with its utility. A major challenge in HUIM is that the utility measure is not monotonic or anti-monotonic, and thus that pruning techniques developed in FIM cannot be directly used in FIM to prune the search space. Many HUIM algorithms such as Two-Phase [13], IHUP [2], BAHUI [11], PB [8], and UPGrowth+ [15] overcome this challenge by using a measure called the *Transaction-Weighted Utilization (TWU)* measure, which provides an upper-bound on the utility of itemsets and is anti-monotonic [2,13,15]. The aforementioned algorithms first identify candidate high utility itemsets by calculating their TWUs. Then, in a second phase, they scan the database to calculate the exact utility of all candidates found in the first phase to eliminate low utility itemsets. The TWU measure is defined as follows. The *transaction utility* (TU) of a transaction T_c is the sum of the utilities of all the items in T_c. i.e. $TU(T_c) = \sum_{x \in T_c} u(x, T_c)$. The *transaction-weighted utilization* (TWU) of an itemset X is defined as the sum of the transaction utilities of transactions containing X, i.e. $TWU(X) = \sum_{T_c \in g(X)} TU(T_c)$. For instance, the TUs of T_0, T_1, T_2, T_3 and T_4 are respectively 25, 20, 8, 22 and 9. The TWU of single items a, b, c, d, e are respectively 55, 54, 84, 53 and 76. $TWU(\{c, d\}) = TU(T_0) + TU(T_1) + TU(T_2) = 25 + 20 + 8 = 53$. The TWU has the following useful property for pruning the search space [13].

Property 1 (Pruning search space using the TWU). Let X be an itemset, if $TWU(X) < minutil$, then X and its supersets are low utility [13].

Recently, algorithms were proposed to mine HUIs directly using a single phase [5,9,12], and were shown to outperform previous algorithms. FHM is to our knowledge the fastest algorithm for mining HUIs [5]. It performs a depth-first search to explore the search space of HUIs, and introduces an additional optimization named EUCP [5] to prune the search space using information about co-occurrences. FHM assign a structure named *utility-list* [5,9,12] to each itemset. Utility-lists allow calculating the utility of an itemset quickly by making join operations with utility-lists of shorter patterns. Utility-lists are defined as follows. Let $\succ$ be any total order on items from I. The *utility-list* of an itemset X in a database D is a set of tuples such that there is a tuple $(tid, iutil, rutil)$ for each transaction T_{tid} containing X. The $iutil$ element of a tuple is the utility of X in T_{tid}. i.e., $u(X, T_{tid})$. The $rutil$ element of a tuple is defined as $\sum_{i \in T_{tid} \wedge i \succ x \forall x \in X} u(i, T_{tid})$. For instance, assume that $\succ$ is the alphabetical order. The utility-list of $\{a\}$ is $\{(T_0, 5, 20), (T_2, 5, 3), (T_3, 10, 12)\}$. The utility-list of $\{d\}$ is $\{(T_0, 6, 3), (T_1, 6, 3), (T_2, 2, 0)\}$. The utility-list of $\{a, d\}$ is $\{(T_0, 11, 3), (T_2, 7, 0)\}$. To discover HUIs, FHM performs a single database scan to create utility-lists of patterns containing single items. Then, longer patterns are obtained by performing the join operation of utility-lists of shorter patterns (see [5,12] for details). Calculating the utility of an itemset using its utility-list and pruning the search space is done as follows.

Property 2 (Calculating utility of an itemset using its utility-list). The utility of an itemset is the sum of $iutil$ values in its utility-list.

Property 3 (Pruning search space using utility-lists). Let X be an itemset. Let the *extensions* of X be the itemsets that can be obtained by appending an item y to X such that $y \succ i$, $\forall i \in X$. If the sum of $iutil$ and $rutil$ values in $ul(X)$ is less than $minutil$, X and its extensions are low utility.

FHM is very efficient. However, it can generate a huge amount of HUIs. This can make the algorithm run out of storage space, and fail to terminate. Furthermore, it is very inconvenient for a user to analyze a large set of HUIs. To discover small and representative subsets of all HUIs, concise representations of HUIs have been proposed such as *closed HUIs* [16], *maximal HUIs* [14], and *generators of HUIs* [6], defined as follows. The *support* of an itemset X in a database D is denoted as $sup(X)$ and defined as $|g(X)|$, the number of transactions containing X. A HUI X is a *closed HUI* (*CHUI*) [16] iff there exists no HUI Y such that $X \subset Y$ and $sup(X) = sup(Y)$. A HUI X is a *maximal HUI* (*MaxHUI*) [14] iff there exists no HUI Y, such that $X \subset Y$. An itemset X is a *generator of high-utility itemsets* (*GHUI*) iff (1) there exists no itemset $Y \subset X$, such that $sup(X) = sup(Y)$, and (2) there exists an itemset Z such that $X \subseteq Z$ and $u(Z) \geq minutil$ [6].

3 The Minimal High Utility Itemsets

CHUIs, MaxHUIs and GHUIs are designed to provide answers to different questions that retailers may have about customer transactions, as outlined in the introduction. A drawback of the representations of CHUIs and MaxHUIs is that they tend to find very long HUIs, containing many items. A problem with these representations is thus that these HUIs often represent rare cases, as generally few customers exactly buy a same large set of items. For marketing purpose, a retailer may be more interested in finding the smallest sets of items generating a high profit, since it is easier to co-promote a small set of items targeted at a many customers rather than a large set of items targeted at few customers. The representation of GHUIs [6] partially addresses this issue by finding the smallest itemsets common to groups of customers having bought a set of items generating a high profit. However, no research has yet considered mining only the smallest HUIs. To address this research gap, we thereafter propose the novel concise representation of *minimal high-utility itemsets* (MinHUIs).

Definition 1 (Minimal HUIs). An itemset X is a *minimal high-utility itemset* (*MinHUI*) iff $u(X) \geq minutil$ and there does not exist an itemset $Y \subset X$ such that $u(Y) \geq minutil$.

This proposed representation is the opposite of maximal HUIs, i.e. it consists of the smallest sets of items that generate a high profit rather than the largest. To better show the relationship between the proposed MinHUIs, and the previously proposed CHUIs, MinHUIs and GHUIs, Fig. 1 presents an illustration of these various types of patterns, for the running example. In this figure, all equivalence classes containing at least a HUI are represented. An *equivalence class* is a set of itemsets supported by the same set of transactions, ordered by the subset relation. For example, $\{\{a, e\}, \{a, c, e\}\}$ is the equivalence class of itemsets appearing in transactions T_0 and T_2. An alternative and equivalent definition of GHUIs and CHUIs is the following. For each equivalence class containing a HUI, the CHUI is the largest itemset (the one having no superset in that equivalence class), while GHUI(s) are the smallest itemsets (those having no subset in that same equivalence class). Note that in the illustration equivalence classes are represented as Hasse diagrams and that low-utility itemsets that are not GHUIs in each equivalence class are not shown. As it can be seen in this example, MaxHUIs can be very long and thus offer few useful information to the user. For example, the only

MaxHUI found in the running example is $\{a, b, c, d, e\}$, and it represents the very specific case of a single customer (T_0). CHUIs are interesting but they also tend to contain very large itemsets. For example, CHUIs include $\{a, b, c, d, e\}$ in the example. GHUIs find the smallest itemsets common to a set of customers. However, a drawback of GHUIs is that some of these itemsets are low-utility such as $\{e\}$ in the example. To address these issues, the proposed MinHUIs are defined as the smallest high-utility itemsets. These itemsets are interesting as they tend to have a high support (represent many customers) as shown in this example, and are all HUIs. MinHUIs in this example are: $\{b, c\}$, $\{b, d\}$, $\{b, e\}$, $\{a, c\}$ and $\{c, e\}$. Formally, the relationship between these various sets of HUIs are the following: $MinHUIs \subseteq HUIs \subseteq 2^I$, $MaxHUIs \subseteq CHUIs \subseteq HUIs \subseteq 2^I$, and $GHUIs \subseteq 2^I$.

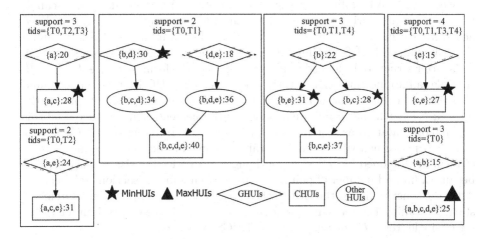

Fig. 1. HUIs and their equivalence classes (represented using Hasse diagrams)

A problem with previous representations is that the number of discovered patterns can still be very large since the number of HUIs, CHUIs, GHUIs and MaxHUIs increases when the *minutil* threshold is decreased. It is interesting to note that this is not necessarily the case for MinHUIs (Property 4).

Property 4 (Influence of minutil on MinHUI count). If *minutil* is lowered, the number of MinHUIs may increase, decrease or stay the same. Moreover, if $minutil = 1$, the set of MinHUIs is equal to I.

The above property is demonstrated using the running example. For $minutil = 20$, there is 3 MinHUIs: $\{a\}$, $\{b\}$, and $\{c, e\}$. For $minutil = 25$, there are 5 MinHUIs: $\{b, c\}$, $\{b, d\}$, $\{b, e\}$, $\{a, c\}$, and $\{c, e\}$. For $minutil = 30$, there are 3 MinHUIs: $\{b, d\}$, $\{b, e\}$, and $\{a, c, e\}$. Another interesting property of MinHUIs is used for pruning the search space in the proposed MinFHM algorithm.

Property 5 (pruning property of minimal high-utility itemsets). If an itemset X is a MinHUI, then supersets of X are not MinHUIs.

4   The MinFHM Algorithm

This section presents the proposed MinFHM algorithm. It first describes the main procedure, which is inspired by the FHM [5] algorithm. This procedure is designed to mine all HUIs. Then, it explains how that procedure is adapted to find only MinHUIs. The resulting algorithm is called MinFHM. The main procedure of MinFHM (Algorithm 1) takes as input a transaction database with utility values and the *minutil* threshold. The algorithm first scans the database to calculate the TWU of each item. Then, the algorithm identifies the set I^* of all items having a TWU no less than *minutil* (other items are ignored since they cannot be part of a high-utility itemsets by Property 3). The TWU values of items are then used to establish a total order $\succ$ on items, which is the order of ascending TWU values (as suggested in [12]). A second database scan is then performed. During this database scan, items in transactions are reordered according to the total order $\succ$, the utility-list of each item $i \in I^*$ is built and a structure named EUCS (Estimated Utility Co-Occurrence Structure) is built [5]. This latter structure is defined as a set of triples of the form $(a, b, c) \in I^* \times I^* \times \mathbb{R}$. A triple (a,b,c) indicates that $TWU(\{a, b\}) = c$. The EUCS can be implemented as a triangular matrix or as a hash map of hash maps where only tuples of the form (a, b, c) such that $c \neq 0$ are kept. In our implementation, we have used this latter representation as it is more memory efficient. Building the EUCS is very fast (it is performed with a single database scan) and occupies a small amount of memory, bounded by $|I^*| \times |I^*|$, although in practice the size is much smaller because a limited number of pairs of items co-occurs in transactions (cf. Sect. 5). After the construction of the EUCS, the depth-first search exploration of itemsets starts by calling the recursive procedure *Search* with the empty itemset $\emptyset$, the set of single items I^*, *minutil* and the EUCS structure.

Algorithm 1. The MinFHM algorithm

---

    input : D: a transaction database, *minutil*: a user-specified threshold
    output: the set of high-utility itemsets

1 Scan D to calculate the TWU of single items;
2 $I^* \leftarrow$ each item i such that $TWU(i) \geq minutil$;
3 Let $\succ$ be the total order of TWU ascending values on I^*;
4 Scan D to build the utility-list of each item $i \in I^*$ and build the $EUCS$;
5 Output each item $i \in I^*$ such that $SUM(\{i\}.utilitylist.iutils) \geq minutil$;
6 **Search** $(\emptyset, I^*, minutil, EUCS)$;

The *Search* procedure (Algorithm 2) takes as input (1) an itemset P, (2) extensions of P having the form Pz meaning that Pz was previously obtained by appending an item z to P, (3) *minutil* and (4) the EUCS. The search procedure operates as follows. For each extension Px of P, if the sum of the *iutil* values of the utility-list of Px is no less than *minutil*, then Px is a high-utility itemset and it is output (cf. Property 4). Then, if the sum of *iutil* and *rutil* values in the utility-list of Px are no less than *minutil*, it means that extensions of Px should be explored. This is performed by merging Px with all extensions Py of P such that $y \succ x$ to form extensions of the form Pxy containing $|Px| + 1$ items. The utility-list of Pxy is then constructed as in HUI-Miner by calling the *Construct* procedure (cf. Algorithm 3) to join the utility-lists of P, Px and Py. This latter procedure is the same as in HUI-Miner [12] and is thus not detailed here. Then, a recursive call to the *Search* procedure with Pxy is done to calculate its utility and explore its extension(s). Since the *Search* procedure starts from single items, it recursively explores the search space of itemsets by appending single items and it only prunes the search space based on Property 5. It can be easily seen based on Property 1, 2 and 3 that this procedure is correct and complete to discover all high-utility itemsets.

Algorithm 2. The *Search* procedure

---

input : P: an itemset, *ExtensionsOfP*: a set of extensions of P, , *minutil*: a user-specified threshold, *EUCS*: the *EUCS*
output: the set of high-utility itemsets

1 **foreach** *itemset* $Px \in$ *ExtensionsOfP* **do**
2   **if** *SUM(Px.utilitylist.iutils)*+*SUM(Px.utilitylist.rutils)* $\geq$ *minutil* **then**
3   *ExtensionsOfPx* $\leftarrow \emptyset$;
4   **foreach** *itemset* $Py \in$ *ExtensionsOfP* *such that* $y \succ x$ **do**
5   **if** $\exists (x, y, c) \in$ *EUCS* *such that* $c \geq minutil$) **then**
6                  $Pxy \leftarrow Px \cup Py$;
7                  $Pxy.utilitylist \leftarrow$ **Construct** (P, Px, Py);
8   *ExtensionsOfPx* $\leftarrow$ *ExtensionsOfPx* $\cup \{Pxy\}$;
9   **if** *SUM(Pxy.utilitylist.iutils)* $\geq$ *minutil* **then** output Px;
10   **end**
11   **end**
12   **Search** $(Px, ExtensionsOfPx, minutil)$;
13   **end**
14 **end**

We now explain how the search procedure is modified to mine only MinHUIs, rather than all HUIs. The first modification is to the main MinFHM procedure (Algorithm 1). During the first database scan, the utility of each single item is now calculated. Then, each item x that is a high-utility itemset is directly output. The reason is that each such item x is a MinHUI, since no smaller itemset can be a HUI. Thereafter, each such item x is removed from the set I (and thus will not be inserted in I^*). Thus, no superset of x will be explored by the *Search* procedure, and item x will be ignored in TWU and remaining utility calculations, afterward. The reason for removing item x from I is that if x is a HUI, then all supersets of x are not MinHUIs according to Property 5. By applying the previous modification, the algorithm will correctly output MinHUIs that are

single items. To find MinHUIs having more than one item, modifications are
made to the *Search* procedure (Algorithm 2) as follows. A new structure called
the *MinHUI-store* is introduced. At any time, this structure stores the itemsets,
which are currently considered to be MinHUIs. When a new HUI Pxy is found,
the modified algorithm checks if there exists an itemset Y in the *MinHUI-store*
such that $Y \subset Pxy$. If there exists such an itemset Y, then Pxy is not a MinHUI.
Thus, Pxy is not output. Moreover, by Property 5, supersets of Pxy are also not
MinHUIs. Thus, Pxy is not added to the set *ExtensionsOfPx*, to ensure that
extension of Pxy will not be considered by the search procedure. If there does
not exist an itemset Y such that $Y \subset Pxy$, then Pxy is assumed to be a MinHUI.
The itemset Pxy is thus inserted into the *MinHUI-store*. Then, the modified
algorithm removes each itemset Z in the *MinHUI-store* such that $Pxy \subset Z$,
because each such itemset Z is no longer a MinHUI, after the discovery of Pxy.
When the algorithm terminates, all MinHUIs in the left-store are output. The
union of these itemsets with the single items that are MinHUIs (which have been
previously output), are the full set of MinHUIs. By the definition and properties
presented in this paper, it can easily be seen that this algorithm is correct and
complete for mining MinHUIs.

Algorithm 3. The Construct procedure

---

    input   : P: an itemset, Px: the extension of P with an item x, Py: the
               extension of P with an item y
    output: the utility-list of Pxy

1 $UtilityListOfPxy \leftarrow \emptyset$;
2 **foreach** *tuple $ex \in Px.utilitylist$* **do**
3   **if** $\exists ey \in Py.utilitylist$ *and* $ex.tid = exy.tid$ **then**
4   **if** $P.utilitylist \neq \emptyset$ **then**
5   Search element $e \in P.utilitylist$ such that $e.tid = ex.tid$.;
6             $exy \leftarrow (ex.tid, ex.iutil + ey.iutil - e.iutil, ey.rutil)$;
7   **end**
8   **else**
9             $exy \leftarrow (ex.tid, ex.iutil + ey.iutil, ey.rutil)$;
10   **end**
11         $UtilityListOfPxy \leftarrow UtilityListOfPxy \cup \{exy\}$;
12   **end**
13 **end**
14 **return** $UtilityListPxy$;

To further optimize the MinFHM algorithm, it is important to implement the
MinHUI-store structure efficiently. In our implementation, it is implemented as a
list of lists of itemsets. More specifically, the *MinHUI-store* structure stores itemsets having the same size in the same list of itemsets. This allows to efficiently
check if an itemset Pxy has proper supersets (subsets) in the *MinHUI-store*, by
only comparing Pxy with larger (smaller) itemsets. Furthermore, to be able to

quickly compare two itemsets, items in itemsets are lexicographically ordered. Another optimization is that it is not necessary to check if a HUI containing two items has a subset in the *MinHUI-Store*, since MinHUIs of size 1 are not used to generate larger itemsets. Thus, HUI of two items found by the search procedure can be directly assumed to be MinHUIs. Finally, the LA-Prune optimization [9] is also incorporated. Moreover, for each MinHUI $\{x, y\}$ of size 2 that is found, the corresponding tuple in the EUCS can be replaced by $(x, y, 0)$ to help prune the search space.

5   Experimental Study

We assessed the performance of MinFHM on a computer with a third generation 64 bit Core i5 processor running Windows 7 and 5 GB of free RAM. We compared the performance of the proposed MinFHM algorithm with FHM [5], CHUD [16], and GHUI-Miner [6], which are respectively the state-of-the-art algorithms for mining HUIs, CHUIs and GHUIs. All memory measurements were done using the Java API. The experiment was carried on four real-life datasets commonly used in the HUIM literature: *mushroom, retail, kosarak* and *foodmart*. These datasets have varied characteristics and represent the main types of data typically encountered in real-life scenarios (dense, sparse and long transactions). Let $|I|$, $|D|$ and A represents the number of transactions, distinct items and average transaction length. *mushroom* is a dense dataset ($|I| = 16,470$, $|D| = 88,162$, $A = 23$). *kosarak* is a dataset that contains many long transactions ($|I| = 41,270$, $|D| = 990,000$, $A = 8.09$). *retail* is a sparse dataset with many different items ($|I| = 16,470$, $|D| = 88,162$, $A = 10,30$). *foodmart* is a sparse dataset ($|I| = 1,559$, $|D| = 4,141$, $A = 4.4$). *foodmart* contains real external and internal utility values. For the other datasets, external utilities for items are generated between 1 and 1,000 by using a log-normal distribution and quantities of items are generated randomly between 1 and 5, as the settings of [2,12,15]. The source code of all algorithms and datasets can be downloaded as part of the SPMF open-source data mining library [7] at http://www.philippe-fournier-viger.com/spmf/. Algorithms were run on each dataset, while decreasing the *minutil* threshold until they became too long to execute, ran out of memory or a clear trend was observed. Figure 2 shows the execution times of MinFHM, FHM, CHUD, and GHUI-Miner. Figure 3 compares the number of MinHUIs, HUIs, CHUIs and GHUIs, respectively generated by these algorithms.

It can first be observed that mining MinHUIs using MinFHM is faster than mining HUIs, CHUIs and GHUIs, using FHM, CHUD and GHUI-Miner. On *mushroom*, MinFHM is up to 824, 44, and 71 times, faster than FHM, CHUD and GHUI-Miner. On *foodmart*, MinFHM is up to 80, 52, and 75 times faster than FHM, CHUD and GHUI-Miner. On *retail*, MinFHM is up to 6, 62, and 63 times faster than FHM, CHUD and GHUI-Miner. On *kosarak*, MinFHM is up to 1.8, 15, and 16 times faster than FHM, CHUD and GHUI-Miner. The reason

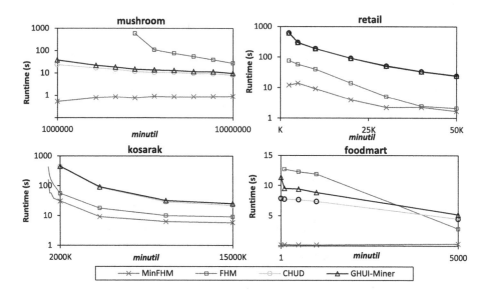

Fig. 2. Execution times

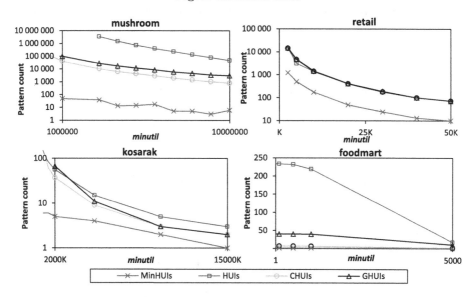

Fig. 3. Number of patterns found

for the excellent performance of MinFHM is that it prunes a large part of the search space by not exploring the transitive extensions[1] of MinHUIs.

---

[1] Recall that for an itemset X, the *extensions* of X are the itemsets that can be obtained by appending an item y to X such that $y \succ i$, $\forall i \in X$.

A second observation is that MinFHM scales well when *minutil* is decreased. For example, on *mushroom*, the runtime of MinFHM does not vary much and remains less than 1 s, while the runtime of FHM increases rapidly as *minutil* is decreased, taking more than 10 min to terminate. MinFHM shows a similar behavior on *foodmart* dataset, where the runtime of MinFHM is very stable while the runtimes of other algorithms increase considerably when *minutil* is decreased. On the *retail*, the runtime of MinFHM increases by a lesser amount compared to the other algorithms when *minutil* is decreased. Finally, on the *Kosarak*, the increase is comparable to the other algorithms. The reason why the runtime of MinFHM is generally very stable is that when *minutil* is decreased, the number of MinHUIs generally increases less rapidly than the number of HUIs, CHUIs and GHUIs (see Fig. 3). As mentioned in Property 4, MinHUIs have the nice property that their number may increase or decrease, as *minutil* is decreased, while the numbers of HUIs, CHUIs and GHUIs cannot decrease, and generally increase very quickly.

It is also interesting to observe that the number of MinHUIs never exceeded 1,300 patterns, while other algorithms generated up to millions of patterns. For example, on the dense *mushroom* dataset and *minutil* = 3, 000, 000, 38 MinHUIs, 3,538,181 HUIs, 10,311 CHUIs, and 27,640 GHUIs, are found. The number of MinHUIs is thus respectively, 931,000, 271, and 727 times less than the number of HUIs, CHUIs and GHUIs. It can thus be concluded that HUIs, CHUIs and GHUIs, generally depend on a very small set of MinHUIs, and that finding these MinHUIs provides a very compact and informative set of results to the user.

6   Conclusion

This paper has studied a novel representation of high-utility itemsets named Minimal High-Utility Itemsets (MinHUIs), its properties, and presented an efficient algorithm named MinFHM to discover MinHUIs. MinFHM includes numerous optimizations to discover MinHUIs efficiently. An extensive experimental study on real-life datasets shows that mining minimal HUIs is almost two orders of magnitude faster than mining HUIs, CHUIs or GHUIs and that it can greatly reduce the result set presented to the user. The source code of all algorithms and datasets can be downloaded as part of the SPMF open-source data mining library [7] at http://www.philippe-fournier-viger.com/spmf/.

For future work, an interesting possibility is to use MinHUIs as a negative border in HUI stream mining and incremental HUI mining, and also to explore the properties of MinHUIs for associative classifiers [18], and the discovery of minimal high-utility sequential patterns [19,20]. Lastly, another possibility is to design a faster algorithm for mining MinHUIs based on EFIM [21], a recently-proposed algorithm that was shown to outperform FHM for the traditional problem of HUI mining.

References

1. Agrawal, R., Srikant, R.: Fast algorithms for mining association rules in large databases. In: Proceedings of International Conference on Very Large Databases, pp. 487–499 (1994)
2. Ahmed, C.F., Tanbeer, S.K., Jeong, B.-S., Lee, Y.-K.: Efficient tree structures for high-utility pattern mining in incremental databases. IEEE Trans. Knowl. Data Eng. **21**(12), 1708–1721 (2009)
3. Deng, Z.: DiffNodesets: an efficient structure for fast mining frequent itemsets. Appl. Soft Comput. **41**, 214–223 (2016)
4. Deng, Z., Lv, S.-H.: PrePost+: an efficient N-lists-based algorithm for mining frequent itemsets via Children-Parent Equivalence pruning. Expert Syst. Appl. **42**(13), 5424–5432 (2015)
5. Fournier-Viger, P., Wu, C.-W., Zida, S., Tseng, V.S.: FHM: faster high-utility itemset mining using estimated utility co-occurrence pruning. In: Andreasen, T., Christiansen, H., Cubero, J.-C., Raś, Z.W. (eds.) ISMIS 2014. LNCS, vol. 8502, pp. 83–92. Springer, Heidelberg (2014)
6. Fournier-Viger, P., Wu, C.-W., Tseng, V.S.: Novel concise representations of high utility itemsets using generator patterns. In: Luo, X., Yu, J.X., Li, Z. (eds.) ADMA 2014. LNCS, vol. 8933, pp. 30–43. Springer, Heidelberg (2014)
7. Fournier-Viger, P., Gomariz, A., Gueniche, T., Soltani, A., Wu, C., Tseng, V.S.: SPMF: a Java open-source pattern mining library. J. Mach. Learn. Res. (JMLR) **15**, 3389–3393 (2014)
8. Lan, G.C., Hong, T.P., Tseng, V.S.: An efficient projection-based indexing approach for mining high utility itemsets. Knowl. Inf. Syst. **38**(1), 85–107 (2014)
9. Krishnamoorthy, S.: Pruning strategies for mining high utility itemsets. Expert Syst. Appl. **42**(5), 2371–2381 (2015)
10. Li, Y.-C., Yeh, J.-S., Chang, C.-C.: Isolated items discarding strategy for discovering high utility itemsets. Data Knowl. Eng. **64**(1), 198–217 (2008)
11. Song, W., Liu, Y., Li, J.: BAHUI: fast and memory efficient mining of high utility itemsets based on bitmap. Int. J. Data Wareh. **10**(1), 1–15 (2014)
12. Liu, M., Qu, J.: Mining high utility itemsets without candidate generation. In: Proceedings of 22nd ACM International Conference on Information and Knowledge Management, pp. 55–64 (2012)
13. Liu, Y., Liao, W., Choudhary, A.K.: A two-phase algorithm for fast discovery of high utility itemsets. In: Ho, T.-B., Cheung, D., Liu, H. (eds.) PAKDD 2005. LNCS (LNAI), vol. 3518, pp. 689–695. Springer, Heidelberg (2005)
14. Shie, B.-E., Yu, P.S., Tseng, V.S.: Efficient algorithms for mining maximal high utility itemsets from data streams with different models. Expert Syst. Appl. **39**(17), 12947–12960 (2012)
15. Tseng, V.S., Shie, B.-E., Wu, C.-W., Yu, P.S.: Efficient algorithms for mining high utility itemsets from transactional databases. IEEE Trans. Knowl. Data Eng. **25**(8), 1772–1786 (2013)
16. Tseng, V., Wu, C., Fournier-Viger, P., Yu, P.: Efficient algorithms for mining the concise and lossless representation of closed+ high utility itemsets. IEEE Trans. Knowl. Data Eng. **27**(3), 726–739 (2015)
17. Uno, T., Kiyomi, M., Arimura, H.: LCM ver. 2: efficient mining algorithms for frequent/closed/maximal itemsets. In: Proceedings of ICDM 2004 Workshop on Frequent Itemset Mining Implementations, CEUR (2004)

18. Nguyen, D., Vo, B., Le, B.: CCAR: an efficient method for mining class association rules with itemset constraints. Eng. Appl. Artif. Intell. **37**, 115–124 (2015)
19. Yin, J., Zheng, Z., Cao, L.: USpan: an efficient algorithm for mining high utility sequential patterns. In: Proceedings of 18th ACM SIGKDD International Conference on Knowledge Discovery and Data Mining, pp. 660–668 (2012)
20. Zida, S., Fournier-Viger, P., Wu, C.-W., Lin, J.C.W., Tseng, V.S.: Efficient mining of high utility sequential rules. In: Proceedings of 11th International Conference on Machine Learning and Data Mining, pp. 1–15 (2015)
21. Zida, S., Fournier-Viger, P., Lin, J.C.-W., Wu, C.-W., Tseng, V.S.: EFIM: a highly efficient algorithm for high-utility itemset mining. In: Sidorov, G., Galicia-Haro, S.N. (eds.) MICAI 2015. LNCS, vol. 9413, pp. 530–546. Springer, Heidelberg (2015)

Authenticity, Privacy, Security, and Trust

Automated k-Anonymization and l-Diversity for Shared Data Privacy

Anne V.D.M. Kayem[1,2]([⊠]), C.T. Vester[1], and Christoph Meinel[2]

[1] Department of Computer Science, University of Cape Town,
Rondebosch, Cape Town 7701, South Africa
akayem@cs.uct.ac.za
http://infosec.cs.uct.ac.za/
[2] Hasso-Plattner-Institute, Potsdam, Germany
http://hpi.de/meinel/lehrstuhl.html

Abstract. Analyzing data is a cost-intensive process, particularly for organizations lacking the necessary in-house human and computational capital. Data analytics outsourcing offers a cost-effective solution, but data sensitivity and query response time requirements, make data protection a necessary pre-processing step. For performance and privacy reasons, anonymization is preferred over encryption. Yet, manual anonymization is time-intensive and error-prone. Automated anonymization is a better alternative but requires satisfying the conflicting objectives of utility and privacy. In this paper, we present an automated anonymization scheme that extends the standard k-anonymization and l-diversity algorithms to satisfy the dual objectives of data utility and privacy. We use a multi-objective optimization scheme that employs a weighting mechanism, to minimise information loss and maximize privacy. Our results show that automating l-diversity results in an added average information loss of 7 % over automated k-anonymization, but in a diversity of between 9–14 % in comparison to 10–30 % in k-anonymised datasets. The lesson that emerges is that automated l-diversity offers better privacy than k-anonymization and with negligible information loss.

Keywords: Automated data anonymization · Multi-objective optimization · k-anonymity · l-diversity · Data outsourcing

1 Introduction

A common challenge faced by law enforcement agencies in developing world regions is that of analyzing large volumes of crime data [7,27]. Recent statistics from the United Nations (UN) and World Bank (WB) [28] estimate that violent crime cost Guatemala an estimated \$2.4 billion or 7.3 % of her Gross Domestic Product (GDP) in 2007, and the Mexican government estimated the costs of violence in 2007 at \$9.6 billion, primarily from lost investment, local business and jobs. The UN and WB also estimated that, in 2007, Jamaica and Haiti could have increased their GDP by 5.4 % merely by bringing down their crime levels to that of Costa Rica [28]. In South Africa for instance, it is estimated that

© Springer International Publishing Switzerland 2016
S. Hartmann and H. Ma (Eds.): DEXA 2016, Part I, LNCS 9827, pp. 105–120, 2016.
DOI: 10.1007/978-3-319-44403-1_7

more than a million of the approximately 2 million crimes reported annually, are never resolved [17,31]. Surveys indicate that corruption and police ineffectiveness fuel fears of disclosure and the general belief that most offenses go unresolved [17]. Challenges faced by the law enforcement authorities include limited "in-house" computational processing power which makes handling large volumes of crime data challenging and perhaps more importantly, the lack of data analytics expertise which is essential in identifying relevant data for crime resolution. Outsourcing the data to a third-party Data Analytics Service Provider (DASP) offers a cost effective management solution to the data analytics problem but the sensitivity of the data makes pre-processing to protect the data a necessary step before the data is transferred to the DASP.

Existing solutions based on encrypting the data before it is transferred to the DASP are time-intensive in terms of query response time which is undesirable when performance as well as data protection are a concern [4,9,11,12,18,30,33]. Data protection alternatives such as anonymization, are a better solution from the performance perspective. Manual anonymization is however, a time-consuming and error-prone procedure that can result in inadvertent disclosures of information. A further concern with manual anonymization is the challenge of preventing new releases of anonymized datasets from being adversarially combined with historical data to provoke linking and inferential attacks.

In this paper, we present an automated anonymization scheme that extends the standard k-anonymization and l-diversity algorithms to satisfy the dual objectives of data utility and privacy. The automated scheme employs a multi-objective optimization approach that uses a weighting mechanism to maximize information utility (minimize information loss) and diversity to maximise privacy by circumventing linking and inference attacks. This is handled via a two pronged approach where in the first step we maximize information utility under a modified k-anonymity algorithm in a manner that ensures security against linking attacks. In the second step, we extend the k-anonymity algorithm based on the concepts of l-diversity to provide protection against inference attacks. Our results indicate that l-diverse datasets incur an average information loss of 7% over k-anonymised datasets, but offer better privacy (protection against linking and inference attacks) with a diversity of between 9–14% in comparison to 10–30% in k-anonymised datasets. The lesson that emerges is that in automated anonymization, augmenting k-anonymization with l-diversity offers better privacy and at a negligible cost to utility.

The outline of the paper is as follows. In Sect. 2 we provide an overview of the literature on privacy preserving data publishing. We proceed in Sect. 3 with a specification of our proposed multi-objective scheme to support k-anonymization and l-diversity in automated data anonymization. In Sect. 4, we present results from experiments conducted on a prototype implementation platform [27]. We offer conclusions and suggestions for future work in Sect. 5.

2   Related Work

Privacy preserving data publishing combines efficient protection with availability in data analytics [6,16,19,22,25,32,36]. There are two tenets to privacy preserving data publishing. The first is to anonymize and then mine the data [2,3,6,16] and the second, to mine and then anonymize the released query results [1–3]. The second approach is better suited to users without the adequate in-house human-capital and computational resources. For this reason, we focus on privacy preserving data publishing schemes where the onus is to anonymize and then share.

Anonymization algorithms can be classified into two main groups namely, syntactic and probabilistic models [10]. Syntactic models have a well defined data output format, such that for small data sets privacy traits can often be confirmed by visually inspecting the data. Privacy violation adversarial models are constructed based on generally available information and generalizations drawn from the syntactic and semantic meaning of the underlying data. k-anonymity [1,6,19], l-diversity [25], and t-closeness [22] algorithms as well as their variants are classified under this category.

On the other hand, probabilistic privacy models employ data perturbations based primarily on noise additions to distort the data [10,34]. Perturbation approaches have been critiqued for being vulnerable to inferential attacks based on adversarial knowledge of the the true underlying distributions of the data [24]. Dwork et al. [15] proposed addressing this caveat with the notion of differential privacy. Differential privacy basically requires that the adversary learns no more from a published data set when one record (or individual) is present in, or removed from, the data set [34]. Attempts have also been made to combine attributes from both syntactic and probabilistic models to form hybrid anonymization approaches. Examples include probabilistic k-anonymity [2], and differential privacy with t-closeness [10]. However, automating these approaches for application on mixed data (categorical and numerical) in ways that minimize information loss and maximize privacy is a challenging problem [16,20].

Since crime data includes a mix of numerical and categorical data, we have opted to focus on syntactic anonymization models, specifically k-Anonymity and l-Diversity. For reasons, centered around high processing costs, we decided against considering the t-closeness scheme. Recall that one of the constraints we mentioned, is the limitation on computational processing power that the organizations face. Work on k-Anonymity was pioneered by Sweeney [29] as an approach to sharing data in plain text without revealing private or sensitive information about individuals. The principle behind k-anonymity is to use the notion of bucketization to create k sets of data (equivalence classes) such that for every tuple there exist at least $k - 1$ tuples that have the same quasi-identifier[1] values. Sweeney's work [29] triggered a plethora of schemes such as [13,14,21,23] aimed at performance improvement and circumventing inferential attacks.

Various l-diversity schemes have been proposed to address this drawback by considering that sensitive attributes are the main reason behind disclosures of information used to provoke inferential attacks [8,23,26]. l-diversity requires in addition, that the most frequent sensitive attribute occurences in an equivalence

---

[1] Quasi-identifiers: Attributes which independently or combined can be used to uniquely identify an individual.

class (EC) should not appear more than $\frac{1}{l}$ times in the EC. So, at least l distinct sensitive values must exist in each EC. As in k-anonymity schemes, efficiently obtaining usable but privacy preserving data sets is provably NP-Hard [35] and so, optimization heuristics have been proposed to improve on the basic l-diversity scheme [13,14,26,35]. We note that l-diversity has the drawback of being dependent on the distribution of sensitive attributes in the data set and so, sensitive attribute values with high probability mass functions (that is some values have a very high frequency and others a very low frequency of occurrence) are prone to provoking high information loss in the anonymized data set. In addition l-diversity only considers the frequency of specific values within independent ECs and not in the dataset as a whole which can result in inadvertent inferential disclosure. t-closeness addresses this caveat but requires a high degree of computational resources. Other issues are centered on the semantics of generalizations and the effect these generalizations have on enabling information disclosures [13,22,25].

In the following section, we propose augmented k-anonymity and l-diversity schemes to support automated data anonymization. The idea is to use the notion of Pareto optimality [5] that has the nice quality of considering that no optimal solution exists for a given problem but rather that the solution space consists of a set of optimal points [5]. This quality, is useful in designing an automated anonymization scheme in that it allows the scheme select the best optimal with respect to data utility and privacy at some given instant and to consider historical data releases. As mentioned before, automated data anonymization is a cost-effective and privacy preserving pre-processing step for data that is outsourced to DASPs. Application examples emerge for law enforcement authorities in developing world countries and organizations lacking the "in-house" computational processing power as well as the data analytics expertise. We now describe our proposed solution in the next section.

3   Multi-Objective Data Anonymization (MOA)

In this section we describe our multi-objective optimization scheme that is geared at supporting automated data anonymization via the k-anonymization and l-diversity algorithms. We begin by providing some basic notation to support our subsequent discussions.

3.1   Information Loss Notation

Let A be the attribute space (columns in a data table) such that $a \in A$ represents a specific attribute (column in the data table) in A and d represents a tuple that contains all the attributes in A.

We denote $T(a)$ as the generalization tree for numerical attributes and $K(a)$ is the generalization tree for categorical attributes. Furthermore, $T(a)_{max}$ and $T(a)_{min}$ denote the upper and lower limits respectively for numerical attribute generalizations while $t_{d,i}(a)_{max}$ and $t_{d,i}(a)_{min}$ represent the upper and lower limits of the generalization of an attribute a in tuple d during the ith iteration of the anonymization algorithm.

Finally, $K(a)_{total}$ is the total number of leaf nodes generated for $K(a)$ and P is the number of nodes created by $K(a)$. $k(a)_p$ is a sub-tree of $K(a)$ rooted at a node $p \in P$ and $k(a)_{p,total}$ is the number of leaf nodes in $k(a)_p$.

3.2 Information Loss and Severity Weighting

Once the data has been processed and generalized, the next step is to find a suitable balance between information loss and privacy. Minimizing information loss is useful in ensuring data usability while maximizing privacy ensures adequate data protection from adversarial access. In line with our goal of multi-objective optimization, we employ a piece-wise function to handle information loss on both categorical and numerical data.

$$IL_{d,i}(a) = \begin{cases} \dfrac{k(a)_{p,total} - 1}{P - 1} & \text{if categorical} \\[2ex] \dfrac{t_{d,i}(a)_{max} - t_{d,i}(a)_{min}}{T(a)_{max} - T(a)_{min}} & \text{if numerical} \end{cases} \tag{1}$$

where the Information Loss Metric is given by:

$$LM_i(a) = \sum_{d \in D} \sum_{a \in A} IL_{d,i}(a) \tag{2}$$

To minimize information loss, we employ a weighting scheme for the loss metric which enables authorized end users to prioritize specific attributes during anonymisation. By this we mean that the data owner can decide to specify the Quasi-Identifiers (QIDs) that should contain more information without negatively impacting on data privacy. The weighting scheme acts as a sort of utility function that can be adjusted dynamically to allow the data owner decide what levels of privacy to sacrifice in favor of query result accuracy without negatively impacting on the overall privacy of the data. The weighted information loss metric ($IL_{weight,i}$) at the ith iteration of the algorithm is computed as follows.

$$IL_{weight,i} = \sum_{d \in D} \sum_{a \in A} w_a \times IL_{d,i}(a) \tag{3}$$

where w_a is the weight assigned to attribute $a \in A$ by the data owner. Finally, to facilitate automated anonymization we use a sensitive attribute severity weighting $S(c)$ where $c \in SA$. SA is the list of sensitive attributes and $S(\cdot)$ maps the sensitive attribute category to its weight.

Example 1. In Table 1, SA denotes the list of offences (sensitive attribute) and $S(\cdot)$ maps the crime category to its weight, which in this case is simply the guideline sentence duration (in time - months, years...) for a given crime. So, $S(\text{Theft}) = 5$ indicates a sentence of 5 years. We note that following this scale, the risk of privacy loss for a tuple containing *"Robbery"* is higher than for a tuple with *"Disorderly Conduct"*.

Table 1. Crime severity weightings

Crime	Severity
Embezzlement	3
Disorderly conduct	3
Theft	5
Drunken driving	5
Robbery	7

We now describe our automated anonymization schemes, namely CG-Kanon and CG-Diverse that are extensions of the k-anonymization and l-diversity algorithms respectively.

3.3   CG-Kanon Scheme

Our proposed CG-Kanon scheme uses the severity weighting and bucketization, to hide tuples with highly sensitive values in larger ECs while tuples of lower sensitivity are classified in smaller ECs. For instance, a tuple concerning a "Robbery" should be classified in a 20-anonymity EC while "theft" could be placed in a lower level EC say, 5-anonymity. This idea of hiding more sensitive values in larger ECs does not affect the absolute level of k-anonymity for different sensitive attribute categories. It is instead a relative statement regarding the level of k-anonymity required for different sensitive attributes in the anonymized dataset. The severity weighting is converted to a severity penalty which is used by the CG-Kanon scheme. To do this, we compute an absolute required minimum level of k-anonymity ($k_{\min}$) for the dataset and use $k_{\min}$ to guarantee a global minimum level of k-anonymity that all ECs must adhere to in the dataset. We compute $k_{\min}$ as follows:

$$k_{min} = \max\left(k_{cons}, \min\left(S_D(\cdot)\right)\right) \tag{4}$$

where k_{cons} is a fixed minimum level of k and $S_D(\cdot)$ is the set of all severities for the dataset D. The definition of k_{min} shows that the global minimum level of k-anonymity is fixed at k_{cons} or at the lowest level of attribute sensitivity in the dataset when $\min\left(S_D(\cdot)\right) > k_{cons}$. If $k_{cons} = 5$ and $\min\left(S_D(\cdot)\right) = 3$ then $k_{min} = 5$. However if $\min\left(S_D(\cdot)\right) = 7$ then $k_{min} = 7$ instead. The CG-Kanon scheme uses $k_{\min}$ as the k-anonymity baseline when deciding on appropriate ECs for tuples based on sensitivity.

Once $k_{\min}$ has been computed, we compute the severity penalty for each classification since the CG-Kanon scheme requires this information to optimize the information loss and privacy cost-benefit trade-off. The severity penalty determines the level of loss of privacy for a single tuple $d \in D(\cdot)$ and is computed as follows.

$$SP_{d,i} = \frac{S_d(c)}{|e_{d,i}|} \tag{5}$$

where $D(\cdot)$ is the dataset, $S_d(c)$ is the severity weight of sensitive attribute $c \in d$, and E is the set of ECs such that $|e_{d,i}|$ is the size of the EC that a tuple d is classified in during the ith iteration of the CG-Kanon scheme.

Example 2. From the severity penalty computation, highly sensitive attributes in small ECs result in high penalties and vice versa. So, if a "murder" report with a severity weighting of 25 were located in a 5-anonymity EC, a penalty of $\frac{25}{5} = 5$ is generated. An incident of "theft" with a severity weighting of 5 generates a severity penalty of 1, indicating that this information is comparatively less sensitive. The CG-Kanon scheme uses the severity penalty as a criterion besides, information utility, to determine tuple placement in ECs to minimize the overall sensitive information exposure risk.

Finally, the CG-Kanon scheme must compute the aggregate severity penalty, $SP_{tot,i}$, for the entire dataset, to determine whether the obtained anonymized dataset satisfies at least the threshold goals of privacy and utility. $SP_{tot,i}$ is computed as follows:

$$SP_{tot,i} = \sum_{d \in D} SP_{d,i} \tag{6}$$

and expresses the total severity penalty for the dataset as the summation of the severity penalties of the individual tuples. The $SP_{tot,i}$ is then feed into a fitness function to decide whether each tuple in D satisfies both objectives. We express the fitness function as follows:

$$FF_i^{CG-Kanon} = \frac{1}{\max\left(SP_{tot,i}, LM_{CG,i}\right)} \tag{7}$$

So, the result for $FF_i^{CG-Kanon}$ at iteration i is the inverse of the maximum of $SP_{tot,i}$ and $LM_{CG,i}$. Recall that a high $SP_{tot,i}$ indicates a strong risk of privacy exposure, while a high $LM_{CG,i}$ indicates a high level of information loss. Therefore, it is desirable that the fitness function generates results that iteratively converge towards a high value for $FF_i^{CG-Kanon}$, expressed by low values of $SP_{tot,i}$ and $LM_{CG,i}$ respectively.

The main drawback here is that, depending on tuple distribution, the diversity of the sensitive attributes in large ECs can be quite low and this negatively impacts on privacy. As well, a large proportion of tuples are suppressed to satisfy the minimum level of k-anonymity which results in high information loss. We addressed this by limiting the size of ECs to a pre-defined threshold size and as we discuss in Sect. 4, found that this reduces the number of suppressions to satisfy $k_{\min}$-anonymity. We still have the caveat of inferential attacks and so augment our CG-Kanon scheme with the CG-Diverse scheme (l-diversity algorithm inspired) to help circumvent these attacks.

3.4 CG-Diverse Scheme

Instead of using $SP_{tot,i}$ to classify tuples into ECs, the CG-Diverse scheme computes the average severity, AS_D, for D as well as the EC average severity weighting AS_e. The AS_D is computed for D and is used to start the anonymization process to ensure that the target level of l-diversity in D is such that $l = AS_D$. We compute AS_D as follows:

$$AS_D = \frac{\sum_{d \in D} S_d(c)}{|D|} \tag{8}$$

A high AS_D implies a higher level of diversity in the entire dataset. As a stopping criterion for deciding when an acceptable level of k_{min} and AS_D has been satisfied by all the ECs, we bound the l-diversity range with the severity weighting scale and use AS_D to compute the fitness of the dataset with respect to privacy and utility. We employ the following modified fitness function, expressed as follows:

$$FF_i^{CG-Diverse} = \frac{1}{\max\left(AS_{D,i}, LM_{CG,i}\right)} \tag{9}$$

However, as mentioned before suppressing the ECs that fail to meet the required levels of AS_D and k_{min} would result in a high level of information loss. Therefore, we alleviate this problem by identifying ECs with a lower average severity (but adequate relative diversity) to avoid high suppression rates. This is achieved by assessing the privacy of individual ECs that do not meet the global AS_D-diversity requirement. To this end the EC average severity weighting AS_e is computed as follows:

$$AS_e = \frac{\sum_{d \in D} S_d(c)}{|e|} \tag{10}$$

The AS_e of an EC is compared to the relative diversity l_e, and if $AS_e > l_e$ the tuples in the EC are generalized to the highest possible level to avoid suppression. Alternatively, when the diversity is higher than AS_e no changes are made. We note that this procedure is computationally inexpensive since it simply requires comparing AS_e with the actual observed diversity of the EC.

Example 3. Table 2 shows the average severity measures calculated for a given sample dataset. The $AS_e = 5$ is calculated as follows: $\frac{5+3+7+5+5}{5}$ using the crime severity weightings given in Table 1. By considering Table 1, and Eqs. (8) and (10), the l-diversity range can be restricted to between 3–25, depending on the underlying dataset. Yet requiring ECs to satisfy the global level of AS_D-diversity might be too restrictive. We alleviate this issue by moving tuples between ECs to minimise the information loss due to suppression. For instance, in Table 2 we observe that in the 5-anonymity EC, "Robbery" has a severity of 7 which implies an inference risk. CG-diverse handles such cases by using the AS_e to move the tuple to the more appropriate 7-anonymity EC as highlighted in Table 2.

We are now ready to discuss our experimental platform, results and analysis.

4   Results and Analysis

We demonstrate the feasibility of our proposed automated data anonymization scheme with results from experiments conducted on a prototype crime data collection application [27]. A host server with an Ubuntu server 12.04 operating

Table 2. Average severity versus diversity

Age	Crime	Diversity (Equivalence Class)	AS_D (Dataset)	AS_e (Equivalence Class)
18 - 22	Theft	4	11	5.0
18 - 22	Embezzlement	4	11	5.0
18 - 22	Robbery	4	11	5.0
18 - 22	Drunken Driving	4	11	5.0
18 - 22	Theft	4	11	5.0
18 - 87	Rape	8	11	7.0
18 - 87	Vandalism	8	11	7.0
18 - 87	Robbery	8	11	7.0
18 - 87	Assault	8	11	7.0
18 - 87	Murder	8	11	7.0

system running on a 64 bit machine with 8 GB RAM and a processor speed of 3.2 GHz (Intel Xeon E3-1230 Quad Core) was used. The algorithms were implemented in Java 1.7.0_65 while Python 2.7.3 was used to run the web server. A PostgreSQL 9.1 database management system and a Postfix email server were used to store the dataset, both plain and anonymized. Our dataset consisted of 10000 records because this is a reasonable bound for daily average crime report rates per police station [17]. The attributes considered included "Age", "Suburb", "Crime" and "Reporter". Sensitive attributes such as "Names" and "Date of Birth" were removed during pre-processing. Quasi-identifiers which more closely match the k-anonymity requirement for CG-Kanon were generated before the anonymization process. This was done by generalizing attributes to the highest node in the generalization hierarchy (tree) for ECs that do not meet the k-anonymity requirement. We qualitatively assessed the anonymized data produced by the CG-Kanon and the CG-Diverse algorithms, by considering aspects such as information loss, classification accuracy and the impact of the weighting scheme on linking and inference attacks. Throughout the discussion of the results we refer to an anonymization based on the weightings of the quasi-identifiers (QIDs) used during the anonymization. This will be denoted as $A_{w_{Age}} : S_{w_{Suburb}} : R_{w_{Reporter}}$. For example where equal weights were assigned to the QIDs this will be denoted as an $A1 : S1 : R1$ anonymization, similarly where we use $A10 : S5 : R1$ weights of 10, 5, and 1 were used for the *Age*, *Suburb*, *Reporter* attributes respectively (Fig. 1). $k_{constant}$ was set to 5 for all results on CG-Kanon anonymization. Our minimum crime severity level for the data was set to 3 and in this case, $k_{min} = 5$. For CG-Diverse, we set our lowest diversity level to 3 for all anonymization runs as a standard minimum privacy level. Since on average, the lower severity crimes were located in such ECs, this was acceptable. All algorithms were allowed to run for 30 min after which the algorithm was stopped. Pre-experiment sampling revealed that running for shorter periods, say 15 min resulted in high severity penalties and information loss for larger ECs, with only between 3–6 % of tuples meeting the minimum anonymity level. Running for much longer resulted in better success rates, but at the price of time. Once stopped the anonymized data was checked for compliance with the desired level of privacy. Tuples not satisfying the privacy criteria on termination were processed further according to the respective CG-Kanon and CG-Diverse algorithms (Fig. 1). Figure 2 shows the CG-Kanon algorithm classifying data using ECs only with no severity weighting support. We note that the crimes are clustered around smaller sized ECs which is good for protection against inference attacks, but bad for information loss. When the severity penalty is applied, we note as shown in Fig. 3 that more severe crimes are classified in larger ECs but this has the caveat of introducing inferential disclosure. For instance, from Fig. 3 one can see directly that more severe crime has a higher frequency with *"Murder"* being as high as 31 %. We address this with the CG-Diverse scheme. As shown in Figs. 4 and 5, based on the $A1 : S1 : R1$ weighting and an average severity level of 11, the global diversity and average severity of each EC is evaluated before suppressing the QIDs. When compared to Figs. 2 and 3, we note that the average diversity in CG-Kanon varies between 10 % and 30 % while that of CG-Diverse is much lower at 9 % to 14 % and consequently lowers inferential risk. The desired lower frequency (i.e. higher diversity) for more

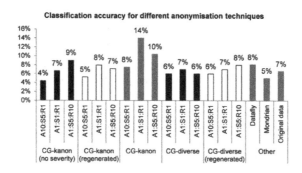

Fig. 1. Classification accuracy of CG-Kanon and CG-Diverse

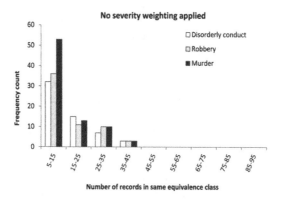

Fig. 2. Severity impact on dataset (no severity weighting)

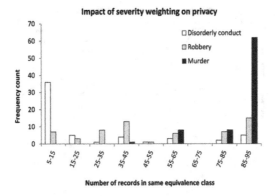

Fig. 3. Impact of severity weighting on privacy

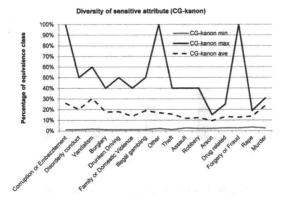

Fig. 4. Sensitive attributes frequency for CG-Kanon using A1:S1:R1

severe crimes is evident in CG-diverse whereas in CG-Kanon there is no such correlation. More severe crimes (Rape and Murder) in this case actually have lower average diversity and consequently less risk of inferential exposure. In addition we see the deviation from the mean frequency for more severe crimes is lower as severity increases. So not only does the average diversity increase as crime severity increases but the variance decreases as well. This gives us more certainty that more severe crimes will be less vulnerable to inference attacks. Finally, we note that l-diversity guarantees at least k-anonymity where $k = l$. The lowest diversity of 3 may appear weak from the privacy perspective when compared to the global diversity of 11 but it is unlikely, practically speaking, that severe crime (sensitive data) will be included in such lower diversity ECs. For instance, if we revisit our earlier results for CG-Kanon where the most serious crime ("Murder") was in an EC of size 90 and still only achieved a 3-diversity. Figures 6 and 7 show the aggregated information losses for different weighting schemes after termination of the algorithm. We selected three weighting schemes to mon-

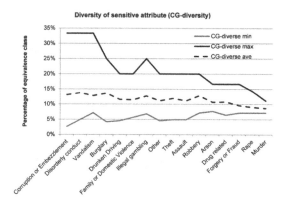

Fig. 5. Sensitive attributes frequency for CG-Diverse using A1:S1:R1

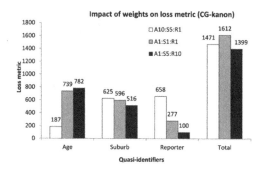

Fig. 6. Information loss for CG-Kanon

itor how the algorithms perform when attributes with varying granularity are weighted differently. For instance the $A10 : S5 : R1$ scheme overweights the *Age* attribute which is highly granular and under weighs the *Reporter* attribute, while $A1 : S5 : R10$ test the opposite scenario and $A1 : S1 : R1$ is equivalent to having no weighting scheme. The marginal increase in information loss for CG-diverse relative to CG-Kanon seems quite acceptable given the improved privacy provided by CG-Diverse. For our results the information loss across the three weighting schemes was on average 7 % higher for CG-diverse. However, this reduced data utility is acceptable given our desire for better anonymized data privacy. One further insight relates to the number of parameters that are used for the fitness function in selecting QIDs. We see from Figs. 8 and 9 that information loss for CG-diverse is a much lower proportion of its starting value than for CG-Kanon. This is attributed to the fact that CG-Kanon searches for solutions that minimize both the information loss and the severity penalty, in addition to satisfying k-anonymity. While CG-diverse only minimizes information loss and endeavours to meet the diversity requirement. The additional parameter (severity penalty) for CG-Kanon increases the search space and reduces the efficiency

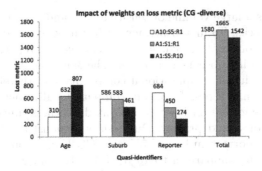

Fig. 7. Information loss for CG-Diverse

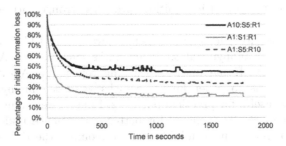

Fig. 8. Information loss reduction versus time (CG-Kanon)

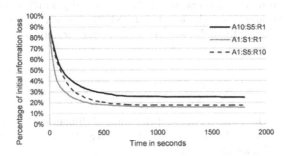

Fig. 9. Information loss reduction versus time (CG-Diverse)

of the algorithm. For instance, at termination the reduction in the initial information loss for $A10 : S5 : R1$ in CG-diverse (Fig. 9) was 74 % compared to 55 % for CG-Kanon (Fig. 8).

5   Conclusions

We presented two algorithms namely, CG-Kanon and CG-diverse that augment the standard k-anonymity and l-diverse algorithms to facilitate automatic classification and anonymization of data. In particular, we considered crime data because it contains a large volume of sensitive data and is vulnerable to linking and inferential attacks. To match privacy with utility, we used a random sampling approach without replacement so, historical released reports were excluded from being selected in subsequent releases. The sampling approach also offers the advantage of reduced computational complexity and therefore runtime for our algorithms which is a plus for use in computationally constrained environments. To reduce information loss, we also used a fitness function to improve classification accuracy, and privacy. Our results demonstrate that CG-diverse incurs an average information loss of 7 % over CG-Kanon, but with a diversity of between 9–14 % in comparison to 10–30 % CG-Kanon. So, we can conclude that, since CG-Diverse offers anonymity levels that are at least equal to CG-Kanon's, the percentage of information loss incurred does not significantly affect query response accuracy and in addition, provides stronger privacy guarantees than CG-Kanon.

Possible avenues for future work include evaluating CG-Kanon and CG-Diverse on de facto anonymization benchmarks such as the Adult's census dataset from the UC Irvine machine learning repository. Additionally, evaluations of robustness to other known attacks against k-anonymization and l-diversity will be useful for practical purposes. Finally, we should also consider parametrizing the t-closeness model for better performance under constrained conditions as an interesting candidate for overcoming the drawbacks of CG-Kanon and CG-Diverse.

Acknowledgements. The authors gratefully acknowledge funding for this research provided by the National Research Foundation (NRF) of South Africa, and the Hasso-Plattner-Institute (HPI). In addition, the authors are grateful for the anonymous reviews.

References

1. Aggarwal, C.C.: On k-anonymity and the curse of dimensionality. In: Proceedings of the 31st International Conference on Very Large Data Bases, VLDB 2005, pp. 901–909. VLDB Endowment (2005)
2. Aggarwal, C.C.: On unifying privacy and uncertain data models. In: Proceedings of the 2008 IEEE 24th International Conference on Data Engineering, ICDE 2008, pp. 386–395. IEEE Computer Society, Washington, DC (2008)
3. Aggarwal, C.C., Yu, P.S.: Privacy-Preserving Data Mining: Models and Algorithms, 1st edn. Springer, New York (2008)
4. Arasu, A., Eguro, K., Kaushik, R., Ramamurthy, R.: Querying encrypted data. In: Proceedings of the 2014 ACM SIGMOD International Conference on Management of Data, SIGMOD 2014, pp. 1259–1261. ACM, New York (2014)
5. Aytug, H., Koehler, G.J.: New stopping criterion for genetic algorithms. Eur. J. Oper. Res. **126**(3), 662–674 (2000)
6. Bayardo, R.J., Agrawal, R.: Data privacy through optimal k-anonymization. In: 21st International Conference on Data Engineering (ICDE 2005), pp. 217–228, April 2005

7. Burke, M., Kayem, A.: K-anonymity for privacy preserving crime data publishing in resource constrained environments. In: 28th International Conference on Advanced Information Networking and Applications Workshops, AINA 2014 Workshops, Victoria, BC, Canada, 13–16 May 2014, pp. 833–840 (2014)
8. Ciriani, V., Vimercati, S.D.C., Foresti, S., Samarati, P.: k-anonymous data mining: a survey. In: Aggarwal, C.C., Yu, P.S. (eds.) Privacy-Preserving Data Mining: Models and Algorithms, pp. 105–136. Springer, Boston (2008)
9. Ciriani, V., De Capitani Di Vimercati, S., Foresti, S., Jajodia, S., Paraboschi, S., Samarati, P.: Combining fragmentation and encryption to protect privacy in data storage. ACM Trans. Inf. Syst. Secur. 13(3), 22:1–22:33 (2010)
10. Clifton, C., Tassa, T.: On syntactic anonymity and differential privacy. Trans. Data Priv. 6(2), 161–183 (2013)
11. De Capitani Di Vimercati, S., Foresti, S., Jajodia, S., Paraboschi, S., Samarati, P.: Encryption policies for regulating access to outsourced data. ACM Trans. Database Syst. 35(2), 12:1–12:46 (2010)
12. De Capitani Di Vimercati S., Foresti, S., Paraboschi, S., Pelosi, G., Samarati, P.: Shuffle index: efficient and private access to outsourced data. ACM Trans. Storage 11(4), 19:1–19:55 (2015)
13. Dewri, R., Ray, I., Ray, I., Whitley, D.: Exploring privacy versus data quality trade-offs in anonymization techniques using multi-objective optimization. J. Comput. Secur. 19(5), 935–974 (2011)
14. Dewri, R., Whitley, D., Ray, I., Ray, I.: A multi-objective approach to data sharing with privacy constraints and preference based objectives. In: Proceedings of the 11th Annual Conference on Genetic and Evolutionary Computation, GECCO 2009, pp. 1499–1506. ACM, New York (2009)
15. Dwork, C., Roth, A.: The algorithmic foundations of differential privacy. Found. Trends Theor. Comput. Sci. 9(3–4), 211–407 (2014)
16. Ghinita, G., Karras, P., Kalnis, P., Mamoulis, N.: Fast data anonymization with low information loss. In: Proceedings of the 33rd International Conference on Very Large Data Bases, VLDB 2007, pp. 758–769. VLDB Endowment (2007)
17. Gould, C., Burger, J., Newham, G.: The saps crime statistics: what they tell us and what they don't. SA Crime Quaterly (2012). https://www.issafrica.org/uploads/1crimestats.pdf
18. Hang, I., Kerschbaum, F., Damiani, E.: ENKI: access control for encrypted query processing. In: Proceedings of the 2015 ACM SIGMOD International Conference on Management of Data, SIGMOD 2015, pp. 183–196. ACM, New York (2015)
19. Iyengar, V.S.: Transforming data to satisfy privacy constraints. In: Proceedings of the Eighth ACM SIGKDD International Conference on Knowledge Discovery and Data Mining, KDD 2002, pp. 279–288. ACM, New York (2002)
20. Kifer, D., Machanavajjhala, A.: No free lunch in data privacy. In: Proceedings of the 2011 ACM SIGMOD International Conference on Management of Data, SIGMOD 2011, pp. 193–204. ACM, New York (2011)
21. Last, M., Tassa, T., Zhmudyak, A., Shmueli, E.: Improving accuracy of classification models induced from anonymized datasets. Inf. Sci. 256, 138–161 (2014). Business Intelligence in Risk Management
22. Li, N., Li, T., Venkatasubramanian, S.: t-closeness: privacy beyond k-anonymity and l-diversity. In: 2007 IEEE 23rd International Conference on Data Engineering, pp. 106–115, April 2007
23. Lin, J.L., Wei, M.C.: Genetic algorithm-based clustering approach for k-anonymization. Expert Syst. Appl. 36(6), 9784–9792 (2009)

24. Liu, K., Giannella, C., Kargupta, H.: A survey of attack techniques on privacy-preserving data perturbation methods. In: Aggarwal, C.C., Yu, P.S. (eds.) Privacy-Preserving Data Mining: Models and Algorithms, pp. 359–381. Springer, Boston (2008)
25. Machanavajjhala, A., Kifer, D., Gehrke, J., Venkitasubramaniam, M.: L-diversity: privacy beyond k-anonymity. ACM Trans. Knowl. Discov. Data **1**(1), 1–52 (2007)
26. Nergiz, M.E., Tamersoy, A., Saygin, Y.: Instant anonymization. ACM Trans. Database Syst. **36**(1), 2:1–2:33 (2011)
27. Sakpere, A.B., Kayem, A., Ndlovu, T.: A usable and secure crime reporting system for technology resource constrained context. In: 29th IEEE International Conference on Advanced Information Networking and Applications Workshops, AINA 2015 Workshops, Gwangju, South Korea, 24–27 March 2015, pp. 424–429 (2015)
28. Seckan, B.: Violent crime in the developing world: research roundup. Journalist's Resource: Research on today's New topics (2012). http://journalistsresource. org/studies/international/development/crime-violence-developing-world-research-roundup
29. Sweeney, L.: K-anonymity: a model for protecting privacy. Int. J. Uncertainty Fuzziness Knowl.-Based Syst. **10**(5), 557–570 (2002)
30. Wang, F., Kohler, M., Schaad, A.: Initial encryption of large searchable data sets using hadoop. In: Proceedings of the 20th ACM Symposium on Access Control Models and Technologies, SACMAT 2015, pp. 165–168. ACM, New York (2015)
31. Website: South Africa's police: something very rotten. In: The Economist: Middle East and Africa (2012). http://www.economist.com/node/21557385
32. Wicker, S.B.: The loss of location privacy in the cellular age. Commun. ACM **55**(8), 60–68 (2012)
33. Wong, W.K., Kao, B., Cheung, D.W.L., Li, R., Yiu, S.M.: Secure query processing with data interoperability in a cloud database environment. In: Proceedings of the 2014 ACM SIGMOD International Conference on Management of Data, SIGMOD 2014, pp. 1395–1406. ACM, New York (2014)
34. Xiao, Q., Reiter, M.K., Zhang, Y.: Mitigating storage side channels using statistical privacy mechanisms. In: Proceedings of the 22nd ACM SIGSAC Conference on Computer and Communications Security, CCS 2015, pp. 1582–1594. ACM, New York (2015)
35. Xiao, X., Yi, K., Tao, Y.: The hardness and approximation algorithms for l-diversity. In: Proceedings of the 13th International Conference on Extending Database Technology, EDBT 2010, pp. 135–146. ACM, New York (2010)
36. Xu, J., Wang, W., Pei, J., Wang, X., Shi, B., Fu, A.W.C.: Utility-based anonymization using local recoding. In: Proceedings of the 12th ACM SIGKDD International Conference on Knowledge Discovery and Data Mining, KDD 2006, pp. 785–790. ACM, New York (2006)

Context-Based Risk-Adaptive Security Model and Conflict Management

Mahsa Teimourikia[✉], Guido Marilli, and Mariagrazia Fugini

Politecnico di Milano, Via Ponzio 34/35, 20133 Milan, Italy
{mahsa.teimourikia,mariagrazia.fugini}@polimi.it,
guido.marilli@mail.polimi.it

Abstract. In dynamic and risk-prone environments, security rules should be flexible enough to permit the treatment of risks, and to manage privileges on resources based on the situation at hand. For this purpose, we define safety-centric contexts based on risk description that is provided by the safety management system. This paper presents a risk-adaptive access control model that adopts hierarchies of contexts and security domains to make adaptations to risks at different levels of criticality. Since various risks may arise simultaneously, two or more security domains might be applicable at the same time incorporating various security rules which might lead to conflicts. Therefore, an approach to analyze conflicts is essential. In this work, we propose a conflict analysis algorithm based on set theory and we illustrate its usage with the proposed risk-adaptive access control model.

Keywords: Attribute-Based Access Control · Security · XACML · Conflict analysis · Context-awareness · Safety management

1 Introduction

Today, Access Control (AC) paradigms are moving from traditional models such as Role-Based Access Control (RBAC) to Attribute-Based Access Control (ABAC) [6], which offers a fine-grained AC over resources by considering relevant attributes for users, resources and the environment, and hence, it enables designing of more expressive security rules. On one hand, access control models are also being applied on physical and virtual resources [3,4,10], specially in smart work environments where the "things" (i.e., machinery and tools) are interconnected to form the Internet of Things (IoT). On the other hand, in a risk-prone environment such as construction and process industries, security policies should be flexible enough to permit the treatment of risks when necessary, and to manage privileges on physical and virtual resources for various authorized users based on the situation at hand.

Context-awareness in security is concerned with adaptation of security rules at run time to the situation at hand. In smart environments, various monitoring data are available to recognize the situational factors, facilitating incorporation

© Springer International Publishing Switzerland 2016
S. Hartmann and H. Ma (Eds.): DEXA 2016, Part I, LNCS 9827, pp. 121–135, 2016.
DOI: 10.1007/978-3-319-44403-1_8

of context-awareness into access control of physical resources. In these environments, security rules should be managed adaptively based on the risks that arise on the fly [3].

Previously, we presented a risk-adaptive AC model based on ABAC where we adopted Access Control Domains (ACDs) that include set of security rules defined for specific risk situations [3]. In the proposed approach, we assumed dealing with only one risk at a time to avoid unpredictable conflicts that may arise if more than one ACD be applicable simultaneously. We adopt XACML that enables usage of combination algorithms at rule and policy levels to avoid conflicts. However, in our case this is not enough and still unpredictable results may arise. Since we consider the ACDs to be predefined, we propose a conflict analysis algorithm to detect conflicting rules at design time and assist the security administrators to resolve them.

Furthermore, in order to avoid repetition of the security rules in the ACDs there is a need to consider hierarchies of ACDs where child ACDs inherit the security rules defined in their parents. In addition, it should be taken into consideration that when there is a case of an emergency or a crisis, there is not enough time to reason about the situation at hand and an approach should be considered for managing the emergency situations at run-time.

This paper, presents our approach to resolve mentioned problems. We define safety-centric contexts based on the risk description. We also propose an approach for calculating the level of criticality of the contexts to form hierarchies of contexts, each related to an ACD that deals with risks with different types and at different levels of criticality. Furthermore, to manage the emergency situations that are time-critical, we consider a "break-glass" security rules that apply when one or more of the contexts' criticality levels indicate an "emergency".

The paper is organized as follows: Sect. 2 reviews the state of the art. In Sect. 3 we introduce the preliminary concepts. Section 4 illustrates a scenario that puts in evidence a typical case of policy conflict. Section 5 shows the architecture of the Risk-Adaptive Access Control System (ACS). Section 6 presents our solution for conflict analysis and its resolution at run-time based on the proposed architectural framework. Section 7 gives some details about how we implemented our solution. And finally, Sect. 8 contains concluding remarks and ideas for future work.

2   Related Work

Access Control (AC) Systems are the first line of defense in the overall security of a system. AC applied to physical and virtual resources mainly protects the access points. Traditional AC models such as Mandatory Access Control (MAC), Discretionary Access Control (DAC) and Role-Based Access Control (RBAC), focus on defining user rights precisely to avoid any violations of the defined security rules [8]. However, traditional access control models demonstrate limited capabilities for adaptations to dynamic changes because of considering static security rules that determine the authorization decisions [8].

Risk-Adaptive Access Control is an emerging topic in the current research [1,14] which mainly concerns with balancing the risk of granting or denying access to resources. While the research has mostly focused on security risks management in access control models [1,2], there are limited works on considering adaptations based on safety risks. In [14], authors propose a criticality-aware ACS based on RBAC, where, according to the critical state of the environment, privileges of users can be dynamically altered by changing the user roles and the Access Control Lists (ACL) associated to the resources. Due to the adoption of RBAC, in the dynamic authorization process, they do not consider relevant attributes such as the location of the person or physical devices but they limit their proposal to the clearance of users.

ABAC has a potential to enable fine-grained access control in IoT applications because of its ability in accommodating changes to various attributes of users and resources to promote fine-grained and dynamic AC [6]. In [3], we described our approach, based on ABAC, to manage the dynamic changes to the security-related attributes of users and resources to dynamically authorize their privileges based on risks that are detected in the environment. In this work, we extend [3] to introduce the use of hierarchies of contexts and ACDs. Contexts in our view are safety-centric and represent a situation based on risks detected in the environment and their level of criticality. With the valuable data gathered in IoT environments, risks concerning the safety of people at various levels of criticality can be identified [5]. To manage risks, different approaches employ break-glass policies in AC [12]. However, more flexible and fine-grained AC can be applied having various, yet finite and manageable sets of security policies (defined as access control domains) for different contexts.

However, during dynamic adaptations of security policies, conflicts may arise. Conflict analysis has attracted great amount of research during current years [9,13,15]. Approaches that adopt translation of rules into first-order logic usually exhibit low performance. In this paper we propose an approach based on set theory in XACML 3.0 and we show that we can achieve an acceptable performance.

3 Preliminary Definitions

In this section, we define preliminary concepts that will be used throughout this paper. Risks related to safety in industrial environments refer to the threats that might endanger the health or life of the workers, which we simply refer to as "risks" from now on. As a policy language we adopt XACML, which its components are described in Table 1.

Table 2 lists the four main rule/policy combining algorithms in XACML, namely: deny overrides; permit overrides; first applicable; only one applicable.

Moreover, Table 3, summarizes the basic definitions regarding the proposed AC model. These include: Subject S, Object O, Environment EN, Privilege P, Access Control Domain ACD, Security Rule RU, Monitoring Device MD, Hazard H, Risk R, and *Consequence*.

Table 1. XACML Components

XACML component	Description
Policy set	It's a set of policies, characterized by a target and a combining algorithm
Policy	It's a set of rules, that apply to a certain target. Its result is computed basing on the chosen combining algorithm
Rule	It's contained in a policy and is composed by a target, condition and an effect
Target	Describes a set (or range) of values for the various categories' attributes, under which the policy/rule is applicable
Condition	It's an expression in a rule that evaluates to true or false and along with the target determines the effect of the rule/policy
Effect	It's the outcome of a rule/policy. The possible allowed values are usually *permit* and *deny*
Combining Algorithm	It's the procedure according to which the results of the policies/rules are combined
Attribute	Characteristic of a subject, resource, action or environment. Each category usually has a set of attributes

Table 2. Combining Algorithms

Algorithm	Description
Deny-overrides	If any evaluation returns deny, then the result must be deny, even if other evaluations have returned permit
Permit-overrides	If any evaluation returns permit, then the result must be permit, even if other evaluations have returned deny
First applicable	Rules are evaluated in their listing order
Only-one-applicable	For all of policies in the policy set, if no policy applies, then the result is NotApplicable. If more than one policy applies, then the result is Indeterminate. If only one policy applies, then the result is the result of evaluating that policy

ABAC is the basis of the AC model, where S, O, and EN and their attributes, SA, OA, and ENA respectively, are evaluated by the AC system for a fine-grained authorization, considering the applied RU. We distinguish two types of attributes ($ATTR = SA \cup OA \cup ENA$): (1) security related $ATTR_{sec}$, (e.g., *security level*, *role*) that are defined by the security administrator; and (2) IoT-based attributes that are dynamically gathered from the ambient using various sensors and monitoring devices $ATTR_{context}$, like *location*, and *time*, that are dynamically set when the relevant data is received. EN is considered to be a Smart Work Environment (SWE), where, the calculated context defines the global safety-centric situation that identifies the risk type and its level of criticality that affects the whole or parts of the environment including the subjects and objects inside it. To adopt XACML, RU is mapped to the Rule component, and ACD is mapped to the Policy component in XACML.

Table 3. Definitions & Notations

Notation	Definition
S	Finite set of entities both needing authorization to access resources (e.g., safety teams) and needing protection against risks (e.g., workers)
O	Finite set of physical resources or "things" (objects), e.g., tools, machinery, devices, that subjects can access or act on
EN	Finite set of environment sections
P	Finite set of privileges that are actions which subjects can perform on objects
ACD	Finite set of access control domains that contain security rules designed for different contexts
RU	Finite set of security rules
MD	Finite set of monitoring devices that sense data from S, O, and/or EN, e.g., sensors, cameras, wearable sensors, etc.
H	Finite set of hazards acknowledged via events in the environment; hazards might turn into risks
R	Finite set of risks identified in EN and endowed with attributes such as *Type*, *Probability*, *Source*, *Location*, and *Consequence*
C	Finite set of consequences which originate from each $r_i \in R$, endowed with attributes such as: *Type*, *Intensity*, *Probability*

4   A Motivating Scenario

In this section, we introduce a motivating scenario to illustrate a use-case in which conflicts may arise. In our approach under "safe" conditions acd_{safe} applies. If we enter into a risk state with a particular intensity, the related $acd_i \in ACD_{context}$ applies that contain security rules previously designed based on the organizations policies and protocols for management of that specific risk which usually relax some otherwise restricted security rules. Considering the possibility of dealing with more than one risk simultaneously, security rules should allow management of all the risks that are present. Therefore, when there are conflicts, rules permitting the execution of preventive or corrective strategies for risk management should prevail. While, in some cases we might want to restrict some permissions for safety reasons, e.g., a machinery is detected to be faulty and we want to restrict access to that machinery to avoid eventual risks. Hence, we cannot always have the assumption of Permit-Overrides when combining several ACDs.

In what follows we make an example for clarifying the issue. Assuming to have following ACDs, defined for two different risk situations:

acd_1 :**Context** $= ShortCircuit$

   ru_1 : **IF** $\{req.o.Type == \text{``FireSprinkler''} \wedge req.p == \text{``TurnOn''}\}$

   THEN $\{effect == Deny\}$

$acd_1.ru_1$ indicates that if anyone tries to activate the fire sprinkler system when there is a risk of electrical short circuit, the effect must be deny (water on a electrical short circuit may cause electric shock).

acd_2 :**Context** $= Fire$

   ru_1 : **IF** $\{req.s.ActiveRole ==$ "$RiskManager$" $\wedge$ $req.o.Type ==$ "$FireSprinkler$"

   $\wedge$ $req.p ==$ "$TurnOn$"$\}$

   THEN $\{effect == Permit\}$

while, $acd_2.ru_1$ indicates that in case of fire, the risk manager should be permitted to turn on the fire sprinkler system if necessary. Therefore, if *electrical short circuit* and *fire* risks are detected simultaneously, ACS have to consider both security rules at the same time. In this case, if the fire sprinkler system starts, the short circuit could intensify the fire and cause electric shock so the proper XACML combining algorithm should be deny-overrides. In conclusion, the hypothesis that the in case of multiple risks the most permissive security rule should prevail is not always correct.

Another interesting case happens if the XACML combining algorithm is First-applicable. Considering the particular nature of our system, the order of applicability of the ACDs is not predictable and therefore, it is possible that the ACDs would be analyzed in two possible orders (acd_1, acd_2 or acd_2, acd_1). In this way, AC system behaves unpredictably. In critical systems concerning with security and safety, the unpredictability of behavior and results is by no means desirable.

In the following sections, we introduce our proposed risk-adaptive AC system followed with the conflict analysis method for tackling the problems mentioned in this scenario.

5   Risk-Adaptive Access Control System (ACS)

In this section, the architecture of the ACS is presented. Its novelty lies in considering hierarchies of contexts and access control domains to manage different risks at different levels of criticality; while adopting break-glass policies in case of a crisis.

5.1   Risk-Adaptive AC Architecture

In this section we shortly describe the architecture of the adaptive AC system. While MD monitor $ATTR_{dynamic}$, data streams are sent to the Safety Management System (SMS), which, in a MAPE loop [3]: (1) monitors the meaningful parameters and identifies the hazards $H_c \subseteq H$ if there are any; (2) analyzes $H_c \subseteq H$ and performs risk assessment to provide the description of the risk and its consequences; (3) plans the preventive strategies; (4) executes the strategies that can be automatically executed by the SMS (e.g., turn on alarms), and supports the execution of human-operated strategies (e.g., evacuation of an area).

Hazard $h_i \in H_c$ may lead to a set of risks $R_c \subseteq R$, each with the following attributes that constitute the "*risk description*": *Type* as a unique name identifying the kind of the risk (e.g., fire); *Source* as the entity in the EN (device, machine, etc.) causing the risk; and *Location* refers to the $en_i \in EN$ affected by the risk. To simplify, we consider risks to be independent, as the dependency between risks can complicate the risk assessment procedure since dependent risks may have effects on one another.

Each $r_i \in R_c$ is connected to a set of consequences $C_{ri} \subseteq C$. In the analysis phase, the SMS also calculates the following attributes for each risk's consequence: *Type*, that is a unique name identifying the kind of the consequence, e.g., damage to the infrastructure, injury, death, etc.; and *Probability*, namely the degree of uncertainty related to the occurrence of such consequence. In the planning phase performed by the SMS, the AC system receives the *risk description* and uses it to identify the *"context"*. We consider a safety-centric approach to model the context, considering different attributes of a risk and its consequences.

5.2 Hierarchical Contexts and Access Control Domains

Figure 1 shows how contexts are mapped to the ACDs considering their hierarchies. Contexts are defined in a hierarchy where the safe context ($context_{safe}$) defines the state of the environment that is considered as safe, namely with no risks. $Context_{risk} = \{context_1, \ldots, context_i\}$, refer to a risk-prone state of the environment, where different risks and consequences are present. Context level $context_i.level$ is considered to represent the criticality of the $context_i$ that allows prioritizing its importance. For instance, $context_1$ representing a risk of fire with $context_1.level = 19$ is considered with a higher priority than $context_2$ corresponding to the risk of injury, with $context_2.level = 5$, hence deserving different priority of actions to face the potential risks. Finally, the emergency context ($context_{emergency}$) is associated with the highest level of criticality and has the highest priority over all the other contexts. To define safety-centric contexts we consider the risk description $rd_i \in RD$:

$$rd_i = \{r_i.Type, r_i.Probability, r_i.Source, r_i.Location, r_i.C_{ri}\} \tag{1}$$

where $r_i \in R_c$, and $r_i.C_{ri}$ represents the set of consequences identified for r_i and $cj \in r_i.C_{ri} = \{c_j.Type, c_j.Probability\}$. We define the context that represents the type of the risk $r_i.Type$ in the environment. The context level is defined based on risk impact value that is defined by safety experts considering rd_i and the organization's specific regulations. Context level illustrates the hierarchy of the contexts with respect to its criticality and is used for defining the priority of security rules in case conflicts were detected between them. Receiving the risk description from the SMS triggers the process of setting the context and calculating its level.

ECA Rules for Dynamically Selecting the Access Control Domains: To set the *Context* that best describes the safety status of the environment we consider a rule-based approach. Event-Condition-Action (ECA) rules have been commonly used for modeling context-aware behavior because of their flexibility and expressiveness. Moreover, adopting ECA rules enables management of contexts based on protocols and overall safety objectives of different organizations.

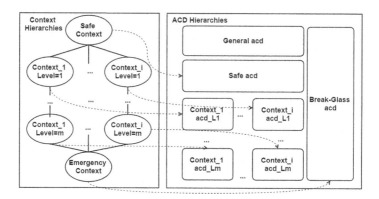

Fig. 1. Hierarchical levels of *ACD*.

Table 4. Notations for formal representations of ECA rules

Notation	Representation
ON	Operator catching an event
IF	Logical conditional operator for checking the conditions represented in the risk description
$\rightarrow$	Logical then operator representing the action which is setting the proper contexts
$\wedge$	Logical AND operator
$\vee$	Logical OR operator
$\sim$	Logical NOT operator
$>$	Greater than
$<$	Less than
$==$	Equivalent to
$=$	Set to
$++$	Increment operator
$ADD\,(param, set)$	Operator for adding a value ($param$) to a set
$REMOVE\,(param, set)$	Operator for removing a value ($param$) to a set
$ISIN\,(param, set)$	A Boolean operator that checks if a value ($param$) exists in a set

We introduce formal notations for representation of ECA rules. Later on we introduce our XML implementation used for implementing and adopting them in applications. Table 4 lists the formal notations defined for representing ECA rules.

Calculating Context Level of Criticality: Security rules are usually defined based on organization regulations and unique needs, and they need to be reviewed and managed by hierarchies of security administrators and managers to guarantee their compliance with specific policies of the organization. Therefore, usually security rules are predefined as dynamic generation of them at run-time can be very time consuming and in addition can pose security risks or do not achieve compliance with existing regulations. To this end, considering that we would have predefined security rules, it is also required to have a limited number of ACDs for feasibility and manageability. This number should indicate a right balance between fine-grained context-awareness and manageability of the security rules. In this work, number of required ACDs has a close relationship with the criticality levels that are considered for contexts, as they indicate the hierarchies required in the ACDs.

To calculate the level of criticality of contexts we refer to common and standard techniques adopted in risk management. Among various techniques used for analyzing and prioritizing risks, calculation of risk impact that leads to the calculation of risk exposure is the most commonly used. Based on the required levels of granularity the likelihood (probability) of the risk is categorized at different levels. For example, low, medium and high can be used for three levels of granularity which is a very rough estimate of the risk likelihood. Another example is to consider five or even ten levels of granularity starting from a very low likelihood and going up to catastrophic levels. In this work, we select the five-level granularity for both the risk likelihood and its impact to enable the manageability of the corresponding ACDs. Table 5, shows these levels, in addition to the numeric values considered to represent them.

On the other hand, the risk impact is an estimate representing the severity of the consequences of the risk [11]. Risk impact is calculated by risk experts and risk managers, considering the costs that the risk consequences have for the specific organization. Various consequences may be considered to have different costs. For example, one organization may consider reputation damage more important than the financial damage and hence attribute a higher cost to it. For the calculation of risk impact various techniques are introduced in the literature [11], however, reviewing and introducing them is out of the scope of this paper. We consider the risk impact as given, and similar to the risk probability we consider five levels of granularity for its value.

Having the risk probability and the risk impact and considering the numerical values assigned to them (see Table 5) risk exposure and hence the context's criticality level is calculated as follows:

$$context_i.level = RiskExposure(r_i) = r_i.Probability \times r_i.Impact \qquad (2)$$

where $r_i \in R_c$ and the $context_i$ corresponds to the $r_i.Type$. Table 5, shows the calculations of context levels for different risk probabilities and impacts. The contexts levels are color coded where the diagonal values show the medium levels of criticality for the given risk, the top triangle indicates high levels of criticality and the down triangle corresponds to low levels of criticality.

To define the safe and emergency contexts, we define two thresholds based on the views and protocols of the organization, namely: T_{safe} which represents a threshold on context level below which is considered safe; and $T_{emergency}$ that is a threshold on context level above which is considered an emergency situation.

Table 5. Qualitative and quantitative scales for calculating context levels

Probability \ Impact	$VeryLow = 1$	$Low = 2$	$Medium = 3$	$High = 4$	$VeryHigh = 5$
$VeryHigh = 5$	5	10	15	20	25
$High = 4$	4	8	12	16	20
$Medium = 3$	3	6	9	12	15
$Low = 2$	2	4	6	8	10
$VeryLow = 1$	1	2	3	4	5

All other *Context Level* values in between T_{safe} and $T_{emergency}$ are considered as the levels indicating the criticality of each risk type as the context. As an example we can define $T_{safe} = 2$ and $T_{emergency} = 20$ and hence $context_i.level \leq 2$ will correspond to the acd_{safe}; and $context_i.level \geq 2$ will correspond to the $acd_{emergency}$.

Mapping Contexts to ACDs: Contexts are mapped into different hierarchies of ACDs that are sets including the security rules applied to control access to objects under safety-related circumstances. As depicted in Fig. 1, we define a hierarchy of $acd_i \in ACD$. The $acd_{General} \in ACD$ defines a set of security rules that is commonly applied regardless of the context. This allows avoiding to repeat the common, shared security rules in different ACDs. The $acd_{safe} \in ACD$ includes a set of security rules that are applied in a safe context. Moreover, $ACD_{Context} = \{acd_{c1,l1}, \ldots, acd_{cn,lm}\}$ as context-specific ACDs are defined where c_i represents a context in $Context_{risk} = \{context_1, \ldots, context_i\}$. Therefore $acd_{c_i,l_j} \in ACD_{Context}$ includes a set of security rules that are applied in $context_i$ with level of criticality of $context_i.level = l_j$. When a acd_{c_i,l_j} with criticality level of l_j is applied, all the context-specific ACDs with the lower levels for the same context will also apply. Moreover, if a context is considered as an "emergency context" then neither the $acd_{General}$ nor the other $acd_{c_i,l_j} \in ACD_{Context}$ would apply and only the $acd_{emergency}$ will be considered that includes a set of break glass security rules that apply in an emergency situation. In the following an example is shown for clarifications.

An Example: Considering a process industry where a risk description is received from SMS which indicates that there is a medium probability of risk of fire, with the risk source of gas pipes, in the warehouse. The consequences are estimated to be injury, with high probability, and damage to infrastructures and resources with high probability. According to the organization protocols, and experts risk analysis based on the risk description, the risk impact is considered high. Also the thresholds are defined as: $T_{safe} = 2$ and $T_{emergency} = 20$. The $context_i = Fire$ and its level is calculated according to (1) and Table 5 which is equal to 12. Since $context_i.level > T_{safe}$ we are not in the safe context and similarly as $context_i.level < T_{emergency}$, we are not in emergency context either. If the calculation of $context_i.level$ indicates a safe or an emergency context we set the context accordingly. In the next step the context should be specified using ECA rules:

$$ON : r_i.Type == \text{``}Fire\text{''}$$
$$IF : context_i.level \leq T_{emergency} \wedge context_i.level \geq T_{safe}$$
$$\rightarrow ADD(\{\text{``}Fire\text{''}, context_i.level\}, ContextList).$$

The above mentioned ECA rule specifies that on the event that there is a risk of fire, on the condition that the context level is in between the safe and emergency thresholds, the pair of the context and its level are added to the list of active contexts.

6   Conflict Analysis of XACML 3.0 Rules

As shown previously, when there are several contexts active at the same time, conflicts may arise with the combination of different security rules in the corresponding ACDs. For conflict analysis we adopt an innovative approach based on set theory. A policy (that maps to our concept of ACD) in XACML 3.0, includes multiple rules (which maps to RU). Each rule in XACML 3.0 has a target and a condition elements. The main idea of this approach is that if the intersection of the targets and conditions of the existing security rules are not empty, then there is a possible conflict between them when their effects are not the same. More formally, considering RU, A, S, and O as the sets of security rules, actions, subjects, and objects respectively. If, for instance, we have two policies (or policy sets) that contain two rules $ru_1 \in RU$ and $ru_2 \in RU$, where ru_1 permits a set of actions $A_1 \subset A$ to set of subjects $S_1 \subset S$ on objects $O_1 \subset O$ and ru_2 denies $A_2 \subset A$ for subjects $S_2 \subset S$ on objects $O_2 \subset O$, then there is a conflict if and only if $S_1 \cap S_2 \neq \emptyset$, $A_1 \cap A_2 \neq \emptyset$ and $O_1 \cap O_2 \neq \emptyset$.

A target in XACML 3.0 includes set of *AnyOff* (i.e., disjunction) and *AllOff* (i.e., conjunction) XML elements. Therefore, the target is represented as follows in our set theory approach:

$$TC = \overbrace{(M_1 \wedge M_2 \wedge \ldots M_n)}^{AllOf} \vee \overbrace{(M_1' \wedge M_2' \wedge \ldots M_n')}^{AllOf} \vee \ldots \vee \overbrace{(M_1'' \wedge M_2'' \wedge \ldots M_n'')}^{AllOf} \quad (3)$$

where M_i represent the conditional statements in the target to which we refer as *match* elements. XACML 3.0 rule conditions include set of *Apply* elements and it is defined as follows:

$$Apply = <AS, D, V, F, S, R> \quad (4)$$

where AS is the set of apply elements contained in this element; D represents the XACML *AttributeDesignator*; V is the *AttributeValue*; F is the *Function*; S represents the *AttributeSelector*; and R is the *AttributeReference*. And therefore, the rule condition is defined as follows:

$$RC = \{Apply_1, Apply_2, \ldots, Apply_n\} \quad (5)$$

To check whether two XACML 3.0 rules ru_1 and ru_2 have a conflict we check the intersection of their targets TC_1 and TC_2. If $TC_1 \cap TC_2 \neq \emptyset$ and the conditions of these rules $RC_1 \cap RC_2 \neq \emptyset$ and the effects of the rules are not the same ($effect_{ru1} \neq effect_{ru2}$) then the rules are considered to have conflicts.

7   Implementation

In this section we shortly describe the implementation details of the proposed risk-adaptive ACS and the conflict analysis algorithm. The ACS is developed employing open source technologies offered by WSO2. More precisely, Balana (an implementation of XACML 3.0) and WSO2 Identity Server, which implement XACML 3.0's data-flow model are used. To be able to use these frameworks for our purposes some modifications have taken place that is described in what follows.

7.1   Customization of the Policy Editor

WSO2 Identity Server is modified to adapt its PAP's (Policy Administration Point) user interface, to enable customization of XACML security rules, policies and policy sets based on our specific requirements. In the original WSO2 PAP Basic Policy Editor user interface, it was not possible to specify a target/condition on a different resource attribute than its identifier. Moreover, we needed to add contexts as the attribute of the Environment and to set it as the target of XACML Policy for defining context-specific ACDs. With the mentioned modifications we are able to build XACML policy language to define ACDs based on any attribute of Subjects, Objects and the Environment.

7.2   ECA Rules Implementation

ECA rules are implemented via the following XML schema shown in Listing 1.4. We adopted similar notations to XACML policies for ECA rules for expressiveness, clarity and simplification of usage.

```
<metarule> :- <when> <if> <then>

<when> :- <when_anyof>
<when_anyof> :- <when_allof>+
<when_allof> :- <risk>+
<risk> :- name,<risk_parameters>
<risk_parameters> :- <risk_parameter>+

<if> :- <if_anyof>
<if_anyof> :- <if_allof>+
<if_allof> :- <condition>+
<condition> :- name,<condition_parameters>
<condition_parameters> :- <condition_parameter>+
<condition_parameter> :- value,type
```

```
<then>  :− <actions>
<actions>  :− <action>⁺
<action>  :− <action_parameters>
<action_parameters>  :− <action_parameter>⁺
<action_parameter>  :− actionType , name , value , type
type  :− "variable"|"immediate"
```

Listing 1.1. Meta-rule XML structure

7.3 Managing Multiple Contexts in XACML 3.0

According to the proposed risk-adaptive AC system, when a new risk is detected and the system is in its "safe" state (i.e., safe context is active), then the string that identifies the context will be assigned to the environment context attribute ($en_i.Context$ where $en_i \in EN$). Otherwise, if the system is already in a risk state, the new risk-specific context will be appended to $en_i.Context$, according to the following notation:

$$en_i.context == \text{"}context_1\#level\&context_2\#level\& \ldots \&context_n\#level\text{"}$$

where *context* is the unique name of the context and *level* is the context's level of criticality. Thanks to this kind of notation, we can keep the environment's context attribute as a simple data type. When needed, we can read all the contexts that are activated, by simply splitting the string. This does not affect the efficiency much as this string does not get very long for each section of the environment assuming that no more that some limited number of risks happen at a time. When a context is no longer active (which is detected by the SMS [3]), the context in the $en_i.Context$ string is removed. If there are no more active risk-specific contexts left, the PDP considers general and safe ACDs.

7.4 Performance Analysis

In this Section the results of experiments on the proposed conflict analyzer are illustrated. The performance analysis is conducted using a notebook running Windows 10 with 8 GB of RAM and an Intel Core i5-4210U dual core processor (3M Cache, up to 2.70 GHz). Since the main objective is to be able to analyze ACDs we have only considered XACML policies and their rules and we did not consider policy sets in this analysis. To run the experiment two well-known policy packages: Continue-a and Synthetic-360 are used. We ran our software on each of these policy packages ten times; the results are shown in Table 6. As shown in Table 6, a positive correlation exists between the number of conflicts detected and the analysis time. By analyzing the policies which compose our test packages, we noticed that Synthetic-360's rules contain a noticeably higher number of match elements and this obviously increases the time necessary to perform the analysis. Considering that XACML 3.0 language has a much more complicated structure than XACML 2.0, we started to compare the results obtained with the ones presented in [7] which concludes that we have been able to maintain the same order of magnitude for the processing time (although our analyzer is not able to analyze policy sets).

Table 6. Sample policy analysis results

Datasets	#Policies	#Rules	Average analysis time (s)	Conflicts detected
Continue-a	266	298	1.07	10483
Synthetic-360	72	360	162.15	26810

8 Concluding Remarks and Future Works

In this paper we have presented our risk-adaptive ACS, based on the ABAC paradigm and XACML 3.0 policy language. For each category of entity involved in the system (*subject, object* and *environment*), two types of attributes are considered: *security related* and *context specific*, where the values of the latter depend on the data received from the monitoring devices and is updated when there is a change. We have realized a hierarchical structure of safety-centric contexts, which lets us manage security rules specified for various risks of different levels of criticality using hierarchies of ACDs. To define the contexts ECA rules are considered that using the risk description and applying the predefined conditions identify the context.

Considering that the dynamic combination of the ACDs may pose unpredictable results, at this point, we have adopted a conflict analysis algorithm based on set theory to detect the potential conflicts at design time so that the security administrators can elaborate on resolving them. The conflict analysis approach is implemented on top of Balana, and performance analysis shows that we can achieve an acceptable response time. To be able to response in a timely manner to emergency situations we have enabled the possibility to adopt Break-Glass security rules. Finally, when the environment is in a safe state the AC system roles back to security rules that are normally applied (the safe ACD).

One of the challenges regarding the security rules and the ECA rules adopted in this work is to automatically check their consistency. Moreover, to evaluate real-time efficiency of the proposed approach use-cases and scenarios should be considered and tested on the implemented AC system. As future works, authors will consider mentioned issues.

References

1. Al-Zewairi, M., Alqatawna, J., Atoum, J.: Risk adaptive hybrid RFID access control system. Secur. Commun. Netw. **8**(18), 3826–3835 (2015)

2. Fall, D., Okuda, T., Kadobayashi, Y., Yamaguchi, S.: Risk adaptive authorization mechanism (RAdAM) for cloud computing. J. Inf. Process. **24**(2), 371–380 (2016)
3. Fugini, M., Teimourikia, M., Hadjichristofi, G.: A web-based cooperative tool for risk management with adaptive security. Future Gener. Comput. Syst. **54**, 409–422 (2016)
4. Gusmeroli, S., Piccione, S., Rotondi, D.: A capability-based security approach to manage access control in the internet of things. Math. Comput. Model. **58**(5), 1189–1205 (2013)
5. Hoyos, C.G., Zimolong, B.: Occupational Safety and Accident Prevention: Behavioral Strategies and Methods. Elsevier, Amsterdam (2014)
6. Hu, V.C., Kuhn, D.R., Ferraiolo, D.F.: Attribute-based access control. Computer **2**, 85–88 (2015)
7. Jebbaoui, H., Mourad, A., Otrok, H., Haraty, R.: Semantics-based approach for detecting flaws, conflicts and redundancies in XACML policies. Comput. Electr. Eng. **44**, 91–103 (2015)
8. Jin, X., Krishnan, R., Sandhu, R.: A unified attribute-based access control model covering DAC, MAC and RBAC. In: Cuppens-Boulahia, N., Cuppens, F., Garcia-Alfaro, J. (eds.) DBSec 2012. LNCS, vol. 7371, pp. 41–55. Springer, Heidelberg (2012)
9. Neri, M.A., Guarnieri, M., Magri, E., Mutti, S., Paraboschi, S.: Conflict detection in security policies using semantic web technology. In: 2012 IEEE First AESS European Conference on Satellite Telecommunications (ESTEL), pp. 1–6. IEEE (2012)
10. Roman, R., Zhou, J., Lopez, J.: On the features and challenges of security and privacy in distributed internet of things. Comput. Netw. **57**(10), 2266–2279 (2013)
11. Sage, A.P., Haimes, Y.Y.: Risk Modeling, Assessment, and Management. Wiley, Hoboken (2015)
12. Schefer-Wenzl, S., Bukvova, H., Strembeck, M.: A review of delegation and break-glass models for flexible access control management. In: Abramowicz, W., Kokkinaki, A. (eds.) BIS 2014 Workshops. LNBIP, vol. 183, pp. 93–104. Springer, Heidelberg (2014)
13. Shamoon, I., Rajpoot, Q., Shibli, A.: Policy conflict management using XACML. In: 2012 8th International Conference on Computing and Networking Technology (ICCNT), pp. 287–291. IEEE (2012)
14. Venkatasubramanian, K.K., Mukherjee, T., Gupta, S.K.: CAAC – an adaptive and proactive access control approach for emergencies in smart infrastructures. ACM Trans. Auton. Adapt. Syst. (TAAS) **8**(4), 20 (2014)
15. Yan, D., Huang, J., Tian, Y., Zhao, Y., Yang, F.: Policy conflict detection in composite web services with RBAC. In: 2014 IEEE International Conference on Web Services (ICWS), pp. 534–541. IEEE (2014)

Modeling Information Diffusion
via Reputation Estimation

Bao-Thien Hoang[(✉)], Kamel Chelghoum, and Imed Kacem

LCOMS EA7306, University of Lorraine, Metz 57000, France
{bao-thien.hoang,kamel.chelghoum,imed.kacem}@univ-lorraine.fr

Abstract. We tackle the problem of predicting information diffusion in social networks. In this problem, we are given social data and would like to infer the diffusion process in the near future. Although this problem has been extensively studied, the challenge of how to effectively combine user activities, network structures and diffused information in social data remains largely open. In addition, no prior work judged the effect of user reputation on the diffusion process. Availability of such reputation score is really important for a user to decide whether he might share information. In this paper, we first devise a novel method for estimating user reputation. Our approach integrates network structure with user features, link features and the content of items shared by the users, then measures the strength of each of these factors. Based on this estimation approach, we develop a model predicting the tendency of a new information item as well as the number of participants of this diffusion process. We conduct several experiments on a snapshot of Twitter which show that our proposed model outperforms other baselines.

Keywords: Predictive model · Information diffusion · Reputation estimation

1 Introduction

Online social networks (OSNs) have become one of the most important media for users to exchange ideas and share interesting information. We could observe plenty of diffusion actions of users (i.e., the action that a user expresses when receiving an information item shared by his/her friends such as "re-shares", "likes", "comments", "re-tweets", "taggings", etc.) in this social media, and the number of diffusion actions could in turn reflects the distinction of the associated information item, in terms of its novelty, popularity and importance. It has been reported that the size of information cascades fits power-law [7], which means that of all the information items posted everyday, only a tiny proportion of them will attract a large number of users to actively participate in propagating them. Therefore, predicting the propensity of a new content item before the outbreak actually happens would give us a great benefit. In addition, the ability to predict

This work has been supported by the ANR INFORSN project.

S. Hartmann and H. Ma (Eds.): DEXA 2016, Part I, LNCS 9827, pp. 136–150, 2016.
DOI: 10.1007/978-3-319-44403-1_9

prospective users that may participate in the diffusion process of an information item is also desirable and valuable in many cases. For example, full understanding of social network and prediction of potential customers allows a mobile phone company to choose a better marketing strategy for their new products.

In addition, to the best of our knowledge, no prior prediction models focus on analyzing all impact factors of user reputation. Generally, the reputation of a user n refers to a unique global trust value that reflects the experiences of all users in the network with user n. It has been shown that the availability of such reputation scores is crucial for many applications, for instance, item ranking, or recommending users to follows. Thus, in the context of the information diffusion problem, that score can be leveraged to evaluate the attractiveness of the item to the user. Specifically, the reputation score of a user can represent his/her ability of diffusing a "good" content item such that another random user not necessarily connected to him/her would give a diffusion action to the item shared by him/her.

Challenges. In this paper, we target at devising a diffusion model that (i) analyzes all factors (including user reputation factor) that affect the diffusion action of users, (ii) predicts the trend of a content item in the near future, e.g., determine whether it becomes a phenomenon, and (iii) predicts (the number of) participants of the diffusion process of this item.

The proposed problem is challenging from at least three points of view. First, the impacts affecting a user's diffusion action originate from different sources: his profile, his relationships, his action logs, and shared items. But the question is how all features interact to generate the decision of user. Second, only diffusion actions are observable, while the underlying influences that trigger the actions are implicit. Consequently, using only the performed diffusion actions, correctly quantifying the weights of these impacts is also challenging. Third, estimating user reputation score from social data is generally difficult because social networks are complex in the sense there can be multiple factors affecting user action such as user profile, user interest, shared items, social influence, etc. As a result, computing the reputation score is as hard as evaluating the weights of factors of a diffusion action.

Moreover, in much prior work on user reputation or influence in social networks, one common approach is to construct an influence graph among users based on how information propagates, then apply methods such as plain PageRank [4] to identify influential users in the network. This leads to the influential users are usually the ones having a high number of neighbors. Recently, Yang *et al.* [26] introduced a method of estimating user reputation from social data by combining biased social data (i.e., the observable or raw data) with aggregated data from an unbiased context (e.g., specific data from LinkedIn Today module). Specifically, they presented a Z-Model using only biased data, and thus, could be used in our problem. However, our experiments (Sect. 3) show that all of these approaches do not well perform when applied to the prediction model.

Contributions. This paper describes our approach to address the above challenges. Our contributions are summarized as follows:

- We propose a novel way to estimate user reputation scores using social data. Our techniques is built upon the combination of the network structure, user and link features, the learned PageRank scores, the sharing levels of users and the attractiveness of the shared items.
- Based on the reputation estimation approach, we devise a model that predicts the tendency of a new information and the perspective users of this diffusion process.
- We compare our predictive model with other baselines by conducting empirical analysis on a snapshot of Twitter. The experimental results show the effectiveness of our approach in terms of prediction quality.

Outline. The rest of this paper is organized as follows. Section 2 defines the notations used throughout the paper, presents the way to estimate user reputation, and describes our predictive model. In Section 3, we present our empirical analysis on a real dataset crawled from Twitter. We review related work in Sect. 4, and conclude our work in Sect. 5.

2 Methodology

2.1 Problem Setting

We consider here a social network of size N modeled as a directed graph $G = (V, E)$ where $V = \{1, 2, 3, \ldots, N\}$ is the set of nodes representing users and $E \subseteq V \times V$ is the set of directed edges between users.

Diffusion Action. When a user sees information diffused (or shared[1]) by other ones, he/she may respond *positively* to the broadcasted content through re-diffusing the item or may just ignore it. More formally, given two nodes n and s, a diffusion action, defined as a tuple (s, n, t), indicates that node n responds positively to the content item diffused by another node s at timestamp t. We refer to the moment at which a node responds positively to some information as its *activation* and that node is called *active*. Conversely, a node that responds *negatively* is called *inactive*.

Item Features. A diffused item may contain some information in its own post or incorporate links to full information of the post. Thus the terms, "information item", "content item" or simply "item", used interchangeably henceforth refer to the main content of the post that the item links to or consists of. We denote the set of topics as $\mathcal{K} = \{k_1, k_2, \ldots, k_{|\mathcal{K}|}\}$. To characterize the content quality of item i, we use a vector $h_i \in \mathbb{R}^L$ where L is the size of the content feature space.

---

[1] We use these terms "diffuse" and "share" interchangeably henceforth.

(For example, if information is textual, vector h_i may be a *tf-idf* vector [15].) We also represent the proximity level of content of item i to various topics as a vector $q_i \in \mathbb{R}^{|\mathcal{K}|}$ where the k-th entry, denoted q_{ik}, is the proximity level on topic k.

User Reputation. The reputation of user n refers to a *unique global trust value* that reflects the experiences of all users in the network with user n. In other words, in the context of information diffusion, it represents his/her ability of propagating "good" information. We study here the user reputation for different content topics. More precisely, given a topic k, we define the reputation μ_{nk} of user n as the propensity of a random user v, who is not necessarily connected to n but interested in topic k, to diffuse a typical (or random) content item on that topic diffused by n. In our approach, all nodes in the network participate in computing these values.

Node and Edge Features. Nodes and edges in the network have rich auxiliary information with a set of features. We assume each node n has a feature vector $\boldsymbol{x}_n$ describing the node (e.g., age, gender, hometown), and each edge (n, v) also has a corresponding feature vector $\boldsymbol{w}_{nv}$ describing the interaction attributes (e.g., the number of messages n and v exchanged, or the similarity of the profiles). Like item features, node and edge features could be related to various topics. Hence, we represent here these features as a distribution of preferences overs $|\mathcal{K}|$ topics.

2.2   Proposed Model

In this section, we present our proposed model for estimating user reputation and predicting information diffusion.

Diffusion Model. Let $y_{sn}(t)$ be a binary value indicating whether at timestamp t user n would be active to any content item diffused by user s. We assume this value follows the Bernoulli distribution with activation probability $p_{sn}(t)$ modeled as a function of node n's interest vector $\boldsymbol{\eta}_n(t)$, node s's reputation score $\boldsymbol{\mu}_s(t)$, and feature vector $\boldsymbol{w}_{sn}(t)$ of edge (s, n):

$$p_{sn}(t) = \sigma(\boldsymbol{\alpha}_{sn}^T \boldsymbol{w}_{sn}(t) + \sum_{k \in \mathcal{K}} \eta_{nk}(t)\mu_{sk}(t) + b) \tag{1}$$

where $\sigma(x) = 1/(1+e^{-x})$ is a logistic function, $\boldsymbol{\alpha}_{sn} \in \mathbb{R}^q$ is a q-dimension vector of regression coefficients, with q the size of the edge feature vector $\boldsymbol{w}_{sn}$, η_{nk} is the user interest for different topics k, μ_{sk} represents the reputation score of node s on topic k, and b is a global bias term [22]. Parameters $\boldsymbol{\alpha}_{sn}, \eta_{nk}, \mu_{nk}$ (for all k) and b will be learned from data. We sometimes omit (t) in the notations for the sake of simplicity.

User Interest Model. Each user may express different levels of interest and expertise for various topics. We represent the interest distribution of a user n, $\boldsymbol{\eta}_n = (\eta_{n1}, \eta_{n2}, \ldots, \eta_{n|\mathcal{K}|})^T$, through a linear form of the user feature vector $\boldsymbol{x}_n$:

$$\boldsymbol{\eta}_n = \boldsymbol{\theta}_n \boldsymbol{x}_n \tag{2}$$

where $\boldsymbol{\theta}_n$ is a $|\mathcal{K}| \times |\boldsymbol{x}_n|$ weighting matrix to be learned from data. For a topic k, the k-th entry η_{nk} is $\eta_{nk} = \boldsymbol{\theta}_{nk}^T \boldsymbol{x}_n$ where $\boldsymbol{\theta}_{nk} \in \mathbb{R}^l$ is the k-th row of $\boldsymbol{\theta}_n$, l is the size of the user feature space.

User Reputation Model. Like user interest, the reputation of a user s is viewed as the distribution of *sharing and interest preferences* over the topics. It is hence modeled as a $|\mathcal{K}|$-dimension latent vector, namely, $\boldsymbol{\mu}_s = (\mu_{s1}, \mu_{s2}, \ldots, \mu_{s|\mathcal{K}|})^T$, where μ_{sk} represents the reputation of user s on topic k and is defined through a linear regression of three factors: s's interest μ_{sk} on topic k, the level of sharing ψ_{sk} on topic k, and the *PageRank score* [4] ϕ_{sk}:

$$\mu_{sk} = \gamma_{sk} \eta_{sk} + \kappa_{sk} \psi_{sk} + \rho_{sk} \phi_{sk} \tag{3}$$

where γ_{sk}, κ_{sk}, ρ_{sk} are coefficients to be learned s.t. $\gamma_{sk} + \kappa_{sk} + \rho_{sk} = 1$.

Sharing Level ψ_{sk} is defined by:

$$\psi_{sk} = \frac{1}{|I_s|} \sum_{i \in I_s} \zeta_{ik} \tag{4}$$

where I_s is the set of items that s has diffused, ζ_{ik} is the *attractiveness of item* i on topic k, determined as a function of the proximity q_{ik} of the content item i to topic k and the average reputation score of the sharers:

$$\zeta_{ik} = q_{ik} \frac{1}{|S_i|} \sum_{s \in S_i} \mu_{sk} \tag{5}$$

Here S_i is a set of users diffusing item i.

PageRank Score. PageRank [4] and its variants like Personalized PageRank [14] and Random Walks with Restarts (RWR) [24] are popular methods for ranking nodes on graphs and evaluating node reputation. The simple idea of the RWR approach would be to start a random walk at node s and compute the proximity of each other node to s by setting up the random walk stochastic transition vector so that the walk may jump back to s and thus restarts the walk. The stationary distribution of such random walk determines the node score expressing the ranking of proximity between s and other nodes in the graph. Note that, the original RWR method only considers the network structure. We here use RWR but combine with other factors such as link feature.

Let us consider a random walk originating from s. We build a transition matrix $\mathbf{A} = (A_{nv})$ demonstrating the probability the walk will traverse an edge (n, v) given that it is currently at node n as follows:

$$A_{nv} = \begin{cases} \frac{a_{nv}}{\sum_u a_{nu}} & \text{if } (n, v) \in E \\ 0 & \text{otherwise} \end{cases} \tag{6}$$

where a_{nv} represents the strength of the interaction (n, v): $a_{nv} = \boldsymbol{\alpha}_{nv}^T \mathbf{w}_{nv}$.

To address RWR, we incorporate the restart probability λ into the transition matrix $\mathbf{A}$, i.e., the random walk jumps back to the seed node s with probability λ and thus "restarts". Accordingly, the transition matrix of the random walk becomes $\mathbf{P} = (P_{nv})$ where

$$P_{nv} = (1 - \lambda)A_{nv} + \lambda \mathbb{1}_{\{v=s\}}. \tag{7}$$

Here $\mathbb{1}_{\{v=s\}}$ is an indicator function. The PageRank score vector $\phi = (\phi_n)$ is then defined by the stationary distribution of the RWR. More precisely, that is the solution of the eigenvector equation $\phi^T = \phi^T \mathbf{P}$. And thus, we obtain the PageRank score ϕ_{nk} of node n on topic k as follows:

$$\phi_{nk} = \boldsymbol{\phi}_k^T \mathbf{P}_{nk} = \sum_j \phi_{jk} P_{jnk}. \tag{8}$$

2.3   Parameter Learning

When representing all parameters as $\boldsymbol{\omega} = \{\boldsymbol{\alpha}, \boldsymbol{\beta}, \boldsymbol{\gamma}, \kappa, \rho\}$, our goal is to find $\boldsymbol{\omega}$ that maximizes the likelihood of observation data. In this section, we present the way of learning these parameters.

Learning Framework. The probability of observing the outcome at time t is:

$$Pr(t) = \prod_{y_{sn}(t) \in \mathbf{Y}} (p_{sn}(t))^{y_{sn}(t)} (1 - p_{sn}(t))^{1 - y_{sn}(t)} \tag{9}$$

where Y is the training data consisting of all users' actions.

When considering all timestamps $t \in [0, \tau]$ in data $\mathbf{Y}$, the likelihood of observing all diffusion actions is:

$$\mathcal{L}(\boldsymbol{\omega}|\mathbf{Y}) = \prod_{c \in \mathcal{C}} \prod_{t_1(c) \leq t \leq t_2(c)} Pr(t) \tag{10}$$

Using natural logarithm of Eq. (10), the problem of modeling how information propagates corresponds to finding a parameter

$$\boldsymbol{\omega}^* = \text{argmax}_{\boldsymbol{\omega}} \log \mathcal{L}(\boldsymbol{\omega}|\mathbf{Y}). \tag{11}$$

Interaction	Dataset	#Tw/Rt	#Articles
Retweeting	Whole dataset	14.8M	789K
	Training set	12.6M	656K
	Testing set	2.1M	124K
Mention	Whole dataset	2.5M	140K
	Training set	2.2M	119K
	Testing set	304K	19K

Fig. 1. Dataset statistics.

Learning Algorithm. The problem tackled here is a logistic regression and non-convex problem, but we can solve it by using *coordinate ascent* (CA) approach. Due to space limitation, we omit this algorithm here.

3 Experiments

We conducted several experiments to evaluate the performance of our proposed model. In this section, we first describe the dataset, then present the experimental setup and performance analysis.

3.1 Dataset

We conduct experiments on a snapshot of Twitter containing a large coverage of all tweets involving French articles and media (such as Le Monde, L'Équipe, TF1, BFMTV, etc.). The dataset was crawled from May 1st to October 31st, 2014 with 1M users and 14.8M tweets related to 789K articles. We also crawled the content of the articles retrieved from the URLs in the tweets. The average lifetime of a diffusion process in the dataset is about 45 h.

We build two networks on the users from the structure of interaction via (i) *retweeting*, and (ii) *mention* (@-messages). In the former network, we create a link from u to v if v retweets at least one post diffused by u. In the latter network, if a tweet of user u mentions user v (i.e., including "@v") in at least one tweet, we include a directed edge from u to v.

For each network, we generate a duplicated dataset from the original dataset. Then we split each dataset into a training set and a testing set with the ratio 5:1 by timestamp. In particular, considering each month as a timestamp, the training set is composed of all data between May 2014 and Sep. 2014 and is used for estimating the weighting parameters. The testing set contains data in the following month, Oct. 2014, and is used for evaluating the model. In the testing set, we excluded all data of the last 2 days of Oct. 2014 because the diffusion process of this data may not have been finished or could not provide a valid grouthtruth for our evaluation. Figure 1 describes the statistics of our datasets.

3.2 Experimental Setup

Topic Classification. Before the evaluation, we first executed some preprocessing on the text of articles and tweets such as removing stop words and low-frequency words, stemming words. We extracted and classified the content of articles into 24 topics.

Parameter Settings. The restart probability λ controls for how "far" the walk wanders from a seed node s before it restarts and jumps back to s. A low value of λ allows the walk to discover new nodes at a high distance from the seed node, while high value gives very short and local random walk. The appropriate value λ depends on the characteristics of the graph. Generally, in social networks, more than half of all edges at the time of creation close a triangle, i.e., a user usually connects to a friend of his friend [19]. In our graphs, more than 90 % of pairs of interacting nodes have a distance of at most 3 hops just before they become friends. According to [2], we set empirically $\lambda = 0.25$.

User Features. Each user n has the following features:

- The total number of items (original tweets and retweets) shared by user, the mention count and hashtag count related to the user.
- The *topical interest* on topic k at timestamp t: that is the fraction between the number of items on that topic n shared at timestamp t and the number of items on topic k that n received at timestamp t.
- The *willingness of diffusion* at timestamp t: that is the fraction between the number of items of all topics that n shared at timestamp t and the number of items of all topics n received at timestamp t.

Note that, for each mentioned feature computation, we assume that if the denominator is zero then the value of that feature is zero.

Link Features. We annotate each edge (n, u) with the following features:

- *Information-based influence* on topic k at timestamp t: This feature indicates the influence of content of item on the diffusion decision of user n. It is defined by the number of items on topic k shared by u and diffused by n at timestamp t, divided by the number of items on that topic shared by u at timestamp t.
- *Relation-based influence*: Users may not be excited about topic k but diffuse information shared by their friends. That action is purely originated from the social, emotional or friendship relations among them. We evaluate the influence of user n on user u at timestamp t as the ratio between the number of items that n has reposted from u at timestamp t and the number of items shared by u at timestamp t.
- *Crow-effect influence* on topic k is the fraction between the number of friends of n sharing items on k and total number of friends of n.
- The number of common friends between n and u.
- *Link age*: as recommended in [1], this feature is computed by the average of $1/(\tau - t_{nu})^{\beta}$, where τ is the time cutoff Sep. 30th, t_{nu} is the time of edge creation, and $\beta = \{0.1, 0.3, 0.5\}$.

Evaluation Metrics. To judge the performance of the proposed method, we use three evaluation metrics: *F1-score*, *Mean Average Precision* (MAP) and Area Under the ROC Curve (AUC). The algorithms are implemented in R.

3.3 Estimating User Reputation

Our first task is estimating user reputation score using the training data, then evaluating the estimated scores with the real influence of user in the testing data. We empirically define a test sharer as (i) the one who shares at least 500 items (including tweets or retweets) in the retweeting-based network, and (ii) the one who are mentioned in at least 200 items in the mention-based network. That results in 423 users and 258 users respectively in two testing sets. We also randomly select other 577 users who diffuse more than 100 items from the testing set of the retweeting-based network, and 442 users who are mentioned in at least 50 items from the testing set of the mention-based network. The computation of evaluation metrics is based on these selected users. We repeat the random selection 20 times, then report the average over these independent runs.

Baselines. We compare our reputation estimation method, denoted by FEP, to the following baselines:

- **PageRank (PR)** [4] **and Plain Random Walks with Restarts (RWR)** [24]. These unsupervised methods can be viewed as simplified forms of our reputation model because the user interest, link features, and the sharing levels are ignored in the computation of reputation score. These baselines are used to prove the need of making use of those additional dimensions for evaluating the reputation score.
- **Feature-Based Reputation Model (FR)** [26]. Yang *et al.* [26] explored the *unbiased* reputation scores of sharer by combining biased social response data with aggregated data from an unbiased context. Specially, they presented a Z-Model where the reputation score is based only on biased social response data. This model can be viewed as a special form of our model in which only user interest profile and sharing levels are considered, while the PageRank score is dismissed.

Results. We present overall results in terms of F1, MAP and AUC metrics in Fig. 2. Generally, the performance of our model is notable. For instance, MAP values are between 0.45–0.48, this means out of all user ranking based on reputation score, more than 45 % of them are correct.

Moreover, according to Fig. 2, FEP outperforms other approaches. Specifically, FEP gives a significant margin over the PR and RWR. It also gains over the FR model. For example, in the retweeting-based network, in terms of AUC we get more than 11 % over RWR and PR, and near 5 % over FR, while in terms of MAP the relative improvement are respectively more than 108 % and 16 %, and in terms of F1-score, FEP gives a 18 % and a 3 % lift over the PageRank-based methods and FR method.

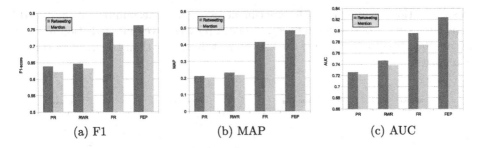

Fig. 2. Performance of different models in estimating user reputation.

The RWR algorithm slightly trails PR approach due to the characteristics of networks: more than 90 % of pairs of nodes have a distance of at most 3 hops, and thus node jumps back to the seed node with high probability. However, both two algorithms are the bottom performers. Recall these PageRank-based baselines do not take into account user interest profile, influences between user and other ones, and the content of items diffused by sharers which play an important role in affecting the diffusion actions of users, and thus, their reputation scores. So it is not surprising that they are worse than the featured-based models like FEP and FR.

Now we discuss the comparison between FEP and FR. By combining user and link features, the item content and the PageRank score in the computation of reputation score, all evaluation metrics of FEP are further improved. This result indicates that the feature-based model and PageRank-based model can be complementary to each other. Only considering one of them is not enough.

Figure 3 visualizes the performance comparison via Precision-recall and ROC curves. We can see that our method obtains a significant precision over the PR and RWR methods, and also outperforms the FR model.

3.4 Predicting Information Diffusion

Another target of our work is to predict the tendency of a new tweet, namely, to identify (i) whether it becomes a trending at a future timestamp, and (ii) the active users of the diffusion process of this tweet.

In the first subtask of predicting the trending tweet, we observe the distribution of the number of retweets in the dataset and define a trending tweet as the one having at least 500 (resp. 50) reposts for the retweeting (resp. mention) network. As a result, we get 281 and 258 trending tweets from the testing sets of the retweeting graph and mention graph, respectively. We also randomly select 719 (resp. 742) tweets having more than 100 (resp. 20) reposts from the testing set of retweeting (resp. mention) graph. We measure the performance of this task for these original tweets.

Likewise, in the second subtask, we randomly select 1000 tweets with the number of retweets exceeding 100 (resp. 20) from the testing set of retweeting (resp. mention) graph. In each experiment, we count the number of predicted active users w.r.t. one selected tweet, then compute the average prediction results for all 1000 tweets. We repeat the experiment 20 times, then the average result is reported.

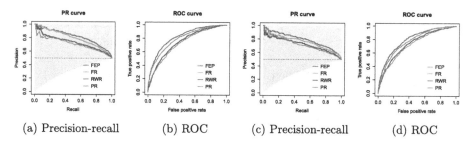

(a) Precision-recall   (b) ROC   (c) Precision-recall   (d) ROC

Fig. 3. Presion-recall and ROC curves in estimating user reputation for the retweeting (a and b) and mention (c and d) networks.

Baselines. We compare the predictive performance of our model, denoted by FIR, to several state-of-the-art models:

- **Featured-Based Model (ETL)** [6]. This method considers the hot emerging topic learner problem. It explores all user and link features for outbreak training and prediction of topic emerging. Thus, the proposed features and learning methods can be applied to predict the information diffusion. This baseline is used to measure the contribution of user and link features.
- **Influence-Based Model (IBP)** [11]. This work analyzes pair-wise influence among users and assigns each pair an influence probability of predicting whether a user performs an action w.r.t. a diffusion process. We consider this model as a restricted form of our model where only link features are taken into account. This baseline is used to measure the contributions of the link features.
- **Supervised Random Walk (SRW)** [1]. Backstrom *et al.* [1] proposed a link prediction algorithm (SRW) that combines user and link features with RWR. In SRW, the transition probability is determined by a function of the attributes of users and links, and the function is adjusted through supervised learning. However, this approach ignores the attractiveness of content items of sharers. Hence, this approach is used to prove the need of analyzing the content of diffused items.

Results of Predicting Trending Tweet. Figure 4 compares the results of various methods on the retweeting and mention networks. First, we can see that the feature-based model ETL has lower performance compared to other

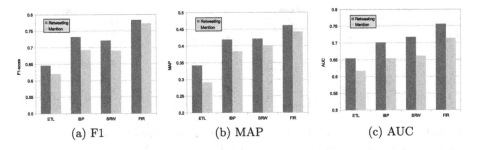

(a) F1   (b) MAP   (c) AUC

Fig. 4. Performance of predicting the trending of a new tweet.

baselines. This can be possibly explained by the fact that ETL uses few amount of training data. Hence, its prediction ability is limited. Second, the influence model IBP outperforms SRW in terms of non-ordered-based measurement like F1-score, but gives lower performance in terms of other ordered-based scores such as MAP and AUC. A possible explanation is as follows: under SRW model, the PageRank scores of users are contributed to the transition probability of users. As we know that, this score brings the order to the network [4]. And in SRW, it is evaluated as a function of user features, link features and influences among users, and is adjusted through supervised learning. This leads to the ranking-based evaluation metrics it produces are more precise than the ones of models considering only features or influence of users. Therefore, the ability of prediction of SRW is better than other influence-based and feature-based models. Moreover, this result indicates that the prediction analysis based on influence of users is not sufficient to obtain a good result combining both the user features, link features and the PageRank score of users.

Finally, our model outperforms each of the baselines by a significant margin in terms of the evaluation metrics. For instance, in the retweeting network, we achieve a F1-measure of 0.77, a MAP score of 0.46 and an AUC score of 0.77, compared to the best baseline's scores of 0.73, 0.42, 0.70 respectively. We give more details concerning the precision of each model at different recalls and the ROC curves in Fig. 5. All of the experimental results confirm that the appropriate combination of user features, influences among users, the PageRank scores and the attractiveness of content items shared by users improves the accuracy of the prediction.

Results of Predicting Active Users. The results of predicting active users are shown in Fig. 6. Similar to the first subtask, the ETL method still remains the worst performer, while our proposed method yields the best results under all evaluation measures. The underlying reason is that when a user makes a decision on diffusion of an item, he/she will consider the characteristics of a network in a comprehensive way, instead of focusing on some single aspect.

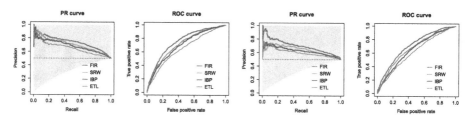

Fig. 5. Presion-recall and ROC curves in predicting tendency of a new tweet for the retweeting (a and b) and mention (c and d) networks.

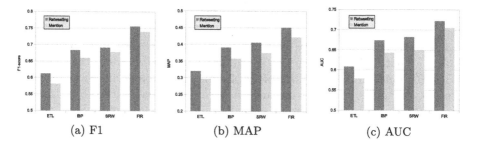

Fig. 6. Performance of predicting active users.

Discussion. From all empirical results, we find that each component in our model has different significance in the prediction model, and the performance drops significantly by removing any of the proposed components. For instance, if we consider only user features and link features, then the model achieves the worst performance. When we add subsequently other components like PageRank score, user interest, influence of users, and the relatedness level of items, the result becomes better. And the best model is the combination of all mentioned components.

4   Related Work

Several approaches related to information diffusion and social influence in networks have been proposed. Leskovec *et al.* [18] described how recommendations of products propagate among users. Gruhl *et al.* [13] addressed the problem of tracking the propagation patterns of topics. However these work either require network characteristic or focus on decompising the topics along two orthogonal axes like chatter and spikes, and this makes them different from our target.

Independent Cascade [8] and Linear Threshold [16] are useful models for simulating the information flow in networks, but cannot be directly applied to predict information diffusion. The *heat-diffusion models* are applied in various domains such as classification, dimensionality reduction [3,17], or can be used as the core of DiffusionRank [25]. Yet all of these studies only take into account the level of influence from one user to another one, while we address all aspects affecting the link between two users. Introduced by Kempe *et al.* [16], the *influence maximization* problem have been tackled by many approaches such as [5,12]. However they are different from our study because they assume the influence probability is known in advance. In [11], the authors proposed a probabilistic model for learning influence probablities by mining the influence in the history logs. In [23], Tang *et al.* studied the topic-level social influence. Ma *et al.* [20], introduced a model of social network marketing for the influence maximization problem. Again, these papers do not consider the edge feature or just focus on the defense against diffusion of the negative information, which are not our consideration.

Another line of related work is the *link prediction problem*. Many approaches have been designed to infer the most plausible links such as NetInf [10], Connie [21] and Netrate [9]. However, all of them concentrate in analyzing the link features, but do not take into account other aspects like user profile interest, content items of sharers.

5   Conclusion

In this paper, we first presented a scheme for evaluating user reputation. This score is measured by combining network structure, user and link features, PageRank scores, sharing levels of users, and the attractiveness of the shared items. In addition, we focused on modeling how information diffuses by presenting a predictive model that identifies the trending of a new content item and the perspective participants of this diffusion process. We furthermore conducted empirical analysis on a snapshot of Twitter which shows the effectiveness of our approach compared to alternative state-of-the-arts. We plan to propose a predictive model in competitive networks, where each network adopts some strategy and the influences under the strategies is considered as payoffs.

References

1. Backstrom, L., Leskovec, J.: Supervised random walks: predicting and recommending links in social networks. In: WSDM, pp. 635–644 (2011)
2. Ballester, C., Vorsatz, M.: A new measure of rank correlation. Rev. Econ. Stat. **3**, 383–401 (2014)
3. Belkin, M., Niyogi, P.: Laplacian eigenmaps for dimensionality reduction and data representation. Neural Comput. **15**(6), 1373–1396 (2003)
4. Brin, S., Page, L.: The anatomy of a large-scale hypertextual web search engine. Comput. Netw. **30**(1–7), 107–117 (1998)
5. Chen, W., Wang, C., Wang, Y.: Scalable influence maximization for prevalent viral marketing in large-scale social networks. In: SIGKDD, pp. 1029–1038 (2010)
6. Chen, Y., Amiri, H., Li, Z., Chua, T.: Emerging topic detection for organizations from microblogs. In: SIGIR, pp. 43–52 (2013)
7. Cui, P., Jin, S., Yu, L., Wang, F., Zhu, W., Yang, S.: Cascading outbreak prediction in networks: a data-driven approach. In: SIGKDD, pp. 901–909 (2013)

8. Goldenberg, J., Libai, B., Muller, E.: Talk of the network: a complex systems look at the underlying process of word-of-mouth. Mark. Lett. **12**(3), 211–223 (2001)
9. Gomez-Rodriguez, M., Balduzzi, D., Schölkopf, B.: Uncovering the temporal dynamics of diffusion networks. In: ICML, pp. 561–568 (2011)
10. Gomez-Rodriguez, M., Leskovec, J., Krause, A.: Inferring networks of diffusion and influence. TKDD **5**(4), 21 (2012)
11. Goyal, A., Bonchi, F., Lakshmanan, L.V.S.: Learning influence probabilities in social networks. In: WSDM, pp. 241–250 (2010)
12. Goyal, A., Bonchi, F., Lakshmanan, L.V.S.: A data-based approach to social influence maximization. PVLDB **5**(1), 73–84 (2011)
13. Gruhl, D., Liben-Nowell, D., Guha, R.V., Tomkins, A.: Information diffusion through blogspace. SIGKDD Explor. **6**(2), 43–52 (2004)
14. Haveliwala, T.H.: Topic-sensitive pagerank: a context-sensitive ranking algorithm for web search. IEEE Trans. Knowl. Data Eng. **15**(4), 784–796 (2003)
15. Jones, K.S.: A statistical interpretation of term specificity and its application in retrieval. J. Documentation **60**(5), 493–502 (2004)
16. Kempe, D., Kleinberg, J.M., Tardos, É.: Maximizing the spread of influence through a social network. In: SIGKDD, pp. 137–146 (2003)
17. Kondor, R., Lafferty, J.D.: Diffusion kernels on graphs and other discrete input spaces. ICML **2002**, 315–322 (2002)
18. Leskovec, J., Adamic, L.A., Huberman, B.A.: The dynamics of viral marketing. TWEB **1**(1), 5 (2007)
19. Leskovec, J., Backstrom, L., Kumar, R., Tomkins, A.: Microscopic evolution of social networks. In: SIGKDD, pp. 462–470 (2008)
20. Ma, H., Yang, H., Lyu, M.R., King, I.: Mining social networks using heat diffusion processes for marketing candidates selection. In: CIKM, pp. 233–242 (2008)
21. Myers, S.A., Leskovec, J.: On the convexity of latent social network inference. In: NIPS, pp. 1741–1749 (2010)
22. Myers, S.A., Zhu, C., Leskovec, J.: Information diffusion and external influence in networks. In: SIGKDD, pp. 33–41 (2012)
23. Tang, J., Sun, J., Wang, C., Yang, Z.: Social influence analysis in large-scale networks. In: SIGKDD, pp. 807–816 (2009)
24. Tong, H., Faloutsos, C., Pan, J.: Fast random walk with restart and its applications. In: ICDM, pp. 613–622 (2006)
25. Yang, H., King, I., Lyu, M.R.: Diffusionrank: a possible penicillin for web spamming. In: SIGIR 2007, pp. 431–438 (2007)
26. Yang, J., Chen, B., Agarwal, D.: Estimating sharer reputation via social data calibration. In: SIGKDD, pp. 59–67 (2013)

Data Clustering

Mining Arbitrary Shaped Clusters and Outputting a High Quality Dendrogram

Hao Huang[1], Song Wang[1], Shuangke Wu[1], Yunjun Gao[2], Wei Lu[3(✉)], Qinming He[2], and Shi Ying[1]

[1] State Key Laboratory of Software Engineering,
Wuhan University, Wuhan, People's Republic of China
{haohuang,xavierwang,wsk9551,yingshi}@whu.edu.cn
[2] College of Computer Science, Zhejiang University,
Hangzhou, People's Republic of China
{gaoyj,hqm}@zju.edu.cn
[3] Key Laboratory of Data Engineering and Knowledge Engineering,
Renmin University of China, MOE, Beijing, People's Republic of China
uqwlu@ruc.edu.cn

Abstract. Hierarchical clustering (HC for short) outputs a dendrogram that offers more topological information than flat clustering (e.g., k-means). However, the existing HC algorithms focus on either the quality of the dendrogram or the ability of mining arbitrary shaped clusters. To address the above two aspects simultaneously, we present HICMEN by adopting (1) the classic agglomerative clustering framework that can generate a complete dendrogram, and (2) a novel similarity measure based on mutual k-nearest neighbors to capture the connectivity of data points and help properly merge up each arbitrary shaped cluster piece by piece. More importantly, we prove that the similarity measure has a nice property called weak monotonicity, which guarantees the quality of the dendrogram generated by HICMEN. Extensive experimental results show that HICMEN is capable of mining arbitrary shaped clusters effectively, and can simultaneously output a high quality dendrogram.

Keywords: Clustering · Arbitrary shaped clusters · Dendrogram

1 Introduction

Hierarchical clustering (abbreviated as HC henceforth) groups data points into a tree hierarchy of clusters, in which every cluster node contains children clusters while sibling nodes of clusters partition data points covered by their common parent according to a similarity measure. Figure 1 illustrates an example of such a process which organizes data points into a tree hierarchy called dendrogram.

A good dendrogram generated in HC should be able to preserve and present the intrinsic proximities of original data points, and offer valuable information in practice. For example, in the area of biology, by performing HC on the physical signs of living being, the dendrogram can be used to find the subspecies of each

© Springer International Publishing Switzerland 2016
S. Hartmann and H. Ma (Eds.): DEXA 2016, Part I, LNCS 9827, pp. 153–168, 2016.
DOI: 10.1007/978-3-319-44403-1_10

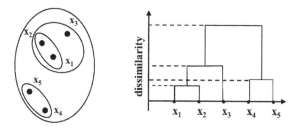

Fig. 1. An illustrative example of HC, and its generated dendrogram.

category, and reveal their taxonomic relations [5]. In the area of Internet, by adopting HC to categorize web pages, the dendrogram can be used to build a catalog of these web pages as a web directory, and facilitate the construction of web-directory-based browse systems [17].

Nonetheless, as a rule the existing HC algorithms only focus on either the quality of the dendrogram, such as traditional HC approaches using linkage metrics [7,23,25] which however are more applicable to compact and spherical clusters, or the ability of mining arbitrary shaped clusters, such as CHAMELEON algorithm [19] which however cannot output a complete dendrogram, let alone guarantee the quality of the dendrogram. In fact, the existing HC algorithms' efforts on either aspect usually sacrifice the performance of the other.

In this paper, we present HICMEN (**HI**erarchical **C**lustering with **M**utual k-n**E**arest **N**eighbors) an HC algorithm that takes both aspects into account. To the best of our knowledge, it is the first time that we explicitly identify and solve the problem of simultaneously mining arbitrary shaped clusters and outputting a high quality dendrogram. HICMEN uses the classic agglomerative clustering framework to generate a complete dendrogram. By adopting a novel similarity measure described by the MkNN (Mutual k-Nearest Neighbors) relationship across two sub-clusters, HICMEN prefers to merge up sub-cluster pairs with similar local densities and close proximities, and aggregates each arbitrary shaped cluster piece by piece. We prove that the proposed similarity measure has a nice property called weak monotonicity, which has the following two advantages, i.e., (1) it can better reflect the real cohesiveness between sub-clusters in an arbitrary shaped cluster and help HICMEN achieve an accurate clustering performance, and (2) with this similarity measure, the dendrogram generated by HICMEN can obtain a high quality, which is quantitated by a commonly used criterion for dendrogram quality called CPCC (Cophenetic Correlation Coefficient) [24].

The remaining sections are organized as follows. We review the related work in Sect. 2, introduce a MkNN-based similarity measure in Sect. 3, and present HICMEN algorithm in Sect. 4. We experimentally verify the effectiveness and efficiency of our approach in Sect. 5 before concluding the paper in Sect. 6.

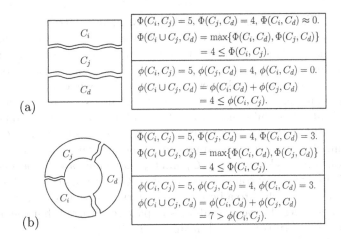

Fig. 2. Comparison of traditional similarity measure $\Phi(\cdot)$ (e.g., single link) and boundary similarity measure $\phi(\cdot)$. (a) Case 1: when C_d is only linked with C_j (i.e., $\phi(C_i, C_d) = 0$), both $\Phi(\cdot)$ and $\phi(\cdot)$ satisfy restrictive monotonicity property; (b) Case 2: when C_d is linked with both C_i and C_j, $\Phi(\cdot)$ still satisfies restrictive monotonicity property, while $\phi(\cdot)$ will be re-calculated according to the changed boundary regions after merging C_i and C_j and the result may not satisfy restrictive monotonicity property any more.

2 Related Work

The related work to the problem of simultaneously mining arbitrary shaped clusters and outputting a high quality dendrogram can be briefly categorized into two groups, namely (1) the dendrogram centered HC algorithms, and (2) the arbitrary shaped clustering algorithms.

2.1 Dendrogram Centered HC Algorithms

Dendrogram centered HC algorithms focus on the completeness and quality of the dendrograms they generate. To this end, they perform the classic agglomerative clustering framework (i.e., beginning from individual data points and recursively merging two most similar sub-clusters) to output a complete dendrogram, and adopt a similarity measure $\Phi(\cdot)$ that satisfies a *restrictive monotonicity property*, i.e., $\Phi(C_i \cup C_j, C_d) \leqslant \Phi(C_i, C_j)$, where C_i and C_j are the merged sub-clusters and C_d is any other disjoint sub-cluster ($\forall\, p \in \{i, j\},\, C_d \cap C_p = \varnothing$). Linkage metrics (such as single link [23], complete link [7] and average link [25]) exemplify this kind of similarity measures. Dendrograms generated by this restrictive monotonic manner were claimed to be true and reflect real cohesiveness of sub-clusters, and have high quality.

The above claim holds true when sub-cluster C_d is linked with either C_i or C_j, such as the situation illustrated in Fig. 2(a). However, in an arbitrary shaped cluster, sub-clusters C_d, C_i and C_j may all be linked with each other, such as the example illustrated in Fig. 2(b). In this situation, the aforementioned claim may not hold. The reason is twofold. (1) Firstly, in an arbitrary shaped cluster, the adjacent sub-clusters are only connected by a small part, i.e., their contact boundaries, while their majority parts are often not so cohesive to each other due to the arbitrary shape. Hence, compared against using a global or average similarity measure, a reasonable estimation on the boundary cohesiveness of the contact boundaries for adjacent sub-clusters helps better reflect the true probability that these sub-clusters are from a same arbitrary shaped cluster structure [18]; (2) on this basis, after merging up adjacent sub-clusters C_i and C_j, if the new sub-cluster $C_i \cup C_j$ has larger contact boundaries to sub-cluster C_d, the boundary cohesiveness between $C_i \cup C_j$ and C_d would increase. Sometimes it may even exceed the boundary cohesiveness between C_i and C_j, i.e., $\phi(C_i \cup C_j, C_d) > \phi(C_i, C_j)$, where $\phi(\cdot)$ refers to a boundary similarity measure that can reflect the boundary cohesiveness between a sub-cluster pair, resulting in that the restrictive monotonicity property is not valid in this case.

In brief, although similarity measures with restrictive monotonicity property were claimed to be helpful in preserving the quality of the dendrogram, they may fail to reflect real cohesiveness of sub-clusters in arbitrary shaped clusters. By contrast, a similarity measure that can dynamically update boundary cohesiveness according to changed contact boundaries helps to better reflect the real cohesiveness between sub-clusters in an arbitrary shaped cluster.

2.2 Arbitrary Shaped Clustering Algorithms

Many HC algorithms have been proposed to identify arbitrary shaped clusters from large data sets. CHAMELEON [19], OPTICS family [1], and CURE [14] are such pioneering examples. CHAMELEON builds a kNN (k-Nearest Neighbors) graph on a given data set, and partitions the graph to a predefined number of sub-graphs, followed by merging up closely linked sub-graph pairs. OPTICS family outputs a cluster ordering which is a linear list of all data points and reflects their density-based clustering structures. CURE first shrinks the data set size by sampling and conducts an agglomerative clustering on the sampled data points. Compared against CHAMELEON, ROCK and OPTICS family, CURE can be much faster since it takes only a few sampled data points for similarity computation. Nonetheless, the clustering results of CURE are sensitive to sampling quality. To mitigate this sensitivities, SPARCL [3] and CLASP [18], two evolutions of CURE, decompose a data set into small local groups via k-means, take group centers as representative data examples, and merge up the representative data examples; some other solutions like ABACUS [4] find the representative data examples by making data points to iteratively glob sufficiently close neighbors from their kNN. Although the above HC algorithms show good capability on mining arbitrary shaped clusters, they can only output an incomplete dendrogram since their HC processes start from sub-graphs or representative data examples rather than the original data set. Moreover, these HC algorithms do not take the dendrogram monotonicity into account, neither the restrictive monotonicity or a weak monotonicity, and lack guarantees for the dendrogram quality.

Besides the above HC algorithms, many non-HC algorithms can also identify arbitrary shaped clusters. For example, spectral clustering [6,20,26] embeds the arbitrary shaped clusters into a low-dimensional space to make cluster structures more distinguishable for clustering; the graph-based approaches [8,16,21] formulate the problem of clustering as a graph partition task; the density-based approaches [9,22,27] detect a cluster by searching a set of density-connected data points. However, non-HC methods do not output any dendrogram, and are inapplicable for the scenario where users require a dendrogram after clustering.

Similar to our proposed HICMEN algorithm, some existing HC and non-HC algorithms also adopt MkNN-based similarity measures [8,13,15,18] to estimate boundary cohesiveness between sub-clusters. Nevertheless, to the best of our knowledge, few of these approaches provide any dendrogram quality guarantee.

In summary, although some existing clustering algorithms show good capability on mining arbitrary shaped clusters, they are not competent to clustering tasks in which a complete and high quality dendrogram is required by users.

3   MkNN-Based Similarity Measure

Before introducing the detailed steps of HICMEN algorithm, in this section, we first present a novel MkNN-based similarity measure for arbitrary shaped clustering, followed by a theoretical analysis on its dendrogram quality guarantee.

3.1   Similarity Measure Definition

Data points have different number of MkNN (see Definition 1). This is because even if a data point x is one of the kNN of another data point y, x may not find y as its kNN when there are enough alternatives around x (i.e., local density around x is high enough). Thus, MkNN relationship tends to appear between sub-cluster pairs that are closely connected (i.e., they have close contact boundaries with similar local densities). This property is of practical significance to reveal the boundary cohesiveness of adjacent sub-clusters, especially those located in an arbitrary shaped cluster structure and only linked by their contact boundaries. Given this property, we introduce an MkNN-based similarity measure (see Definition 2) for arbitrary shaped clustering.

Definition 1. *Given a data set D and a positive integer k, the mutual k-nearest neighbors of a data point $x \in D$, denoted by $MkNN(x)$, is defined as $MkNN(x) = \{y \in D \mid x \in kNN(y) \wedge y \in kNN(x)\}$, where $kNN(x)$ denotes the k-nearest neighbors of data point x.*

Definition 2. *Given disjoint sub-clusters C_i and C_j, let S_{ij} be the set of data points that participate in the MkNN relationship across C_i and C_j, i.e., $S_{ij} = \{x \cup y \mid x \in C_i, y \in C_j, x \in MkNN(y)\}$. Then, the similarity between C_i and C_j, denoted by $\phi(\cdot)$, is defined as*

$$\phi(C_i, C_j) = \max \left\{ \frac{|S_{ij} \cap C_i|}{|C_i|}, \frac{|S_{ij} \cap C_j|}{|C_j|} \right\}.$$

This similarity measure $\phi(\cdot)$ is symmetric (i.e., $\phi(C_i, C_j) = \phi(C_j, C_i)$). It refers to the maximum ratio of connecting points (i.e., data points that have MkNN relationship across a given sub-cluster pair) to their host sub-cluster. A high value of $\phi(C_i, C_j)$ indicates that sub-clusters C_i and C_j are tightly contacted with each other by their contact boundaries, and it is very likely that they are within a same arbitrary shaped cluster.

3.2 Guarantee for High Quality Dendrogram

MkNN-based similarity measures are commonly used in arbitrary shaped clustering [8,13,15,18] since they can often better reflect the real cohesiveness between adjacent sub-clusters in arbitrary shaped clusters, and help conduct a more accurate clustering work. Nonetheless, few of the existing MkNN-based similarity measures take the dendrogram quality into account when they are used in HC.

In contrast, our similarity measure $\phi(\cdot)$ is able to preserve the quality of the dendrogram generated by HC, i.e., as much as possible, it helps HC organize the dendrogram in a monotonic manner so that original data points with closer proximities would be merged as preferred. Note that the HC mentioned here performs the classic agglomerative clustering framework, i.e., starting from individual data points, it recursively merges two most similar sub-clusters until there is only one cluster left at the end. This progress can be described as follows. Given disjoint sub-clusters C_i, C_j, and C_d, if C_i and C_j are merged first, then the merging criterion ensures the inequality below.

$$\max \left\{ \phi(C_i, C_d), \phi(C_j, C_d) \right\} \leqslant \phi(C_i, C_j). \tag{1}$$

With this inequality, we can prove the dendrogram quality guarantee of our similarity measure $\phi(\cdot)$ in the following two situations.

(1) When C_d is linked with either C_i or C_j (i.e., $\phi(C_i, C_d) > 0 \wedge \phi(C_j, C_d) = 0$, or $\phi(C_i, C_d) = 0 \wedge \phi(C_j, C_d) > 0$, such as the situation illustrated in Fig. 2(a)), similar to traditional similarity measures, our similarity measure $\phi(\cdot)$ also helps HC organize the dendrogram in a restrictively monotonic manner.

Theorem 1. *Given disjoint sub-clusters C_i, C_j, and C_d, if C_i and C_j are merged first and relationships $\phi(C_i, C_d) = 0$ and $\phi(C_j, C_d) > 0$ hold, then the proposed similarity measure $\phi(\cdot)$ satisfies restrictive monotonicity property, i.e.,*

$$\phi(C_i \cup C_j, C_d) \leqslant \phi(C_i, C_j).$$

Proof. As $\phi(C_i, C_d) = \max \left\{ \frac{|S_{id} \cap C_i|}{|C_i|}, \frac{|S_{id} \cap C_d|}{|C_d|} \right\} = 0$, we have $S_{id} \cap C_i = S_{id} \cap C_d = \varnothing$, $(S_{id} \cup S_{jd}) \cap C_d = S_{jd} \cap C_d$. Based on the definition of $\phi(\cdot)$,

$$\phi(C_i \cup C_j, C_d) = \max \left\{ \frac{|S_{id} \cap C_i| + |S_{jd} \cap C_j|}{|C_i| + |C_j|}, \frac{|(S_{id} \cup S_{jd}) \cap C_d|}{|C_d|} \right\}$$

$$= \max \left\{ \frac{|S_{jd} \cap C_j|}{|C_i| + |C_j|}, \frac{|S_{jd} \cap C_d|}{|C_d|} \right\}.$$

As inequality $\frac{|S_{jd} \cap C_j|}{|C_i|+|C_j|} \leqslant \frac{|S_{jd} \cap C_j|}{|C_j|}$ holds naturally, we have

$$\phi(C_i \cup C_j, C_d) \leqslant \max\left\{\frac{|S_{jd} \cap C_j|}{|C_j|}, \frac{|S_{jd} \cap C_d|}{|C_d|}\right\} = \phi(C_j, C_d).$$

Combining with Inequality (1), we have $\phi(C_i \cup C_j, C_d) \leqslant \phi(C_i, C_j)$, and the proof completes. ∎

Therefore, when C_d is linked with either C_i or C_j, $\phi(\cdot)$ also has the restrictive monotonicity property to ensure the monotonicity of dendrogram, and thus helps the dendrogram to preserve the intrinsic proximities of original data points.

(2) When C_d has contact boundaries with both C_i and C_j (i.e., $\phi(C_i, C_d) > 0 \wedge \phi(C_j, C_d) > 0$, such as the situation illustrated in Fig. 2(b)), as mentioned in Sect. 2.1, similarity measures with restrictive monotonicity property may fail to reflect the real cohesiveness between the sub-clusters in this situation. Our similarity measure $\phi(\cdot)$ is not restricted by the restrictive monotonicity property in this situation. Instead, it re-evaluates the similarity $\phi(C_i \cup C_j, C_d)$ between C_d and the newly merged sub-cluster $C_i \cup C_j$ based on the changed contact boundaries. But this re-evaluated similarity $\phi(C_i \cup C_j, C_d)$ is bounded. It will not be significantly greater than the similarity $\phi(C_i, C_j)$ between the previously merged sub-clusters C_i and C_j, and prevent a large distortion for the monotonicity between two successive levels in dendrogram. We refer to this property as *weak monotonicity*, which can be proven by the following lemma and theorem.

Lemma 1. *Given disjoint sub-clusters C_i, C_j, and C_d, if C_i and C_j are merged first and relationships $\phi(C_i, C_d) > 0$ and $\phi(C_j, C_d) > 0$ hold, then the proposed similarity measure $\phi(\cdot)$ satisfies the following relationship, i.e.,*

$$\phi(C_i \cup C_j, C_d) \leqslant \max\left\{\phi(C_i, C_d), \phi(C_j, C_d), \frac{|S_{id} \cap C_d| + |S_{jd} \cap C_d|}{|C_d|}\right\}.$$

Proof. Without loss of generality, assuming $\frac{|S_{id} \cap C_i|}{|C_i|} \leqslant \frac{|S_{jd} \cap C_j|}{|C_j|}$, then we have

$$\frac{|S_{id} \cap C_i|}{|C_i|} \leqslant \frac{|S_{id} \cap C_i| + |S_{jd} \cap C_j|}{|C_i| + |C_j|} \leqslant \frac{|S_{jd} \cap C_j|}{|C_j|}.$$

On the other hand, the following inequality holds naturally.

$$\frac{|S_{id} \cap C_d|}{|C_d|}, \frac{|S_{jd} \cap C_d|}{|C_d|} \leqslant \frac{|(S_{id} \cup S_{jd}) \cap C_d|}{|C_d|} \leqslant \frac{|S_{id} \cap C_d| + |S_{jd} \cap C_d|}{|C_d|}.$$

Combining the above inequalities, we have

$$\max\left\{\frac{|S_{id} \cap C_i|}{|C_i|}, \frac{|S_{id} \cap C_d|}{|C_d|}\right\} \leqslant \max\left\{\frac{|S_{id} \cap C_i| + |S_{jd} \cap C_j|}{|C_i| + |C_j|}, \frac{|(S_{id} \cup S_{jd}) \cap C_d|}{|C_d|}\right\}$$

$$\leqslant \max\left\{\frac{|S_{jd} \cap C_j|}{|C_j|}, \frac{|S_{jd} \cap C_d|}{|C_d|}, \frac{|S_{id} \cap C_d| + |S_{jd} \cap C_d|}{|C_d|}\right\}.$$

By Definition 2, the above inequality can be re-written as

$$\phi(C_i, C_d) \leqslant \phi(C_i \cup C_j, C_d) \leqslant \max\left\{\phi(C_j, C_d), \frac{|S_{id} \cap C_d| + |S_{jd} \cap C_d|}{|C_d|}\right\}.$$

By moving the leftmost item into the rightmost item, we get

$$\phi(C_i \cup C_j, C_d) \leqslant \max\left\{\phi(C_i, C_d), \phi(C_j, C_d), \frac{|S_{id} \cap C_d| + |S_{jd} \cap C_d|}{|C_d|}\right\},$$

and the proof completes.   ∎

With Lemma 1, we have the theorem below.

Theorem 2. *Given disjoint sub-clusters C_i, C_j, and C_d, if C_i and C_j are merged first and relationships $\phi(C_i, C_d) > 0$ and $\phi(C_j, C_d) > 0$ hold, then the proposed similarity measure $\phi(\cdot)$ satisfies the following relationship, i.e.,*

$$\phi(C_i \cup C_j, C_d) \leqslant 2 \cdot \phi(C_i, C_j).$$

Proof. The following three inequalities holds naturally.

$$\frac{|S_{id} \cap C_d| + |S_{jd} \cap C_d|}{|C_d|} \leqslant 2 \cdot \max\left\{\frac{|S_{id} \cap C_d|}{|C_d|}, \frac{|S_{jd} \cap C_d|}{|C_d|}\right\},$$

$$\frac{|S_{id} \cap C_d|}{|C_d|} \leqslant \max\left\{\frac{|S_{id} \cap C_i|}{|C_i|}, \frac{|S_{id} \cap C_d|}{|C_d|}\right\} = \phi(C_i, C_d),$$

$$\frac{|S_{jd} \cap C_d|}{|C_d|} \leqslant \max\left\{\frac{|S_{jd} \cap C_j|}{|C_j|}, \frac{|S_{jd} \cap C_d|}{|C_d|}\right\} = \phi(C_j, C_d).$$

Combining the above three inequalities, we have

$$\frac{|S_{id} \cap C_d| + |S_{jd} \cap C_d|}{|C_d|} \leqslant 2 \cdot \max\left\{\phi(C_i, C_d), \phi(C_j, C_d)\right\}.$$

Based on Inequality (1), we get

$$\max\left\{\phi(C_i, C_d), \phi(C_j, C_d), \frac{|S_{id} \cap C_d| + |S_{jd} \cap C_d|}{|C_d|}\right\} \leqslant 2 \cdot \phi(C_i, C_j).$$

Combining Lemma 1, we have $\phi(C_i \cup C_j, C_d) \leqslant 2 \cdot \phi(C_i, C_j)$, and the proof completes.   ∎

Theorem 2 shows that with $\phi(\cdot)$, when HC merges a sub-cluster C (e.g., $C_i \cup C_j$) with any other disjoint sub-cluster (e.g., C_d), the similarity between them will not exceed twice of the similarity between the parent sub-cluster pair (e.g., C_i and C_j) of C. In other words, this theorem provides the theoretical upper bound for the distortion of two successive levels in the dendrogram generated by HC, and roughly guarantees the dendrogram's quality (i.e., its monotonicity).

Algorithm 1. HICMEN Algorithm

    Input   : data set $D = \{x_1, x_2, \ldots, x_N\}$; parameter k.
    Output: dendrogram root R.
1 $\Gamma = \varnothing; \Lambda = \varnothing;$ //set Γ of clusters and set Λ of outliers
2 $[\Gamma, \Lambda] =$ remove_outliers(D);
3 **while** $|\Gamma| > 1$ **and** $\phi(C_K, C_L) > 0$ **do**
4 | $[C_K, C_L] = \arg\max\{\phi(C_i, C_j)\}$ where $(C_i \text{ and } C_j \in \Gamma)$;
5 | $C_M = C_K \cup C_L$; children(C_M) $= \{C_K, C_L\}$;
6 | $\Gamma = \Gamma \cup \{C_M\} \setminus \{C_K\} \setminus \{C_L\}$;
7 **for** each $x_i \in \Lambda(1 \leqslant i \leqslant |\Lambda|)$ **do**
8 | $x_{nn} = \arg\min_x\{dist(x_i, x) \mid x \in D \setminus \Lambda\}$;
9 | $C_{nn} = C_{nn} \cup \{x_i\}$ where $x_n \in C_{nn}$ and $C_{nn} \in \Gamma$;
10 $R =$ average_link(Γ);   //run average link with Euclidean distance

---

4 HICMEN Algorithm

4.1 Algorithm Description

With the similarity measure $\phi(\cdot)$, we propose HICMEN algorithm to mine arbitrary shaped clusters and output a high quality dendrogram. Before presenting the detailed steps, we would like to clarify that there may be no MkNN relationship across natural clusters in a data set. In this situation, $\phi(\cdot)$ will regard the similarities between the natural clusters as zeros, and hinder HC from carrying out a complete agglomerative clustering. Hence, when the maximal similarity evaluated by $\phi(\cdot)$ is zero and the work of agglomerative clustering has not been finished (i.e., there is still more than one cluster), we adopt average link algorithm with Euclidean distance to complete the HC process.

The pseudo-code of HICMEN is presented in Algorithm 1. It takes as inputs a data set D containing N data points, and a parameter k for calculating $\phi(\cdot)$.

HICMEN is carried out by three phases, namely the initialization phase (lines 1–2), the merging phase (lines 3–6), and the ending phase (lines 7–10). (1) In the initialization phase, to prevent outliers from affecting the identification of real clustering structures, HICMEN first removes outliers Λ from data set D via an efficient outlier detection algorithm proposed by Bay and Schwabacher [2]. The rest of data points are classified into set Γ for the succeeding merge process (line 4). (2) In the merging phase, starting from individual data points in Γ, HICMEN recursively merges the sub-clusters by using the proposed similarity measure $\phi(\cdot)$. In each iteration, it searches and merges two most similar sub-clusters (line 4) and labels the parent-child relationship in dendrogram (line 5). The merging process does not stop until every data point is in a single cluster or the maximal similarity is zero. (3) In the ending phase, HICMEN firstly assigns each outlier to its nearest cluster, and adopts average link algorithm with Euclidean distance to merge the rest of sub-clusters (if any) before returning the root of dendrogram.

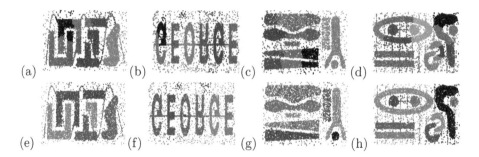

Fig. 3. The clustering results of (a)–(d) average link and (e)–(h) HICMEN (with $k = 22$) on $D1.1$–$D1.4$ (in which # of final clusters are 9, 15, 8, and 12, respectively).

4.2   Complexity Analysis

In the initialization phase, after kNN search which takes $O(dN^{2-1/d} + N \log N + kN)$ time by building a k-d tree [11], the outlier detection algorithm proposed by Bay and Schwabacher [2] can achieve a linear time complexity.

In the merging phase, MkNN relationship can be found by searching the kNN list with $O(kN)$ time. At each iteration, we update the merged sub-clusters' hash tables with $O(k)$ time since the average number of MkNN-connected sub-clusters is $O(k)$. We also update the hash tables of other sub-clusters that contain the merged sub-clusters. Given ν such sub-clusters, it takes $O(\nu k)$ time to update the hash tables, and $O(\nu \log N)$ time to find the next two most similar sub-clusters to be merged with the help of a maximum heap. In summary, the merging phase takes about $O(\nu N \log N + \nu kN)$ time since there are at most $(N - 1)$ iterations.

In the ending phase, suppose that the number of outliers is ω. It takes $O(\omega N)$ time to search their nearest non-outlier neighbors. Supposing that there still exist κ sub-clusters after merging phase, it takes $O(\kappa N)$ time to merge them by average link algorithm. In summary, the ending phase takes $O(\omega N + \kappa N)$ time.

According to extensive experimental results, ν, ω and κ are far less than N. Hence, the overall time complexity of HICMEN is about $O(dN^{2-1/d} + N \log N)$.

5   Experimental Evaluation

In this section, we first verify the effectiveness of HICMEN by comparing it with (1) single link, complete link, and average link algorithms, which are the most famous dendrogram centered HC algorithms, (2) CHAMELEON algorithm, which is the paradigm of arbitrary shaped clustering centered HC algorithms, (3) ABACUS algorithm, which is a state-of-the-art evolution of CURE, and (4) DBSCAN, which is a high-performance representative of non-HC algorithms. We then conduct an efficiency study for the algorithms followed by a discussion on the impact of parameter k to HICMEN's clustering performance and execution time. Note that as the parameters of CHAMELEON and DBSCAN often affect the clustering performance of these two algorithms, we give them a privilege, i.e.,

Table 1. Description of data sets $D2.1$–$D2.6$

Data set	Name (domain application)	N	d	c
$D2.1$	Iris	150	4	3
$D2.2$	Breast Cancer Wisconsin	683	9	2
$D2.3$	Vehicle Silhouettes	846	18	4
$D2.4$	Image Segmentation	2,100	16	7
$D2.5$	Landsat Satellite	6,435	36	6
$D2.6$	Letter Recognition	20,000	16	26

Table 2. NMI scores of HC algorithms on $D2.1$–$D2.6$

Algorithm	$D2.1$	$D2.2$	$D2.3$	$D2.4$	$D2.5$	$D2.6$
Single Link	0.72	0.01	0.01	0.35	0.62	0.40
Complete Link	0.72	0.64	0.18	0.50	0.48	0.39
Average Link	0.81	0.68	0.17	0.49	0.64	0.40
CHAMELEON	0.70	0.77	0.12	0.59	0.61	0.31
ABACUS	0.79	0.70	0.16	0.56	0.61	0.40
DBSCAN	0.73	0.74	0.15	0.52	0.58	0.29
HICMEN	**0.82**	**0.84**	**0.21**	**0.68**	**0.69**	**0.43**

we vary their parameters at each execution and report their best performance from 20 times of run. All tested algorithms are implemented in C++, running on a desktop PC with 8GB RAM and Intel Core i7-2600 CPU at 3.40 GHz.

5.1 Effectiveness Evaluation

We first evaluate the effectiveness of HICMEN from the following two aspects, namely (1) clustering performance, and (2) dendrogram quality.

(1) Clustering Performance Study. In this experiment, we first demonstrate HICMEN's ability of mining arbitrary shaped clusters, and compare HICMEN with the other tested algorithms in terms of clustering accuracy.

(a) Effectiveness of Mining Arbitrary Shaped Clusters: We run HICMEN on four commonly used 2D data sets $D1.1$–$D1.4$ (8,000 points in $D1.1$–$D1.3$ and 10,000 points in $D1.4$), which were also used to evaluate CHAMELEON, ABACUS and DBSCAN algorithms. As shown in [4,19], CHAMELEON and ABACUS have good performance on these data sets, while DBSCAN's performance on them often has minor flaws.

Figure 3 illustrates the clustering results of HICMEN and average link algorithm, from which we can observe that HICMEN can effectively identify arbitrary shaped clusters, while average link fails. Single link and complete link algorithms have similar failures to average link.

Table 3. Description of data sets $D3.1$–$D3.8$

Data set	N	d	c	Data set	N	d	c
$D3.1$	3,000	2	20	$D3.5$	2,701	4	10
$D3.2$	5,250	2	35	$D3.6$	4,051	6	10
$D3.3$	7,500	2	50	$D3.7$	5,401	8	10
$D3.4$	2,026	3	10	$D3.8$	6,751	10	10

Table 4. CPCC scores of HC algorithms

Algorithm	$D3.1$	$D3.2$	$D3.3$	$D3.4$	$D3.5$	$D3.6$	$D3.7$	$D3.8$
Single Link	0.61	0.58	0.52	0.79	0.86	0.91	0.92	0.88
Complete Link	0.72	**0.71**	0.67	0.79	0.90	0.84	0.92	0.87
Average Link	0.74	0.70	0.66	0.82	**0.92**	**0.92**	**0.94**	0.91
HICMEN	**0.75**	**0.71**	**0.69**	**0.84**	**0.92**	**0.92**	**0.94**	**0.92**

(b) Accuracy Comparison: We compare HICMEN (with parameter $k = 22$) with the other tested approaches on six real data sets $D2.1$–$D2.6$ from UCI machine learning repository [10]. The data set properties are described in Table 1, in which N, d, and c indicate the number of points, data set dimensions, and the number of real clusters. The accuracy performance of each algorithm is measured by NMI (Normalized Mutual Information). Each NMI score falls in the range $[0, 1]$. A greater NMI score indicates a more accurate clustering result.

Table 2 lists the NMI scores of clustering results of each algorithm, from which we can observe that our HICMEN algorithm outperforms the other tested HC algorithms in accuracy performance.

(2) Dendrogram Quality. In this experiment, we compare the quality of the dendrogram generated by each algorithm on data sets $D3.1$–$D3.8$ [12], of which the properties are summarized in Table 3. Data sets $D3.1$–$D3.3$ have the same dimension with increased numbers of clusters, while data sets $D3.4$–$D3.8$ have the same cluster number with increased dimensions. We adopt CPCC (Cophenetic Correlation Coefficient) as dendrogram quality criterion, which describes how faithfully a dendrogram preserves the intrinsic proximities between the original data points. The definition of CPCC is as follows.

$$CPCC = \frac{\frac{1}{M}\sum_{i=1}^{N-1}\sum_{j=i+1}^{N} d(i,j)c(i,j) - \mu_P\mu_C}{\sqrt{\left(\frac{1}{M}\sum_{i=1}^{N-1}\sum_{j=i+1}^{N} d^2(i,j) - \mu_P^2\right)\left(\frac{1}{M}\sum_{i=1}^{N-1}\sum_{j=i+1}^{N} c^2(i,j) - \mu_C^2\right)}}$$

where $d(i,j)$ and $c(i,j)$ are the ordinary Euclidean distance and dendrogrammatic distance between points i and j respectively, N is the number of points, $M = \frac{1}{2}N(N-1)$, and μ_P and μ_C are defined as

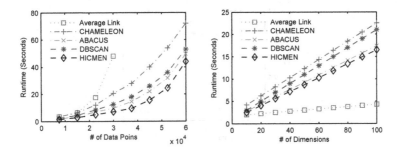

Fig. 4. Efficiency performance (runtime) of HC algorithms on data sets with different number of data points and different number of dimensions.

$$\mu_P = \frac{1}{M} \sum_{i=1}^{N-1} \sum_{j=i+1}^{N} d(i,j), \quad \mu_C = \frac{1}{M} \sum_{i=1}^{N-1} \sum_{j=i+1}^{N} c(i,j).$$

A greater CPCC score indicates a better dendrogram quality.

Table 4 presents the CPCC score of the dendrogram generated by HIC-MEN (with parameter $k = 22$) and HC using linkage metrics. Note that CHAMELEON, ABACUS and DBSCAN have no CPCC score since they cannot output a full dendrogram. As shown in the table, HICMEN has almost the greatest CPCC scores among all HC algorithms, indicating that it can most faithfully preserve the intrinsic proximities between the original data points, better than traditional similarity measures with restrictive monotonicity property.

5.2   Efficiency Evaluation

In this experiment, we evaluate the efficiency of the tested algorithms with various data set sizes and dimensions. To investigate their scalability to the number of data points, we generate $D4.1$–$D4.8$ based on $D3.3$ by creating new data points nearby the original data points from the original number $7,500$ to the number of $60,000$ (with interval $7,500$). To investigate their scalability to the dimensionality of data sets, we generate $D5.1$–$D5.8$ based on $D3.3$ by creating new dimensions from 10 to 100 (with interval 10).

Figure 4 shows the execution time of the tested algorithms on $D4.1$–$D4.8$ and $D5.1$–$D5.8$ respectively (The parameter k of HICMEN is set to 22). We skip to plot the runtime curves of single link and complete link algorithms since they are very similar to that of average link algorithm.

From the figure, we can have the following observations. (1) With the growth of data set size, the execution time of our HICMEM algorithm increases slower than that of the other tested algorithms, indicating that our HICMEM algorithm shows better scalability to large data sets; (2) with the growth of dimensions, the execution time of average link algorithm increases very slowly, whereas the execution time of HICMEN, CHAMELEON, ABACUS and DBSCAN increase linearly with regard to the dimensions. Nonetheless, our HICMEN algorithm still keeps faster than CHAMELEON, ABACUS and DBSCAN.

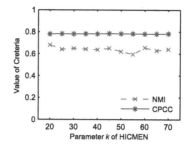

Fig. 5. Impact of parameter k to HICMEN's clustering result accuracy (in terms of NMI scores) and dendrogram quality (in terms of CPCC scores).

5.3   Impact of Parameter

HICMEN has only one parameter k for calculating the MkNN-based similarity measure $\phi(\cdot)$. In this experiment, we evaluate the impact of this parameter k to the effectiveness of HICMEN. By varying the value of k, we report the NMI score of HICMEN's clustering result, and the CPCC score of its generated dendrogram.

Figure 5 depicts the corresponding results on $D2.4$ (similar observations can be obtained on the rest of UCI data sets used in this paper), from which we can observe that the clustering performance and dendrogram quality of our HICMEM algorithm keep relatively stable with various k values, indicating that the effectiveness of HICMEN is relatively insensitive to the value of parameter k.

6   Conclusion

In this paper, we have defined an MkNN-based similarity measure for HC, and proven its weak monotonicity which enables HC to accurately express arbitrary shaped data sets with little distortion on the dendrogram. Based on this similarity measure, we have proposed HICMEN a simple yet effective HC algorithm for accurately identifying arbitrary shaped clusters and with a complete and high quality dendrogram as the output. Experimental results on both real and synthetic data sets have verified the effectiveness and efficiency of our approach.

Acknowledgements. This work was supported in part by NSFC Grants (61502347, 61502504, 61522208, 61572376, 61472359, 61379033, 61373038, and 61364025), the Fundamental Research Funds for the Central Universities (2015XZZX005-07, 2015XZZX004-18, and 2042015kf0038), and the Research Funds for Introduced Talents of WHU.

References

1. Ankerst, M.: OPTICS: ordering points to identify the clustering structure. In: SIGMOD, pp. 49–60 (1999)
2. Bay, S.D., Schwabacher, M.: Mining distance-based outliers in near linear time with randomization and a simple pruning rule. In: KDD, pp. 29–38 (2003)
3. Chaoji, V., Hasan, M.A., Salem, S., Zaki, M.J.: SPARCL: an efficient and effective shape-based clustering. Knowl. Inf. Syst. $21(2)$, 201–229 (2009)
4. Chaoji, V., Li, G., Yildirim, H., Zaki, M.J.: ABACUS: mining arbitrary shaped clusters from large datasets based on backbone identification. In: SDM, pp. 295–306 (2011)
5. Chen, Y.-A., Tripathi, L.P., Dessailly, B.H., Nyström-Persson, J., Ahmad, S., Mizuguchi, K.: Integrated pathway clusters with coherent biological themes for target prioritisation. Plos One $9(6)$, e99030 (2014)
6. Correa, C.D., Lindstrom, P.: Locally-scaled spectral clustering using empty region graphs. In: KDD, pp. 1330–1338 (2012)
7. Defays, D.: An efficient algorithm for a complete link method. Comput. J. $20(4)$, 364–366 (1977)
8. Ertöz, L., Steinbach, M., Kumar, V.: Finding clusters of different sizes, shapes, and densities in noisy, high dimensional data. In: SDM, pp. 47–58 (2003)
9. Ester, M., Kriegel, H.-P., Sander, J., Xu, X.: A density-based algorithm for discovering clusters in large spatial databases with noise. In: KDD, pp. 226–231 (1996)
10. Frank, A., Asuncion, A.: UCI machine learning repository (2010)
11. Friedman, J.H., Bentley, J.L., Finkel, R.A.: An algorithm for finding best matches in logarithmic expected time. ACM Trans. Math. Softw. $3(3)$, 209–226 (1977)
12. SIPU Clustering datasets. http://cs.joensuu.fi/sipu/datasets/
13. Guha, S., Rastogi, R., Shim, K.: ROCK: a robust clustering algorithm for categorical attributes. In: ICDE, pp. 512–521 (1999)
14. Guha, S., Rastogi, R., Shim, K.: CURE: an efficient clustering algorithm for large databases. Inf. Syst. $26(1)$, 35–58 (2001)
15. Houle, M.E.: The relevant-set correlation model for data clustering. In: SDM, pp. 775–786 (2008)
16. Hu, T., Liu, C., Tang, Y., Sun, J., Song, H., Sung, S.Y.: High-dimensional clustering: a clique-based hypergraph partitioning frameworks. Knowl. Inf. Syst. $39(1)$, 61–88 (2014)
17. Huang, H., Gao, Y., Chen, L., Li, R., Chiew, K., He, Q.: Browse with a social web directory. In: SIGIR, pp. 865–868 (2013)
18. Huang, H., Gao, Y., Chiew, K., Chen, L., He, Q.: Towards effective and efficient mining of arbitrary shaped clusters. In: ICDE, pp. 28–39 (2014)
19. Karypis, G., Han, E.H., Kumar, V.: CHAMELEON: hierarchical clustering using dynamic modeling. IEEE Comput. $32(8)$, 68–75 (1999)
20. Li, J., Xia, Y., Shan, Z., Liu, Y.: Scalable constrained spectral clustering. IEEE Trans. Knowl. Data Eng. $27(2)$, 589–593 (2015)
21. Mok, P.K., Huang, H.Q., Kwok, Y.L., Au, J.S.: A robust adaptive clustering analysis method for automatic identification of clusters. Pattern Recogn. $45(8)$, 3017–3033 (2012)
22. Alex, R., Alessandro, L.: Clustering by fast search and find of density peaks. Science $344(6191)$, 1492–1496 (2014)
23. Sibson, R.: SLINK: an optimally efficient algorithm for the single-link cluster method. Comput. J. $16(1)$, 30–34 (1973)

24. Sokal, R.R., Rohlf, F.J.: The comparison of dendrograms by objective methods. Taxon **11**(2), 33–40 (1962)
25. Voorhees, E.M.: Implementing agglomerative hierarchic clustering algorithms for use in document retrieval. Inf. Process. Manag. **22**(6), 465–476 (1985)
26. Yang, Y., Ma, Z., Yang, Y., Nie, F., Shen, H.T.: Multitask spectral clustering by exploring intertask correlation. IEEE Trans. Cybern. **45**(5), 1069–1080 (2015)
27. Kim, Y., Shim, K., Kim, M.-S., Lee, J.S.: DBCURE-MR: an efficient density-based clustering algorithm for large data using MapReduce. Inf. Syst. **42**, 15–35 (2014)

Hierarchically Clustered LSH for Hierarchical Outliers Detection

Konstantinos Georgoulas[(✉)] and Yannis Kotidis

Department of Informatics, Athens University of Economics and Business,
Patission 76, 10434 Athens, Greece
{kgeorgou,kotidis}@aueb.gr

Abstract. In this work we introduce hierarchical outliers that extend the notion of distance-based outliers for handling hierarchical data domains. We present a novel framework that permits us to detect hierarchical outliers in a consistent manner, providing a desired monotonicity property, which implies that a data observation that finds enough support so as to be disregarded as an outlier at a level of the hierarchy, will not be labelled as an outlier when examined at a more coarse-grained level above. This way, we enable users to grade how suspicious a data observation is, depending on the number of hierarchical levels for which the observation is found to be an outlier. Our technique utilizes an innovative locality sensitive hashing indexing scheme, where data points sharing the same hash value are being clustered. The computed centroids are maintained by our framework's scheme index while detailed data descriptors are discarded. This results in reduced storage space needs, execution time and number of distance evaluations compared to utilizing a straightforward LSH index.

1 Introduction

An outlier is an observation that differs so much from others so as to arouse suspicion that it was generated by a different process than the rest of the data. In order to put this intuition into a context where outliers can be formally defined and computed many alternative definitions have been proposed. One of the most commonly used approach is the distance-based outlier definition, which suggests that given a dataset P, a positive integer N and a positive real number r, a data object p of P is a $O(N, r)$-outlier, if less than N objects in P lie within distance r from p, for some appropriate distance metric.

Outliers detection is critical for many modern applications such as decision support (OLAP), customer behavior analysis and network management. However, none of the well known outlier detection techniques takes into consideration the hierarchical nature of the data domains that is inherent in such applications. The natural aggregation of atomic values along a domain hierarchy is a critical summarization technique that can be used to detect different *grades* of abnormal behavior by looking at all levels of the hierarchy.

© Springer International Publishing Switzerland 2016
S. Hartmann and H. Ma (Eds.): DEXA 2016, Part I, LNCS 9827, pp. 169–184, 2016.
DOI: 10.1007/978-3-319-44403-1_11

As an example, we consider the case of an electronic store. There are several ways to categorize products (*ProductId, Group, Class* categories) that a customer purchases. Table 1 presents an example of customers and the products they purchased. Distance-based computations of outliers in this example can be performed by mapping each customer into a point in a high-dimensional domain (e.g. dimensions being the productIds). The values of the coordinates on each dimension (i.e. productId) can be boolean values (indicating whether the user has purchased the product), or may be derived from different statistics (e.g. number of times the customer purchased a product, her rating, etc.).

Independently of the details of this mapping, if we compare customers based on the productIds of the products they purchased, then John and Mary show no apparent similarity. However, if we look at the *Group* category of the products, it is obvious that they both purchased Smart Phones. Similarly, John and Jim look dissimilar until they are observed at the upper level of the product's hierarchy (*Class* category). Consequently, distance-based outliers derived by looking at the data domain that corresponds to the leaves of the product's domain hierarchy (Product→Class→Group→ProductID) may find support when these observations are aggregated further up the hierarchy.

Table 1. Product purchases

User	ProductId	Group	Class
John	Samsung Galaxy S4	Smart phones	Computers
John	Apple iPhone 6	Smart phones	Computers
Tim	Nikon Camera D750	Cameras	Tvs-cameras
Jim	Apple iPad Air 2	Tablets	Computers
Mary	LG Nexus 5	Smart phones	Computers

Given that domain hierarchies are commonly used in many applications, in this work we first look at the problem of deriving an intuitive definition that extends the notion of distance-based outliers over hierarchical domains. A straightforward independent computation of distance-based outliers over all hierarchical levels may yield inconsistent results that complicate data analysis. As an example, depending on the selected threshold values N and r, an observation that is not an outlier at the leaves of the hierarchy may be deemed as such at an intermediate level. This goes against intuition, which suggests that as atomic values are being aggregated via the hierarchy, data observations tend to look similar.

In this work, we introduce the notion of hierarchical outliers for handling hierarchically organized data domains. Our proposed definition computes outliers in a consistent manner, which implies that a data observation that finds enough support so as to be disregarded as an outlier at a level of the hierarchy, can not be labelled as an outlier when examined at a more coarse-grained level above. This intended monotonicity property not only leads to conclusions that are not surprising to the user analyst but also enable us to grade how suspicious a data observation is, depending on the number of hierarchy levels for which the observation is found to be an outlier.

In addition to providing a proper definition of hierarchical outliers, in this work we also look at efficient techniques that enable us to compute such outliers in large datasets. Locality sensitive hashing (LSH) is a popular technique that partitions a high-dimensional dataset into buckets so as to avoid performing all-pair computation of item distances. Direct application of LSH for hierarchical outliers identification is prohibitively expensive as independent indexes need to be constructed for each level of the hierarchy, leading to increased computational and storage overhead.

Thus, we propose an innovative LSH index scheme, termed as hierarchically clustered LSH (cLSH), which instead of storing the data items at an independent index for every level of the hierarchy, it only maintains the centroids of clusters, which are constructed performing a clustering technique among the data items that share a common hash value at every index. As a result, both the computational and storage overhead for the cLSH is reduced compared to the original LSH structure.

The contributions of our work are:

- We introduce the notion of hierarchical outliers and provide an intuitive framework for detecting hierarchical outliers over hierarchically organized data domains. Our framework assigns a simple and intuitive statistic called *grade* for every data item identified as hierarchical outlier, which is a positive integer referring to the number of levels for which the specific item is outlier. The higher the *grade*, the more erroneous the item is.
- We propose an innovative indexing scheme based on locality sensitive hashing. This scheme maintains centroids of data clusters at LSH indexes of hierarchical levels, making it less space demanding compared to the case of independently created original LSH indexes at every level.
- We introduce a bottom up computation via the hierarchy of data domain in order to detect hierarchical outliers. At each level, our method utilizes results from previously performed computations resulting in faster computation of outliers.
- We present an experimental evaluation for our framework measuring the accuracy and the efficiency (in terms of space and time) of our proposed techniques.

2   Related Work

Many previous works in different areas of data management have studied the problem of outlier detection. Different approaches for the definition of an outlier have been presented in case of multidimensional data. In [2,9,15] distance based outliers are discussed, while [3,13] consider density-based outliers as well. In the first case, data items are considered as outliers based on the distances from their neighbors. In the second case data items are studied by computing the density of data around their local neighbors. The relative density of a data item compared to its neighbors is computed as an outlier score. Different approaches have discussed different variants for computing this score [8,16]. A different outlier definition presented in [10] suggests that angles between data vectors are more stable than distances in high dimensional spaces. In this case items are compared using angle-based similarity metrics, like the cosine similarity metric. A data item is not considered as outlier if most of the objects are located in similar directions with it.

Many of the aforementioned solutions exploit well known indexing techniques (like the R-tree and its variants) in order to perform range or NN queries that are necessary for outlier detection. Conventional multidimensional indexes are inapplicable in large data domains. For instance, the cardinality of the product dimension in a data warehouse can be in the order of tens of thousands. Instead, our technique utilizes a probabilistic indexing method termed LSH [4] that approximates the results of NN and range queries in high dimensional spaces and extends it in order to handle efficiently the hierarchical structure that data follows. The LSH scheme that we present in this work can be extended to support different distance metrics, including the cosine similarity for angle-based computation of outliers [6].

3   Motivational Example

Suppose that we would like to detect hierarchical outliers in the data warehouse of an electronic store. A set of data items could be derived by projecting every customer's purchases at the hierarchically organized *Product* domain space. Figure 1 shows a data item representing a customer's purchases. The hierarchy of the *Product* domain consists of four levels. The lowest level l_4 contains all *ProductIds*, which are used for the unique identification of the products. At level l_3, the *Group* category of the products is represented (i.e. *Home Theatre, TVs, Cameras, Smart Phones, Tablets, Laptops*). Level l_2 depicts the *Class* category of products (i.e. *Audio, Tvs & Cameras, Computers*), while l_1 contains the *Top* level representing all products. In this example, without loss of generality, at the

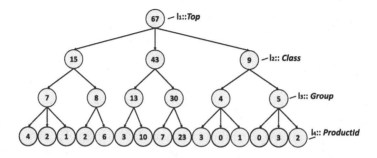

Fig. 1. Product domain hierarchy

lowest level of the hierarchy the values represent cumulative purchases of different productIds for this customer. Aggregated values at upper levels are obtained by the utilization of *sum* function (e.g. as in a typical roll-up aggregation).

Given this setting of data, we may focus on detecting outliers at every level of this hierarchy. We could try to identify outliers based on the productIds that the customer bought, or according to her aggregated purchases over product *Groups*. Someone else may take into consideration customers purchases at more abstracted summarization levels provided by the *Class* or *Top* category. We suggest a holistic approach that considers all abstraction levels of product purchases, based on the specified hierarchy. We introduce the notion of *hierarchical outlier* that takes into consideration the whole hierarchical tree that represents her complete purchasing history, supporting a more intuitive decision whether a customer is an outlier, or not.

Moreover, our framework provides an succinct measure termed *hierarchical outlier grade* that denotes the number of levels a customer is identified as an outlier. For instance, if a customer is regarded as hierarchical outlier with $grade = 2$, this would suggest that her purchases based on the productIds and their Groups are significantly different from other people on the dataset. On the other hand, this result implies that her purchases when aggregated at the *Class* level are similar to many other customers.

4   Hierarchical Outliers

The hierarchical nature of the data domain motivates us to examine data at every level of the hierarchy they follow, in order to be identified as outliers. In our motivational example, if we check all customers at level l_4 and identify a specific customer as an outlier, we have no evidence to regard her as an outlier at upper levels too. It is likely that only few customers purchase the same exactly products as she does (in terms of productIds), while there are many who purchase similar quantities of products at the *Group* level of products categorization.

This observation leads us to propose a framework for the outlier detection problem that takes into consideration the hierarchical structure of the data domain. An obvious solution would be to compute distance-based outliers at all different abstracted levels in a completely separate way. Given the fact that data items are high dimensional, someone could construct an index for every hierarchical level, in order to retrieve the nearest neighbors of the queried item at every level and then according to the distance-based outlier definition she could decide whether it is outlier or not. However, it is quite possible a specific item in question to be identified as an outlier at some levels and not to be considered as an outlier at some others lower or higher to previous ones, depending on the selected distance thresholds.

This lack of coherence stems from the main drawback of an independent evaluation of distance-based outliers: it handles the different abstraction levels of a data item as independent observations, rather than different abstractions of the same data item, obtained through the hierarchy. By manipulating data in this manner, there could be no *consistent* results in order to characterize a customer's behavior in total.

In order to alleviate this inconsistency of results for hierarchically organized data we introduce the notion of *hierarchical outlier* $HO(N, r)$.

Definition 1 (Hierarchical Outlier $HO(N, r)$). *Given a dataset P over a hierarchically organized data domain with h hierarchical levels, a positive integer N (threshold) and a positive real number r, a data item $p \in P$ is a $HO(N, r)$-Hierarchical Outlier with grade L, if there are L levels of data hierarchy, at which less than N objects in P lie within distance r_i from p, where $1 \le i < h$. $r_i = \sqrt{(2 * max(F^i) - 1)} * r_{i+1}$, $r_h = r$ and $max(F^i)$ denotes the maximum fanout of those hierarchical tree's nodes belong to hierarchical level l_i.*

Intuitively, the definition utilizes a certain method for computing the distance thresholds at the different levels. As will be explained in what follows this ensures that outliers' grade can be computed in a consistent manner following the desired monotonicity property.

We first present some preliminaries that we utilize to better describe our hierarchical outlier definition.

Lemma 1. *Given a data item $X = \{x_1, x_2, x_3, \ldots, x_{d_{i+1}}\}$ and a query point $q = \{q_1, q_2, q_3, \ldots, q_{d_{i+1}}\}$ in a domain organized by a hierarchy H, it holds that $D_i(q, X) \le \sqrt{(2 * max(F^i) - 1)} * D_{i+1}(q, X)$ where $D_{i+1}(q, X), D_i(q, X)$ are the Euclidean distance between q and X at hierarchical levels l_{i+1} and l_i, respectively, where $1 \le i < h$ and $max(F^i)$ is the maximum fanout of those hierarchical tree's nodes belonging to hierarchical level l_i.*

Lemma 1 ensures the consistency of the results that the proposed hierarchical outlier detection process provides. Based on this property, if a data item q has N items that lie within distance r_{i+1} from it at level l_{i+1}, then it will also have at least the same N items in distance $r_i = \sqrt{(2 * max(F^i) - 1)} * r_{i+1}$ at higher level l_i. By utilizing the popular distance-based outlier definition, we disregard q as an outlier at a level l_{i+1} and furthermore we also do not consider it as outlier at any upper level l_i with the condition of defining distance thresholds r_i based on Lemma 1. In Fig. 2, we graphically depict how the distance threshold r_4 at the lowest level l_4 is "expanded" at the upper levels of hierarchy H. When thresholds are increased in a manner consistent to Lemma 1, the computation of distance-based outliers provides the desired consistency.

Proof. Here, we prove that $D_i(q, X) \le \sqrt{(2 * max(F^i) - 1)} * D_{i+1}(q, X)$. For every level l_i, where $1 \le i < h$ we know that

$$D_i^2(q, X) = \sum_{j=1}^{d_i} \left((q_{k_j+1} - x_{k_j+1}) + \cdots + \right.$$

$$\left. + (q_{k_j+f_j} - x_{k_j+f_j}) \right)^2 = \sum_{j=1}^{d_i} Value(j).$$

where f_j is the fanout of j-th node at level l_i of hierarchical tree. d_i is dimensionality of level l_i and $k_j = \sum_{w=0}^{j-1} f_w$.

$$Value(j) = (q_{k_j+1} - x_{k_j+1})^2 + \cdots + (q_{k_j+f_j} - x_{k_j+f_j})^2 +$$
$$+2 * \sum_{w=k_j+1}^{k_j+f_j-1} \sum_{y=w+1}^{k_j+f_j} (q_w - x_w)(q_y - x_y)$$

and thus,

$$D_i^2(q, X) = D_{i+1}^2(q, X) + \sum_{j=1}^{d_i} extra(j) \tag{1}$$

where $extra(j) = 2 \sum_{w=k_j+1}^{k_j+f_j-1} \sum_{y=w+1}^{k_j+f_j} (q_w - x_w)(q_y - x_y)$.

Bounding the $\sum_{j=1}^{d_i} extra(j)$ of Eq. 1, we are able to express distance $D_i(q, X)$ as a factor of $D_{i+1}(q, X)$. It is

$$\sum_{j=1}^{d_i} extra(j) \leq 2 * \sum_{j=1}^{d_i} \sum_{w=k_j+1}^{k_j+f_j-1} \sum_{y=w+1}^{k_j+f_j} |q_w - x_w||q_y - x_y|$$
$$\leq 2 * (max(F^i) - 1) \sum_{j=1}^{d_i} \sum_{w=k_j+1}^{k_j+f_j} (q_w - x_w)^2$$
$$\leq 2 * (max(F^i) - 1) \sum_{j=1}^{d_{i+1}} (q_j - x_j)^2$$
$$\leq 2 * (max(F^i) - 1) * D_{i+1}^2(q, X)$$

and thus we prove that: $D_i^2(q, X) \leq (2max(F^i) - 1) * D_{i+1}^2(q, X)$

Although our techniques are tailored for the popular Euclidean metric, they can be adapted appropriately for different distance metrics and aggregation functions applied to the data domain's hierarchy.

In the following sections, we present in detail our adopted LSH indexing structure that is tailored to identify hierarchical outliers, as well as our algorithm for their efficient detection based on the proposed index.

5   Hierarchically Clustered LSH Indexing

Given that we need to compare high-dimensional data when looking for hierarchical outliers, we adapt a powerful dimensionality reduction technique called LSH [1]. LSH generates an indexing structure by evaluating multiple hashing functions over each data item. Using the LSH index, we can identify the nearest neighbors of each customer and compute outliers based on the distances from her neighbors.

We utilize hash functions that are based on 2-stable distributions and create several different hash tables in order to increase the effectiveness of the LSH indexing schema. There have been many proposals on how to tune and increase performance of LSH (e.g. [6,11]), however such techniques are orthogonal to the work we present here.

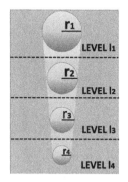

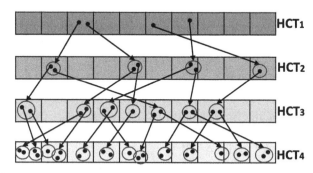

Fig. 2. Bounding r for different hierarchical levels.

Fig. 3. Hierarchical clustering of our LSH scheme.

A direct approach for indexing a data set over a hierarchical domain would be the construction of independent LSH hash schemes (one per hierarchical level). Each hash scheme would contain T hash tables, named $HT_1^1 \ldots HT_1^T$, $HT_2^1 \ldots HT_2^T$, ..., $HT_h^1 \ldots HT_h^T$, which would maintain the data items of levels $l_1, l_2, \ldots, l_h$, respectively. As we have already denoted, storing the whole dataset at independent LSH indexes for every hierarchical level is not an efficient way of indexing. Instead, we introduce a more space-saving index by creating T_i hash tables at every level l_i, named $HCT_1^1 \ldots HCT_1^{T_1}$, $HCT_2^1 \ldots HCT_2^{T_2}$, $HCT_h^1 \ldots HCT_h^{T_h}$. Each HCT_i is a hierarchically clustered hash table and contains a small number of centroids, which are computed by clustering the data items that falls in the bucket with the same id for the HCT_{i+1} hash table of the immediate lower level l_{i+1}.

At most k centroids (to be stored in HCT_i) are computed by clustering the data items belonging to the same bucket of HCT_{i+1}. k is a user-defined parameter, which affects the space cost of our proposed LSH scheme. Its value could vary from one to the exact number of items stored every time at a bucket. The more centroids per bucket we maintain, the higher the space requirements of the index would be. A more flexible option that we apply in our framework, is the derivation of different values for k at every bucket so as a target space reduction ratio rr is achieved for the whole index.

Thus, we compute k_{B^j} (where $k_{B^j} = \frac{|B^j|}{rr}$) centroids for every B^j bucket of HCT_{i+1}. We compute the hash values for the computed centroids and store them in the appropriate bucket of HCT_i. We have to notice that, both the number T_i

of HCT tables that are created at every level l_i and the hash functions utilized to hash the centroids, are selected following the same parametrization process [14] as in the case of building independent hash schemes for every level. Similarly, for hash table HCT_{i-1}, we compute the centroids after the clustering of the centroids maintained to each bucket of HCT_i. As a result, HCT_{i-1} maintains centroids of the clusters constructed over the centroids stored at each bucket of HCT_i. Following the same procedure, we create the HCT tables for all the remaining levels up to l_1, in a bottom-up process. The higher the hierarchical level, the fewer centroids need to be indexed to its corresponding hash table HCT. The aforementioned procedure can also be performed at the lowest level l_h. In this case a primary LSH scheme for level l_h is constructed. The clusters and their centroids for every bucket are computed and stored to newly created HCT hash tables, while the primary LSH scheme is not required any more and is, thus, discarded.

In more detail, our LSH indexing structure construction requires the following steps:

- We initially construct a temporary LSH scheme for indexing the real data items of level l_h. These hash tables are auxiliary (i.e. used for the construction of the HCT hash tables at level l_h) and they are discarded immediately after the next step of the process is completed.
- We compute $k_{B^j} = \frac{|B^j|}{rr}$ centroids for every bucket B^j of the hash table HT_h^1. In our framework, we utilize k-means for clustering, however this choice is orthogonal to our scheme. This set of centroids are hashed to a set of T_h hash tables using hash functions $g_h^1 \ldots g_h^{T_h}$, where g_h^i for $1 \leq i \leq T_h$ is a family of a 2-stable distribution functions [5].
- We repeat the previous step for every level l_i, with $1 \leq i \leq h - 1$. Each time, we perform a k-means clustering at the centroids stored to the buckets of the HCT_{i+1}^1 hash table. These centroids are abstracted to the upper hierarchical level l_i, forming a much smaller dataset (in terms of cardinality) than the real dataset, for the level l_i. Based on the centroids' hash values, they are stored at the corresponding buckets of $HCT_i^1 \ldots HCT_i^{T_i}$ hash tables.

In Fig. 3, we show an instance of our indexing structure for the case of our running example's hierarchy. For ease of presentation, we create only one table, instead of T_i, at every hierarchical level l_i. The hierarchy consists of 4 levels and, thus, we create four tables $HCT_4, HCT_3, HCT_2, HCT_1$, one per level. Every arrow links a centroid, that is maintained at HCT_{i-1}, with the cluster of a HCT_i hash table at level l_i which members it represents. There can be one or multiple clusters in the same bucket, for example the two clusters at the first bucket of HCT_4. At hash table HCT_3, we can see thirteen centroids derived from HCT_4's data. As we mentioned previously, these centroids are assigned to buckets of HCT_3, based on their hash value during index's construction. Only these thirteen centroids are maintained in the hierarchically clustered hash table HCT_3 of level l_3. Similarly, at HCT_2 seven centroids are constructed based on the hash values of the centroids of the seven clusters created at HCT_3. Finally at the top HCT_1 hash table, we notice only four entries.

It is clear that the number of data items/centroids stored at each level are quite fewer than the data maintained in the case of storing the whole dataset at every hash table of every hierarchical level. Consequently, our index requires significantly smaller space compared to the original LSH scheme. It maintains hash tables, consisted only of a small number of centroids performing a per-bucket clustering of the data items. These centroids, as we will explain in Sect. 6, are utilized in order to compute the support score to a query point during the hierarchical outlier detection process. This evaluation leads to reduced number of distance computations when querying the index resulting to even faster outlier identification compared to the baseline approach. In our experimental analysis, we depict several figures proving our aforementioned claims. Encapsulated information in centroids, such as the number of data items that the cluster contains and the cluster range (i.e. distance of the centroid to its furthest cluster member), is a key factor for the reduction of the computation cost of hierarchical outlier identification, as we show in the next section.

6   Efficient Identification of Hierarchical Outliers

Given the proposed cLSH indexing scheme, we are able to identify hierarchical outliers $HO(N,r)$ based on Definition 1 and compute their *grade* according to the procedure described below.

Formally, given a query point q, we would like to compute its grade. Notice that q may be part of the data set, or an arbitrary point (e.g. a new customer). The identification process begins at the lowest level l_h of the hierarchy. Firstly, a nearest neighbor (NN) query is executed for the query point q utilizing the HCT_h hash tables created for indexing data at level l_h. Following the original LSH scheme's way of NN evaluation [7], we compute the hash value of q by applying the g_h^i hash function for every one of the T_h hash tables at level l_h, where $1 \leq i \leq T_h$. We retrieve from every HCT_h hash table the content from those buckets which id value is the same with the computed hash value of q. A set of items is returned from each bucket. These sets are merged, removing any duplicates, forming a resulted set, named $SupCand_h(q)$. $SupCand_h(q)$ set contains all the centroids of the data clusters containing data items that are candidates to lie within distance r from q.

A query item q *gains* support (i.e. increases its support score), if a centroid lies within distance r_i from it at level l_i. In order to compute the support that a centroid *gives* to a query item we proceed to an approximation technique, based on which a centroid c, with radius r_c and $rep(c)$ (where $rep(c)$ are the number of data items a cluster contains), gives support $sup(c)_q$ to a query point q according to the following formula:

$$sup(c)_q^i = rep(c) \times \frac{V(Sphere(c, r_c) \bigcap Sphere(q, r_i))}{V(c, r_c)} \tag{2}$$

where $Sphere(q, r_i)$ is the the hyper-sphere having as center the point q and radius r_i. Figure 4 provides a visualization of this process.

Algorithm 1. $HO\_Query(q, L, support, grade)$

Input: q is the query point

l_i is i-th level of hierarchy H

$support$ is the support score of q at level l_{L+1}

$cur\_support$ is the support score of q at level l_L

$grade$ is Hierarchical Outlier Grade for q

1: $SupCand_L(q) = \emptyset$ $cur\_support = 0$
2: **for** $j = 1 \ldots T_L$ **do**
3:      $SupCand_L(q) = SupCand_L(q) \cup lsh(q, HCT_L^j)$
4: **end for**
5: **for** $\forall c \in SupCand_L(q)$ **do**
6:      $sup(c)_q^L = ComputeSupport(q, c)$
7:      $cur\_support = cur\_support + sup(c)_q^L$
8:      **if** $pred(c) \notin bucket(q, HCT_{L+1}^1)$ **then**
9:          $support = support + sup(c)_q^L$
10:     **end if**
11: **end for**
12: **if** $support < N$ OR $cur\_support < N$ **then**
13:     $grade + +$
14:     **if** $L \geq 2$ **then**
15:         $HO\_Query(q, L - 1, support, grade)$
16:     **end if**
17: **end if**

Algorithm 1 shows the algorithm for computing whether q is a hierarchical outlier and return its grade. Firstly, we compute the set $SupCand_L(q)$ of candidate centroids (Lines 2–4), that may give support to q at level L. Function lsh returns those centroids from all the hierarchical hash tables at level l_L, which have the same hash value with q. We then update (Lines 5–10) the support score of q at level l_L based on every centroid c that belongs to $SupCand_L(q)$. Function $ComputeSupport(q, c)$ (Line 6) approximates (as it implied by Eq. 2) the support that c provides to q. In Line 7 the support $sup(c)_q^L$ increases the $cur\_support$ of q at level l_L. In case the centroid c represents a cluster of centroids (this information is derived by function $pred(c)$) that belong to a bucket HCT_{L+1}^1 of level l_{L+1} that it has not been processed during the query evaluation at level l_{L+1} (i.e. members of c's cluster do not fall in the same bucket of HCT_{L+1}^1 with the one that q's hash value implies - $bucket(q, HCT_{L+1}^1)$), its providing support to q is also added to the support that q has already gained by the previous levels (Lines 8–10). If the $support$ or $cur\_support$ do not exceed threshold N the q's grade is increased by one and we recursively call the algorithm for level l_{L-1}. Our process terminates whenever the obtained support at a level l_i exceeds threshold N or level l_1 is reached.

In our running example, given a query point q, we first compute its hash values for hierarchical levels l_4, l_3, l_2, l_1 and then we assign it to the corresponding buckets that our index maintains. As it is depicted in Fig. 5, q falls in buckets $B4_{10}, B3_8, B2_7$ and $B1_3$ of HCT_4, HCT_3, HCT_2 and HCT_1 respectively. For

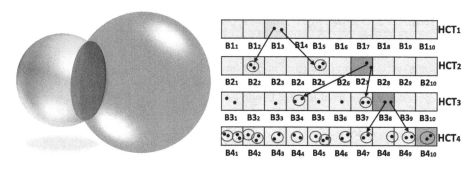

Fig. 4. Computing support that a cen-   **Fig. 5.** Buckets visited during query execution.
troid c "gives" to a query point q

ease of presentation, we only depict one table per level, instead of T^i copies
for every level l_i that our method suggests. However, query execution utilizing
T^i tables per level is straightforward to what we discuss here. We only need to
merge the sets of items retrieved from the buckets of T^i tables of a specific level
and then proceed as we describe below.

For a given threshold value $N = 10$, our method starts at level l_4 evaluating
the Euclidean distance between q and centroids stored in bucket $B4_{10}$ of HCT_4.
Based on these evaluations we approximate q's support score at hierarchical level
l_4. In case support value is greater than N, we terminate query's evaluation and
answer that q is not a hierarchical outlier, otherwise we set its grade value to 1
and we continue checking q at level l_3. Assuming in this example that support
score at hierarchical level l_4 is two, we continue at level l_3 retrieving the two
centroids (c_{310}, c_{311}) stored at bucket $B3_8$ of HCT_3 based on q's hash value. For
each one of these two centroids we approximate the support that they provide to
q by utilizing Eq. 2. Suppose that $sup(c_{310})^3_q = 3$ and $sup(c_{311})^3_q = 2$ we conclude
that q is also an outlier at level l_3 and increase its grade by one. Continuing at
level l_2 we obtain centroids c_{25}, c_{26} from $B2_7$ bucket, which represent clusters
that its members are stored in bucket $B3_4$ and $B3_7$ respectively, that has not
been processed during the lower levels query evaluation and thus could be added
to the already computed support score of q. The support that c_{25} provides (e.g.
$sup(c_{25})^2_q = 6$) is added both to the *cur_support* for level l_2 and *support* that q
have already gained from levels l_3 and l_4. *support* exceeds threshold N and query
execution terminates (without accessing bucket $B1_3$ of level l_1). As a result, our
algorithm replies that q is identified as a hierarchical outlier with grade = 2.

Concluding, we should notice that a hierarchical outlier detection query
involves processing of several buckets of the HCT tables for levels l_h up to
l_1. However, the higher the hierarchical level our method examines, the lower is
the number of centroids obtained by these buckets, as the number of centroids
at higher levels is reduced as an effect of the recursive clustering over the hierarchy during index construction. Moreover, our algorithm retains the value of
support from previous (lower) levels, in order to expedite processing. Consider
level l_1, where we have already processed three buckets ($B4_{10}$, $B3_8$, $B2_7$), that

give the necessary support to q at l_2 and so the $B1_3$ *bucket* is not need to be accessed. Finally, the monotonicity property of hierarchical outliers, permits us to terminate the query, when enough support is gained at a specific level.

7 Experimental Evaluation

In this section, we present an experimental evaluation of the proposed hierarchical outlier detection framework. All algorithms are implemented in Java and the experiments run on a desktop PC with an i7 CPU (4 cores, 3.4 GHz), 8 GB RAM, and a 128 GB SSD.

7.1 Experimental Setup

Data Sets. In the experimental study, we employ two data sets. In the first dataset, we created a hierarchy of products consisting of six levels with dimensions (cardinality) 2654, 380, 51, 13, 3, 1 from the leaves to the root of hierarchical tree, respectively. We generated data for 50000 customers with their purchases over the 2654 different products at the lowest level of the hierarchy. In order to generate the purchases of a customer, we first set the number of cumulative purchases for every customer by selecting uniformly from the range 30000–80000. We then selected randomly 20 % of the 2654 products belonging at the lowest level of product's domain hierarchical tree. These 20 % of products are considered as high interest products for customers and 80 % of her total purchases are uniformly distributed to these products. The remaining 20 % of a customer's purchases are distributed randomly to the rest of products (that span 80 % of the produce domain) that are considered as low interest. We created several clusters of customers where customers of the same cluster have the same sets of high and low interest products.

We also used the OLAP Council APB-1 benchmark generator [12] to create a second dataset which contains 5300 customers. For every customer, the generator produced a vector representing her cumulative purchases over a period of 17 months on a domain of 6050 products. The products' domain hierarchy consists of six hierarchical levels.

Algorithms. We evaluate our hierarchical outlier detection algorithm that utilizes the hierarchically clustered LSH (cLSH) index and we compared it to an alternative implementation of the same algorithm that utilizes independent LSH indices for every level of the hierarchy. All indices are parametrized as described in [14].

Metrics. Our main metrics include: (a) the average number of distance evaluations for a hierarchical outlier detection query, (b) the average query execution time, (c) the average number of candidates points that the indices return per query execution, (d) the storage needs for both implementations, and (e) the precision of the results of both techniques computed as

$$precision_{level(i)} = \frac{|customers\ retrieved_{level(i)} \cap real\ outliers_{level(i)}|}{|customers\ retrieved_{level(i)}|}$$

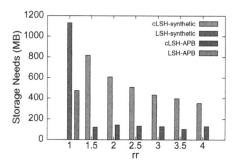

Fig. 6. Storage requirements

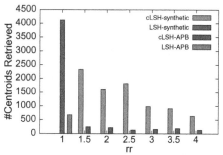

Fig. 7. Index points retrieved per query

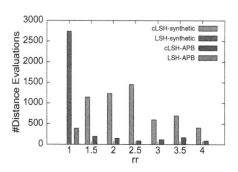

Fig. 8. Average distance evaluations per query

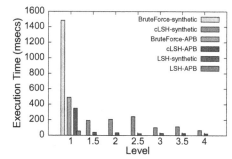

Fig. 9. Average execution time per query

Queries. We present average values over 100 queries, where all query points are outliers at lowest level of products hierarchy on both datasets. For the remaining levels the number of outliers range from 4 to 74. The higher the hierarchical level, the smaller the number of queries that are outliers. For instance at fifth level there are 74 queries identified as outliers for APB dataset and 13 for the synthetic one while at the highest hierarchical level there are only 6 and 4 outliers respectively.

Parameters. We conduct experiments varying the reduction ratio rr (1.5–4) that defines $k_i = \frac{|B_i|}{rr}$ for the k-means clustering evaluation on every bucket B_i.

7.2 Experimental Results

Space Cost. In Fig. 6, we depict the storage needs of our technique for various values of rr. The higher the requested value of rr the lower the space cost because larger clusters are constructed and, thus, fewer centroids are stored at the cLSH index. Given that only centroids are maintained by cLSH it is expected that we gain in terms of space compared to the original LSH scheme.

Distance Evaluations and Index Points Retrieved. In Fig. 7, we show the average number of the points (LSH)/centroids (cLSH) returned as candidates by each index to provide support to a query point. This number is significantly

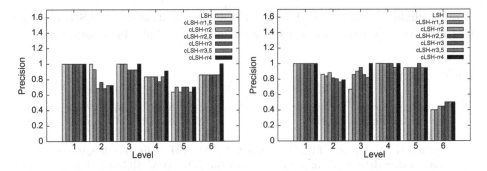

Fig. 10. Synthetic dataset   **Fig. 11.** APB dataset

smaller for cLSH, as a result of the recursive clustering over the hierarchy and the way these centroids are used to increase the support of a query point.

Figure 8 depicts the average number (over 100 queries) of distance evaluations required in order to compute the hierarchical outliers and their grade. Our method using cLSH performs up to 80 % fewer distance evaluations in order to detect a hierarchical outlier, compared to the straightforward LSH scheme. This significant reduction is attributed to the use of centroids in order to calculate the support from a whole cluster of points to the query point in a single step, instead of a per-data-item calculation.

Execution Time. Figure 9 shows that our method is up to 75 % faster compared to the original LSH scheme and 15 times faster than a brute-force method that does not use any index. The main factor that increases the execution time is the number of distance evaluations. This is evident by the fact that the evaluations and execution time graphs follow the same trend.

Precision. Figs. 10 and 11 depict the precision of both indices (LSH/cLSH) in hierarchical outlier identification. Both techniques are very accurate and provide high precision results. We do not provide a similar graph for the recall because it was 100 % for all levels of the hierarchy, in these experiments (i.e. our approximation technique – summarized in Formula 2 – overestimates the support score). Even though cLSH is significantly more condense than the LSH index, it provides equally accurate results.

8   Conclusions

In this work we introduced a framework for detecting outliers in hierarchically organized domains. Key to our method is a monotonicity property that enables us to grade in an intuitive manner how erroneous a data item seems with respect to the rest of the data. We also discussed a novel indexing scheme that computes hierarchical clusters of data items and embeds them in a LSH index. Using this index we can quickly identify hierarchical outliers with reduced storage and computation cost compared to using a straightforward LSH index. The benefits of our techniques stem from the hierarchical organization of the LSH buckets that permits us to reuse distance computations while exploring a data item.

References

1. Andoni, A., Indyk, P.: Near-optimal hashing algorithms for approximate nearest neighbor in high dimensions. In: FOCS, pp. 459–468. IEEE Computer Society (2006)
2. Bhaduri, K., Matthews, B.L., Giannella, C.R.: Algorithms for speeding up distance-based outlier detection. In: Proceedings of the 17th ACM SIGKDD International Conference on Knowledge Discovery and Data Mining, pp. 859–867. ACM (2011)
3. Breunig, M., Kriegel, H., Ng, R., Sander, J., et al.: LOF: identifying density-based local outliers. Sigmod Rec. **29**(2), 93–104 (2000)
4. Charikar, M.: Similarity estimation techniques from rounding algorithms. In: STOC (2002)
5. Datar, M., Immorlica, N., Indyk, P., Mirrokni, V.S.: Locality-sensitive hashing scheme based on p-stable distributions. In: Proccedings of SCG (2004)
6. Georgoulas, K., Kotidis, Y.: Distributed similarity estimation using derived dimensions. VLDB J. **21**(1), 25–50 (2012)
7. Indyk, P., Motwani, R.: Approximate nearest neighbors: towards removing the curse of dimensionality. In: STOC (1998)
8. Jin, W., Tung, A.K.H., Han, J., Wang, W.: Ranking outliers using symmetric neighborhood relationship. In: Ng, W.-K., Kitsuregawa, M., Li, J., Chang, K. (eds.) PAKDD 2006. LNCS (LNAI), vol. 3918, pp. 577–593. Springer, Heidelberg (2006)
9. Knorr, E., Ng, R., Tucakov, V.: Distance-based outliers: algorithms and applications. VLDB J. **8**(3), 237–253 (2000)
10. Kriegel, H., Zimek, A., et al.: Angle-based outlier detection in high-dimensional data. In: Proceeding of ACM SIGKDD, pp. 444–452 (2008)
11. Lv, Q., Josephson, W., Wang, Z., Charikar, M., Li, K.: Multi-probe LSH: efficient indexing for high-dimensional similarity search. In: VLDB, pp. 950–961 (2007)
12. OLAP Council APB-1 Benchmark. http://www.olapcouncil.org/research/resrchly. htm
13. Papadimitriou, S., Kitagawa, H., Gibbons, P., Faloutsos, C.: LOCI: fast outlier detection using the local correlation integral. In: Proceedings of ICDE, pp. 315–326 (2003)
14. Slaney, M., Lifshits, Y., He, J.: Optimal parameters for locality-sensitive hashing. In: Proceedings of the IEEE, pp. 2604–2623 (2012)
15. Sugiyama, M., Borgwardt, K.: Rapid distance-based outlier detection via sampling. In: Advances in Neural Information Processing Systems, pp. 467–475 (2013)
16. Tang, J., Chen, Z., Fu, A.W., Cheung, D.W.: Enhancing effectiveness of outlier detections for low density patterns. In: Chen, M.-S., Yu, P.S., Liu, B. (eds.) PAKDD 2002. LNCS (LNAI), vol. 2336, pp. 535–548. Springer, Heidelberg (2002)

Incorporating Clustering into Set Similarity Join Algorithms: The *SjClust* Framework

Leonardo Andrade Ribeiro[1]([✉]), Alfredo Cuzzocrea[2],
Karen Aline Alves Bezerra[3], and Ben Hur Bahia do Nascimento[3]

[1] Instituto de Informática, Universidade Federal de Goiás, Goiânia, Goiás, Brazil
laribeiro@inf.ufg.br
[2] DIA Department, University of Trieste and ICAR-CNR, Trieste, Italy
alfredo.cuzzocrea@dia.units.it
[3] Departmento de Ciência da Computação, Universidade Federal de Lavras,
Lavras, Brazil
karen.bezerra@posgrad.ufla.br, bhn@computacao.ufla.br

Abstract. *Data cleaning and integration* found on *duplicate record identification*, which aims at detecting duplicate records that represent the same real-world entity. *Similarity join* is largely used in order to detect pairs of similar records in combination with a subsequent clustering algorithm meant for grouping together records that refer to the same entity. Unfortunately, the clustering algorithm is strictly used as a post-processing step, which slows down the overall performance, and final results are produced at the end of the whole process only. Inspired by this critical evidence, in this paper we propose and experimentally assess *SjClust*, a framework to integrate similarity join and clustering into a single operation. The basic idea of our proposal consists in introducing a variety of cluster representations that are smoothly merged during the set similarity task, carried out by the join algorithm. An optimization task is further applied on top of such framework. Experimental results, which are derived from an extensive experimental campaign, we retrieve are really surprising, as we are able to outperform the original set similarity join algorithm by an order of magnitude in most settings.

1 Introduction

Data cleaning and integration (e.g., [7,15]) found on *duplicate record identification* (e.g., [8,24]), which aims at detecting duplicate records that represent the same real-world entity. This is becoming more and more relevant in emerging *big data research* (e.g., [14,22,25]), as a plethora of real-life applications are characterized by the presence of multiple records representing the same real-world entity, which practically plagues every large database. Such records are often referred to as *fuzzy duplicates* (duplicates, for short), because they might not be exact copies of one another. Duplicates arise due to a variety of reasons, such as typographical errors and misspellings, different naming conventions, and as a result of the integration of data sources storing overlapping information.

S. Hartmann and H. Ma (Eds.): DEXA 2016, Part I, LNCS 9827, pp. 185–204, 2016.
DOI: 10.1007/978-3-319-44403-1_12

Duplicates degrade the quality of the data delivered to application programs, thereby leading to a myriad of problems. Some examples are misleading data mining models owing to erroneously inflated statistics, inability of correlating information related to a same entity, and unnecessarily repeated operations, e.g., mailing, billing, and leasing of equipment. Duplicate identification is thus of crucial importance in data cleaning and integration.

Duplicate identification is computationally very expensive and, therefore, typically done offline. However, there exist important application scenarios that demand (near) real-time identification of duplicates. Prominent examples are data exploration [10], where new knowledge has to be efficiently extracted from databases without a clear definition of the information need, and virtual data integration [7], where the integrated data is not materialized and duplicates in the query result assembled from multiple data sources have to be identified — and eliminated — on-the-fly. Such scenarios have fueled the desire to integrate duplicate identification with processing of complex queries [2] or even as a general-purpose physical operator within a DBMS [6].

An approach to realize the above endeavor is to employ *similarity join* in concert with a *clustering algorithm* [9]. Specifically, similarity join is used to find all pairs of records whose similarity is not less than a specified threshold; the similarity between two records is determined by a *similarity function*. In a post-processing step, the clustering algorithm groups together records using the similarity join results as input. For data of string type, *set similarity join* is an appealing choice for composing a duplicate identification operator. Set similarity join views its operands as sets — strings can be easily mapped to sets. The corresponding similarity function assesses the similarity between two sets in terms of their overlap and a rich variety of similarity notions can be expressed in this way [6]. Furthermore, a number of optimization techniques have been proposed over the years [3,6,18,19,23,24] yielding highly efficient and scalable algorithms.

The strategy of using a clustering algorithm strictly for post-processing the results of set similarity join has two serious drawbacks, however. First, given a group of n, sufficiently similar, duplicates, the set similarity join performs $\binom{n}{2}$ similarity calculations to return the same number of set pairs. While this is the expected behavior considering a similarity join in isolation, it also means that repeated computations are being performed over identical subsets. Even worse, we may have to perform much more additional similarity calculations between non-duplicates: low threshold values are typically required for clustering algorithms to produce accurate results [9]. Existing filtering techniques are not effective at low threshold values and, thus, there is an explosion of the number of the comparisons at such values. Second, the clustering is a blocking operator in our context, i.e., it has to consume all the similarity join output before producing any cluster of duplicates as result element. This fact is particularly undesirable when duplicate identification is part of more complex data processing logic, possibly even with human interaction, because it prevents pipelined execution.

In this paper, we propose and experimentally assess *SjClust*, a framework to integrate set similarity join and clustering into a single operation, which addresses the above issues. The main idea behind our framework is to represent groups of similar sets by a *cluster representative*, which is incrementally updated during the set similarity join processing. Besides effectively reducing the number similarity calculations needed to produce a cluster of n sets to $O(n)$, we are able to fully leverage state-of-the-art optimization techniques at high threshold values, while still performing well at low threshold values where such techniques are much less effective. Indeed, the resulting composed algorithm is even up to an order of magnitude faster than the original set similarity join algorithm for low threshold values. Moreover, we exploit set size information to identify when no new set can be added to a cluster; therefore, we can then immediately output this cluster and, thus, avoid the blocking behavior. Furthermore, there exists a plethora of clustering algorithms suitable for duplicate identification and no single algorithm is overall the best across all scenarios [9]. Thus, versatility in supporting a variety of clustering methods is essential. Our framework smoothly accommodates various cluster representation and merging strategies, thereby yielding different clustering methods for each combination thereof.

2   Related Work

The duplicate identification problem has a long history of investigation conducted by several research communities spanning databases, machine learning, and statistics, frequently under different names, including record linkage, de-duplication, and near-duplicate identification [8,12]. Over the last years, there is growing interest in realizing duplicate identification on-the-fly. In [1], a query-driven approach is proposed to reduce the number of cleaning steps in simple selections queries over dirty data. The same authors presented a framework to answer complex Select-Project-Join queries [2]. Our work is complementary to these proposals as our algorithms can be encapsulated into physical operators to compose query evaluation plans.

There is long line of research on (exact) set similarity joins [3,6,18,19,23,24]. Aspects most relevant to our work are discussed at length in Sect. 3. To the best of our knowledge, integration of clustering into set similarity joins has not been previously investigated in the literature—the general idea of our proposal was presented in [17]. In [15], the authors employ the concept of proximity graph to cluster strings without requiring a predefined threshold value. The algorithm to automatically detected cluster borders was improved later in [11]. However, it is not clear how to leverage state-of-the-art set similarity joins in these approaches to improve efficiency and deal with large datasets. In [9], a large number of clustering algorithms are evaluated in the context of duplicate identification. These algorithms use similarity join to produce their input, but can start only after the complete similarity join execution.

By looking at the innovative context of duplicate detection over big data repositories, which is really emerging at now, some relevant state-of-the-art proposals are the following ones. [25] proposes a data cleaning algorithm based on *MapReduce* that extracts relations from nodes in the target Cloud environment, and then cleans data based on *an innovative weighted-based knowledge model*. [14] evidences the relevance of data cleaning methodologies in big data scenarios, and harnesses both *context* and *usage patterns* of data entities to determine relationships among objects that are recognized as similar. Finally, [22] focuses the attention on the specific case of *big RDF data cleaning*, by also considering *semi-automatic methods*.

3   Fundamental Concepts and Background Knowledge

3.1   Basic Concepts and Definitions

We map strings to *sets of tokens* using the popular concept of *q-grams*, i.e., sub-strings of length q obtained by "sliding" a window over the characters of an input string v. We (conceptually) extend v by prefixing and suffixing it with $q-1$ occurrences of a special character "$\$$" not appearing in any string. Thus, all characters of v participate in exact q q-grams. For example, the string *"token"* can be mapped to the set of *2*-gram tokens $\{\$t, to, ok, ke, en, n\$\}$. As the result can be a multi-set, we simply append the symbol of a sequential ordinal number to each occurrence of a token to convert multi-sets into sets, e.g., the multi-set $\{a,b,b\}$ is converted to $\{a\circ1, b\circ1, b\circ2\}$. In the following, we assume that all strings in the database have already been mapped to sets.

We associate a weight with each token to obtain *weighted sets*. A widely adopted weighting scheme is the Inverse Document Frequency (*IDF*), which associates a weight $idf(tk)$ to a token tk as follows: $idf(tk) = ln(1 + N/df(tk))$, where $df(tk)$ is the *document frequency*, i.e., the number of strings a token tk appears in a database of N strings. The intuition behind using IDF is that rare tokens are more discriminative and thus more important for similarity assessment. The weight of a set r, denoted by $w(r)$, is given by the weight summation of its tokens, i.e., $w(r) = \sum_{tk \in r} w(tk)$.

We consider the general class of set similarity functions. Given two sets r and s, a set similarity function $sim(r,s)$ returns a value in $[0,1]$ to represent their similarity; larger value indicates that r and s have higher similarity. Popular set similarity functions are defined as follows.

Definition 1 (Set Similarity Functions). *Let r and s be two sets. We have:*

- *Jaccard similarity:* $J(r,s) = \frac{w(r \cap s)}{w(r \cup s)}$.
- *Dice similarity:* $D(r,s) = \frac{2 \cdot w(r \cap s)}{w(r) + w(s)}$.
- *Cosine similarity:* $C(r,s) = \frac{w(r \cap s)}{\sqrt{w(r) \cdot w(s)}}$

We now formally define the set similarity join operation.

Definition 2 (Set Similarity Join). *Given two set collections $\mathcal{R}$ and $\mathcal{S}$, a set similarity function sim, and a threshold τ, the set similarity join between $\mathcal{R}$ and $\mathcal{S}$ returns all scored set pairs $\langle (r,s), \tau\prime \rangle$ s.t. $(r,s) \in \mathcal{R} \times \mathcal{S}$ and $sim(r,s) = \tau\prime \geq \tau$.*

In this paper, we focus on self-join, i.e., $\mathcal{R} = \mathcal{S}$; we discuss the extension for binary inputs in Sect. 3.3. For brevity, we use henceforth the term similarity function (join) to mean set similarity function (join). Further, we focus on the Jaccard similarity and the IDF weighting scheme, i.e., unless stated otherwise, $sim\,(r, s)$ and $w\,(tk)$ denotes $J\,(r, s)$ and $idf\,(tk)$, respectively.

Example 1. Consider the sets r and s below

$$r = \{\textbf{A, B, } C\textbf{, D, E}\}$$
$$s = \{\textbf{A, B, D, E, } F\}$$

and the following token-IDF association table:

tk	A	B	C	D	E	F
$idf\,(tk)$	1.5	2.5	2	3.5	0.5	2

Thus, we have $w\,(r) = w\,(s) = 10$ and $w\,(r \cap s) = 8$; thus $sim\,(r, s) = \frac{8}{10+10-8} \approx 0.66$.

3.2 Optimization Techniques

Similarity functions can be equivalently represented in terms of an *overlap bound* [6]. Formally, the overlap bound between two sets r and s, denoted by $O\,(r, s)$, is a function that maps a threshold τ and the set weights to a real value, s.t. $sim\,(r, s) \geq \tau$ iff $w\,(r \cap s) \geq O\,(r, s)$[1]. The similarity join can then be reduced to the problem of identifying all pairs r and s whose overlap is not less than $O\,(r, s)$. For the Jaccard similarity, we have $O\,(r, s) = \frac{\tau}{1+\tau} \cdot (w\,(r) + w\,(s))$.

Further, similar sets have, in general, roughly similar weights. We can derive bounds for immediate pruning of candidate pairs whose weights differ enough. Formally, the weight bounds of r, denoted by $min\,(r)$ and $max\,(r)$, are functions that map τ and $w\,(r)$ to a real value s.t. $\forall s$, if $sim\,(r, s) \geq \tau$, then $min\,(r) \leq w\,(s) \leq max\,(r)$ [19]. Thus, given a set r, we can safely ignore all other sets whose weights do not fall within the interval $[min\,(r), max\,(r)]$. For the Jaccard similarity, we have $[min\,(r), max\,(r)] = \left[\tau \cdot w\,(r), \frac{w(r)}{\tau}\right]$. We refer the reader to [20] for definitions of overlap and weight bounds of several other similarity functions, including Dice and Cosine.

We can prune a large share of the comparison space by exploiting the *prefix filtering principle* [6,19], which allows discarding candidate pairs by examining only a fraction of the input sets. We first fix a global order $\mathcal{O}$ on the universe $\mathcal{U}$ from which all tokens are drawn. A set $r' \subseteq r$ is a prefix of r if r' contains the first $|r'|$ tokens of r. Further, $pref_\beta\,(r)$ is the shortest prefix of r, the weights of whose tokens add up to more than β. The prefix filtering principle is defined as follows.

---

[1] For ease of notation, the parameter τ is omitted.

Definition 3 (Prefix Filtering Principle [6]). Let r and s be two sets. If $w\,(r \cap s) \geq \alpha$, then $pref_{\beta_r}(r) \cap pref_{\beta_s}(r) \neq \varnothing$, where $\beta_r = w\,(r) - \alpha$ and $\beta_s = w\,(s) - \alpha$, respectively.

We can identify all candidate matches of a given set r using the prefix $pref_\beta(r)$, where $\beta = w\,(r) - min\,(r)$. We denote this prefix simply by $pref\,(r)$. It is possible to derive smaller prefixes for r, and thus obtain more pruning power, when we have information about the set weight of the candidate sets, i.e., if $w\,(s) \geq w\,(r)$ [3] or $w\,(s) > w\,(r)$ [18]. Note that prefix overlap is a condition necessary, but not sufficient to satisfy the original overlap constraint: an additional verification must be performed on the candidate pairs. Finally, the number of candidates can be significantly reduced by using the *inverse document frequency ordering*, $\mathcal{O}_{idf}$, as global token order to obtain sets ordered by decreasing IDF weight[2]. The idea is to minimize the number of sets agreeing on prefix elements and, in turn, candidate pairs by shifting lower frequency tokens to the prefix positions — recall that higher IDF weights are associated to low-frequency tokens.

Example 2. Consider the sets r and s in Example 1 and $\tau = 0.6$. We have $O\,(r, s) = 7.5$; $[min\,(r), max\,(r)]$ and $[min\,(s), max\,(s)]$ are both $[6, 16.7]$. By ordering r and s according to O_{idf} and the IDF weights in Example 1, we obtain:

$$r = [D, B, C, A, E]$$
$$s = [D, B, F, A, E].$$

We have $pref\,(r) = pref\,(s) = [D]$.

3.3   Similarity Join Algorithms: Definitions and Usage

Similarity join algorithms based on inverted lists are effective in exploiting the previous optimizations [3,18,19,24]. Most of such algorithms have a common high-level structure following a filter-and-refine approach.

Algorithm 1 formalizes the steps of a similarity join algorithm. The algorithm receives as input a set collection sorted in increasing order of set weights, where each set is sorted according to $\mathcal{O}_{idf}$. An inverted list I_t stores all sets containing a token t in their prefix. The input collection R is scanned and, for each *probe set* r, its prefix tokens are used to find *candidate sets* in the corresponding inverted lists (lines 4–10); this is the *candidate generation phase*, where the map M is used to associate candidates to its accumulated overlap score os (line 3). Each candidate s is dynamically removed from the inverted list if its weight is less than $min\,(r)$ (lines 6–7). Further filters, e.g., filter based on overlap bound, are used to check whether s can be a true match for r, and then the overlap score is accumulated, or not, and s can be safely ignored in the following processing (lines 8–10). In the *verification phase*, r and its matching candidates, which are stored in M, are checked against the similarity predicate and those pairs satisfying the

---

[2] A secondary ordering is used to break ties consistently (e.g., the lexicographic ordering).

Algorithm 1. Similarity join algorithm

Input: A set collection $\mathcal{R}$ sorted in increasing order of the set weight; each set is sorted according to $\mathcal{O}_{idf}$; a threshold τ

Output: A set S containing all pairs (r, s) s.t. $Sim\,(r, s) \geq \tau$

1   $I_1, I_2, \ldots I_{|\mathcal{U}|} \leftarrow \varnothing, S \leftarrow \varnothing$

2   **foreach** $r \in \mathcal{R}$ **do**

3     $M \leftarrow$ empty map from set id to overlap score (os)

4   **foreach** $t \in pref\,(r)$ **do**   // can. gen. phase

5   **foreach** $s \in I_t$ **do**

6   **if** $w\,(s) < min\,(r)$

7   Remove s from I_t

8   **if** $filter\,(r, s, M\,(s))$

9           $M\,(s)\,.os \leftarrow -\infty$ // invalidate s

10   **else** $M\,(s)\,.os = M\,(s)\,.os + w\,(t)$

11    $S \leftarrow S \cup Verify\,(r, M, \tau)$ // verif. phase

12   **foreach** $t \in pref\,(r)$ **do**   // index. phase

13      $I_t \leftarrow I_t \cup \{r\}$

14 **return** S

---

predicate are added to the result set. To this end, the *Verify* procedure (not shown) employs a merge-join-based algorithm exploiting token order and the overlap bound to define break conditions (line 11) [18]. Finally, in the *indexing phase*, a *pointer* to set r is appended to each inverted list I_t associated with its prefix tokens (lines 12 and 13).

Algorithm 1 is actually a self-join. Its extension to binary joins is trivial: we first index the smaller collection and then go through the larger collection to identify matching pairs. For simplicity, several filtering strategies such positional filtering [24] and min-prefixes [18], as well as inverted list reduction techniques [3,18] were omitted. Nevertheless, these optimizations are based on bounds and prefixes and, therefore, our discussion in the following remains valid.

4 Our Proposal: The Innovative *SjClust* Framework

We now present *SjClust*, a general framework to integrate clustering methods into similarity joins algorithms. The goals of our framework are threefold: (1) *flexibility and extensibility* by accommodating different clustering methods; (2) *efficiency* by fully leveraging existing optimization techniques and by reducing the number of similarity computations to form clusters; (3) *non-blocking behavior* by producing results before having consumed all the input, preferably much earlier.

The backbone of *SjClust* is the similarity join algorithm presented in Sect. 3. In particular, *SjClust* operates over the same input of sorted sets, without requiring any pre-processing, and has the three execution phases present in Algorithm 1, namely, candidate generation, verification, and indexing phases. Nevertheless, there are, of course, major differences.

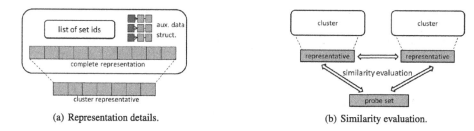

(a) Representation details.   (b) Similarity evaluation.

Fig. 1. Cluster representation.

First and foremost, the main objects are now cluster of sets, or simply clusters. Figure 1 illustrates strategy adopted for cluster representation. The internal representation contains a list of its set element's ids, an (optional) auxiliary structure, and the cluster's *complete representation*, a set containing all tokens from all set elements. A cluster exports its external representation as the so-called *cluster representative* (or simply representative) (Fig. 1(a)). Representatives are fully comparable to input sets and similarity evaluations are always performed on the representatives, either between a probe set and a cluster or between two clusters (Fig. 1(b)). In the following, we use the term cluster and representative interchangeably whenever the distinction is unimportant for the discussion.

Figure 2 depicts more details on the *SjClust* framework. In the candidate generation phase, prefix tokens of the current probe set are used to find cluster candidates in the inverted lists (Fig. 2(a)). Also, there is a *merging phase* between verification and indexing phases (Fig. 2(b)). The verification phase reduces the number of candidates by removing false positives, i.e., clusters whose similarity to the probe set is less than the specified threshold. In the merging phase, a new cluster is generated from the probing set and the clusters that passed through the verification are considered for merging with it according to a *merging strategy*. In the indexing phase, references to the newly generated cluster are stored in the inverted lists associated with its prefix tokens. Finally, there is the so-called *Output Manager*, which is responsible for maintaining references to all clusters — a reference to a cluster is added to the Output Manager right after its generation in the merging phase (Fig. 2(b)). Further, the Output Manager sends a cluster to the output as soon as it is identified that no new probing set can be similar to this cluster. Clusters in such situation can be found in the inverted lists during the candidate generation (Fig. 2(a)) as well as identified using the weight of the probe set (not shown in Fig. 2).

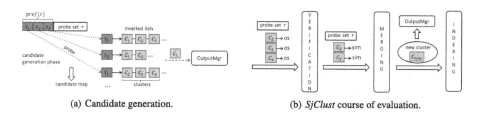

(a) Candidate generation.   (b) *SjClust* course of evaluation.

Fig. 2. *SjClust* framework components.

The aforementioned goals of *SjClust* are met as follows: flexibility and extensibility are provided by different combinations of cluster representation and merging strategies, which can be independently and transparently plugged into the main algorithm; efficiency is obtained by the general strategy to cluster representation and indexing; and non-blocking behavior is ensured by the Output Manager. Next, we provide details of each *SjClust* component.

5 *SjClust* Architecture and Components

5.1 Cluster Representation

Cluster representatives are used to compactly represent a cluster, while capturing the most significant features of its elements. In our context, there is the additional requirement that cluster representatives must be fully comparable with the original sets. Also, we want flexibility in obtaining different representation strategies.

We start by defining the complete representation of a cluster of sets, from which we extract the corresponding representative. Intuitively, the complete representation of a cluster is given by the union of its sets. We then order all tokens according to a *cluster ordering*, denoted by $\mathcal{O}_{cl}$. While $\mathcal{O}_{idf}$ is used to increase prefix filtering effectiveness (recall Sect. 3.2), $\mathcal{O}_{cl}$ is used to improve quality by sorting the tokens in the complete representation in decreasing order of importance. We formally define the concept of complete representation of a cluster in the following.

Definition 4 (Complete Representation). *Let* $C = \{r_1, r_2, \ldots, r_n\}$ *be a cluster of sets. The complete representation of* C*, denoted by* $CompR_C$*, is the union of its elements, i.e.,* $CompR_C = r_1 \cup r_2 \cup \ldots \cup r_n$*, sorted in decreasing order according to* $\mathcal{O}_{cl}$*.*

Given that the tokens in $CompR_C$ are sorted according to some notion of importance, we can use the prefix concept to derive the representative containing the most important tokens in C. To this end, we need to first define the size of the prefix. A natural choice is to use average weight of sets in C. Then, we define a slight variation of the prefix concept: given a sorted set r, $pref\prime_\alpha(r)$ is the shortest prefix of r, the weights of whose tokens add up to *not less* than α. Finally, to be comparable with probe sets, we need to further sort the tokens in the representative according to $\mathcal{O}_{idf}$. We are now ready to formally define the concept of cluster representative in our context.

Definition 5 (Cluster Representative). *Let* $C = \{r_1, r_2, \ldots, r_n\}$ *be a cluster of sets and* $CompR_C$ *its complete representation according to Definition 4. The cluster representative of* C*, denoted by* CR_C*, is the following prefix of* $CompR_C$*:*

$$CR_C = pref\prime_\alpha(CompR_C), \text{ where } \alpha = \sum_{i=1}^{n} \frac{w(r_i)}{n},$$

ordered according to $\mathcal{O}_{idf}$*.*

We can now derive cluster representation strategies by instantiating $\mathcal{O}_{cl}$. Specifically, $\mathcal{O}_{cl}$ can be defined by associating weights to tokens using a weighting scheme in the same fashion as $\mathcal{O}_{idf}$. A suitable weighting scheme to our context is the TF (Term Frequency), where the weight $tf\,(tk)$ of a token tk in a cluster $\mathcal{C}$ is directly given by the frequency of tk in $\mathcal{C}$. The intuition behind using TF-based ordering is to represent a cluster by its most frequent tokens. Such strategy requires the maintenance of a token-tf table for each cluster. Two sorting operations are needed after a merging between a probe set and one or more clusters: the first on the complete representation using the updated token-TF table and the second on the representative using IDF weights. Note that merging always occurs on the complete representation and the new representative is generated afterward.

Furthermore, except for $\mathcal{O}_{idf}$ and the more general definition in terms of weighted sets, this cluster representation is the same as the one used in [15]. A clear drawback of using TF is that frequent tokens in the collection, consequently with low IDF weights, tend to be also frequent within clusters. As opposed to rare tokens, such highly frequent tokens are often unimportant for similarity assessment, but can nevertheless appear in the representative owing to the TF-based ordering.

An alternative is to simply make $\mathcal{O}_{cl} = \mathcal{O}_{idf}$, i.e., the complete representation follows the same ordering of the input sets. A representative now is composed by tokens with the highest IDF values. This approach has a lower computational cost as compared to the previous representation: it does not require maintenance of an extra data structure such as the token-TF table, nor further sorting after merging clusters to probe sets. However, we now have the drawback that the representative may contain tokens that appear in only a few sets in the cluster.

Finally, we can avoid the issues of the previous strategies, while keeping their advantages, by using the TF-IDF weighting scheme: the weight $tf\text{-}idf\,(tk)$ of a token tk is given by $tf\text{-}idf\,(tk) = (1 + ln\,(tf\,(tk))) \cdot idf\,(tk)$. Henceforth, we refer to the proposed representation strategies by their adopted weighting scheme, i.e., as TF, IDF, and TF-IDF representations.

Example 3. Consider the sets (or cluster representatives) r and s in Example 1. After the union of r and s, the resulting token-TF and token-TF-IDF association tables are as follows (for simplicity, we have not taken the logarithm of TF in the latter table).

tk	A	B	C	D	E	F
$tf\,(tk)$	2	2	1	2	2	1
$tf\text{-}idf\,(tk)$	3	5	2	7	1	2

The complete representation and the corresponding cluster representative for strategies TF, IDF, and TF-IDF are shown in Table 1—for the complete representation using TF and TF-IDF, ties are broken using the IDF-based ordering.

Table 1. Cluster representations from Example 3 for strategies TF, IDF, and TF-IDF.

TF	Complete representation
	D B A E C F
	Representative
	D B C A E —
IDF	Complete representation
	D B C F A E
	Representative
	D B C F — —
TF-IDF	Complete representation
	D B A C F A
	Representative
	D B A C F —

5.2 Merging Strategies

We now discuss strategies for the merging phase. The output of the verification phase is the current probe set r and a set S of similar clusters. After generating a new (singleton) cluster C_r from the probe set, and before sending it to the indexing phase, there are three cases to consider.

(1) S is empty: the probe set is similar to no cluster and C_r goes directly to the indexing phase.
(2) S contains a single cluster: this single cluster is merged into C_r .
(3) S contains more than one cluster: we apply a merging strategy, considering that the elements in S were not identified as similar to one another in previous *SjClust* iterations.

The new cluster C_r is also sent to the Output Manager in cases (2) and (3), and in case (1) if singleton clusters are allowed to appear in the result.

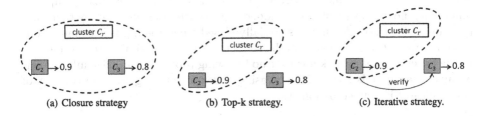

(a) Closure strategy   (b) Top-k strategy.   (c) Iterative strategy.

Fig. 3. Cluster merging strategies.

Now, we present three strategies for case (3) as depicted in Fig. 3.

– *Closure*: the simplest strategy is to merge all clusters in S into C_r (Figure 3(a)). This strategy corresponds to calculating the *transitive closure* of the similarity graph induced by input sets. The main problem of this strategy is that it tends to produce bigger clusters with several sets representing non-duplicates, thereby leading to poor precision in the results.

- *Top-K*: in this strategy, we first sort the elements in S according to their similarity to the probe set. Then, we take the K closest clusters and merge them into C_r (Figure 3(b)). An issue with this strategy is choosing the value of K: we can have the same issue of poor precision as Closure if K is too large; conversely, we can face the opposite problem if K is too small, i.e., smaller clusters are formed with duplicates in different clusters, thereby leading to poor recall.
- *Iterative*: this strategy is an specialization of Top-K, which aims at allowing the use of a small K value to maintain precision, while avoiding drop in recall. First, the K closest clusters are merged into C_r; afterward the algorithm proceeds iteratively, evaluating the similarity between C_r and the following $K+1, K+2, \ldots, |S|$ clusters, in decreasing order of similarity to the original probe set. If the similarity between the current cluster C_{K+i} and C_r is greater than the original threshold τ, then C_{K+i} is merged into C_r and the algorithm proceeds to the next representative; otherwise it stops. This strategy is similar in spirit to the merge-and-refine strategy to duplicate identification [4], which exploits the insight that a merging operation can lead to new matches.

A last point is that clusters are invalidated after being merged into another cluster, i.e., they are ignored in the subsequent processing. Since there are several references to clusters in the inverted lists and Output Manager, we need some *garbage collection* mechanism. To this end, we use a simple attribute in the cluster object to indicate whether it is valid or not. This attribute is checked every time a reference to a cluster is found in the candidate evaluation phase (see line 5 in Algorithm 1) and references to invalid ones are promptly discarded.

5.3   The Output Manager

At each *SjClust* iteration, a probe set is converted into a new cluster and indexed. Afterward, this cluster may progressively become part of bigger clusters, up to a point when no new element can be added to the current cluster; we say then that this cluster is closed. A cluster is trivially closed when the input is exhausted or when the weight of the current probe set is too large to be similar to this cluster. In the latter situation, we know that no following probe set can be similar either, because the input is sorted in increasing order of set weights. The concept of closed cluster is defined as follows.

Definition 6 (Closed Cluster). *Let C be a cluster and CR_C its representative; let r be the current probe set. C is a closed cluster if $w(CR_C) < min(r)^3$.*

---

[3] This definition can be made consistent when the input is exhausted by defining a conceptual probe set of infinite weight after the last input set.

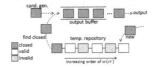

Fig. 4. The Output Manager.

The Output Manager is the *SjClust* component in charge of sending closed cluster to the output. The Output Manager is illustrated in Fig. 4. It contains two data structures: a temporary repository and an output buffer. Clusters generated in the merging phase are first stored in the temporary repository. It is a kind of priority queue, which maintains clusters sorted in increasing order of their representative weights. We use a simple but highly efficient implementation based on linked lists. Incoming clusters are usually larger than the majority of the stored ones, and, therefore, we only need to scan at most a few positions from the tail to the head of the list to find the point of insertion.

We search the temporary repository for closed clusters at end of each *SjClust* iteration using the $min(r)$ value of the current probe cluster. The search is performed from the head to the tail of the list. Closed clusters are sent to the output buffer as they are found and the search is stopped when the first non-closed cluster is met. Note that the temporary repository may also contain invalid clusters, e.g., clusters that were merged into other clusters. Entries to invalid clusters are removed both during insertion and search time and, thus, the size of the temporary repository is kept to a minimum.

The output buffer is a queue , which can be used to deliver clusters in a pipelined execution either in pull- or push-model. Besides from the temporary repository, the output buffer can also receive closed clusters from candidate evaluation phase. This occurs when references to closed clusters are found in the inverted lists (see line 6 in Algorithm 1; in this part, the algorithm is extended with a call to the Output Manager to send the closed cluster to the output buffer). References to closed clusters also need to be garbage collected. They are set as invalid after being sent to the output buffer, so their references can be removed afterward from inverted lists or the temporary repository.

6   Experimental Assessment and Analysis

We used publicly available datasets from the Stringer Project[4], which have been extensively used to evaluate duplicate identification algorithms [9]. Starting with a clean dataset as source, duplicates were generated by performing controlled transformations, such as character-level modifications (insertions, deletions, and substitutions), word swapping, and domain specific abbreviations, such replacing Incorporated with Inc. The "dirtiness level" of a generated dataset is determined by the percentage of duplicates to which transformations are applied (*erroneous duplicates*) and the extent of transformations applied to each erroneous duplicate (*errors in duplicates*).

---

[4] http://dblab.cs.toronto.edu/project/stringer/clustering/.

Table 2. Information about duplicate datasets.

Group	Name	Percentage of	
		Dirty duplicates	Errors in duplicates
High error	H1	90	30
	H2	50	30
Average error	M1	30	30
	M2	10	30
	M3	90	10
	M4	50	10
Low error	L1	30	10
	L2	10	10

In this evaluation, we used datasets generated from two sources: *Company*, containing company names and *DBLP* containing information about titles of computer science publications (dblp.uni-trier.de/xml). The average string length of Company is 21.03 characters, while of DBLP 33.55 characters. The parameters used in the generation processes are shown in Table 2; the percentage of token swap and abbreviations for all datasets were 20 % and 50 %, respectively. There are 8 datasets for each source, grouped according to their "dirtiness", i.e., high, average, and low error. Finally, the number of clusters in each dataset is 500.

We evaluate the quality of the results using two metrics: the pairwise F1 measure, denoted by *pF1*, and the closest cluster F1 measure, denoted by *ccF1*. The first metric is based on counting the number of pairs of duplicates correctly identified and is defined as follows [16]. Let G be set of ground truth clusters, i.e., the clusters whose duplicates have been all correctly identified, and D be the set of clusters returned by some cluster algorithm. Further, given a set of clusters P, let *pairs* (P) be a function that returns the set of distinct pairs of elements that are in the same cluster. For example, if $P = \{\langle a, b \rangle, \langle d, e, f \rangle\}$, then *pairs* $(P) = \{(a, b), (d, e), (d, f), (e, f)\}$. Thus, the pairwise precision and recall are defined as

$$pPr\,(G, D) = \frac{|pairs\,(G) \cap pairs\,(D)\,|}{pairs\,(D)}, \text{ and } pRe\,(G, D) = \frac{|pairs\,(G) \cap pairs\,(D)\,|}{pairs\,(G)}.$$

Therefore:

$$pF1\,(G, D) = \frac{2 \cdot pPr\,(G, D) \cdot pRe\,(G, D)}{pPr\,(G, D) + pRe\,(G, D)}.$$

The closest cluster F1 measure is based on summing up the pairwise Jaccard similarity of (unweighted) clusters. The corresponding pairwise precision and recall are defined as follows [4].

$$ccPr\,(G, D) = \frac{\sum_{d \in D} max_{g \in G} J\,(d, g)}{|D|}, \text{ and } ccRe\,(G, D) = \frac{\sum_{g \in G} max_{d \in D} J\,(g, d)}{|G|}.$$

Therefore:

$$ccF1\,(G, D) = \frac{2 \cdot ccPr\,(G, D) \cdot ccRe\,(G, D)}{ccPr\,(G, D) + ccRe\,(G, D)}.$$

We ran our experiments on an Intel Xeon E3-1240 Quad-core, 3,3 GHz, 8 MB CPU cache, and 8 GB of main memory. All algorithms were implemented using Java JDK 8 (Oracle). The processing cost of the algorithms is measured in average wall-clock time over repeated runs.

We implemented the set similarity join described in [18]; this implementation provided the basis for the own *SjClust* implementation as well as was used directly in the performance experiments. *SjClust* was set to not produce singletons. We used Jaccard as similarity function in all evaluations and used $K = 1$ for Top-K and Iterative merging strategies. We converted strings to upper-case letters, eliminated repeated white spaces, and generated the corresponding weighted token sets using q-grams of size 3 and IDF weighting scheme. We did not perform any further data pre-processing, such as removal of stop words.

In the experimental charts, the representation strategy TF-IDF is abbreviated to TI and the merging strategies Closure, Top-K, and Iterative are abbreviated to C, T, and I, respectively. The combination of representation and merging strategies is represented by their abbreviated form connected by a hyphen, i.e., TI-T represents the combination of TF-IDF and Top-K.

6.1 Accuracy Results

We first report accuracy results. We show the average accuracy value for each group of datasets. To calculate the average, we executed each algorithm 7 times with threshold value varying from 0.2 to 0.8 and took the best result value for each metric. As an important observation, the highest threshold leading to the best accuracy result was never higher than 0.3 on DBLP and lower than 0.4 in 67 % of the evaluations on Company.

Figure 5 shows the results on Company datasets. The F1 values are above 0.85 for all algorithms on lower error datasets (Fig. 5(a)). The strategies based on TF-IDF representation performed the best, while those based on IDF the worst. The IDF-based ordering on the complete representation shifts low-frequency

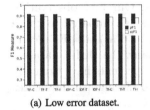
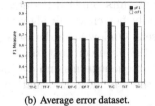
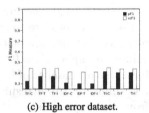

(a) Low error dataset.   (b) Average error dataset.   (c) High error dataset.

Fig. 5. Accuracy results on the Company dataset.

tokens to the cluster representative (recall to the discussion in Sect. 5.1). In this context, representatives contain tokens that appear in a few set elements in the corresponding cluster. As a result, objects representing duplicates, i.e., probe sets and representatives, cannot be easily identified because they have fewer tokens in common and, thus, lower similarity.

On the other hand, the explanation for the worse results of TF as compared to TF-IDF is that the Company dataset is characterized by the presence of high frequency and similar words. For example, the sub-string "Comp" appears in one third the strings. The token distribution contains a large number of high-frequency tokens accordingly, which are then shifted to the cluster representative in the TF strategy. Clusters representing distinct strings can then be evaluated as similar and erroneously merged due to common high-frequency tokens in their representatives, thereby hurting precision.

Differences in performance are less pronounced when comparing merging strategies. In general, the Closure strategy leads to slight better results when combined with IDF and TF-IDF, whereas Top-K and Iterative perform better with TF. Closure merges more clusters than the other strategies, and therefore, favors recall over precision. Hence, it tends to compensate the relative low-recall of IDF (and TF-IDF), while causing drop in precision when combined with TF. As expected, accuracy results degrade as we move to average error datasets (Fig. 5(b)). However, such accuracy degradation is only moderated on the algorithms based on TF and TF-IDF, which show better robustness to quality decrease of the underlying datasets. On the other hand, F1 values are less than 0.45 for all algorithms on high error datasets (Fig. 5(c)). Correctly identifying duplicates on such low-quality datasets is a challenge as some generated duplicates are hardly identified even by manual inspection.

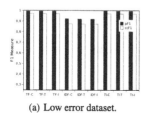 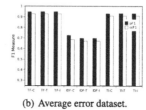 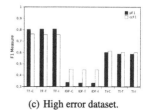

(a) Low error dataset.   (b) Average error dataset.   (c) High error dataset.

Fig. 6. Accuracy results on the DBLP dataset.

Figure 6 shows accuracy results on DBLP datasets. Our first observation is that F1 values are markedly better for all algorithms as compared to those obtained on Company datasets. On low error datasets (Fig. 6(a)), all pF1 (ccF1) values are higher than 0.92 (0.87). In contrast to Company, the DBLP datasets have a small number of high-frequent terms and distinct elements have typically low similarity. As a result, there is more information to distinguish duplicates from non-duplicates.

Algorithms using TF performed even better than those using TF-IDF on average and high error datasets (Figs. 6(b) and 6(c), respectively); particularly, pF1 values are still higher than 0.8 for these algorithms on the latter. In contrast to Company, the token distribution derived from DBLP is highly skewed, with a large number of rare tokens. As result, the problem of low recall caused by IDF-based ordering is exacerbated.

6.2   Performance Results

We now report and analyze performance results. For this experiment, we generated a dataset from DBLP containing 20k groups of 5 duplicates (totaling 100k strings). Besides using a larger dataset, we also increased the average string length to 69 characters by appending the corresponding author names to each publication title. We only used lower threshold values from 0.2 to 0.5, because we observed poor accuracy with higher values in the previous experiment.

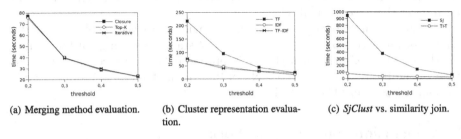

(a) Merging method evaluation.   (b) Cluster representation evaluation.   (c) *SjClust* vs. similarity join.

Fig. 7. Performance results on the DBLP dataset.

The results are show in Fig. 7. We first compared the performance of the merging strategies; we fixed the cluster representation to TF-IDF. There is no noticeable performance difference among them, as shown in Fig. 7(a). Even with the tendency to performing more merging, Closure exhibits nearly the same performance as compared to Top-K and Iterative. The underlying algorithm exploits token ordering to optimize the merging process and, therefore, the negative impact on performance is reduced. Also, closed inspection revealed that merging of more than 2 clusters is quite rare.

Further, we compared the three proposed representation strategies using Top-K as merging strategy. Figure 7(b) shows the results. Now, while there is relatively little difference between IDF and TF-IDF, TF is about 3× slower than them. Because TF uses more frequent tokens in the cluster representatives, the corresponding prefixes have more incidences of such tokens even with the posterior IDF-based ordering. As a result, token collisions in the prefixes of dissimilar objects are more frequent, reducing pruning power because those objects need to be verified.

Our next experiment compared *SjClust* with similarity join. Recall that for duplicate identification, similarity join is followed by clustering algorithm, which only starts after the similarity joins has completed. Hence, the results showed here for similarity join are only a (loose) lower bound for the scenario of sequential composition of similarity join and clustering. For *SjClust*, we used TF-IDF

Table 3. Input consumed before starting producing results.

Threshold	0.2	0.3	0.4	0.5
# of sets proc.	10080	479	105	51

and Top-K as representation and merging strategies, respectively. Figure 7(c) shows the results. Remarkably, *SjClust* is dramatically faster than similarity join. For the threshold value of 0.2, *SjClust* is 12× times faster. The reason is that prefix filtering is ineffective for low threshold values, which causes an explosion in the number of candidates and, consequently, in the number of similarity calculations. This limitation may prevent the use of similarity joins in duplicate identification in large datasets, because lower threshold are often required to obtain accurate results. In contrast, *SjClust* drastically reduces the number of similarity calculations by restricting them to cluster representatives, which are much fewer than the original sets.

Finally, we illustrate the non-blocking behavior of *SjClust*. Table 3 shows the number of input sets processed before *SjClust* starts producing cluster results. For threshold of 0.5, the first cluster is produced before processing less than 0.1 % of the input. Even for threshold of 0.2, the first result is produced after consuming less than 11 % of the input.

7   Conclusions and Future Work

In this paper, we presented *SjClust*, a framework to integrate clustering into set similarity join algorithms. We demonstrated the flexibility of *SjClust* in incorporating different clustering methods by proposing several cluster representation and merging strategies. *SjClust* is an order of magnitude faster than the original set similarity join algorithm for lower thresholds, which are often needed in practice to obtain accurate results in duplicate identification. Furthermore, our proposal produces results earlier, thereby avoiding blocking behavior. We described *SjClust* and its main components in detail and experimentally evaluated its accuracy and efficiency using different datasets. Future work is mainly oriented towards enriching our framework with advanced features such as *uncertain data management* (e.g., [13]), *adaptiveness* (e.g., [5]), and *execution time prediction* (e.g., [21]).

Acknowledgments. This research was partially supported by the Brazilian agencies CNPq and CAPES.

References

1. Altwaijry, H., Kalashnikov, D.V., Mehrotra, S.: Query-driven approach to entity resolution. PVLDB **6**(14), 1846–1857 (2013)
2. Altwaijry, H., Mehrotra, S., Kalashnikov, D.V.: Query: a framework for integrating entity resolution with query processing. PVLDB **9**(3), 120–131 (2015)

3. Bayardo, R.J., Ma, Y., Srikant, R.: Scaling up all pairs similarity search. In: Proceedings of the WWW Conference, pp. 131–140 (2007)
4. Benjelloun, O., Garcia-Molina, H., Menestrina, D., Qi, S., Whang, S.E., Widom, S.: Swoosh: a generic approach to entity resolution. VLDB J. **18**(1), 255–276 (2009)
5. Cannataro, M., Cuzzocrea, A., Mastroianni, C., Ortale, R., Pugliese, A.: Modeling adaptive hypermedia with an object-oriented approach and xml. In: WebDyn 2002 (2002)
6. Chaudhuri, S., Ganti, V., Kaushik, R.: A primitive operator for similarity joins in data cleaning. In: Proceedings of the 22nd International Conference on Data Engineering, p. 5 (2006)
7. Doan, A., Halevy, A.Y., Ives, Z.G.: Principles of Data Integration. Morgan Kaufmann, Burlington (2012)
8. Elmagarmid, A.K., Ipeirotis, P.G., Verykios, V.S.: Duplicate record detection: a survey. TKDE **19**(1), 1–16 (2007)
9. Hassanzadeh, O., Chiang, F., Miller, R.J., Lee, H.C.: Framework for evaluating clustering algorithms in duplicate detection. PVLDB **2**(1), 1282–1293 (2009)
10. Idreos, S., Papaemmanouil, O., Chaudhuri, S.: Overview of data exploration techniques. In: Proceedings of the SIGMOD Conference, pp. 277–281 (2015)
11. Kazimianec, M., Augsten, N.: PG-Skip: proximity graph based clustering of long strings. In: Yu, J.X., Kim, M.H., Unland, R. (eds.) DASFAA 2011, Part II. LNCS, vol. 6588, pp. 31–46. Springer, Heidelberg (2011)
12. Koudas, N., Sarawagi, S., Srivastava, D., Record linkage: similarity measures and algorithms. In: Proceedings of the SIGMOD Conference, pp. 802–803 (2006)
13. Leung, C.K.-S., Cuzzocrea, A., Jiang, F.: Discovering frequent patterns from uncertain data streams with time-fading and landmark models. In: Küng, J., Wagner, R., Cuzzocrea, A., Dayal, U., Hameurlain, A. (eds.) TLDKS VIII. LNCS, vol. 7790, pp. 174–196. Springer, Heidelberg (2013)
14. Liu, H., Ashwin Kumar, T.K, Thomas, J.P.: Cleaning framework for big data - object identification and linkage. In: Proceedings of the Big Data Congress, pp. 215–221 (2015)
15. Mazeika, A., Böhlen, M.H.: Cleansing databases of misspelled proper nouns. In: Proceedings of the VLDB Workshop on Clean Databases (2006)
16. Menestrina, D., Whang, S., Garcia-Molina, H.: Evaluating entity resolution results. PVLDB **3**(1), 208–219 (2010)
17. Ribeiro, L.A., Cuzzocrea, A., Bezerra, K.A.A., do Nascimento, B.H.B.: SJClust: Towards a framework for integrating similarity join algorithms and clustering. In: Proceedings of the ICEIS Conference, pp. 75–80 (2016)
18. Ribeiro, L.A., Härder, T.: Generalizing prefix filtering to improve set similarity joins. Inf. Syst. **36**(1), 62–78 (2011)
19. Sarawagi, S., Kirpal, A.: Efficient set joins on similarity predicates. In Proceedings of the SIGMOD Conference, pp. 743–754 (2004)
20. Schneider, N.C., Ribeiro, L.A., de Souza, A., Inácio, H.M., Wagner, A., von Wangenheim. SimDataMapper: An architectural pattern to integrate declarative similarity matching into database applications. In: Proceedings of the SBBD Conference, pp. 967–972 (2015)
21. Sidney, C.F., Mendes, D.S., Ribeiro, L.A., Härder, T.: Performance prediction for set similarity joins. In: Proceedings of the SAC Conference, pp. 967–972 (2015)
22. Tang, N.: Big RDF data cleaning. In: Proceedings of the ICDE Conference Workshops, pp. 77–79 2015)

23. Wang, J., Li, G., Feng, J.: Can we beat the prefix filtering?: an adaptive framework for similarity join and search. In: Proceedings of the SIGMOD Conference, pp. 85–96 (2012)
24. Xiao, C., Wang, W., Lin, X., Yu, J.X., Wang, G.: Efficient similarity joins for near-duplicate detection. TODS **36**(3), 15 (2011)
25. Zhang, F., Xue, H.-F., Xu, D.-S., Zhang, Y.-H., You, F.: Big data cleaning algorithms in cloud computing. iJOE **9**(3), 77–81 (2013)

Distributed and Big Data Processing

"Overloaded!" — A Model-Based Approach to Database Stress Testing

Jorge Augusto Meira[1,2(✉)], Eduardo Cunha de Almeida[2], Dongsun Kim[1],
Edson Ramiro Lucas Filho[2], and Yves Le Traon[1]

[1] SnT Research Center, University of Luxembourg, Luxembourg City, Luxembourg
{jorge.meira,dongsun.kim,yves.letraon}@uni.lu
[2] C3SL - Federal University of Paraná, Curitiba, Brazil
{jmeira,eduardo,erlfilho}@inf.ufpr.br

Abstract. As a new era of "Big Data" comes, contemporary database management systems (DBMS) introduced new functions to satisfy new requirements for big volume and velocity applications. Although the development agenda goes at full pace, the current testing agenda does not keep up, especially to validate non-functional requirements, such as: performance and scalability. The testing approaches strongly rely on the combination of unit testing tools and benchmarks. There is still a testing methodology missing, in which testers can model the runtime environment of the DBMS under test, defining the testing goals and the harness support for executing test cases. The major contribution of this paper is the MoDaST (**Mo**del-based **Da**tabase **S**tress **T**esting) approach that leverages a state transition model to reproduce a runtime DBMS with dynamically shifting workload volumes and velocity. Each state in the model represents the possible running states of the DBMS. Therefore, testers can define state goals or specific state transitions that revealed bugs. Testers can also use MoDaST to pinpoint the conditions of performance loss and thrashing states. We put MoDaST to practical application testing two popular DBMS: PostgreSQL and VoltDB. The results show that MoDaST can reach portions of source code that are only possible with non-functional testing. Among the defects revealed by MoDaST, when increasing the code coverage, we highlight a defect confirmed by the developers of VoltDB as a major bug and promptly fixed.

1 Introduction

Scalable and high performance data processing is one of the key aspects for successful business operations as the volume of incoming transactions is getting larger for most application areas. Over the last 40 years traditional "one-size-fits-all" Database Management Systems (DBMS), such as DB2, Oracle, PostgreSQL, have been successful in processing transactions. However, the recent growth of the transaction workload (e.g., Internet, Cloud computing, Big Data) is challenging these DBMS requiring revisiting their kernel. Even new DBMS are being designed ground up to better tackle these workloads.

© Springer International Publishing Switzerland 2016
S. Hartmann and H. Ma (Eds.): DEXA 2016, Part I, LNCS 9827, pp. 207–222, 2016.
DOI: 10.1007/978-3-319-44403-1_13

Although the development agenda goes at full pace, with the recent appearance of a great deal of new DBMS, the testing agenda does not keep up, specially for validating non-functional requirements, such as performance, robustness and scalability. With the increasing demand in volume and velocity of transactions, many load conditions challenge the DBMS in unexpected ways that the state of the art in testing tools cannot exercise. Different bugs can be found in the literature describing that the root cause is linked to different conditions of transient load shifts or sudden spikes. The result in the DBMS can be treacherous leading to a number of non-functional failures, such as: poor performance query plans [1], backpressure[1], lock escalation (for lock-based mode) [2], poor performance estimations [3], performance degraded mode and load shedding [4]. Many of these failures are also called "Heisenbugs" [5], because the root cause are not easy to detect and may elude the bug-catcher for years of execution.

1.1 Motivation

The bug-catching task becomes even harder when the existing testing methodologies for transaction processing only validate functional requirements [6,7]. The validation of non-functional requirements is still an open issue and strongly rely on the combination of unit testing tools[2] (e.g., Jepsen, JUnit, Jmeter, PeerUnit [8]) and benchmarks to reproduce specific workloads (e.g., TPC-like, YCSB).

The main problem with this combination is that it is strictly based on tools and does not adhere to a general methodological testing approach. In general, this combination has to be conservative to eke out the "ideal" testing environment: test cases mimic any benchmark workload and then execute on top of an unit testing tool. However, the expected environment grounds testing with a proper methodological approach to define the testing goals and the harness support for executing test cases. Writing and executing test cases come later.

The major contribution of this paper is such a methodological approach that can eventually be implemented on top of any unit testing tool with the benchmark of your choice. Figure 1 shows the impact on PostgreSQL of executing test cases with and without a testing methodology. The impact is measured by the code coverage ratio of our methodological approach and the same test case reproducing the TPC-C benchmark workload on top of a unit test tool without following any testing methodology. First, we see that the impact of shifting the transaction load in PostgreSQL can only be analyzed when testing is driven by a methodological approach. Second, we notice that the load shifting exercises PostgresSQL in different code portions. More interestingly, when the DBMS is upon heavy loads (rightmost bar), the throughput goes down, but exercising almost 60 % of the source code of the kernel (12 % more than the steady condition).

This result shows that even the kernel of a mature DBMS, such as PostgreSQL, is not acquainted to non-functional testing, which would reveal the

---

[1] https://voltdb.com/docs/UsingVoltDB/DesignAppErrHandling.php.

[2] VoltDB testing: https://voltdb.com/blog/how-we-test-voltdb.

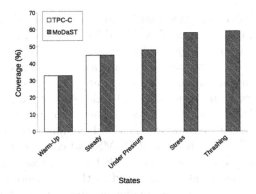

Fig. 1. Example of inherent limitations of the existing testing tools.

bugs that we discuss in this paper (see Fig. 2). To come up with a general non-functional testing approach for transaction processing in the Big Data era, it is important to define a running model of the Database System Under Test (DUT) that allow reproducing and harnessing load shifts.

1.2 Contribution

This paper presents, MoDaST (**Mo**del-based **Da**tabase **S**tress **T**esting), a novel methodological approach to DBMS stress testing. This approach focuses on testing scalability and performance of DBMS with dynamically changing load levels by using a test model for database systems. The approach leverages a state machine model with observable runtime states: warm-up, steady, under-pressure, stress, and thrashing. The model allows us to infer and explore internal states of the DUT even if black-box testing is only available. The observable states can basically be used for guiding the testing goals with test cases forcing the state transitions. More importantly, MoDaST allows reproducing the state transitions for regression or to figure out what is the exact condition that revealed a bug.

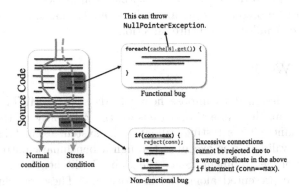

Fig. 2. Conceptual execution paths under normal and stress conditions.

To evaluate MoDaST and show that it can be put to practical application, we applied it to two different classes of real-world DBMS (i.e., SQL and NewSQL) collected from open source projects: PostgreSQL and VoltDB. We designed a distributed stress testing environment by using a cluster facility and a distributed testing driver to shift the submission volume and velocity of the workload. We collected performance monitoring data and code coverage to figure out whether there is any potential defect. We also conducted a comparative study between MoDaST and unit testing running the TPC-C benchmark to find out which one can correctly test different behaviors of DBMS and cover more source code.

The results of the experiments showed that our approach successfully drove the DUTs into the different states specified in the test model. MoDaST found out that DUTs actually follows the model by observing performance data. In addition, the results revealed that our approach explored different performance behaviors and increased test coverage up to 20 % in certain code packages of PostgreSQL and 12 % for VoltDB compared to the baseline technique. Newly covered lines by MoDaST exposed three new bugs. In particular, one of the bugs had significant impact on VoltDB by affecting not just non-functional but also functional requirements upon heavy load conditions. This bug was confirmed and promptly fixed by the VoltDB hackers after our reporting[3].

Overall, this paper makes the following contributions: **1- Database state model**: We designed a running model to infer the internal states of DBMS based on performance observations. Among different possible states, our model detects performance loss and thrashing states at runtime. **2- MoDaST, a model-based DB stress testing approach**: We introduce a novel testing approach to force state transitions in the model by shifting the transaction loads. The state transitions allow exercising different source code portions of the DBMS that would never be exercised by single test cases of unit testing tools. **3- Empirical evaluation**: We present empirical evaluation results by applying MoDaST to popular open-source DBMS. Based on the evaluation results, we identified and reported potential bugs. One of them was confirmed as "major bug" and promptly fixed by the core developers.

The remainder of this paper is organized as follows. Section 2 we discuss the related work. Section 3 describes our model-based approach to database stress testing. Section 4 we present empirical results stress testing two popular DBMS. Finally, Sect. 5 concludes with future directions.

2   Related Work

Stress testing is designed to impose heavy loads such as HTTP requests or database queries at the same time to ensure the reliability of the system In DBMS, performance/stress testing validates the system from different angles. Commonly, this validation is executed through a benchmark pattern to reproduce a production environment. Since the DebitCredit benchmark [9], several benchmarks were presented along the last decades. These benchmarks focus

---

[3] https://issues.voltdb.com/browse/ENG-6881.

on comparing metrics (e.g., response time, throughput, and resource consumption) [3]. The TPC-like benchmarks offer different workload levels to evaluate databases from two perspectives: OLTP or OLAP. In contrast, the Yahoo Cloud Serving Benchmark (YCSB) [10] is designed to evaluate four specific features of distributed databases: Performance, Scalability, Availability and Replication. There are another type of benchmarks focusing on DBMS availability: R-cubed [11], DBench-OLTP [12], Under Pressure Benchmark [13].

Some of existing performance testing tools attempt to test database systems under different levels of workload. Jepsen[4], Hammerora[5], AppPerfect[6] and Oracle Application Testing Suite[7] provide a test driver to build up test cases on top of TPC-like benchmarks. Agenda [7] provides its own methodology and test driver, but this tool can only generate functional test cases. JMeter is also a well known and widely applied load testing tool for different applications, including DBMS. But, it was designed to load test functional requirements.

The main disadvantage of these tools is the lack of a high level testing methodology, like MoDaST. In software testing, the testing methodology is the foundation over which the tools are used [14]. Otherwise, test cases will be narrowed to reproduce specific load conditions that cannot reflect a far more aggressive real-world production environment with load spikes and shifts after a while in steady condition state [1,4]. In addition, these tools cannot correlate performance loss and related defects to specific its root causes.

Finally, techniques to generate test cases can be used to boost testing results of MoDaST for specific testing goals. For instance, [15,16] presents a technique to generate queries with cardinality constraints for validating multidimensional histograms A complementary technique to generate test databases is presented in [17]. Although MoDaST is a testing model, rather than a data/query generation tool, it was built for validating write-mostly, while the mentioned techniques are meant to read-mostly database systems assessments.

3   Approach: MoDaST

This section describes our Model-based Database Stress Testing (MoDaST) approach. Figure 3 shows an overview of this approach. MoDaST consists of the Database State Machine (DSM) and a test driver. The DSM represents a set of observable states of a DBMS and its transition function. The test driver defines the load model of each state and commences performance testing by giving a specific load to the DUT. Then, the driver observes the current performance data of the DUT and figures out state transitions by giving the data to DSM. The remainder of this section details MoDaST.

---

[4] https://aphyr.com/tags/jepsen.

[5] http://hammerora.sourceforge.net/.

[6] http://www.appperfect.com/.

[7] http://www.oracle.com/technetwork/oem/app-test/index.html.

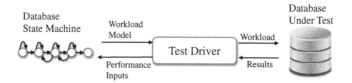

Fig. 3. Architectural overview of MoDaST.

3.1   The Database State Machine (DSM)

The DSM models how a DUT behaves at given workload levels. In particular, DSM focuses on representing observable states of a DUT with respect to performance (i.e., performance behaviors). The behaviors of a DUT can be represented by the following states: Warm-up (s_1), Steady (s_2), Under Pressure (s_3), Stress (s_4), Thrashing (s_5). We formally define the DSM and its corresponding states in Definition 1. Figure 4 depicts the DSM, the running states and transitions.

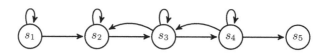

Fig. 4. The Database State Machine (DSM) and the observable states.

Definition 1. *The Database State Machine (DSM) denoted as T, is a 5-tuple* $(\mathcal{S}, s_1, \mathcal{F}, \beta, \tau)$ *where:*

- $\mathcal{S} = \{s_1, s_2, s_3, s_4, s_5\}$ *is a set of states,*
- $s_1 \in \mathcal{S}$ *is the initial state,*
- $\mathcal{F} \subset \mathcal{S}$ *is the set of final states, where* $\mathcal{F} = \{s_5\}$ *in DSM,*
- β *is the set of performance input defined by Definition 2,*
- τ *a state transition function defined by Definition 6.*

The detailed information about every state is available in Sect. 3.1. To describe each state in detail, it is necessary to define the performance input, β. Based on the performance input, the DSM determines state transitions, τ.

Performance Input: DSM takes three different performance input from a DUT to infer its current internal state. The input, β, is the set of (1) the performance variation, (2) the transaction efficiency, and (3) the performance trend.

Definition 2. *The* **performance Input**, *denoted by* β, *is a tuple of three performance variables:* $\beta = < \Delta, \delta, \varphi_{\delta}$, *where* Δ *is the performance variation (Definition 3),* δ *is the transaction efficiency (Definition 4), and* φ *is the performance trend (Definition 5), respectively.*

The **performance variation** represents the stability of transactions treated by a DUT. This is denoted by Δ as shown in Definition 3. MoDaST makes n observations ($n > 1$) and computes the number of transactions treated per second (y) for each observation. For example, if $\Delta \to 0$, the DUT is processing a steady number of incoming transactions.

Definition 3. *The* **performance variation**, Δ, *is the dispersion of the number of treated transactions per second and formally defined as:*

$$\Delta = \sqrt{\frac{1}{n-1} \sum_{i=1}^{n} (y_i - \mu)^2}, \tag{1}$$

where $\mu = \frac{1}{n} \sum_{i=1}^{n} (y_i)$

The **transaction efficiency**, δ, is the proportion between the number of transactions treated by a DUT and requested by clients. This enables to define the upper bound number of transactions in concurrent execution with steady behavior. For example, if $\delta \to 1$ across a number of performance observations, the DUT is successfully treating most of transactions requested by clients.

Definition 4. *The* **transaction efficiency**, *denoted by* δ, *is the proportion of the transactions treated per second* (y) *by the number of transactions requested per second* (z):

$$\delta = \frac{y}{z} \tag{2}$$

The **performance trend**, φ, is a metric explaining the expected performance slope of a DUT within a certain size of a sliding window as described in Definition 5. As shown in Fig. 5, φ can be computed by the distance between the

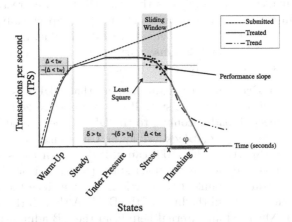

Fig. 5. DSM and its performance input. The X-axis is time in seconds and the Y-axis represents transactions per second. This shows relationships between performance input and states in DSM.

current time (observation) and expected time when the transaction efficiency of the DUT converges to 0 (i.e., $\delta = y/z = 0$, where $z \neq 0$). Section 3.2 describes how to compute the distance in detail.

Definition 5. *The* **performance trend** *is a function defined as*

$$\varphi = x' - x \tag{3}$$

where x is the current time and x' represents the point that the tangent line crosses the time axis.

States: The DSM models, with a state machine, the database performance states found in the literature in which non-functional bugs were reported [1–5]. The following paragraphs describe each state in detail.

State 1 — Warm-up: This state is the startup process of the DUT. In this state, the DUT initializes internal services such as transaction management service. Although some transactions can be treated during this state, performance is not stable since the DUT focuses on filling memory caches. DSM defines the Warm-up state by using performance variations (Δ in Definition 3).

The DSM infers that a DUT is in the *Warm-up* state if Δ is not converging to 0 after the startup of the DUT. In other words, $\neg(\Delta < t_w)$, where t_w is the warm-up threshold value. Otherwise (i.e., $\Delta < t_w$ holds), the transition to the next state (i.e., **Steady**) is triggered. Each DBMS has a unique t_w value. Section 4 explains how to determine the value.

State 2 — Steady: The DSM infers this state if the performance variation, Δ, is converging to 0. Once the DUT is in this state, it never comes back to the *Warm-up* state again since all the internal services are already initialized and running. In addition, the memory cache of the DUT is filled to provide the expected performance, which indicates that the DUT can correctly treat most of incoming transaction requested by clients in time. Specifically, this can be represented as $\delta > t_s$, where t_s is the steady threshold value. Each DBMS has a different value for the threshold that may vary based on the type of the expected workload and the available hardware environment.

State 3 — Under Pressure: This state implies that a DUT is on the limit of performance. The DUT goes to the state if δ approaches to zero, which means that a set of unexpected load is coming to the DUT. Unexpected loads include shifts and sudden spikes (e.g., Black Friday or Christmas) that affect performance [1,4,13]. In this state, the DUT can still deal with the similar amount of transactions processed in the previous state (*Steady*). However, it cannot properly treat a certain amount of transactions in time since the total amount requested by clients is beyond the limit of the DUT. Although this situation can be transient, it might need an external help from the DB administrator (DBA) to go back to *Steady*. For example, DBA can scale up the DUT's capability or set up the DBMS to reject a certain amount of the incoming transactions until the load decreases to an acceptable amount (i.e., $z \to y$ and $\delta > t_s$).

State 4 — Stress: a DUT goes into this state when the number of transactions requested by clients is beyond the performance limit. This state is different from the *Under Pressure* state since the performance variation (i.e., Δ) increases. The DUT in this state is highly vulnerable to crash if no external help is available. For example, the DBA should consider additional solutions such as adopting database replica, adding more cluster machines, or killing long running transactions (normally triggered by bulk loads). If an appropriate solution is performed, the DUT can go back to the *Under Pressure* state and $\Delta < t_{st}$, where t_{st} is the stress threshold value.

State 5 — Thrashing: This state represents that the DUT uses a large amount of computing resources for a minimum number of transactions. The DUT experiences resource contention and cannot deal with any new transaction in this state. In this state, it is no longer possible to come back to the previous one as any external intervention is useless. The DSM detects the transition to the *Thrashing* state if $\varphi < t_{th}$, where t_{th} is the thrashing threshold value. Predicting the thrashing state is explained in Sect. 3.2.

State Transitions: The state transition function, τ, determines whether the DUT changes its internal state based on observed performance data. This function takes *performance input* ($< \Delta, \delta, \varphi >$) from the test driver and gives the next state $s \in S$ as described in Definition 6. In each state, the DSM examines the current values of performance input and compares the values with threshold values[8] (i.e., t_w, t_s and t_{st}). Table 1 summarizes the threshold values.

Definition 6. *The state transition function, τ, is defined as:*

$$\tau : \mathcal{S} \times \beta \to \mathcal{S} \tag{4}$$

where $\forall s \in \mathcal{S}, \exists p \in \beta$ *and* $\exists s' \in \mathcal{S} | (s, p) \to (s')$.

Table 1. Threshold values for state transitions.

States	Target state				
	s_1	s_2	s_3	s_4	s_5
s_1	$\neg(\Delta < t_w)$	$\Delta < t_w$	-	-	-
s_2	-	$\delta > t_s$	$\neg(\delta > t_s)$	-	-
s_3	-	$\delta > t_s$	$\neg(\delta > t_s)$	$\Delta > t_{st}$	-
s_4	-	-	$\neg(\Delta > t_{st})$	$\Delta > t_{st}$	$\varphi < t_{th}$
s_5	-	-	-	-	$\varphi = 0$

---

[8] The values used in the experiments are specified in the Sect. 4 since it is variable depending on the DUT.

3.2 Predicting the Thrashing State

In addition to stress testing, MoDaST can predict crashes before a DUT actually goes into the *Thrashing* state. This indicates the time remaining until *out-of-service* and allows DBA to react before service failure.

The first step to predict the *Thrashing* state is computing the performance slope. MoDaST uses the Least Squares method [18] that approximates the relationship between independent (x) and dependent (y) variables in the form of $y = f(x)$. The testing time in seconds is denoted by x and y denotes the corresponding throughput at time x. It allows the computation of the three required coefficients (i.e., a_0, a_1 and a_2) for the quadratic function (i.e., performance slope): $f(x) = a_0 x^2 + a_1 x + a_2$

Our approach computes the coefficients by using recent p observations[9] of (x, y) (i.e., sliding window). The quadratic function estimates the performance slope as shown in Fig. 5. Once the performance slope is identified it is possible to calculate the derivative $f'(x)$ (i.e., tangent line), considering the current observation x_i. By using the tangent projection in the axis x, MoDaST can estimate the performance trend, φ, according to Definition 5. If the value is converging to the thrashing threshold (t_{th}), we assume that DUT may crash at any moment (i.e., transition from the stress to thrashing state).

3.3 The Test Driver

The goal of the test driver is to generate different load conditions and collect the performance input for the DSM. The test driver is built on top of the PeerUnit distributed testing framework [8]. PeerUnit allows building, coordinating and executing distributed test cases, which are key features for stress testing.

Since the performance of a DBMS can be affected by both the number of connections and transactions, it is necessary to test both the connection and transaction management modules by using two workload cases as follows: **Case #1**: The goal of this case is to submit a heavy load to the **connection** module of the DUT. The number of connections is gradually increased for each step. In this case, the driver submits only one transaction per connection; **Case #2**: The goal of this case is to submit a heavy load to the **transaction** module of the DUT instead of the connection module. The number of transactions is gradually increased for each step. In this case, the driver submits an increasing number of transactions per connection (i.e., fixed number of connections).

4 Empirical Evaluation

We applied our approach to two DBMS running TPC-C: VoltDB 4.5 and PostgreSQL 9.3. These subjects are selected for several reasons. First, both of them are ACID open-source RDBMS. In addition, they have representative characteristics of each category: PostgreSQL is a centralized disk-oriented DBMS and

---

[9] p is defined by the least squares correlation coefficient [18].

VoltDB is a distributed in-memory DBMS. The experiment procedure has four steps: (1) Submit the load condition, (2) Analyze the execution, (3) Collect the code coverage data, and (4) Proceed to the comparative study.

The experiments are executed on a HPC platform [19]. We used two different configurations: (1) 11 machines for PostgreSQL (one DBMS server and ten testers) and (2) 13 machines for VoltDB (three DBMS server and ten testers). Each machine has dual Xeon X5675@3.07 GHz with 48 GB of RAM running Debian GNU/Linux and connected by the Infiniband QDR (40 Gb/s) network. Our approach is implemented in Java 7. To collect the code coverage information of PostgreSQL, we used the GNU/Gcov, which is supported by default by the DBMS. For VoltDB, the code coverage is measured by Eclemma JaCoCo, since the DBMS is implemented in Java.

The threshold values are specified in Table 2. They were set based on the available hardware, the workload of our choice and the architecture of the DBMS. For instance, VoltDB does not need threshold values for the warm-up and thrashing states. Since VoltDB is an in-memory database, the warm-up process is basically instant. The thrashing state was not observed on the VoltDB. The t_s threshold is limited by 90 % of the transaction rate acceptance and the t_{st} is limited by 10 % of the transaction acceptance rate compared to the previous state "tps_{up}" (i.e., *Under Pressure*). For the t_{th}, we used one second. The slide window is set to 60 observations (i.e., $p = 60$).

Table 2. Threshold values for the state transitions. VoltDB does not need values for the warm-up and thrashing states since this DBMS does not experience these states.

	PostgreSQL	VoltDB
t_w	0.1	–
t_s	0.9	0.9
t_{st}	0.1 * tps_{up}	0.1* tps_{up}
t_{th}	1	–

The remainder of this section is guided by four research questions that are, respectively, related to: 4.1 performance results, 4.2 code coverage, 4.3 defects, and 4.4 thrashing prediction.

4.1 Does DSM Properly Reflect Performance Behaviors of a DUT?

This is the baseline question since our approach assumes that the DUT follows the DSM as designed. PostgreSQL experienced all the states of DSM as shown in Fig. 6. It presented an unexpected behavior concerning the ability to maintain a stable performance. During the execution of the workload case #1, the test driver increased the number of connections sequentially as described in Sect. 3.3. According to the specification of PostgreSQL, it can process 2,000 concurrent connections (i.e., defined by the MAX_CONNECTION configuration). However,

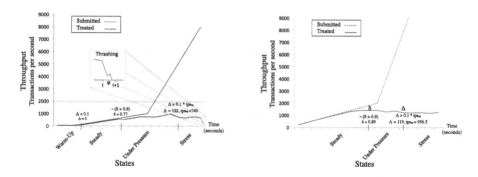

Fig. 6. Performance results of PostgreSQL.   **Fig. 7.** Performance results of VoltDB.

the DUT could not deal with any workload greater than 1,000 concurrent connections as shown in Fig. 6[10]. For the workload case #2, the test driver increased the number of transactions with a fixed number of connections. PostgreSQL's behavior was more stable in this case and did not reach the thrashing state. However, it stayed either in the under pressure or stress states.

VoltDB presented consistent results in terms of throughput stability. Thus, the DUT was mostly either in the steady or under pressure states for both workload cases (see Fig. 7). However, the connection module was forced into stress state triggering a backpressure condition when applying the workload case #1. This condition occurs when a burst of incoming transactions was sustained for a certain period of time. This period must be sufficient to make the planner queue full. More information about this condition will be described in Sect. 4.3.

A curious reader may ask what would happen if instead of using MoDaST, we execute stress testing with a combination of standard benchmark on top of a popular testing tool, like jepsen or jmeter. We call this combination as baseline approach. By definition, this baseline approach considers performance constraints to ensure the DBMS on the steady state during measurement time. For example, one of the constraints defined by TPC-C as "Response Time" is: "At least 90 % of all transactions of each type must have a Transaction RT less than 5 s...". Thus, it is not possible to explore any stress condition of the DUT, since it never reached the under pressure nor the stress states. In both workload cases, the baseline approach only contemplates the steady state.

Answer: *MoDaST drove a DUT into each state of the DSM while the baseline technique can only explore two initial states.*

4.2   How Much Does Our Approach Cover the Source Code of a DUT (i.e., Code Coverage)?

This question is about our assumption that some execution paths in DBMS source code can only be explored when a certain amount of workload is requested.

---

[10] The thrashing state is only observable in the workload case #1.

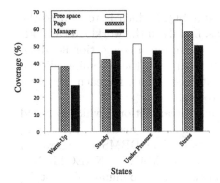
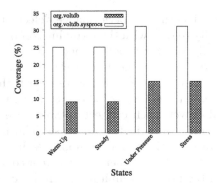

Fig. 8. Code coverage results of Post-greSQL. This focuses on three major modules: Free Space, Page, and Manager.

Fig. 9. Code coverage results of VoltDB. These packages are related to the concurrency control and server management

Figure 8 shows the code coverage results of PostgreSQL. Three packages presented a significant impact on the following modules: (i) *Freespace* that implements the seek for free space in disk pages; (ii) *Page* that initializes pages in the buffer pool; and (iii) *Manager* that implements the shared-row-lock. The coverage increased mainly during the two last states: *stress* and *thrashing*. It occurs because those packages are responsible for managing disk page allocation and the transaction lock mechanism. PostgreSQL needed to execute functions dedicated to stress conditions to allocate extra resources and to deal with the high concurrency when the number of transactions increases.

VoltDB showed a notable improvement of code coverage results as shown in Fig. 9, even though it was not significant compared to that of PostgreSQL. We observed the improvement in two packages: *org.voltdb* and *org.voltdb.sysproc*. These packages manage the maximum number of connections and concurrent transactions. The package "org.voltdb.sysproc" is related to the basic management information about the cluster. The above-mentioned VoltDB classes were not covered when applying the baseline approach. Basically, the warm-up and steady states did not generate any concurrent condition. We see the same result for PostgreSQL testing.

Answer: *MoDaST allows to explore a larger part of source code of DUTs than the baseline technique since a certain part of source code can be executed only when pushing the DUT to heavy loads.*

4.3 Does Our Approach Find Bugs?

This question is correlated to the previous one; if MoDaST can explore more lines of code by submitting different load conditions, we may find new defects located in functions dealing with stress conditions. During the experiments, we found two potential defects (one from PostgreSQL and one from VoltDB) and one new unreported major bug (VoltDB).

We identified a performance defect of PostgreSQL, which is related to the inability to deal with the incoming connections, mainly in the workload case #1. Actually, the defect can be triggered either by transaction or connection flooding. PostgreSQL implements a backend process to deal with the incoming clients. Each client keeps one connection with the database. Thus, for each incoming connection, the DUT starts a new backend process.

Moreover, each connection holds one or more transactions, which proceed to modifications in the database. The modifications are made by *insert* and *update* operations that compose each transaction. The DBMS configuration allows to set the number of concurrent connections (i.e., MAX_CONNECTIONS) up to the resources limitations. In our experiments, the maximum value set for MAX_CONNECTIONS was 2,000. Despite of the limit, PostgreSQL was not able to reach the number of 2,000 open connections at any time. As the number of connections/transactions increases, PostgreSQL spends most of the computational power dealing with the table locks instead of creating new backend processes. From the testing point of view, we consider a potential *defect*.

VoltDB experienced a backpressure condition by applying the workload case #2. The increasing number of submitted transactions, via JDBC interface, fulfills the planner queue limit (i.e., 250) and raised up the message below: (**GRACE-FUL_FAILURE**): '**Ad Hoc Planner task queue is full. Try again.**' This can be considered a potential defect[11], once the planner queue is waiting for VoltDB planner. The planner became full and started to reject new operations.

The code coverage also enabled to reveal a functional bug. Figure 10 shows the code fragment where the defect was identified (Line 426). We reported this bug[12] to the developer community of VoltDB. This branch of code is responsible for ensuring that the DUT does not accept more concurrent connections than the maximum constraint allowed by the server resources. The bug rose up when our approach led the DUT to the stress state, which exposed it to a race condition in the connection module. The solution for this bug is to ensure that, even during race conditions, the number of concurrent connections never goes beyond the limit. Basically it should be guaranteed in the condition statement (i.e., IF) by replacing "==" by ">=". VoltDB developers created a bug report[13] as a major bug and promptly fixed after our reporting.

Answer: *The MoDaST found and reproduced three potential defects and one of them is confirmed as a major bug by the developers of VoltDB.*

```
    ...
426   if (m_numConnections.get() == MAX_CONNECTIONS.get()) {
427     networkLog.warn(''Rejected connection from '' +
    ...
```

Fig. 10. Example of bug only identified under stress conditions.

---

[11] http://zip.net/bmps8J.

[12] http://zip.net/byptRy.

[13] https://issues.voltdb.com/browse/ENG-6881.

4.4 Can Our Approach Predict Performance Degradation (e.g., the Thrashing State)?

This question is necessary because performance prediction is one of the advantages when using MoDaST. If the prediction is available, the DBA can apply several solutions for preventing DBMS crashes. Our approach could predict the thrashing states of PostgreSQL (see Fig. 6). However, due the instability of PostgreSQL, it crashed immediately after detecting $\varphi < t_{th}$. Thus, it is almost impossible to take any action to avoid such a state. VoltDB never went to the thrashing state under the two workload cases. This implies that $\varphi \gg t_{th}$ and VoltDB was highly stable. It does not mean that our approach was not effective. Rather, MoDaST correctly performed thrashing prediction for a stable DBMS. Due to our limited resources, we could not significantly scale up the number transactions. This remains as future work.

Answer: *The thrashing prediction showed to be precise, even with: (1) Performance instability of PostgreSQL; (2) Resources limitations to crash VoltDB.*

5 Conclusion

In this paper, we presented a novel model-based approach to database stress testing, MoDaST. It leverages a state machine to figure out the internal state of DBMS at run time. We evaluated MoDaST on two popular DBMS: PostgreSQL and VoltDB. Our results show that MoDaST can successfully infer their current internal state based on the state model. We also found out that submitting a high workload can lead to exercising the kernel in many different ways that is not possible by the current testing tools. Consequently, we identified new bugs in both DBMS. In particular, one of the bugs is already confirmed as "major bug" and promptly fixed by the VoltDB hackers. Our future work includes applying MoDaST to NoSQL and Streaming DBMS, since they implement different levels of concurrency control for transaction processing and, therefore, require different testing assumptions.

Acknowledgments. Supported by the Digital Inclusion Project: Ministry of Communication of Brazil, National Research Fund of Luxembourg and CNPq grant 441944/2014-0.

References

1. Soror, A.A., Minhas, U.F., Aboulnaga, A., Salem, K., Kokosielis, P., Kamath, S.: Automatic virtual machine configuration for database workloads. ACM Trans. Database Syst. **35**(1), 7:1–7:47 (2008)
2. Chang, J.W., Whang, K.Y., Lee, Y.K., Yang, J.H., Oh, Y.C.: A formal approach to lock escalation. Inf. Syst. **30**(2), 151–166 (2005)
3. Jain, R.: The Art of Computer Systems Performance Analysis - Techniques for Experimental Design, Measurement, Simulation, and Modeling. Wiley Professional Computing. Wiley, Hoboken (1991)

4. Storm, A.J., Garcia-Arellano, C., Lightstone, S.S., Diao, Y., Surendra, M.: Adaptive self-tuning memory in DB2. In: Proceedings of the 32nd International Conference on Very Large Data Bases. VLDB 2006, pp. 1081–1092. VLDB Endowment (2006)
5. Gray, J.: Why do computers stop and what can be done about it? (1985)
6. Willmor, D., Embury, S.M.: An intensional approach to the specification of test cases for database applications. In: Proceedings of the 28th International Conference on Software Engineering, pp. 102–111. ACM (2006)
7. Deng, Y., Frankl, P., Chays, D.: Testing database transactions with agenda. In: Proceedings of the 27th International Conference on Software Engineering. ICSE 2005, pp. 78–87. ACM, New York (2005)
8. de Almeida, E.C., Marynowski, J.E., Sunyé, G., Valduriez, P.: Peerunit: a framework for testing peer-to-peer systems. In: Proceedings of the IEEE/ACM International Conference on Automated Software Engineering. ASE 2010, pp. 169–170. ACM, New York (2010)
9. DeWitt, D.J., Levine, C.: Not just correct, but correct and fast: a look at one of Jim Gray's contributions to database system performance. SIGMOD Rec. $37(2)$, 45–49 (2008)
10. Cooper, B.F., Silberstein, A., Tam, E., Ramakrishnan, R., Sears, R.: Benchmarking cloud serving systems with YCSB. In: Proceedings of the 1st ACM Symposium on Cloud Computing. SoCC 2010, pp. 143–154. ACM, New York (2010)
11. Zhu, J., Mauro, J., Pramanick, I.: R-cubed (r3): rate, robustness, and recovery - an availability benchmark framework. Technical report, CA, USA (2002)
12. Vieira, M., Madeira, H.: A dependability benchmark for OLTP application environments. In: Proceedings of the 29th International Conference on Very Large Data Bases, VLDB 2003, vol. 29, pp. 742–753. VLDB Endowment (2003)
13. Fior, A.G., Meira, J.A., de Almeida, E.C., Coelho, R.G., Fabro, M.D.D., Traon, Y.L.: Under pressure benchmark for DDBMS availability. JIDM $4(3)$, 266–278 (2013)
14. Ghezzi, C., Jazayeri, M., Mandrioli, D.: Fundamentals of Software Engineering. Prentice Hall, Upper Saddle River (1991)
15. Bruno, N., Chaudhuri, S., Thomas, D.: Generating queries with cardinality constraints for DBMS testing. IEEE Trans. Knowl. Data Eng. $18(12)$, 1721–1725 (2006)
16. Lo, E., Binnig, C., Kossmann, D., Tamer Özsu, M., Hon, W.K.: A framework for testing DBMS features. VLDB J. $19(2)$, 203–230 (2010)
17. Binnig, C., Kossmann, D., Lo, E.: Towards automatic test database generation. IEEE Data Eng. Bull. $31(1)$, 28–35 (2008)
18. Radhakrishna Rao, C., Helge Toutenburg, S., Heumann, C.: Linear models and generalizations, least squares and alternatives, 3rd edition. AStA Adv. Stat. Anal. $93(1)$, 121–122 (2009)
19. Varrette, S., Bouvry, P., Cartiaux, H., Georgatos, F.: Management of an academic HPC cluster: The UL experience. In: Proceedings of the 2014 International Conference on High Performance Computing & Simulation (HPCS 2014), Bologna, Italy. IEEE, July 2014

A Cost Model for DBaaS Storage

Djillali Boukhelef[1], Jalil Boukhobza[2(✉)], and Kamel Boukhalfa[1]

[1] University of Science and Technology Houari Boumediene, Algiers, Algeria
{dboukhelef,kboukhalfa}@usthb.dz
[2] UMR 6285, Lab-STICC, Univ. Bretagne Occidentale, Brest, France
boukhobza@univ-brest.fr

Abstract. Cloud infrastructures employ hybrid storage systems that incorporate various types of devices (flash memory solid-state and hard disk drives). Dealing with such heterogeneity makes the use of data placements strategies necessary. These strategies generally rely on cost modeling techniques. In this paper, we propose a cost model for the storage of database objects in a Cloud infrastructure. Our cost model increments the existing work by including: (1) storage cost, which comprises the occupation, the energy and the endurance costs, (2) the penalty cost that could arise from the SLA (Service Level Agreement) violation, and (3) the migration cost resulting from the object movement between storage systems. We also evaluate the relevance of our model and its usability throughout examples.

Keywords: Cloud · DBaaS · Cost model · Storage system · Database · SLA

1 Introduction

Over the last few years, there was a high demand of outsourcing database to cut IT costs. As a result, many Cloud service providers (CSP) started offering Database-as-a-Service (DBaaS) to their customers, such as Amazon RDS and Microsoft Azure. It is then challenging for the CSP to satisfy customers' requirements while minimizing the overall cost. To address this challenge, several technologies and strategies have been employed. From a storage perspective, the efficient use of hybrid storage systems (flash memory solid-state and hard disk drives) [1–8] is one of the most critical issues.

In a DBaaS context, each customer is associated with a Service Level Agreement (SLA) that defines the amount of allocated resources, and the expected quality of service (QoS). The infrastructure SLA includes the metrics that quantify resources allocated, for instance CPU, RAM, and storage (IOPS). The violation of SLA terms results in customer dissatisfaction and is subject to penalty for the CSP. The penalty amount depends on many factors such as the violation degree and time. Therefore, the CSP must allocate the resources in a manner to satisfy the agreed upon SLA. In our study, we focus on storage system I/O throughput metric (IOPS).

I/O operations have always been considered as the main computer system bottleneck [9]. In addition, as reported in [10], the I/O access time represents almost 90 % of the transaction execution time in some database requests. Flash memory technology came to partly bridge the performance gap between main memory and Hard Disk

S. Hartmann and H. Ma (Eds.): DEXA 2016, Part I, LNCS 9827, pp. 223–239, 2016.
DOI: 10.1007/978-3-319-44403-1_14

Drives (HDD). It has very appealing properties: low access latency, low energy consumption, light weight, and shock resistance [8]. But still, HDD are not dead, as flash memory based drives (Solid State Drive –SSD) are about 10 times more expensive than HDDs. There are mainly two ways for architecting hybrid storage system: (1) a vertical integration, or tiered architecture in which SSDs tier is used as a cache for the HDD tier [2, 3, 11], and (2) a horizontal integration where HDDs and SSDs are put at the same level with HDDs [1, 4, 7, 8, 12–14]. We use this latter architecture in the scope of our study.

In order to optimize I/O performances in such hybrid systems, CSP needs to investigate objects placement strategies to reduce the overall operation cost while satisfying customers SLA. To do so, an accurate evaluation of the storage cost of database object placement over the available storage devices is required.

Several state-of-the-art work have proposed models to evaluate the storage system cost and are summarized in Table 1. We structured state-of-the-art contributions according to four parameters that, in our opinion, must be considered: (1) the storage cost, taking into account occupation, endurance (or wear out), and energy, (2) the penalty cost in case of SLA violation, (3) the migration cost in case a migration happens because of a workload change for instance, and (4) the database application specificities. While we think that considering penalty is crucial in cost modeling in a Cloud context, there are few contributions of cost models taking this parameter into account. We can also observe that very few cost models are related to a DBaaS context, they mainly focus on Virtual Machine (VM) placement. In a DBaaS context, each customer has a set of database objects (e.g. a table, an index) that can be independently stored. Note that there exist other related work that proposed to overcome the data migration issue from a performance perspective [5, 6, 15–17]. Nevertheless, they do ignore the cost induced by such operation, for instance the storage, penalty, and even migration costs.

In this paper, we propose a detailed cost model to evaluate the overall cost of data placement in a DBaaS Cloud context. Our cost model includes the *storage system cost, the workload cost (energy cost, endurance cost, penalty cost) and the migration cost.* Note that our cost model takes into account the impact of migration order.

The rest of paper is organized as follows: in Sect. 2, we formulate the problem. Then we present our cost model in Sect. 3 and we evaluate it in Sect. 4. In Sect. 5, we introduce the related work. Finally, Sect. 6 concludes the paper.

Table 1. Related work

Cost Model	Storage Cost			Penalty Cost	Migration		Database
	Occupation	Energy	Endurance		Cost	Order	
[12–14, 18]	x	x	-	-	-	-	-
[4]	x	x	-	-	-	-	x
[1]	-	-	x	-	-	-	-
[2]	x	x	x	-	-	-	-
[19]	x	-	-	-	-	-	-
[7]	-	x	x	x	x	-	-
[5, 6, 15–17]	-	-	-	-	x	-	-
Our Cost Model	x	x	x	x	x	x	x

2 Problem Definition

CSP manages a set of customers $U = \{u_1, u_2, \ldots, u_k\}$. Each customer u_k has: (1) a database instance, which consists of a set of objects O_{u_k} (2) an I/O workload w_{u_k}, (3) a requested I/O performance in terms of IOPS noted $iops_{sla,u_k}$, and (4) a penalty function pn_{u_k}. The database objects are placed in a hybrid storage system with devices of different characteristics.

A Cloud administrator needs to find the right object placement on the hybrid storage system to meet customers' requirements. Two issues need to be considered: (1) the initial placement for storing new customer's objects, (2) a dynamic placement (migration) where the CSP can move objects from an initial placement, to a new one. The first case being a subcase of the second, we focus on the more general case of dynamic placement.

In our model, we suppose that periodically, the Cloud administrator takes decisions about object migration. We defined an **evaluation period T:** which is the period of time over which the overall cost for a placement is evaluated and decision about object migration can be taken.

We assume that migration process is run as a background process without affecting performance of foreground workloads (I/O operations on database objects). To this end, we considered that migration process consumes only the remaining I/O resource performance as in [12, 16]. We also consider that migrations are serialized with respect to a given storage device, which means that each object migration consumes all the available I/O resource performance on a given device. Note that when applying a given number of migrations in a period T, even though each migration cost for a specific object is constant, the overall storage cost of migrations depends on the order upon which the migrations are performed. If a given migration is performed in the beginning of a period, the overall storage cost will be mainly related to the destination device, while if the migration is performed at the end, the storage cost will be mainly related to the source device.

In Appendix A, we declared variables used in this paper.

In the following, we define the main aspects of our problem: Cloud customer, database objects, customer workloads, I/O system, customer requirements, placement, migration, evaluation period and monitoring.

Cloud Customer: We call a Cloud customer u_k any entity that hosts externally their database in a Cloud. An SLA agreement must be concluded between the customer and CSP to define the expected QoS.

Object: Each customer u_k has a database that consists of a set of objects $O_{u_k} = \{o_{1,u_k}, \ldots, o_{i,u_k}\}$. We denote $s_{o_{i,u_k}}$ the size of object o_{i,u_k}. In this work, an object is any logical entity of a database such as index, table, materialized view, log, sequence, etc. A database object can be composed of one or more physical files. We consider the data placement and the data migration at a granularity of a database object. Using a smaller granularity (e.g. block) would increase the volume of metadata, the processing time, and the complexity, while using a virtual machine (VM) granularity as in [7] is not optimal for DBaaS context.

Workload: Each customer u_k generates a workload w_{u_k} that is modeled by a set of query sequence, $w_{u_k} = \{q_{1,u_k}, \ldots, q_{n,u_k}\}$. The customer workload generates a set of I/O operations. In our work, we distinguish four types of I/O operations: random read (rr), sequential read (sr), random write (rw), and sequential write (sw). We denote $OP = \{rr, sr, rw, sw\}$ the set of I/O operation types.

I/O System: We assume that the I/O system consists of set of storage devices $D = \{d_1, d_2, d_3, \ldots, d_j\}$. Each storage device d_j has a capacity c_{d_j}, purchase cost p_{d_j}, and wear out wo_{d_j}. We denote t_{op,d_j} (respectively $iops_{op,d_j}$) the response time (respectively maximum throughput) of device d_j for I/O operations of type op. Note that the I/O system is shared between the cloud customers, and the customer objects can be distributed among several storage devices.

Customer Requirements: Each customer u_k has a storage related performance requirement. As a case of study, we used the frequent IOPS metric. We denote $iops_{sla,u_k}$ the expected IOPS of customer u_k. Let us denote $iops_{offered}(u_k)$ the IOPS delivered by the CSP to customer u_k. Our approach can easily be extended to other performance metrics such as data transfer rate. The violation of customer requirements implies a penalty calculated using a function pn_{u_k} that takes as input the $iops_{offered}(u_k)$.

Placement: We define a placement as a mapping from a set of objects $O, O = \bigcup_{k=1}^{K} O_{u_k} = \{o_{i,u_k} | \exists! O_{u_k}, o_{i,u_k} \in O_{u_k}\}$ to set of devices D. We note that the mapping function $pl(o_{i,u_k})$ indicates the storage device d_j which stores the object o_{i,u_k}. A valid object placement must conform to the capacity and the performance constraints of each storage devices d_j in D. We denote O_{d_j} the set of objects hosted on device d_j, $O_{d_j} = \{o_{i,u_k} | pl(o_{i,u_k}) = d_j, d_j \in D, o_{i,u_k} \in O\}$.

Migration: The migration is the process that consists in moving a subset of objects among the storage devices D to achieve the target placement from an initial placement. Let us denote $O_{mv}, O_{mv} \subseteq O$ the set of objects to move. We use o_{mv_m,d_s,d_d} (respectively M) to refer the mth object (respectively the cardinality) of $O_{mv}, O_{mv} = \{o_{mv_0,d_s,d_d}, o_{mv_1,d_s,d_d}, \ldots, o_{mv_m,d_s,d_d}\}$. We note that d_s, d_d represent respectively source and destination device, $d_s, d_d \in D, d_s \neq d_d$. We assume that objects are moved following an order defined by the administrator.

Monitoring and Trace: To evaluate the storage cost of placement the Cloud administrator needs to extract the I/O access pattern of each object o_{i,u_k} of the previous period. The monitoring phase outputs the average number of I/O requests of type op delivered to object o_{i,u_k} in a period T. Let us denote it $req_{op,o_{i,u_k}}$.

Problem Formulation: In the following, we formally define our problem:

Input: (1) set of storage devices D, where each device d_j has a capacity c_{d_j}, purchase cost p_{d_j}, wear out wo_{d_j}, response time t_{op,d_j}, maximum IOPS $iops_{op,d_j}$ (2) a set of customers $U = \{u_1, \ldots, u_k\}$, where each customer u_k has a set of objects $O_{u_k} = \{o_{1,u_k}, \ldots o_{i,u_k}\}$, a workload $w_{u_k} = \{q_{0,u_k}, \ldots, q_{n,u_k}\}$, an agreed upon IOPS $iops_{sla,u_k}$, a penalty function pn_{u_k} (3) a Target placement, with an initial placement (4) a

Period T, (5) for each object o_{i,u_k}, the average number of I/O requests $req_{op,o_{i,u_k}}$ (6) a set of objects to move O_{mv} according to a well-defined order.

Storage Constraint: Capacity: $\forall d_j \in D, \left(\sum_{o_{i,u_k} \in O_{d_j}} s_{o_{i,u_k}} \right) \leq c_{d_j}$

Performance: $\forall d_j \in D, \left(\sum_{op \in OP} \dfrac{\sum_{o_{i,u_k} \in O_{d_j}} req_{op,o_{i,u_k}}}{iops_{op,d_j}} \right) \leq 1$

Output: The overall monetary cost of placement for a given period T ($Cost_{pl,T}$). So far, we formulated our problem. In the next section, we define our cost model.

3   Cost Model

In the following, we provide a cost model to evaluate the overall cost of a placement for a given period T ($Cost_{pl,T}$). We compute the cost by assuming that target placement is obtained by moving M objects from current placement. Note that, serialize the migration of M objects generates sequence of $M+1$ placements. Thus, we compute the $Cost_{pl,T}$ from the sum of placement costs ($Cost_{pl,t_m}$) as shown in (1). We called $Cost_{pl,t_m}$ (respectively t_m) the placement cost in the period t_m (respectively migration time of mth object).

$$Cost_{pl,T} = \sum_{m=0}^{M} Cost_{pl,t_m} \tag{1}$$

As shown in Fig. 1, we assume that the placement cost is obtained from storage system cost ($Cost_{stg,T}$), penalty cost ($cost_{pnl,T}$), and migration cost ($Cost_{mgr,T}$), see (2). Note that for an initial placement, the migration cost is nil. All the notations used are described in the Appendix.

$$Cost_{pl,T} = Cost_{stg,T} + Cost_{pnl,T} + Cost_{mgr,T} \tag{2}$$

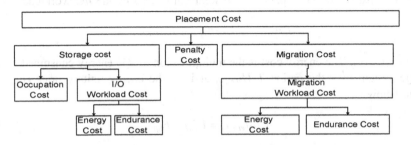

Fig. 1. Our cost model hierarchy

3.1 Storage System Cost

The storage system cost includes three components: (1) the occupation cost, (2) the workload cost, and (3) the management cost. The occupation cost ($Cost_{occp,T}$) represents the cost of storing the database objects over all the storage system. The workload cost ($Cost_{w,T}$) is related to the required performance to handle I/O requests issued to database objects.

Note that there are other additional costs which are not closely related to the database objects placement such as maintenance cost, human resources cost, air-conditioning costs. These costs are integrated in one global cost called management cost ($Cost_{mng,T}$). We used a fixed estimation for the $Cost_{mng,T}$ as in [2, 7]. Equation (3) shows the total cost of storage system.

$$Cost_{stg,T} = Cost_{occp,T} + Cost_{w,T} + Cost_{mng,T} \tag{3}$$

Occupation Cost. We calculate the occupation cost of storage system as the sum of I/O device occupation costs as shown in Eq. (4). We compute the I/O device occupation for a given period T as the multiplication of the amortized cost of device d_j over that period, by the fraction of capacity occupied by the objects.

$$Cost_{occp,T} = \sum_{j=1}^{J} \left(\left(\frac{\sum_{o_{i,u_k} \in O_{d_j}} s_{o_{i,u_k}}}{c_{d_j}} \right) * \left(Cost_{amz,T}(d_j) \right) \right) \tag{4}$$

We obtain the amortized cost of device d_j for a given period T as the multiplication of amortized cost over one unit of time ($Cost_{amz,1}$) by period T as shown in (5). Note that we calculate $Cost_{amz,1}$ by distributing the purchase cost p_{d_j} of I/O device d_j over amortization period representing the guarantee period (e.g. 5 years).

$$Cost_{amz,T}(d_j) = T * Cost_{amz,1}(d_j) \tag{5}$$

Workload Cost. Processing I/O requests consumes energy and affects the devices endurance. Therefore, we summarize the workload cost as the sum of energy and endurance costs as shown in (6). The energy cost ($Cost_{erg,T}$) represents the energy consumed by the storage system to handle the I/O requests in a period T, while the endurance cost ($Cost_{edr,T}$) represents the wear out caused by the I/O workload.

$$Cost_{w,T} = Cost_{erg,T} + Cost_{edr,T} \tag{6}$$

Energy Cost. The energy cost is the multiplication of: (1) energy consumed by the storage system over the period T ($E_{ss,T}$), and (2) the energy unitary price (E_{up}), as shown below.

$$Cost_{erg,T} = E_{ss,T} * E_{up} \tag{7}$$

The energy unitary price is defined by local electricity authority. In this paper, we assume it to be fixed. We consider the energy consumed by the storage system ($E_{ss,T}$) as the sum of the energy consumed by the storage devices ($E_{d,T}$).

$$E_{ss,T} = \sum_{j=1}^{J} \left(E_{d,T}(d_j) \right) \qquad (8)$$

We compute the device storage energy $(E_{d,T})$ for a given period T as the integral of device power P over this period:

$$E_{d,T}(d_j) = \int_T P(d_j) = P(d_j) * T \qquad (9)$$

A storage device has several power states which change according to its activity. We distinguish three states for HDD (standby, idle, active), with the respective power consumptions: $P_{atv}, P_{idl}, P_{sdb}$. Where $P_{atv} > P_{idl} > P_{sdb}$ because of the mechanical movements induced. Several studies noticed that spinning down an idle disk is not very effective in server workloads [20–22]. Thus, we only distinguish two states: active and idle. Note that SSDs have similar number of states, but have no moving parts. Equation (10) shows the power consumption of device d_j. The values of P_{atv,d_j}, P_{idl,d_j} are obtained from device specifications data sheets.

$$P(d_j) = \begin{cases} P_{atv,d_j} & where\,state = active \\ P_{idl,d_j} & where\,state = idle \end{cases} \qquad (10)$$

In this paper, we used a power model that considers constant power values for a given device that do not depend on the workload pattern like it was proved in [7]. Note that from (9) and (10), we calculate $E_{d,T}(d_j)$ as:

$$E_{d,T}(d_j) = \left(P_{atv,d_j} * t_{atv}(d_j) \right) + \left(P_{idl,d_j} * t_{idl}(d_j) \right) \qquad (11)$$

t_{atv} represents the time needed to run the I/O workload issued to the device for all contained objects, while t_{idl} represents the inactivity time. We obtain t_{idl} by subtracting t_{atv} from T $(t_{idl}(d_j) = T - t_{atv}(d_j))$. Therefore, by substitution, we formulate the energy cost consumed by a storage system as follows:

$$Cost_{erg,T} = \left[\sum_{j=1}^{J} \left[\left[P_{atv,d_j} * t_{atv}(d_j) \right] + \left[P_{idl,d_j} * t_{idl}(d_j) \right] \right] \right] * E_{up} \qquad (12)$$

As shown in (13), we calculate the device active time from the device response time t_{op,d_j} and the number of issued I/O operations.

$$t_{atv}(d_j) = \sum_{o_{i,u_k} \in O_{d_j}} \left(\sum_{op \in OP} \left(t_{op,d_j} * req_{op,o_{i,u_k}} * T \right) \right) \qquad (13)$$

The value of $req_{op,o_{i,u_k}}$ is obtained from a monitoring phase, while the value t_{op,d_j} is measured in a calibration phase (see Sect. 4.1).

Endurance Cost. We calculate the storage system endurance cost for a given period T as the sum of storage devices endurance costs $(Cost_{edr,d,T})$ shown in Eq. (14).

$$Cost_{edr,T} = \sum_{j=1}^{J} Cost_{edr,d,T}(d_j) \tag{14}$$

The $Cost_{edr,d,T}(d_j)$ represents the monetary cost of workload impact on the lifetime of the device d_j as shown in (15). Let us define the workload impact by the ratio $\left(\frac{wo_w(d_j)}{wo_{d_j}}\right)$. Therefore, we compute $Cost_{edr,d,T}(d_j)$ by multiplying this ratio by the purchase cost of a device as in [2].

$$Cost_{edr,d,T}(d_j) = p_j * \left(\frac{wo_w(d_j)}{wo_{d_j}}\right) \tag{15}$$

Note that the values wo_{d_j}, $wo_w(d_j)$ are specific to each device d_j. We evaluated SSD lifetime (wo_{d_j}) by the total amount of write operations that it can sustain as in [2]. However, it is more complex to define the HDD wear out. Historically, several studies used the number of duty cycle to represent the HDD endurance [2, 7, 23]. Nowadays, HDD manufacturers replaced the concept of start-stop cycle with the readily quantifiable workload, which is defined as the total amount of data read from (or written to) the device. Therefore, we calculate wo_{d_j} (respectively $wo_w(d_j)$) of device d_j as shown in (16) (respectively (17)).

$$wo_{d_j} = \begin{cases} total\_bytes\_written & if\ d_j \in \{SDD\} \\ workload\_limit & if\ d_j \in \{HDD\} \end{cases} \tag{16}$$

$$wo_w(d_j) = \begin{cases} \sum_{o_{i,u_k} \in O_{d_j}} \left(\sum_{op \in \{rw,sw\}} \left(io_{size} * req_{op,o_{i,u_k}} * T \right) \right) & if\ d_j \in \{SDD\} \\ \sum_{o_{i,u_k} \in O_{d_j}} \left(\sum_{op \in OP} \left(io_{size} * req_{op,o_{i,u_k}} * T \right) \right) & if\ d_j \in \{HDD\} \end{cases} \tag{17}$$

The *total_byte_written* and *workload_limit* are available in the disks datasheet.

Note that both the occupation cost defined in (4) and the endurance cost introduced in (14) evaluate data placement impact on the device lifetime. Thus, the Cloud administrator should consider only the maximum value.

3.2 Penalty Cost

We calculate the penalty cost of a storage system for a given period T $(Cost_{pnl,T})$ as the sum of customers penalties $(Cost_{pnl,u,T}(u_k))$ in this period as shown below.

$$Cost_{pnl,T} = \sum_{k=1}^{K} (Cost_{pnl,u,T}(u_k)) \tag{18}$$

Nowadays, pricing models and SLA templates become more complex. Three types of penalty have been proposed in the literature as in [24]: (1) *Fixed Penalty*: a fixed penalty is applied if the SLA clauses are violated (2) *Delay-dependent Penalty*: The amount of penalty is related to the violation time (3) *Proportional Penalty*: the penalty is proportional to the delay and the degree of SLA violation. Note that the amount of penalty will be deduced from the customer's bill. We used proportional penalty in this work as it considers more parameters related to the violation. The amount of penalty will be proportional to rate of violation $\left(\frac{iops_{offered,u}(u_k)}{iops_{sla,u_k}}\right)$ and the period T. Therefore, for each customer u_k we calculate the penalty $Cost_{pnl,u,T}(u_k)$ as follows:

$$Cost_{pnl,u,T}(u_k) = pn_k\left(\frac{iops_{offered,u}(u_k)}{iops_{sla,u_k}}, T\right) \tag{19}$$

The penalty function pn_{u_k} and $iops_{sla,u_k}$ value are defined in the SLA. While the $iops_{offered,u}(u_k)$ value is calculated from the time needed to handle the I/O workload of customer u_k ($t_{exe,u}(u_k)$) and the total number of I/O requests issued to their objects ($io_{tot,u}(u_k)$):

$$iops_{offered,u}(u_k) = \frac{io_{tot,u}(u_k)}{t_{exe,u}(u_k)} \tag{20}$$

We obtain $t_{exe,u}(u_k)$ from the number of I/O requests issued to the objects and the device response time as shown in (21). Note that $x_{d_j} = 1$ if $pl(o_{i,u_k}) = d_j$, 0 *else*

$$t_{exe,u}(u_k) = \sum_{j=1}^{J} \sum_{o_{i,u_k} \in O_{u_k}} \sum_{op \in OP} \left(x_{d_j} * t_{op,d_j} * req_{op,o_{i,k}} * T\right) \tag{21}$$

We calculate $io_{tot,u}(u_k)$ from the sum of I/O requests issued to the customer's objects:

$$io_{tot,u}(u_k) = \sum_{o_{i,u_k} \in O_{u_k}} \left(\sum_{op \in OP} T * req_{op,o_{i,u_k}}\right) \tag{22}$$

3.3 Migration Cost

Migration consists of reading the object o_{i,u_k} from a source disk d_s and writing it to a destination disk d_d. This workload causes additional I/O traffic that decreases the lifetime of both disks d_s, d_d and increases their energy consumption. Then, we calculate the migration cost as the sum of the endurance cost and energy cost for d_s and d_d.

$$Cost_{mgr,T} = Cost_{edr,d,T}(d_s) + Cost_{edr,d,T} + ((E_{d,T}(d_s) + E_{d,T}(d_d)) * E_{up}) \tag{23}$$

In order to calculate this cost, we only need to determine the migration time as the equations for the endurance and energy have already been introduced. The $Cost_{edr,d,T}$, $E_{d,T}$ are calculated as shown in Eqs. (15) and (12) respectively. We assume that the migration workload consists of x I/O sequential read operations from source disk d_s and x I/O sequential write operations to the destination disk d_d. Indeed, fragmentation of objects is supposed to be low which corresponds to reality:

$$x = \frac{S_{O_{i,u_k}}}{io_{size}} \tag{24}$$

We calculate the migration time (t_m) as the maximum between the read and write time of object as the write and reads are pipelined.

$$t_m = \max\left(x * t_{sr,d_s}, x * t_{sw,d_d}\right) \tag{25}$$

4   Experimentation

In this section, we evaluated the effectiveness of the proposed cost model. First, we show the usability of our cost model by explaining how to obtain the input values. Second, we discuss the relevance of each sub cost used in our cost model and compare our results with the output of state-of-the-art cost models for databases built using TPC-H and TPC-C benchmarks.

Experimental Platform Architecture. As shown in Fig. 2, our VMs run on top of VMware hypervisor. We built a simple storage system consisting of an HDD and an SSD. We developed a kernel module, like in [27], that captures the I/O requests issued to the different I/O devices in order to infer I/O access patterns. The module consolidates the information collected at object level and logs the result in a file. Note that our tool traces below the RDBMS buffer and kernel page cache to avoid the caching impact.

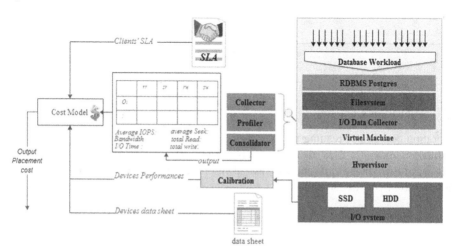

Fig. 2. Experimental platform architecture

Experimental Setup. All of experiments were performed in VMs ran on top of VMware ESXi-5.1.0 server. Each VM is configured with 8 vCPU and 8 GB of RAM. Note that only one VM is running at a time. The hypervisor ESX hosted on a server with a CPU Intel Xeon 2.4 GHz and 12 GB of RAM. We used Debian 7.5 (kernel 3.19.5) and RDBMS PostgreSQL 9.3.5. Table 2 details the storage devices characteristics.

4.1 Cost Model Usability

This part presents a use case of the proposed cost model. We show first how to obtain the input values used in our cost model, and then we detail the calibration phase required to get some specific input values.

Cost Model Input Values. Several input parameters are employed in our cost model (see Sect. 2). We classify them into four categories based on from where one can gather their values: (1) *storage device datasheet:* contains p_{d_j}, c_{d_j}, wo_{d_j}, P_{atv,d_j}, and P_{idl,d_j} as shown in Table 2, (2) *I/O tracer:* as will be explained farther, we developed an I/O tracer allowing to output $req_{op,o_{i,u_k}}$ and $seek_{d_j}$, (3) *Calibration phase:* is executed so that to get accurate devices performance characteristics (t_{op,d_j}, $iops_{op,d_j}$) for each operation type $op \in OP$, this phase will be discussed in more detail in the next section. (4) *Cloud administrator:* has access to system metadata to obtain customers and objects information such as (u_k, $iops_{sla,u_k}$, pn_{u_k}, O_k, o_{i,u_k}, and $s_{o_{i,u_k}}$).

Calibration Phase. We follow the same method described in [4, 25] to measure storage devices response time (t_{op,d_j}). Note that we experimented using the default I/O database block size. Table 3 shows our results. We observe that SSD response time is stable, and the measured values are very close to the average. Thus, we compute t_{op,d_j} from the average response time from measurements. Note that we used a device fill rate value of less than 50 %, as in [13]. HDD response time is rather stable for sequential access due the impact of data location (outer/inner tracks). In our cost model, we used the average response time. However, for random access on HDD, we observe a high instability due the impact of seek distance. To overcome this situation, we have conducted a study to explore the impact of seek distance on the device response time. We confirmed that response times increase linearly with average seek distance. Thus, we

Table 2. Devices specifications

	HDD	SDD
Model	ST91000640SS	850 PRO
Price ($) [$p_{d_j}$]	230	200
Price (GB/cent/h) [$Cost_{amz,1}$]	4.79119E-03	3.56735E-02
Capacity [c_{d_j}]	1000 GO	120 Go
warranty (years)	5	5
Perfomance	Seek Time : 8.5 ms Read/Write :9.5 ms	SR: 540 MB/s, SW:520 MB/s, RR 10 000 IOPS, RW: 40 000 IOPS
Latency (ms)	4.16	
Idle power [P_{idl,d_j}]	3.27	Max. 3.7W
Active power [P_{atv,d_j}]	5.6W	Max. 3.7W
Workload imit[wo_{d_j}]	550 TB/yr	120/250GB: 75TBW

Table 3. Devices response times

	HDD [min, max], [avg, stdev]	SDD [min, max], [avg, stdev]
SR(ms)	[0.6, 1] [0.81, 0.014]	[0.031, 0.031] [0.031, 0.0002]
RR(ms)	[5.29, 15.03] [8.42, 2.49]	[0.32, 0.39] [0.035, 0.024]
SW(ms)	[0.6, 1] [0.80, 0.012]	[0.22, 0.26] [0.023, 0.01]
RW(ms)	[5.08, 14.68] [8.31, 2.37]	[0.53, 0.57] [0.055, 0.013]

used linear regression to define response time based on average seek distance as shown in Eq. (26) for the used disk. This simple model showed an average error of ± 3.11 % and a maximum of ± 11 %.

$$\varphi\left(seek_{j,t}\right) = t_{res,j,op} = \begin{cases} 4.98\,ms\,if\,seek_{j,t} < 0.5\,GB \\ 0.058\left(seek_{j,t}\right) + 5.56\,seek_{j,t} \in [0.5, 150]\,GB \\ 15\,ms\,if\,seek_{j,t} > 150\,GB \end{cases} \qquad (26)$$

4.2   Cost Model Evaluation

Experiments Description. We applied our cost model to estimate the placement cost of several databases built using the conventional TPC-H and TPC-C benchmarks. We varied the size and the workload of the databases to simulate real environment. In this evaluation, the amortized cost shown in Table 4 is computed by distributing the purchase cost of the storage device over a period of 5 years as in [18], and the energy cost is computed using a cost of 0.1$ per kWh as in [13]. Obviously, the migration cost depends on the migration frequency and the amount of the data migrated. We assume that the migration is performed at frequency of twice per day as in [6]. The amount of penalty is calculated as thirty percent (30 %) of the total charges paid by client as in Amazon Cloud. Note that we present results based on the real purchase cost of storage devices. The TPC-H databases are populated using DBGen with a scale factor of 30, 100, and 300. The workload is generated randomly from the 22 TPC-H query template as in [4]. The TPC-C databases were populated using scale factor of 12, 350, and 800 warehouses. We varied query arrival rates for TPC-H workload and the number of terminals/warehouse (or think time) for TPC-C workload to generate different workloads. We used our I/O tracer to extract the I/O access pattern of databases objects. We estimated the HDD response time for random access using Eq. (26). Table 3 below shows the databases access pattern of one hour and the different monthly costs.

Discussion. We show, from Table 4, that all the evaluated costs are in the same order of magnitude and the dominant costs vary according to the database size, the workload pattern, and storage device. We observe that occupation cost increases the operating cost for large databases infrequently accessed (see DB4, DB6). We take as example OLAP databases occasionally requested. However, the endurance cost becomes the most significant for small databases which have I/O intensive workloads (see DB2, DB3). One example is OLTP databases. We observed that the energy cost constitutes 5 to 28 % of the overall cost for the databases placed in the HDD (see DB1 and DB5). The endurance cost can represent up to 90 % of the total cost for databases with intensive write workload placed in SSD. Moreover, the Cloud administrator should be attentive to the migration cost that can easily explode when large quantity of data is selected to move (see DB4). Note that the migration order affects the overall cost. Our experimentations show that the impact of move order represents between 0.5 to 8.25 % of the overall cost. On the other hand, we see that penalty cost is very large and it can achieve 80 % of the overall cost. That is, it becomes one of the biggest concern of the CSP [26].

Table 4. Cost simulation (for one month in $)

	TPC-C			TPC-H		
	DB1	DB2	DB3	DB4	DB5	DB6
Size(GB)	32	1,2	60	147	34	381
Device	HDD	SSD	SSD	HDD	HDD	HDD
RR (op/h)	136800	540000	601200	10800	32400	1080
SR (op/h)	46800	68400	104400	216000	216000	8280
RW (op/h)	108000	468000	543600	3600	7200	360
SW (op/h)	32400	82800	147600	18000	10800	1800
$Cost_{occp,T}$	0.1104	0,0308	1.5411	**0.507**	0.117	**1.314**
$Cost_{edr,T}$	0.1454	**3.9396**	**4.9438**	0.111	0.120	0.005
$Cost_{erg,T}$	**0.3559**	0.0349	0.0402	0.242	**0.253**	0.236
$Cost_{stg,T}$	0.5013	3.9746	4.9841	0.749	0.372	1.550
$Cost_{pnt,T}$	0.96	0.036	1.8	4.410	1.020	0.000
$Cost_{mgr,T}$	0.0854	0.0032	0.1602	**0.393**	0.091	0.000
Total cost	**1.5468**	**4.0138**	**9.2843**	**5.552**	**1.483**	**1.550**
State-of-the-art cost models						
[2]	0.5013	3.9746	4.9841	0.3537	0.3723	0.2412
[4]	0.4663	0.0657	1.5813	0.7494	0.3701	1.5504
[7]	1.5468	4.0138	6.9443	5.1562	1.4831	0.2412
[12–14, 18]	0.3559	0.0349	0.0402	0.242	0.253	0.236

From the discussion above, we demonstrated that for an accurate estimation of the placement cost all the considered parts of our cost model can have a significant impact depending on the overall configuration. Note that the aforementioned evaluation is far smaller in scale than real data centers.

Comparison with State-of-the-Art Cost Model. In the second part of Table 4, we show a comparison between our results and the output costs of the previous cost models. We see that [2, 4, 12–14, 18] propose cost model to accurately estimate the operating cost by combining occupation cost, energy cost, and endurance. Unfortunately, they ignore penalty cost which is a very impacting parameter. We observe that using such cost models leads to inaccurate cost estimation which can reach ±80% difference in the worst case. In [7], the authors propose a cost model that includes penalty and migration cost. However, it ignores the occupation cost and the impact of migration order. The proposed cost model results in an estimation error (compared to ours) between 7 % and 80 % as shown in Table 4. The error increases linearly with database size. Note that this cost model ignores also the impact of migration order with can constitutes 8.25 % of the overall cost in the examples discussed above.

5 Related Work

Our work relies upon previous research on storage system cost models, database objects placement, hybrid storage system management and data migration policy. We classify the related work in three classes.

Storage System Cost Model. Many researchers have made great efforts to estimate the cost of hybrid HDD/SSD storage systems or, more generally, the cost of I/O system [1, 2, 4, 12, 13, 18, 19] as illustrated in Table 1. Most of them represent the overall cost of the storage system by the sum of occupation cost (or purchase cost) and energy cost [4, 12, 13, 18], while others only use endurance cost [1], or the purchase cost [19].

In [7], the authors use the cost arising from I/O workload issued. Other work consider the overall cost of the storage system as the sum of purchase cost, energy cost, and endurance cost as in [2], however, they do not consider customer's SLA and thus cannot be applied as is.

Migration. A large literature explores the data migration issue in hybrid storage [5, 6, 15–17]. The objective of those works is choosing the data to move to optimize performance and moving it without impact on foreground I/O performance. Unfortunately, they do not use a detailed storage cost and do ignore the penalty cost. In [7], authors attempt to propose cost model for VM placement by including both migration cost and storage cost. However, they propose a too simple model for migrations.

Penalty. Few related work include the storage system penalty in the evaluation of the overall cost of placement. The work that is the most related to ours is [7]. They define a cost model for VM Storage in Cloud which includes the storage penalty caused by the violation of I/O constraints defined in SLA. However, this cost model is not applicable in a data base context.

6 Conclusion and Future Work

In this paper, we introduced a new accurate storage cost model to evaluate database objects placement in a DBaaS applications. Our model complement existing model taking into account: (1) the customers SLA constraints, (2) the extra penalty cost, (3) Migration cost, and (4) the impact of the objects migration order in the estimation of overall cost.

The proposed cost model can be used in number of key areas such as: (1) *Pricing strategy* to evaluate precisely the overall cost of storage system; (2) *Placement strategies* to find the optimal placement of objects and (3) *Resources allocation* to effectively allocate the I/O resource to Cloud's customers. While we have focused on HDD/SSD storage system in this paper, our model can be easily generalized to be applied for other new storage class devices. In our experimentations, we demonstrated the usability and relevance of our cost model.

In future work, we plan to use our cost model within Framework based on the MAPE-K autonomic loop. The framework aims to place and automatically migrate the objects across the different storage devices (HDD/SSD) to adapt to the workload fluctuation.

We would like to investigate some issues related to our cost model in future works. The proposed cost model has not considered the case of distributed storage system, as result it ignores network related costs. So the cost model can be updated to suit large distributed storage systems.

Acknowledgement. This work is supported by the PHC (Partenariat Hubert Curien) Tassili GHEEMaS project (number 16MDU964).

Appendix A: Notations

Variable	Description
Customers	
U, u_k	The set of customer in cloud, kth customer in cloud $k \in \{1, K\}$
w_{u_k}	The workload of customer u_k
pn_{u_k}	The penalty of customer u_k
$iops_{sla,u_k}$	The IOPS SLA of customer u_k
$iops_{offered,u}(u_k)$	The IOPS offered to customer u_k in the period T
$t_{exe,u}(u_k)$	The time need to handle the I/O workload of customer u_k
$io_{req,u}(u_k)$	The total number of the IO requests issued from customer u_k
Objects	
O, O_{u_k}	The set of cloud objects, The set of objects of customer u_k
O_{d_j}	The set of objects hosted in device d_j
$o_{i,u_k}, s_{o_{i,u_k}}$	The ith object of customer $u_k i \in \{1, I\}$, its size
$req_{op,o_{i,u_k}}$	The average IOPS of type op issued to the object $o_{i,k}$
Migration	
O_{mv}, o_{mv_m,d_s,d_d}	The set of objects to move, the mth objects to move
$pr_{o_{mv_m}}$	The priorty of mth objects to move
Devices	
D, d_j	The set of device, The jth device $j \in \{1, J\}$
$p_{d_j}, c_{d_j}, wo_{d_j}$	The price, The capacity, The wear out of device d_j
$iops_{op,d_j}$	The max throughput of device d_j for operation type op
t_{op,d_j}	The response time of device d_j for operation type op
$t_{atv}(d_j), t_{idl}(d_j)$	The active time, the idle time of device
$seek_{d_j}$	The average seek distance of device dj
msr_{op,d_j}	The set of experimental measures taken for the device d_j and I/O operations of type op
wo_w	The impact of workload on the lifetime of device
$E_{ss,t}, E_{d,t}, E_{up}$	storage system energy, device energy, energy unit price
$P, P_{atv,d_j}, P_{idl,d_j}$	The power, the active power, the idle power of device d_j
General	
io_{size}	The size of I/O block (database block)
T	Period of time
op	$op \subseteq OP, OP = \{rr, sr, rw, sw\}, rr$:random read. sr: sequential read, rw: random write, and sw:sequential write.
Cost notations	
$Cost_{pl,T}$	The placment cost for given period T
$Cost_{stg,T}$	The storage cost for given period T
$Cost_{pnl,T}$	The penalty cost for given period T
$Cost_{pnl,u,T}(u_k)$	The penalty cost of customer u_k for given period T

(Continued)

Variable	Description
$Cost_{mgr,T}$	The Migration Cost for given period T
$Cost_{mng,T}$	The Management Cost for given period T
$Cost_{occp,T}$	The occupation cost for given period T
$Cost_{w,T}$	The workload cost for given period T
$Cost_{amz,T}$	The amortized cost for given period T
$Cost_{amz,1}$	The amortized cost for one unite of time
$Cost_{erg,T}$	The energy cost for given period T
$Cost_{edr,T}$	The Endurance cost of stotage system for given period T
$Cost_{edr,d,T}(d_j)$	The endurence cost of device d_j for given period T

References

1. Cheng, Y., Iqbal, M.S., Gupta, A., Butt, A.R.: Pricing games for hybrid object stores in the cloud: provider vs. tenant. In: 7th USENIX Workshop on Hot Topics in Cloud Computing (HotCloud 2015) (2015)
2. Li, Z., Mukker, A., Zadok, E.: On the importance of evaluating storage systems' $costs. In: 6th USENIX Workshop on Hot Topics in Storage and File Systems (HotStorage 2014) (2014)
3. Cheng, Y., Iqbal, M.S., Gupta, A., Butt, A.R.: CAST: tiering storage for data analytics in the cloud. In: Proceedings of the 24th International Symposium on High-Performance Parallel and Distributed Computing (2015)
4. Zhang, N., Tatemura, J., Patel, J.M., Hacigümüş, H.: Towards cost-effective storage provisioning for DBMSs. Proc. VLDB Endow. 5, 274–285 (2011)
5. Tai, J., Sheng, B., Yao, Y., Mi, N.: Live data migration for reducing SLA violations in multi-tiered storage systems. In: 2014 IEEE International Conference on Cloud Engineering (IC2E), pp. 361–366 (2014)
6. Zhang, G., Chiu, L., Liu, L.: Adaptive data migration in multi-tiered storage based cloud environment. In: 2010 IEEE 3rd International Conference on Cloud Computing (CLOUD), pp. 148–155 (2010)
7. Ouarnoughi, H., Boukhobza, J., Singhoff, F., Rubini, S.: A cost model for virtual machine storage in cloud IaaS context. In: 2016 24th Euromicro International Conference on Parallel, Distributed, and Network-Based Processing (PDP), pp. 664–671 (2016)
8. Boukhobza, J.: Flashing in the Cloud: Shedding some Light on NAND Flash Memory Storage Systems. IGI Global, Hershey (2013)
9. Shriver, E.: Performance modeling for realistic storage devices (1997)
10. Sharaf, M.A., Chrysanthis, P.K., Labrinidis, A., Amza, C.: Optimizing I/O-intensive transactions in highly interactive applications. In: Proceedings of the 2009 ACM SIGMOD International Conference on Management of Data, pp. 785–798. ACM, New York (2009)
11. Oh, Y., Choi, J., Lee, D., Noh, S.H.: Caching less for better performance: balancing cache size and update cost of flash memory cache in hybrid storage systems. In: FAST (2012)
12. Guerra, J., Pucha, H., Glider, J.S., Belluomini, W., Rangaswami, R.: Cost effective storage using extent based dynamic tiering. In: FAST, pp. 20–20 (2011)

13. Kim, Y., Gupta, A., Urgaonkar, B., Berman, P., Sivasubramaniam, A.: HybridStore: a cost-efficient, high-performance storage system combining SSDs and HDDs. In: 2011 IEEE 19th International Symposium on Modeling, Analysis Simulation of Computer and Telecommunication Systems (MASCOTS), pp. 227–236 (2011)

14. Kim, Y., Gupta, A., Urgaonkar, B., Berman, P., Sivasubramaniam, A.: HybridPlan: a capacity planning technique for projecting storage requirements in hybrid storage systems. J. Supercomput. **67**, 277–303 (2013)

15. Lin, L., Zhu, Y., Yue, J., Cai, Z., Segee, B.: Hot random off-loading: a hybrid storage system with dynamic data migration. In: 2011 IEEE 19th International Symposium on Modeling, Analysis Simulation of Computer and Telecommunication Systems (MASCOTS), pp. 318–325 (2011)

16. Lu, C., Alvarez, G.A., Wilkes, J.: Aqueduct: online data migration with performance guarantees. In: Proceedings of the 1st USENIX Conference on File and Storage Technologies. USENIX Association, Berkeley (2002)

17. Sundaram, V., Wood, T., Shenoy, P.: Efficient data migration in self-managing storage systems. In: IEEE International Conference on Autonomic Computing, 2006, ICAC 2006, pp. 297–300 (2006)

18. Dutta, A.K., Hasan, R.: How much does storage really cost? Towards a full cost accounting model for data storage. In: Altmann, J., Vanmechelen, K., Rana, O.F. (eds.) GECON 2013. LNCS, vol. 8193, pp. 29–43. Springer, Heidelberg (2013)

19. Moore, R.L., D'Aoust, J., McDonald, R.H., Minor, D.: Disk and tape storage cost models. In: Archiving Conference, pp. 29–32. Society for Imaging Science and Technology (2007)

20. Gurumurthi, S., Sivasubramaniam, A., Kandemir, M., Franke, H.: DRPM: dynamic speed control for power management in server class disks. In: Proceedings of 30th Annual International Symposium on Computer Architecture, 2003, pp. 169–179 (2003)

21. Hylick, A., Sohan, R., Rice, A., Jones, B.: An analysis of hard drive energy consumption. In: IEEE International Symposium on Modeling, Analysis and Simulation of Computers and Telecommunication Systems, 2008, MASCOTS 2008, pp. 1–10 (2008)

22. Son, S.W., Chen, G., Kandemir, M.: Disk layout optimization for reducing energy consumption. In: Proceedings of the 19th Annual International Conference on Supercomputing, pp. 274–283. ACM, New York (2005)

23. Pinheiro, E., Weber, W.-D., Barroso, L.A.: Failure trends in a large disk drive population. In: FAST, pp. 17–23 (2007)

24. Garg, S.K., Gopalaiyengar, S.K., Buyya, R.: SLA-based resource provisioning for heterogeneous workloads in a virtualized cloud datacenter. In: Xiang, Y., Cuzzocrea, A., Hobbs, M., Zhou, W. (eds.) ICA3PP 2011, Part I. LNCS, vol. 7016, pp. 371–384. Springer, Heidelberg (2011)

25. Canim, M., Mihaila, G.A., Bhattacharjee, B., Ross, K.A., Lang, C.A.: An object placement advisor for DB2 using solid state storage. Proc. VLDB Endow. **2**, 1318–1329 (2009)

26. Du, L.: Pricing and resource allocation in a cloud computing market. In: Proceedings of the 2012 12th IEEE/ACM International Symposium on Cluster, Cloud and Grid Computing (Ccgrid 2012), pp. 817–822. IEEE Computer Society, Washington (2012)

27. Ouarnoughi, H., Boukhobza, J., Singhoff, F., Rubini, S.: A multi-level I/O tracer for timing and performance storage systems in IaaS cloud. In: REACTION (2014)

A Query Processing Framework for Array-Based Computations

Leonidas Fegaras[(✉)]

University of Texas at Arlington, Arlington, USA
fegaras@cse.uta.edu

Abstract. Current scientific applications must analyze enormous amounts of array data using complex mathematical data processing methods. This paper describes a distributed query processing framework for large-scale scientific data analysis that captures array-based computations using SQL-like queries and optimizes and evaluates these computations using state-of-the-art parallel processing algorithms. Instead of providing a library of concrete distributed algorithms that implement certain matrix operations efficiently, we generalize these algorithms by making them parametric in such a way that the same efficient implementations that apply to the concrete algorithms can also apply to their generic counterparts. By specifying matrix operations as generic algebraic operators, we are able to perform inter-operator optimizations, such as fusing matrix transpose with matrix multiplication, resulting to new instantiations of the generic algebraic operators, without having to introduce new efficient algorithms on the fly. We evaluate the effectiveness of our framework by measuring the performance improvement of matrix factorization when evaluated with inter-operator optimization.

1 Introduction

In recent years, it has become easier and cheaper than ever to collect data but harder to turn these data into value. In computational science, the explosion in scientific data generated by experiments and simulations has created a major challenge for many scientific projects. For data scientists who need to analyze vast volumes of data, data-intensive processing is fast becoming a necessity. They need algorithms capable of scaling to petabytes and faster tools that are more sophisticated, more reliable, and easier to use.

As datasets grow larger, new frameworks in distributed Big Data analytics have become essential tools to large-scale machine learning and scientific discoveries. Among these frameworks, the Map-Reduce programming model [3] has emerged as a generic, scalable, and cost effective solution for Big Data processing on clusters of commodity hardware. The Map-Reduce paradigm is a scale-out solution that brings computations to the data, rather than data to the computations. This is a drastic departure from high-performance computing models, which make a clear distinction between processing and storage nodes. Currently, most programmers prefer to use a higher-level declarative language to code their Map-Reduce applications, such as Apache Hive [11] and PigLatin [18], instead of coding them directly in an algorithmic language, such as Java. For instance, Hive is

S. Hartmann and H. Ma (Eds.): DEXA 2016, Part I, LNCS 9827, pp. 240–254, 2016.
DOI: 10.1007/978-3-319-44403-1_15

used for over 90 % of Facebook Map-Reduce jobs. Most Map-Reduce query languages though provide a limited syntax for operating on data collections, in the form of simple relational joins and group-bys. They cannot express complex data analysis tasks, such as PageRank, data clustering, and matrix factorization, using SQL-like syntax exclusively. Because of these limitations, these languages enable users to plug-in custom scripts into their queries for those jobs that cannot be declaratively coded in their query language. This nullifies the benefits of using a declarative query language and may result in platform-dependent, suboptimal, error-prone, and hard-to-maintain code. Furthermore, some of these languages are inappropriate for complex scientific and graph analysis applications, because they do not directly support iteration in declarative form and are not able to handle complex scientific data. But there are some recent query systems, such as Apache MRQL [17], which are powerful enough to express complex data analysis tasks.

In the past, large-scale data processing was mainly done in the realm of scientific computing. In recent years, the volume of data generated by scientists through experiments and simulations has been steadily increasing at an unprecedented rate. For example, the Large Hadron Collider at CERN and astronomy's Pan-STARRS5 array of celestial telescopes are capable of generating several petabytes of data per day, which need to be made available and analyzed by scientists on worldwide grids of computers. Data-intensive scientific computing shares some of the key ingredients of cloud computing. Just like in cloud computing, scientific computing is driven to use the most efficient computing techniques available, including high-performance computing and low-level data management. Since most of the data generated by scientists are in array form, current scientific applications must analyze enormous amounts of array data using complex mathematical data processing methods. Scientists are typically comfortable with numerical analysis tools, such as MatLab, but are not familiar with the intricacies of Big Data analysis and distributed computing. A declarative distributive query language capable of expressing complex mathematical operations on arrays could help them develop their data analysis applications without any prior knowledge of distributed computing.

The goal of this paper is to support large-scale scientific data analysis by (1) extending an existing distributed query language, namely Apache MRQL [17], with array operations that can capture most array-based computations in declarative form and (2) by developing a query processing framework that can optimize and evaluate these computations using state-of-the-art parallel processing algorithms. Other proposed systems [1,8,20,22] focus on storage structures and indexing techniques for arrays, such as chunking and tiling, to achieve better performance on certain parallel array computations. Although such storage layouts may speed up the processing of individual array operations, they produce results in a certain layout that may need to be restructured before it is used for the next matrix operation. Furthermore, such schemes do not address inter-operation optimization, which is the focus of our work. Our approach is to accept any kind of array representation and storage but at the same time be able to recognize certain array operations in a query and translate them into efficient parallel array processing algorithms. For example, matrix multiplication $X \times Y$ between two sparse matrices X and Y can be implemented efficiently in a distributed environment using a 2D mesh of processors [7,23] by distributing the data to worker nodes in the form of a grid of partitions, where each partition contains only those rows from X and those columns from Y needed to compute a single grid partition of the resulting matrix. If a query language were to adopt a certain matrix representation and provide a fixed number of matrix operations in the form of predefined operators or library functions, then the task of recognizing these operations and mapping them to efficient algorithms would have become easy. Such an approach though does not leave many opportunities of inter-operator optimization, such as fusing matrix transpose with matrix multiplication, because the resulting fused operation would have been a new operation that requires the introduction of a new efficient algorithm on the fly. Instead of looking at concrete algorithms that implement specific mathematical operations, our objective is to generalize these algorithms by making them parametric in such a way that the same efficient implementations that apply to the concrete algorithms can also apply to their generic counterparts.

The most effective method of making an algorithm parametric is to make it higher-order by abstracting parts of its computations into its functional parameters. Such a higher-order operation must capture the essence of the concrete algorithm it generalizes by facilitating an equivalent data distribution and by supporting a similar parallel processing method. To generate such a higher-order operation from a query, a query evaluator must be able to recognize certain syntactic patterns in the query, in their most generic form, that can be mapped to this operation. This task can become more feasible if it is done at the algebraic operation level, rather than at the syntactic level. That is, instead of introducing source-to-source transformations to match parts of a query with certain generic syntactic patterns that correspond to a generic operation, our approach is to translate queries into algebraic forms and then normalize and rewrite these forms into these algorithms using algebraic rewrite rules. We believe that this approach will be very effective when applied, not only to mathematical operations, but also to a wide spectrum of queries whose functionality is in essence equivalent to these mathematical operations.

The contribution of this work can be summarized as follows:

- We introduce a new higher-order operator, called *GroupByJoin*, that generalizes many algorithms that correlate two data sources using an equi-join followed by a group-by with aggregation (Sect. 5).
- We provide an efficient implementation of GroupByJoin in Map-Reduce based on an algorithm that generalizes the SUMMA parallel algorithm for matrix multiplication (Sect. 6).
- We have extended the query optimization framework in MRQL to generate physical plans that use this operator. This is accomplished with algebraic rewrite rules that recognize certain patterns in the algebraic terms derived from MRQL queries that are equivalent to a GroupByJoin operation. We show how these rewrite rules can be used, in conjunction with the existing

algebraic optimization rules in MRQL, to minimize the number of Map-Reduce operations for queries that contain consecutive matrix operations (Sect. 7).
– We report on a prototype implementation of our framework using MRQL running on top of Hadoop Map-Reduce. We show the effectiveness of our method through experiments on two queries, a simple query that combines matrix multiplication with matrix transpose, and the very complex query for matrix factorization, that is both iterative and contains many matrix operations in every iteration (Sect. 8).

2   Related Work

One of the major drawbacks of the Map-Reduce model is that, to simplify reliability and fault tolerance, it does not preserve data in memory between the map and reduce tasks of a Map-Reduce job or across consecutive jobs, which imposes a high overhead to complex workflows and graph algorithms, such as PageRank and matrix factorization, which require repetitive Map-Reduce jobs. To achieve better performance for such complex workflows, it is crucial to minimize the required number of Map-Reduce jobs, mostly because of the high overhead of dumping the intermediate results between consecutive Map-Reduce jobs to the HDFS. As an alternative solution, some recent systems for cloud computing use distributed memory for inter-node communication, such as the main memory Map-Reduce (M3R [21]), Apache Spark [19], Apache Flink [6], and distributed GraphLab [15]. Another alternative framework to the Map-Reduce model is the Bulk Synchronous Parallelism (BSP) programming model [23]. The best known implementations of the BSP model for data analysis on the cloud are Google's Pregel [16] and Apache Hama [10].

Most other array-processing systems use special storage techniques, such as regular tiling, to achieve better performance on certain array computations. SciDB [22] is a large-scale data management system for scientific analysis based on an array data model with implicit ordering. The SciDB storage manager decomposes arrays into a number of equal sized and potentially overlapping chunks, in a way that allows parallel and pipeline processing of array data. Like SciDB, ArrayStore [20] stores arrays into chunks, which are typically the size of a storage block. One of their most effective storage method is a two-level chunking strategy with regular chunks and regular tiles. SystemML [8] is an array-based declarative language to express large-scale machine learning algorithms, implemented on top of Hadoop. It supports many array operations, such as matrix multiplication, and provides alternative implementations to each of them. SciHadoop [1] is a Hadoop plugin that allows scientists to specify logical queries over arrays stored in the NetCDF file format. Their chunking strategy, which is called the Baseline partitioning strategy, subdivides the logical input into a set of partitions (sub-arrays), one for each physical block of the input file. Finally, MLlib, which is part of MLbase [13], is a machine learning library built on top of Spark and includes algorithms for fast matrix manipulation.

3   Background: The MRQL Query Language

Apache MRQL [17] is a query processing and optimization system for large-scale, distributed data analysis. MRQL was originally developed by the author [4,5], but is now an Apache incubating project with many developers and users worldwide. MRQL (the Map-Reduce Query Language) is an SQL-like query language for large-scale data analysis on computer clusters. The MRQL query processing system can evaluate MRQL queries in four modes: in Map-Reduce mode using Apache Hadoop [9], in BSP mode (Bulk Synchronous Parallel model) using Apache Hama [10], in Spark mode using Apache Spark [19], and in Flink mode using Apache Flink (previously known as Stratosphere) [6]. The MRQL query language is powerful enough to express most common data analysis tasks over many forms of raw in-situ data, such as XML and JSON documents, binary files, and CSV documents. MRQL is more powerful than other current high-level Map-Reduce languages, such as Hive [11] and PigLatin [18], since it can operate on more complex data and supports more powerful query constructs, thus eliminating the need for using explicit procedural code. With MRQL, users are able to express complex data analysis tasks, such as PageRank, k-means clustering, matrix factorization, etc., using SQL-like queries exclusively, while the MRQL query processing system is able to compile these queries to efficient Java code.

For example, the following MRQL query that calculates the k-means clustering algorithm (Lloyd's algorithm), by deriving k new centroids from the old (the stopping condition has been omitted):

```
repeat centroids = ...
  step select < X: avg(s.X), Y: avg(s.Y) >
       from s in Points
       group by k: ( select c from c in centroids
                     order by distance (c, s ))[0]
```

where Points is the input data set of points on a plane, centroids is the current set of centroids (k cluster centers), and distance is a function that calculates the distance between two points. The initial value of centroids (the ... value) can be a bag of k random points. The select-query in the group-by part assigns the closest centroid to a point s (where [0] returns the first tuple of an ordered list). The select-query in the repeat step clusters the data points by their closest centroid, and, for each cluster, a new centroid is calculated from the average values of its points.

4   Our Framework

One of the objectives of our work is to accept any kind of array representation but at the same time be able to recognize certain array operations in a query and translate them into efficient parallel array processing algorithms. Sparse vectors and matrices can be captured as regular collections in MRQL. For example, a sparse matrix M can be represented as a collection of triples, (v, i, j), for

$v = M_{ij}$. Then, the matrix multiplication between two sparse matrices X and Y can be expressed as follows in MRQL:

```
select ( sum(z), i, j )
  from (x,i,k) in X, (y,k,j) in Y, z = x*y
  group by i, j
```

that is, we retrieve the values $X_{ik} \in X$ and $Y_{kj} \in Y$ for all i, j, k, and we set $z = X_{ik} * Y_{kj}$. The group-by operation in MRQL lifts each non-group-by variable defined in the from-part of the query from some type T to a bag of T, indicating that each such variable must now contain multiple values, one for each group. Consequently, after we group by the indexes i and j, the variable z will be lifted to a bag of numerical values $X_{ik} * Y_{kj}$, for all k. Hence, sum(z) in the query header will sum up all these values, deriving $\sum_k X_{ik} * Y_{kj}$ for the ij element of the resulting matrix.

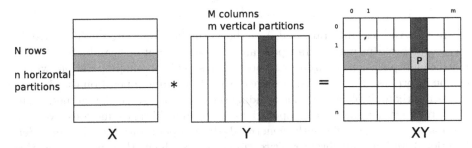

Fig. 1. Matrix multiplication: each partition P requires N/n rows from X and M/m columns from Y

Matrix multiplication is an important operation, used frequently in scientific computations and machine learning. Suppose that X is an $N * K$ matrix and Y is an $K * M$ matrix. If the previous matrix multiplication query for $X \times Y$ is evaluated naively using an equi-join followed by a group-by, the intermediate result of the join would have been of size $N * K * M$, which would have to be shuffled to cluster nodes for the group-by operation. Instead, one may use the SUMMA algorithm for matrix multiplication [7], which has been adapted for the BSP distributed model [23] and later for Map-Reduce [2]. This algorithm distributes the data as a grid of $m * n$ partitions, so that each partition contains N/n full rows from X and M/m full columns from Y (Fig. 1). That is, the X elements are replicated m times and the Y elements are replicated n times. Then, each partition is assigned to a single node in a cluster, which must have enough free memory to multiply the associated submatrices of size $N/n * K$ and $K * M/m$. The goal of this method is to minimize replication (m and n) so that the memory of each worker node in the cluster is fully utilized by performing the submatrix multiplication in memory. When implemented using Map-Reduce, this algorithm requires only one Map-Reduce job: the map task replicates and distributes the data to reducers, while each reducer multiplies its submatrices in memory using a hash join.

How can such algorithm be incorporated into the evaluation engine of a query language? One solution is to provide a library of predefined functions for various matrix operations, using their most efficient implementation. But such an approach does not leave any opportunities for inter-operation optimization. Consider, for example, Matrix Factorization using Gradient Descent [12], used in machine learning applications, such as for recommender systems. The goal of this computation is to split a matrix R of dimension $n \times m$ into two low-rank matrices P and Q of dimensions $n \times k$ and $k \times m$, for small k, such that the error between the predicted and the original rating matrix $R - P \times Q^T$ is below some threshold, where $P \times Q^T$ is the matrix multiplication of P with the transpose of Q and '$-$' is cell-wise subtraction. Matrix factorization can be done using an iterative algorithm that repeatedly applies the following rules to minimize the error matrix E:

$$E \leftarrow R - P \times Q^T$$
$$P \leftarrow P + \gamma(2E \times Q^T - \lambda P)$$
$$Q \leftarrow Q + \gamma(2E \times P^T - \lambda Q)$$

where γ is the learning rate and λ is the normalization factor used in avoiding overfitting. But matrix transpose and cell-wise operations can be fused with matrix multiplication, because they both correspond to a map operation, which can be incorporated into the map stage of the Map-Reduce operation that implements matrix multiplication, thus avoiding the extra map stage all together. That is, instead of defining matrix operations as opaque library functions, we can express them using sufficiently generic algebraic operations (i.e., higher-order functions) and use algebraic rewrite rules to fuse them, thus minimizing the number of processing stages and eliminating intermediate results. That way, in addition to offering more opportunities for optimization, application developers will not be forced to represent their data matrices in the single fixed representation used by the underlying implementation of the concrete matrix algorithms. Instead, they will be free to use any representation, thus focusing only on the computation logic. In addition, by generalizing these algorithms, one can optimize a wider spectrum of queries that resemble matrix multiplication, such as calculating the shortest distance between all pairs of nodes in a graph G:

```
repeat S = G
  step select (x,z,min(d))
       from (x,y,d1) in S, (y,z,d2) in S, z = d1+d2
       group by x, z
```

(assuming for simplicity that $(x, x, 0) \in G$ for every node x).

5   The GroupByJoin Operation

In this section, we generalize matrix multiplication using an algebraic operation, called a *Group-By Join*. Let X and Y be bags of types $\{\alpha\}$ and $\{\beta\}$, respectively, for arbitrary types α and β. The generic MRQL query

```
select h( k, reduce(acc, zero, z) )
  from x in X, y in Y, z = (x,y)
  where jx(x) = jy(y)
  group by k: ( gx(x), gy(y) )
```

which generalizes matrix multiplication, returns a value of type $\{\delta\}$, where

- jx is the left join key function of type $\alpha \to \kappa$,
- jy is the right join key function of type $\beta \to \kappa$,
- gx is the left group-by function of type $\alpha \to \kappa_1$,
- gy is the right group-by function of type $\beta \to \kappa_2$,
- h is the head function of type $((\kappa_1, \kappa_2), \gamma) \to \delta$.
- reduce(acc,zero,s) reduces the elements of a bag s of type $\{(\alpha, \beta)\}$ into a value of type γ, using an accumulator acc of type $((\alpha, \beta), \gamma) \to \gamma$ and a zero value of type γ. That is, $\mathsf{reduce}(\mathsf{acc}, \mathsf{zero}, \{z_1, z_2, \ldots, z_n\}) = \mathsf{acc}(z_1, \mathsf{acc}(z_2, \ldots, \mathsf{acc}(z_n, \mathsf{zero})))$.

To preserve bag semantics, we must have $\mathsf{acc}(x, \mathsf{acc}(y, s)) = \mathsf{acc}(y, \mathsf{acc}(x, s))$, for all x, y, and s.

The previous generic query is captured by the higher-order physical operation:

$$\mathsf{GroupByJoin}(\mathsf{jx}, \mathsf{jy}, \mathsf{gx}, \mathsf{gy}, \mathsf{acc}, \mathsf{zero}, \mathsf{h}, X, Y)$$

which generalizes the SUMMA algorithm by distributing X and Y into a grid of $n * m$ partitions based on their group-by and join key functions.

For example, matrix multiplication, which corresponds to the MRQL query

```
select ( sum(z), i, j )
  from (x,i,k) in X, (y,k,j) in Y, z = x*y
  group by i, j
```

is captured by the operation:

GroupByJoin(λ(x,i,k). k, λ(y,k,j). k, λ(x,i,k). i, λ(y,k,j). j, λ((x,y),c). c+x*y, 0, λ((i,j),c).
       (c,i,j), X, Y)

6   The Implementation of GroupByJoin in Map-Reduce

The GroupByJoin operation distributes the data to worker nodes in the form of a $n * m$ grid of partitions, where each partition contains only those rows from X and those columns from Y needed to compute a single partition of the resulting matrix.

Figure 2 shows the pseudo-code for the implementation of GroupByJoin in Map-Reduce, where flush(H) is:

```
for each (key, value) in H
    emit h(key, value)
clear H
```

```
1    mapLeft ( x ):
2      for each i in 0..m−1
3        emit ( ((hashCode(gx(x)) % n)*m+i, jx(x), 1), (1,x) )
4
5    mapRight ( y ):
6      for each i in 0..n−1
7        emit ( ((hashCode(gy(y)) % m)+m*i, jy(y), 2), (2,y) )
8
9    reduce ( ( partition ,joinkey,tag), values ):
10     if ( partition != current_partition )
11       flush (H)
12       current_partition ← partition
13     // (1,x) tuples arrive before (2,y) tuples in values
14     for each leading (1,x) tuple in values
15       insert x into xs
16     for each (2,y) tuple in the rest of values
17       for each x in xs
18         key ← (gx(x),gy(y))
19         if (H[key] is null)
20           H[key] ← zero
21         H[key] ← acc( (x,y), H[key] )
22
23   cleanup ( ):
24     flush (H)
```

Fig. 2. Map-Reduce pseudo-code for GroupByJoin(jx, jy, gx, gy, acc, zero, h, X, Y)

which applies the function h to each key-value pair in the key-value map H and emits the results to the output. Similar to a regular reduce-side join on Map-Reduce [14], our group-by join uses two mappers, mapLeft and mapRight, for each of the inputs, X and Y, respectively. Both mappers emit pairs of key-values. A mapper value takes the form (tag,data), where data is the input data and tag is the source number 1 or 2, to specify the input source (X or Y). A mapper key is a triple (partition,joinkey,tag), where partition is one of the $n * m$ partitions, and joinkey is the join key value, jx(x) or jy(y). The partition number of a partition (i, j) in the grid of $n * m$ partitions is equal to $i * m + j$. The two mappers replicate the X values m times and the Y values n times (associated with different partition numbers). A value $x \in$ X is sent to all the row partitions (gx(x) mod n, $*$) and a value $y \in$ Y is sent to all the column partitions ($*$, gy(y) mod m). Hadoop Map-Reduce supports custom partitioning, grouping, and sorting functions that control the shuffling of the map results to the reducers. In our Hadoop Map-Reduce implementation,

- the partition function returns the partition value of the mapper key,
- the grouping function returns the pair (partition,joinkey), and
- the sorting is based on partition (major order), joinkey (minor order), and tag (sub-minor order).

That is, each partition will contain multiple reduce groups, one for each join key. For each partition p and for each different join key value v, the grouping values in the reducer method, reduce, will contain all the tuples from $x \in X$ and $y \in Y$ that are shuffled to this partition and satisfy $jx(x) = jy(y) = v$. For matrix multiplication, when X is an $N * K$ matrix and Y is an $K * M$ matrix, the size of values will be $N/n + M/m$ (one column from the X horizontal partition and one row from the Y vertical partition), while the size of hash table H will be $(N * M)/(n * m)$. The number of partitions may be larger than the number of worker nodes (the reducers). That is, each reducer may receive multiple partitions, and each partition may contain multiple groupings. Each grouping is handled separately by the reduce method, and the results of processing each partition is emitted by flush(H) at the end of each partition (when the partition number changes). The result of processing each partition are stored in the hash table H, of maximum size $(N * M)/(n * m)$. That is, we must select n and m to be the minimum values so that H can fit in memory. That is, if there is available memory to fit $\mathcal{T}$ tuples, then $(N * M)/(n * m) = \mathcal{T}$. Our goal is to minimize data replication, which is equal to $N * K * m + K * M * n$. That is, we want to minimize $N/n + M/m$ (if we divide by the constants K and $n * m$). This is possible, when $N/n = M/m = \sqrt{\mathcal{T}}$. Internally though, done implicitly by Hadoop Map-Reduce, each reducer node sorts and groups its entire partition (which contains $N * K/n + K * M/m$ tuples) before reduction, which is done with external sorting at each reducer.

7   Translating Queries to GroupByJoin Operations

Based on the discussion in the Introduction, it would be hard to use source-to-source transformations to put queries, such as matrix multiplication and shortest distance, into an algebraic form, such as GroupByJoin, because query syntax may take many different equivalent forms, which have to be recognized by these source-to-source transformations. Instead, our approach is to translate queries into their default algebraic forms and then normalize and rewrite these forms using algebraic rules.

The MRQL algebra used in this section has already been described in our previous work [4]. The most important algebraic operation in the MRQL algebra is cMap (also known as concat-map or flatten-map in functional programming languages), which generalizes the select, project, join, and unnest operators of the nested relational algebra. Given two arbitrary types α and β, the operation cMap(f, X) maps a bag X of type $\{\alpha\}$ to a bag of type $\{\beta\}$ by applying the function f of type $\alpha \rightarrow \{\beta\}$ to each element of X, yielding one bag for each element, and then by merging these bags to form a single bag of type $\{\beta\}$. Using a set former notation on bags, it is expressed as follows:

$$\text{cMap}(f, X) = \{ z \,|\, x \in X, z \in f(x) \} \tag{1}$$

Given an arbitrary type κ that supports value equality $(=)$, an arbitrary type α, and a bag X of type $\{(\kappa, \alpha)\}$, the operation groupBy(X) groups the elements of the bag X by their first component and returns a bag of type $\{(\kappa, \{\alpha\})\}$, where the first component of each tuple is a unique group-by key and the second is the group (a bag) that contains all values that correspond to this key. For example, groupBy($\{(1, \text{"A"}), (2, \text{"B"}), (1, \text{"C"})\}$) returns $\{(1, \{\text{"A"}, \text{"C"}\}), (2, \{\text{"B"}\})\}$. Although any join $X \bowtie_{j_x(x)=j_y(y)} Y$ can be expressed as a nested cMap, to facilitate the creation of physical plans for joins, the MRQL algebra provides a special join operator:

$$
\begin{aligned}
\text{join}&(j_x, j_y, h, X, Y) \\
&= \{\, h(x, y) \mid x \in X,\ y \in Y,\ j_x(x) = j_y(y) \,\} \\
&= \text{cMap}(\lambda x.\, \text{cMap}(\lambda y.\, \textbf{if } j_x(x) = j_y(y)\, \textbf{then } \{h(x, y)\}\, \textbf{else } \{\ \}, Y),\ X)
\end{aligned}
$$

where an anonymous function $\lambda x.\, e$ specifies a unary function (a lambda abstraction) f such that $f(x) = e$. This operation joins two bags, X of type $\{\alpha\}$ and Y of type $\{\beta\}$, using the join functions, j_x of type $\alpha \to \kappa$ and j_y of type $\beta \to \kappa$, and combines the joining values using the function h of type $(\alpha, \beta) \to \gamma$, deriving a bag of type $\{\gamma\}$. Finally, aggregations are captured by the operation reduce($acc, zero, X$), which reduces the elements of a bag X of type $\{\alpha\}$ into a value of type β, using an accumulator acc of type $(\alpha, \beta) \to \beta$ and a zero value $zero$ of type β. For example, reduce($\lambda(x, s).\, x + s, 0, \{1, 2, 3\}$) = 6.

The algebraic terms derived from MRQL queries can be normalized using rewrite rules, such as:

$$
\text{cMap}(f, \text{cMap}(g, S)) \to \text{cMap}(\lambda x.\, \text{cMap}(f, g(x)), S) \tag{2}
$$

that fuses two cascaded cMaps into a nested cMap, thus avoiding the construction of the intermediate bag. This rule can be proven directly from the cMap definition in Eq. (1):

$$
\begin{aligned}
\text{cMap}&(f, \text{cMap}(g, S)) \\
&= \{\, z \mid w \in \{\, y \mid x \in S,\ y \in g(x) \,\},\ z \in f(w) \,\} \\
&= \{\, z \mid x \in S,\ y \in g(x),\ z \in f(y) \,\} \\
&= \{\, z \mid x \in S,\ z \in \{\, w \mid y \in g(x),\ w \in f(y) \,\} \,\} \\
&= \text{cMap}(\lambda x.\, \text{cMap}(f, g(x)), S)
\end{aligned}
$$

In addition, a cMap can be fused with a join resulting to a join:

$$
\begin{aligned}
\text{join}(\,&j_x,\ j_y,\ h,\ X,\ \text{cMap}(\lambda y.\, \{f(y)\}, Y)\,) \\
&\to \text{join}(\,j_x,\ \lambda y.\, j_y(f(y)),\ \lambda(x, y).\, h(x, f(y)),\ X,\ Y\,) \tag{3} \\
\text{cMap}(\,&\lambda v.\, \{f(v)\},\ \text{join}(\,j_x, j_y, h, X, Y\,)) \\
&\to \text{join}(\,j_x,\ j_y,\ \lambda(x, y).\, f(h(x, y)),\ X,\ Y\,)) \tag{4}
\end{aligned}
$$

In our framework, GroupByJoin operations are derived from algebraic forms with the help of the following rule:

```
cMap( λ(k,s). { h(k,reduce(acc,zero,s)) },
        groupBy( join( jx, jy,
                        λ(x,y). ( (gx(x),gy(y)), (x,y) ),
                        X, Y ) ) )
    → GroupByJoin( jx, jy, gx, gy, acc, zero, h, X, Y )
```

which rewrites an equi-join followed by a group-by to a GroupByJoin. For example, the MRQL query that captures matrix multiplication $X \times Y$:

```
select ( sum(z), i, j )
  from (x,i,k) in X, (y,k,j) in Y, z = x*y
group by i, j
```

is translated into the following algebraic form:

```
cMap( λ((i,j),s). {( reduce(λ(v,c). c+v, 0, s), i, j )},
      groupBy( join( λ(x,i,k). k, λ(y,k,j). k,
                     λ((x,i,k),(y,l,j)). ( (i,j), x*y ),
                     X, Y ) ) )
```

while the MRQL query that captures matrix transpose Y^T:

```
select (y,j,i) from (y,i,j) in Y
```

is translated into the following algebraic form:

```
cMap( λ(y,i,j). {(y,j,i)}, Y )
```

Hence, using Eq. 3, the two cMaps in the composition $X \times Y^T$ are fused into:

```
cMap( λ((i,j),s). {( reduce(λ(v,c). c+v, 0, s), i, j )},
      groupBy( join( λ(x,i,k). k, λ(y,j,k). k,
                     λ((x,i,k),(y,j,l)). ( (i,j), x*y ),
                     X, Y ) ) )
```

which is translated to the following algebraic operation:

```
GroupByJoin( λ(x,i,k). k, λ(y,j,k). k, λ(x,i,k). i, λ(y,j,l). j, λ((x,y),c). c+x*y, 0, λ((i,j),c).
             (c,i,j), X, Y )
```

that combines matrix multiplication with matrix transpose.

8 Performance Evaluation

The platform used for our evaluations is a small cluster of 9 nodes, built on the Chameleon cloud computing infrastructure, www.chameleoncloud.org. This cluster consists of nine m1.medium instances running Linux, each one with 4 GB RAM and 2 VCPUs at 2.3 GHz. For our experiments, we used Hadoop 2.6.0 (Yarn) and MRQL 0.9.6. The cluster frontend was used exclusively as a NameNode/ResourceManager, while the rest 8 compute nodes were used as DataNodes/NodeManagers. For our experiments, we used all the available 16 VCPUs of the compute nodes for Map-Reduce tasks.

We have experimentally validated the effectiveness of our methods using two MRQL queries: Matrix factorization using gradient descent, shown in Fig. 3, and the simple query: multiply(Pmatrix,transpose(Qmatrix)), where multiply and transpose are also given in Fig. 3. Given a matrix R, our matrix factorization query in Fig. 3 calculates the error matrix $E = R - P \times Q^T$ and the factor matrices P and Q, so that R is approximately equal to $P \times Q^T$. For our experiments, we set this query to iterate 10 times and used the learning rate $a = 0.002$ and the normalization factor $b = 0.02$. The matrix to be factorized, Rmatrix, was an $n \times m$

```
1   macro transpose ( X ) {        /* matrix transpose */
2     select (x, j, i )
3       from (x, i, j ) in X
4   };
5   macro multiply ( X, Y ) {      /* matrix  multiplication */
6     select (sum(z), i, j )
7       from (x, i, k) in X, (y, k, j) in Y, z = x*y
8       group by (i, j)
9   };
10  macro mult ( a, X ) {          /* multiplication  by a constant */
11    select ( a*x,  i,  j )
12      from (x, i, j ) in X
13  };
14  macro Cadd ( X, Y ) {          /* cell – wise addition */
15    select ( x+y,  i,   j )
16      from (x, i, j ) in X, (y, i, j ) in Y
17  };
18  macro Csub ( X, Y ) {          /* cell – wise subtraction */
19    select ( x–y, i,   j )
20      from (x, i, j ) in X, (y, i, j ) in Y
21  };
22  macro factorize ( R, Pinit , Qinit ) {    /* matrix  factorization */
23    repeat (E,P,Q) = (R, Pinit , Qinit )
24      step ( Csub(R,multiply(P,transpose(Q))),
25             Cadd(P,mult(a,Csub(mult(2,multiply(E,transpose(Q))),mult(b,P)))),
26             Cadd(Q,mult(a,Csub(mult(2,multiply(E,transpose(P))),mult(b,Q)))) )
27      limit 10
28  };
29  let (E,P,Q) = factorize (Rmatrix,Pmatrix,Qmatrix)
30    in  multiply (P,transpose(Q));
```

Fig. 3. Matrix factorization using gradient descent in MRQL

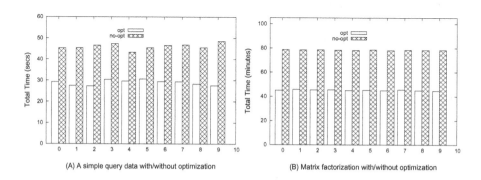

(A) A simple query data with/without optimization

(B) Matrix factorization with/without optimization

Fig. 4. Evaluation of a simple matrix query (A) and matrix factorization (B)

sparse matrix with random integer values between 1 and 5 (resembling the 5-star rating in Netflix) in which only the 10 % of the elements were provided (the rest were zero). The size of m was always kept equal to $10 * n$, while $n * m$ was equal to $100000 + i * 50000$ elements, for $i \in [0, 9]$. That is, $n * m$ took the following values: 100*1000, 122*1220, 141*1410, 158*1580, 173*1730, 187*1870, 200*2000, 212*2120, 223*2230, 234*2340. The initial factor matrices, Pmatrix and Qmatrix, had sizes $n * k$ and $m * k$, respectively, where $k = 10$ for all experiments (a low rank), and were initialized with random values between 1 and 5.

For both MRQL queries, we perform our evaluations in two modes: with and without inter-operation optimization. With inter-operation optimization means that matrix operations were defined using macros so that compositions of operations are fused into one operation, thus avoiding the creation of intermediate results (which Hadoop Map-Reduce must store in the HDFS). Without inter-operation optimization means that the matrix operations were defined as opaque functions, which have to be evaluated as is, thus offering no opportunities for optimization. The results for the simple query multiply(Pmatrix,transpose(Qmatrix)) are shown in Fig. 4A. The results look very similar for different data sizes (100K through 145K tuples) because all matrices (including the intermediate results) are split into 16 files (one for each compute node in the HDFS) and each file can fit into one HDFS block (64MBs) regardless of its size. We can see in Fig. 4A that there is improvement even for just two operations: matrix multiplication and transpose. With inter-operation optimization, these two operations are fused into a single one, a GroupByJoin, which runs in about the same time as matrix multiplication alone. The results for matrix factorization are shown in Fig. 4B. Here, the improvement is even more substantial (the optimized query takes about half the time of the non-optimized one) since the results of all these optimizations are aggregated and repeated at each iteration step.

9 Conclusion

We have presented a general framework for optimizing SQL-like queries that capture array-based computations on sparse arrays. In contrast to related work, we do not provide a library of predefined array operations. Instead, we are letting programmers express their array operations using normal SQL-like syntax, but, at the same time, we provide an optimization framework that translates these queries into efficient distributed array operations. That way, we are able to achieve inter-operation optimization that would be infeasible if these operations were expressed as black boxes.

Acknowledgments. This work is supported in part by the National Science Foundation under the grant CCF-1117369. Our performance evaluations were performed at the Chameleon cloud computing infrastructure, www.chameleoncloud.org, supported by NSF.

References

1. Buck, J., Watkins, N., Lefevre, J., Ioannidou, K., Maltzahn, C., Polyzotis, N., Brandt, S.A.: SciHadoop: array-based query processing in hadoop. In: SC 2011
2. Das, A., Afrati, F.N., Salihoglu, S., Ullman, J.D.: Upper and lower bounds on the cost of a map-reduce computation. In: VLDB 2013
3. Dean, J., Ghemawat, S.: MapReduce: simplified data processing on large clusters. In: OSDI 2004
4. Fegaras, L., Li, C., Gupta, U.: An optimization framework for map-reduce queries. In: EDBT 2012
5. Fegaras, L., Li, C., Gupta, U., Philip, J.J.: XML query optimization in map-reduce. In: International Workshop on the Web and Databases (WebDB) (2011)
6. Apache Flink. http://flink.apache.org/
7. Geijn, R.A., Watts, J.: SUMMA: scalable universal matrix multiplication algorithm. Concurrency: Pract. Experience **9**(4), 255–274 (1997)
8. Ghoting, A., Krishnamurthy, R., Pednault, E., Reinwald, B., Sindhwani, V., Tatikonda, S., Tian, Y., Vaithyanathan, S.: SystemML: declarative machine learning on MapReduce. In: IEEE International Conference on Data Engineering (ICDE) (2011)
9. Apache Hadoop. http://hadoop.apache.org/
10. Apache Hama. http://hama.apache.org/
11. Apache Hive. http://hive.apache.org/
12. Koren, Y., Bell, R., Volinsky, C.: Matrix factorization techniques for recommender systems. In: IEEE Computer, August 2009
13. Kraska, T., Talwalkar, A., Duchi, J., Griffith, R., Franklin, M., Jordan, M.I.: MLbase: a distributed machine learning system. In: Conference on Innovative Data Systems Research (2013)
14. Lin, J., Dyer, C.: Data-intensive text processing with MapReduce. Book preproduction manuscript, April 2010
15. Low, Y., Gonzalez, J., Kyrola, A., Bickson, D., Guestrin, C., Hellerstein, J.M.: Distributed GraphLab: a framework for machine learning and data mining in the cloud. In: VLDB 2012
16. Malewicz, G., Austern, M.H., Bik, A.J.C., Dehnert, J.C., Horn, I., Leiser, N., Czajkowski, G.: Pregel: a system for large-scale graph processing. In: PODC 2009
17. Apache MRQL (incubating). http://mrql.incubator.apache.org/
18. Olston, C., Reed, B., Srivastava, U., Kumar, R., Tomkins, A.: Pig Latin: a not-so-foreign language for data processing. In: SIGMOD 2008
19. Apache Spark. http://spark.apache.org/
20. Soroush, E., Balazinska, M., Wang, D.: ArrayStore: a storage manager for complex parallel array processing. In: SIGMOD 2011
21. Shinnar, A., Cunningham, D., Herta, B., Saraswat, V.: M3R: increased performance for in-memory Hadoop jobs. In: VLDB 2012
22. The SciDB Development Team: overview of SciDB: large scale array storage, processing and analysis. In: SIGMOD 2010
23. Valiant, L.G.: A bridging model for parallel computation. Commun. ACM **33**(8), 103–111 (1990)

Decision Support Systems, and Learning

Creative Expert System: Result of Inference and Machine Learning Integration

Bartlomiej Sniezynski[1]([✉]), Grzegorz Legien[1], Dorota Wilk-Kołodziejczyk[1,2],
Stanislawa Kluska-Nawarecka[2], Edward Nawarecki[1], and Krzysztof Jaśkowiec[2]

[1] AGH University of Science and Technology, Al. Mickiewicza 30,
30-059 Krakow, Poland
bartlomiej.sniezynski@agh.edu.pl

[2] Foundry Research Institute in Krakow, Zakopianska Street 73, Krakow, Poland

Abstract. This paper presents an idea of a creative expert system. It is based on inference and machine learning integration. Execution of learning algorithm is automatic because it is formalized as applying a complex inference rule. Firing such a rule generates intrinsically new knowledge: rules are learned from training data, which consists of facts stored already in the knowledge base. This new knowledge may be used in the same inference chain to derive a decision. Complex rules may also represent other procedural activities, like searching databases. Such a solution makes the reasoning process more creative and allows to continue reasoning in cases when the knowledge base does not have appropriate knowledge explicit encoded. In the paper appropriate model and inference algorithm are proposed. The idea is tested on a decision support system in a casting domain.

Keywords: Web-based expert system · Logic of plausible reasoning · Knowledge representation and processing

1 Introduction

Traditional reasoning techniques applied in AI (e.g. classical logic, rule-based systems based on classical logic [11,17], fuzzy logic [23], Bayesian Networks [15]) offer convergent interpretation of the stored knowledge, which does not provide new knowledge. Machine learning techniques may be creative and provide diversity but are not integrated with inference process. In this paper a method to integrate these two approaches is proposed. Execution of learning algorithm is defined as a complex inference rule executed in an inference chain if the reasoning process is not able to continue classical reasoning. Training data consists of facts stored already in the knowledge base. The new knowledge may be used in the same inference chain to derive a decision.

This work is a continuation of [10] in which machine learning was not applied. We have chosen the same knowledge representation and reasoning formalism: the Logic of Plausible Reasoning (LPR) [6] and added complex inference rules. As a

© Springer International Publishing Switzerland 2016
S. Hartmann and H. Ma (Eds.): DEXA 2016, Part I, LNCS 9827, pp. 257–271, 2016.
DOI: 10.1007/978-3-319-44403-1_16

result, our implementation combines many knowledge manipulation techniques during reasoning. It is able to use a background knowledge, simple proof rules (such as Modus Ponens or generalization) or complex ones (machine learning or searching algorithms) to infer a decision.

The solution we propose allows to create a creative expert system, which, in a case there is no appropriate knowledge in the knowledge base, instead of getting stuck, automatically creates intrinsically new knowledge to continue the reasoning process.

In the following sections related research is discussed, the Creative Reasoning Model and inference algorithm are presented. Next, LPR basics and the software are described. Results of experiments in a domain of material choice support for casting (knowledge base and reasoning scenarios) conclude the work.

2   Related Research

Integration of expert systems and machine learning was analyzed some time ago. A system presented in [5] is based on Neural logic networks corresponding to three-valued logic. System allows for adaptive learning of new rules from its experience.

In [22] neural network was also applied to overcome brittleness of classical expert systems. It is used for choosing the most appropriate questions for the current case. Description of user's interaction with system is collected as training data for the network.

In [21] adaptive expert system is proposed for aircraft maintenance. It recommends the most accurate action for symptoms reported by user. Like in examples above, learning uses historical data (in this case repairs register) to update association weights between symptoms and actions. Certainty of suggested diagnosis is increased in case of successful prediction or decreased in the other case. Symptoms may be also combined using generalization.

In the solutions presented above machine learning algorithms are not part of the formal reasoning system. Therefore the integration of machine learning and reasoning is not complete.

In CoMES system [4] authors attempted to join many popular techniques from Artificial Intelligence and Software Engineering. Machine learning is used for updating the knowledge base, which can be accessed by few algorithms in parallel. The system uses agent architecture to integrate knowledge from human experts and other expert systems.

Our solution is based on the Inferential Theory of Learning (ITL). This theory was created by Michalski [12]. Michalski et al. also developed ITL partial implementation - an INTERLACE system [3]. The system can generate sequences of knowledge operations that will enable the derivation of a target trace from the input hierarchies and traces. Machine learning was not integrated in this system.

Logic of Plausible Reasoning (LPR) used in our system was proposed by Collins and Michalski [6]. The goal of that work was to identify reasoning patterns used by humans and create a formal system, which would be able to represent these patterns. The objective set by the creators has caused that LPR is significantly different from other known knowledge representation methods: there are many inference rules in LPR and many parameters are specified for representing the uncertainty of knowledge.

3   Creative Inference Model

The proposed *Creative Inference Model* assumes that the knowledge representation and reasoning method, can be formalized as a Labeled Deductive System (LDS) [8]. Knowledge is represented by formulas. Inference process consists of a sequence of *Knowledge transmutation* applications.

Knowledge transmutation can be represented as the following triple:

$$kt = (p, c, a), \tag{1}$$

where p is a (possibly empty) premise or precondition, c is a consequence (pattern of formula(s) that can be generated) and a is an action (empty for simple transmutations) that should be executed to generate consequence if premises are true according to the knowledge base. As a result we can represent three types of inference rules:

- simple (e.g. Modus Ponens proof rule);
- complex (using machine learning, e.g. rule induction algorithms or clustering methods);
- search (database or web searching procedures).

We assume that every transmutation has its cost assigned ($cost_{kt}$). The cost should represent its computational complexity and (or) other important resources that are consumed (e.g. database access or search engines fees). Usually, simple transmutations have a low cost, search transmutations have a moderate cost and complex ones have a high cost.

To manage uncertainty a *label algebra* may be used:

$$\mathcal{A} = (A, \{f_{kt}\}). \tag{2}$$

A is a set of labels which estimate uncertainty of formulas. *Labeled formula* is a pair $f : l$ where f is a formula and $l \in A$ is a label. A finite set of labeled formulas can be considered as a knowledge base. Functions f_{kt} are used to calculate labels of reasoning results. If $kt = (p, c, a)$ and p is a conjunction of premises α_i (of length n) then the plausible label l of its conclusion c is calculated using f_{kb} : $A^n \rightarrow A$, hence $l = f_{kt}(l_1, ..., l_n)$, where l_i is a label of α_i.

Creative Inference Algorithm (see Algorithm 1) is an adaptation of LPR proof algorithm [19], where proof rules are replaced by more general knowledge transmutations. It is based on AUTOLOGIC system developed by Morgan [14]. To limit the number of nodes and to generate optimal inference chains, algorithm A* [9] is used.

Input data is a set of labeled formulas KB – a knowledge base and a hypothesis (question) represented by the formula φ, which should be derived from KB.

Input: φ – formula, KB – finite set of labeled formulas
Output: If $\exists l \in A$ such that $\varphi : l$ can be inferred from KB: success, P –
    inference chain of $\varphi : l$ from KB; else: failure
$T :=$ tree with one node (root) $s = [\varphi]$;
$OPEN := [s]$;
while $OPEN$ *is not empty* **do**
 | $n :=$ the first element from $OPEN$;
 | Remove n from $OPEN$;
 | **if** $n = []$ **then**
 | | Generate proof P using path from s to n;
 | | Exit with success;
 | **end**
 | **if** *the first formula of n represents action* **then**
 | | Execute action;
 | | **if** *action was successfull* **then**
 | | | add action's results to KB;
 | | | E:=nodes generated by removing from n action formula;
 | | **end**
 | **else**
 | | $K :=$ knowledge transmutations, which consequence can be unified with
 | | first formula of n;
 | | $E :=$ nodes generated by replacing the first formula of n by premises
 | | and action of transmutations from K and applying substitutions from
 | | unifier generated in the previous step;
 | | **if** *the first formula from n can be unified with element of KB* **then**
 | | | Add to E node obtained from n by removing the first formula and
 | | | applying substitutions from unifier;
 | | **end**
 | **end**
 | Remove from E nodes generating loops;
 | Append E to T connecting nodes to n;
 | Insert nodes from E to $OPEN$;
end
Exit with failure;

Algorithm 1: Creative Inference Algorithm

If there exist a label $l \in A$ such, that $\varphi : l$ can be inferred from KB, appropriate inference chain is returned, else procedure exits with failure.

Agent's experience and the context description should be also stored in KB as LPR formulas.

This algorithm generates a tree T, which nodes (N) are labeled by sequences of formulas. Every edge of T is labeled by a knowledge transmutation, which consequence can be unified with the first formula of a parent node or is labeled by the term $kb(l)$ if the first formula of a parent node can be unified with $\psi : l \in KB$. s is the root of T. It is labeled by $[\varphi]$. The goal is to generate a node labeled by empty set of formulas.

As it was mentioned, to limit the number of nodes expanded, A* algorithm may be used. Therefore nodes in the $OPEN$ sequence can be ordered according to the values of evaluation function $f : N \to \mathbb{R}$, which is defined as follows:

$$f(n) = g(n) + h(n), \tag{3}$$

where $g : N \to \mathbb{R}$ represents the actual cost of the inference chain (sum of $cost_{kt}$ for transmutations applied), using knowledge transmutation costs and label of φ that can be generated, and $h : N \to \mathbb{R}$ is a heuristic function which estimates the cost of the path from n to the goal node (e.g. minimal knowledge transmutation cost multiplied by the length of n can be used).

4   LIIS System

In this section the LPR Intelligent Information System (LIIS) used in experiments is described. At the beginning LPR is introduced, next main features and implementation details of system are described. Finally, inference algorithms and label algebra are presented.

4.1   Introduction to LPR

We have chosen LPR for basic knowledge representation and reasoning. If needed, instead of LPR another technique, which can be formulated using LDS, may be used.

The language used by LPR consists of a countable set of constants C, variables X, the seven relational symbols, and logical connectives $\to$ and $\wedge$. Formally, it is a quadruple: $L = (C, X, \{V, H, B, E, S, P, N\}, \{\to, \wedge\})$. The relational symbols (V, H, B, E, S, P, N) are used for defining the following relationships:

- H defines the hierarchy between concepts; expression $H(o_1, o, c)$ means that o_1 is o in a context c;
- B is used to present the fact that one object is placed below another one in a hierarchy;
- V is used for representing statements: $V(o, a, v)$ is a representation of the fact that object o has an attribute a equal to v;
- E is used for representing relationships; the notation $E(o_1, a_1, o_2, a_2)$ means that values of attribute a_1 of object o_1 depend on attribute a_2 of the second object o_2;
- S determines similarity between objects; $S(o_1, o_2, c)$ represents the fact that o_1 is similar to o_2 in a context c;
- P represents order between concepts: $P(o_1, o_2)$ means that concept o_1 precedes concept o_2;
- N is used for comparing the concepts; $N(o_1, o_2)$ means that concept o_1 is different from the concept o_2. This relation do not appear in the knowledge base only as a premise of some implication.

To represent vagueness of the knowledge it is possible to extend statement definition and allow to use composite value $[v_1, v_2, \ldots, v_n]$ (list of elements of C). It can be interpreted that object o has an attribute a equal to v_1 or v_2, ..., or v_n. If $n = 1$ instead of $V(o, a, [v_1])$ notation $V(o, a, v_1)$ is used.

In statements, value should be placed below an attribute in a hierarchy: if $V(o, a, [v_1, v_2, \ldots, v_n])$ is in a knowledge base, there should be also $H(v_i, a, c)$ for any $1 \leq i \leq n, c \in C$.

LPR formula means every atomic formula: $H(o_1, o_2, c), B(o_1, o_2), V(o, a, v), E(o_1, a_1, o_2, a_2), S(o_1, o_2, c), P(o_1, o_2)$, where $o, o_1, o_2, a, a_1, a_2, c, v \in C$, a conjunction of atomic formulas and implications in the form of $\alpha_1 \wedge \alpha_2 \wedge \ldots \alpha_n \rightarrow V(o^\alpha, a^\alpha, v^\alpha)$, where $n \in \mathbb{N}, n > 0$. It is assumed that α_i has the form of $V(o_i^\alpha, a_i^\alpha, v_i^\alpha)$, $P(v_i^\alpha, w_i^\alpha)$ or $N(v_i^\alpha, w_i^\alpha)$, and $o^\alpha, o_i^\alpha, a^\alpha, a_i^\alpha, v^\alpha, v_i^\alpha, w_i^\alpha \in C \cup X$ for $1 \leq i \leq n$.

The most commonly used proof rules operate on the statement (others can be found in [6]). Index attached to the name of the rule tells us what is transformed: o is an object, and v is the value. These rules are shown in Table 1. GENo and SPECo are generalization and specialization of objects in statements, respectively, while GENo and SPECo are similar transformations of values. SIMo represents reasoning by analogy (similarity) between objects, while SIMv represents analogy of values. MP is the classical Modus Ponens inference rule.

Table 1. Rules transforming object-attribute-value triples

$$GEN_o \; \frac{\begin{array}{c} H(o_1, o, c) \\ E(o, a, o, c) \\ V(o_1, a, v) \end{array}}{V(o, a, v)} \qquad SPEC_o \; \frac{\begin{array}{c} H(o_1, o, c) \\ E(o, a, o, c) \\ V(o, a, v) \end{array}}{V(o_1, a, v)} \qquad SIM_o \; \frac{\begin{array}{c} S(o_1, o_2, c) \\ E(o_1, a, o_1, c) \\ V(o_2, a, v) \end{array}}{V(o_1, a, v)}$$

$$GEN_v \; \frac{\begin{array}{c} H(v_1, v, c) \\ E(a, o, a, c) \\ H(o_1, o, c_2) \\ B(v, a) \\ V(o_1, a, v_1) \end{array}}{V(o_1, a, v)} \qquad SPEC_v \; \frac{\begin{array}{c} H(v_1, v, c) \\ E(a, o, a, c) \\ H(o_1, o, c_1) \\ V(o_1, a, v) \end{array}}{V(o_1, a, v_1)} \qquad SIM_v \; \frac{\begin{array}{c} S(v_1, v_2, c) \\ E(a, o, a, c) \\ H(o_1, o, c_1) \\ B(v_1, a) \\ V(o_1, a, v_2) \end{array}}{V(o_1, a, v_1)}$$

$$MP \; \frac{\begin{array}{c} \alpha_1 \wedge \ldots \wedge \alpha_n \rightarrow \\ V(o, a, v) \\ \alpha_1 \\ \vdots \\ \alpha_n \end{array}}{V(o, a, v)}$$

4.2 LIIS Main Features

The main function of the system is to perform inference based on the knowledge base available and user-defined hypothesis (also known as query). This function is based on the MILS framework.

The query is a statement which can include constants, variables and numeric values. The hypothesis is verified by the inference engine. If it can be proved to have a non-zero probability, a proof is submitted. If there are several proofs, all of them are presented to the user with the relevant information concerning their credibility.

The user can choose maximum depth of the proof tree and knowledge trans-mutations that are to be used in the process.

Machine learning algorithms may be automatically executed during inference process. Currently two complex knowledge transmutations are defined. Both apply rule induction using AQ [13] or C4.5 [16] algorithms. Consequence c (see (1)) has a form of statement for both transmutations. To limit the computation time, the user may limit the learning for a set of attributes called *category set*. In premise p it is checked if it is possible to generate enough examples from the knowledge base. Examples are divided into training and testing data. Second one is used to estimate strength of the rules learned.

The system provides engine for performing decision support. This function-ality is extensively used in the case study described in the next section. Expert system scenarios use both formulas from base and information provided by a user executing decision support procedure. The user can supply knowledge by selecting answer from list or filling in inputs with numerical values. Questions can be skipped. Scenarios can be developed using GUI.

The application supports edition of knowledge base elements, providing tools facilitating this process. Single formula is created with a form, suggesting object names basing on ones already used in system. Formulas of different types are placed in their own tables, where they can be filtered by the name of one of their object.

4.3   Description of LIIS Implementation

In the description of the developed application an attempt was made to charac-terize the solution, heading for the widest possible use of existing development tools, while providing the functionality needed to effectively meet the functional requirements. As a result, LPR Intelligent Information System is a web applica-tion created with Google Web Toolkit, solution supporting the development of browser-based applications. Technology affects system architecture, dividing it into three logical parts.

Server-side part of system brings realization of main features. Reasoning engine, using LPR-Library, implements inference, machine learning and exper-tise conduction. The engine is used by services responsible for providing these functions to users – Reasoning Service and Expert System Service. LPR Service allows storing and obtaining knowledge base elements from MySQL database with use of Data Access Objects. Persistence layer takes advantage of Hibernate object-relational mapping and stores formulas, expert system scenarios, user data and knowledge bases' metadata. System management is provided via User Service and App Service.

Client part of application contains JavaScript views, compiled from Java classes. It communicates with server-side with Remote Proxy Calls, where data is transported via HTTP as Data Transfer Objects – plain, serializable Java classes shared by both sides. LIIS is build with Maven, dependencies management tool. It works on Apache Tomcat web server.

In current implementation of reasoning engine the following formulas can be used: statements (V), hierarchy (H), similarity (S), implications, P and N. Applied transformations includes rules of object and value generalization, specialization, similarity, Modus Ponens, ordering and hierarchy transitivity.

4.4   Label Algebra

In our research we use a simplified version of label algebra defined in [18]. The system uses the following coefficients representing certainty, which are real numbers from the range [0,1]: for formulas V – confidence, for formulas S – similarity rate, for formulas H – typicality and dominance, for formulas B – confidence, for formulas I (implications) – strength. Formulas P and N are certain (have label equal to 1.0).

The certainty label of the statement, which is conclusion of proof rule r_i is a product of the label of each of the premises:

$$f_{r_i}(l_1, l_2, \ldots, l_{n_i}) = \prod_{i=1}^{n_i} l_i \tag{4}$$

If premise represents hierarchy, typicality is used in object transformation and dominance is used in value transformation.

5   Experimental Results

In order to examine the functionality of the system, various scenarios have been developed. Below the selected three scenarios are presented. The description of the scenarios contains the hypotheses, user responses, the proof obtained, the proof rules used during inference, and the description the inference process.

The goal of the decision support system is to find a material which fits the requirements and reduce the cost of the casting through the use of a new material.

5.1   Knowledge Base

The model was tested before on two small domains [1,2]. To show advantages of the proposed solution in a larger scale, a decision support system was developed in a domain, which is complex enough, contain hierarchies of objects, and is characterized by a number of parameters of an intuitive nature, difficult to measure. The system supports the choice of metal products manufacturing technology, casting technology included. Knowledge base consists of more than 700 formulas.

Often the choice of technology for the manufacture of metal item and of the material from which this item is to be produced stems from the experience and knowledge of the engineer designing this item. These human aspects are difficult to represent using formal languages. When the task of designing machine parts is undertaken, parameters that the item should have and the related operational and utility functions must be taken into account. This also applies to the case of the conversion of material. A new type of material must provide at least the same mechanical properties and reliability as the original one. The choice of the method of manufacture is affected by the batch size, dimensional accuracy, dimensions, complexity, the type of the necessary machining and heat treatment, etc. All these factors also create costs. In this situation, the problem which the designer of a particular product (machine part) has to face and solve consists in selecting the material and the technology of its manufacture, which will ensure that the specific technical requirements are satisfied, while allowing the maximum reduction of production costs.

In the application of LIIS system considered here it is very important to indicate the appropriate material, which could replace the traditional materials (forged steel, cast steel). This material can be Austempered Ductile Iron (ADI), which has a favorable relationship between the tensile strength (Rm) and elongation (A), offering at the same time significantly lower manufacturing costs (savings of approx. 20 %). The decision about the possible use of ADI must be based, however, on more detailed analysis of requirements imposed on a particular product and its characteristics, to mention as an example the damping capacity, corrosion resistance, dimensions, the batch size, and the weight of a single item.

It is accepted that the low-volume production includes up to 50 pieces of castings weighing between 0 and 25 kg. Low-volume production also comprises up to 10 cast pieces if the casting weight is 25–500 kg. If the casting weight exceeds 500 kg, the small-volume production comprises 1 cast piece. Medium-volume production covers 50–5000 pieces for the weight range between 0 and 25 kg, 10–100 pieces for the weight range of 25–500 kg, and 2–10 pieces for the total weight of more than 500 kg. All values above this level stand for the large-lot production.

The batch size (production volume) is dependent on the weight of product for each of the three type ranges. This helps to better understand the comparison of prices for the same product made from ADI and carburized steel for different batch sizes and product weights.

Core of the knowledge base are hierarchies. They were defined during consultations with experts. They represent facts that ADI is a kind of cast_iron and define its 63 subtypes (ADI_GSJ-1400-1, ADI_1, ADI_2, ..., ADI_31, ..., ADI_34, ADI_41, ..., ADI_44, ADI_51, ..., ADI_68, ...). Context is related to cost, production volume, application and mechanical properties. The first label value (typicality) is high (often equal to 1.0), which means that certainty of specialization of objects and values (SPECo and SPECv) will be also high. The second label (dominance) is low.

1. H(adi, cast_iron, cost):0.8:0.1
2. H(adi, cast_iron, volume_production):0.8:0.1

3. H(adi_gsj-1400-1, adi, application):1.0:0.1
4. H(adi_4, adi, application):1.0:0.1
5. H(adi_42, adi, application):1.0:0.1

In statements minimum elongation and tensile strength of selected steel grades are expressed. Labels representing certainty have high values. Similar statements are prepared for other types of ADI (like ADI_4, ADI_42, ADI_52 etc.). Some parameters are not known and corresponding statements are missing.

1. V(adi, application, rake):1.0
2. V(adi_gsj-1400-1, minimal_elongation_A, 1):1.0
3. V(adi_gsj-1400-1, tensile_strength_Rm, 1400):1.0
4. V(engjs_14001, chemical_composition_c, 3.462-3.524):1.0
5. V(adi_gsj-1400-1, austenization_time, 105-inf):1.0
6. V(adi_gsj-1400-1, austenization_temp, 867.5-895):1.0
7. V(adi_gsj-1400-1, hardening_time, 187.5-inf):1.0

The rest of formulas have form of implication. Four of them allow to recommend a material for production (see below). They have conclusion V(casting, material_alternative, X). The more parameters are checked (and more premises the rule has), the more certain the answer is. The first implication checks application, costs, tensile strength and minimal elongation and it has certainty 1.0. Fourth rule checks only application, therefore its certainty is equal to 0.25 Other rules allow to predict the production costs assuming a particular batch size and product weight.

1. V(casting, application_required, A) ∧ V(X, application, A) ∧
   V(casting, cost_required, COST_MAX) ∧ V(X, cost, COST_CALCULATED)
   ∧ P(COST_CALCULATED, COST_MAX) ∧
   V(casting, tensile_strength_Rm_required, STRENGTH_MIN)
   ∧ V(X, tensile_strength_Rm, C) ∧ P(STRENGTH_MIN, C)
   ∧ V(casting, minimal_elongation_A_required, ELONG_MIN)
   ∧ V(X, minimal_elongation_A, E) ∧ P(ELONG_MIN, E)]
   →V(casting, material_alternative, X):1.0
...
4. V(casting, application_required, A) ∧ V(X, application, A)] →V(casting, material_alternative, X):0.25

5.2   Scenario 1

The first scenario illustrates a simple case, in which all the knowledge necessary for reasoning is given explicitly in knowledge base. Application of the material is a rake[1], the maximum cost limit is equal to 15, product weight is heavy, the

---

[1] A rake is a tool used in sewage-treatment plants. Its main task is to mix organic materials such as straw, grass, hay, etc. with semi-liquid material obtained from the municipal waste-water treatment after suitable processing, and with soil and refining additives to obtain mineral fertilizer used in agriculture.

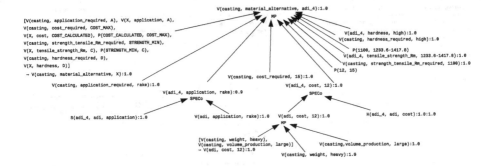

Fig. 1. Graphical presentation of the proof found in scenario 1

batch size is large, minimal tensile strength Rm is equal to 1100 and hardness is high. As a result, the system recommends ADI_4 with confidence 1.0.

The proof was obtained by double application of the Modus Ponens (MP) rule and double object specialization (SPECo) rule. It is presented in Fig. 1. In the first step, the MP rule was applied to implication no. 1, which means that if the required application of casting under consideration is equal to A (premise 1) and is the same as the application allowed for an alternative material in the rule marked by variable X (premise 2), the required maximum cost is equal to COST_MAX (premise 3), and the cost calculated for an alternative material is equal to COST CALCULATED (premise 4) and is lower than the maximum cost (premise 5), the required minimum tensile strength Rm is STRENGTH_MIN (premise 6), and for an alternative material it is C (premise 7) and is higher than STRENGTH_MIN (premise 8), and required hardness described as HARDNESS (premise 9) is the same as for alternative material (premise 10), then the alternative material (X) should be used with confidence 1.0.

Premises 1 and 3 can be adapted to the knowledge base elements or answers to questions. Premise 2 (application acceptable for ADI_4) was inferred using SPECo object specialization rule because ADI_4 is a typical ADI in terms of application, and it is known that ADI may be used to produce rakes. Similarly, premise 4 was derived using SPECo specialization rule and knowing that ADI_4 is a typical ADI in terms of the cost of obtaining it and calculating this cost for ADI based on the mass of the casting and using the implication no. 13 as above. Premises 5–10 can be unified with the knowledge base elements or answers to questions.

5.3 Scenario 2

User requirements in this scenario are the following. Application is also rake, the maximum allowable cost is 15, casting weight (diameter) is medium this time, batch size is large, minimum tensile strength Rm is lower: 1000, hardness is high.

The last parameter is problematic because there are some materials for which it is not measured. It is the case for ADI_42 which matches the other criteria.

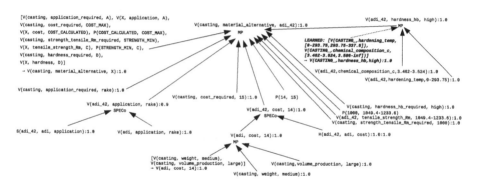

Fig. 2. Graphical presentation of the proof found in scenario 2

However, due to presence of other examples, a classifier predicting hardness may be learned and applied to this case. Therefore, expert system recommends ADI_42 with confidence 1.0.

The proof was obtained by triple application of the Modus Ponens (MP) rule and double object specialization (SPECo). Most of the inference steps look similar to the first scenario. At the beginning, the MP rule was applied to implication no. 1. Its premises 1, 3, 6, 9 are user's responses. Proof is presented as a diagram in Fig. 2.

Exactly like in first scenario, premises 2 and 4 were inferred using SPECo object specialization rule because ADI_42 is a typical ADI in terms of application and cost. Premises 5 and 8 expressed that inferred cost (variable COST_CALCULATED) or given by user tensile strength (variable C) fits demanded range. Premise 7 was unified with knowledge base. The last premise related to hardness was missing and the system was not able to infer it. Therefore a complex knowledge transmutation was applied. 14 examples described by all available attributes were prepared. One of the rules checked hardening temperature and carbon content:

$$V(CASTING, hardening\_temp, [0 - 293.75, 293.75 - 337.5])$$
$$\wedge V(CASTING, chemical\_composition\_c, [3.462 - 3.524, 3.586 - inf])$$
$$\rightarrow V(CASTING, hardness\_hb, high) : 1.0. \quad (5)$$

Its premises were true for ADI_42 and it allowed to derive its hardness. Therefore ADI_42 was supposed to have high hardness and it was recommended to the user.

5.4   Scenario 3

Requirements in this scenario are the same as in the previous one. However, other implication from machine learning results was selected to predict the hardness.

Proof structure is shown in Fig. 3. Inference is similar to one from the second scenario. The only difference is replacing the learned implication in the last step with implication:

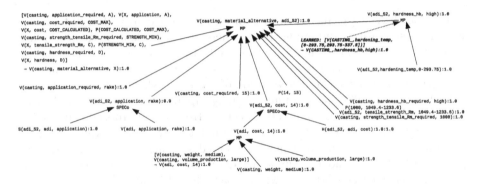

Fig. 3. Graphical presentation of the proof found in scenario 3

$$V(CASTING, hardening\_temp, [0 - 293.75, 293.75 - 337.5])$$
$$\rightarrow V(CASTING, hardness\_hb, high) : 1.0. \quad (6)$$

It has a shorter form – premise related to carbon content is omitted. As a result, casting ADL52 was recommended material with confidence 1.0. In the previous scenario this material was not selected because it was not matching the condition related to the carbon content.

5.5   Time Complexity

We have checked how adding complex knowledge transmutation influences the inference time. We have used the same knowledge base and a scenario for which the answer can be found with and without machine learning. Three attributes were allowed as categories: *hardness*, *application* and *cost*. For maximal proof tree depth equal to 10 the inference time was doubled, while for the depth equal to 80 the time was increased by 33 %.

5.6   Summary

Domain experts reviewed results of presented scenarios. Achieved answers and their certainties was evaluated as correct. Second and third scenario show that MILS model allows to continue reasoning even in cases not covered by the KB. In such cases, recommendations depend on the learned implications. The increase of the inference time after adding complex transmutations is acceptable.

6   Conclusions and Further Works

The presented LIIS system based on LPR allowed to create a web-based expert system for material and technology recommendation in a casting process. The system was tested in many scenarios, three examples of which are described above. Technologists confirmed that the answers are right and the proofs are valid and easy to follow.

The knowledge base created consists of various types of formulas: statements, hierarchies, similarities and implications. Therefore, in the knowledge processing various types of inference patterns are applied (deductive reasoning, generalization, and similarity). As a result, the knowledge and reasoning reflect human way of thinking, what makes the creation of the knowledge base more natural.

Further works will concern adding learning capabilities to the system. Learning module is already implemented (see Sect. 4). However, appropriate knowledge base and use cases are still under construction. Other works concern application in a system in other domains. Knowledge-based systems for telemetry-oriented applications [20] and money laundering detection [7] are under investigation. Also consistency check of the Knowledge Base should be added.

Acknowledgments. The research reported in the paper was supported by the grant of The National Centre for Research and Development (LIDER/028/593/L-4/12/NCBR /2013) and by the Polish Ministry of Science and Higher Education under AGH University of Science and Technology Grant 11.11.230.124.

References

1. Szczepaniak, P., Kacprzyk, J., Niewiadomski, A.: Advances in Web Intelligence. LNCS, vol. 3528. Springer, Heidelberg (2005)
2. Sniezynski, B.: Integration of inference and machine learning as a tool for creative reasoning. In: AAAI Fall Symposium Series, North America, September 2014
3. Alkharouf, N.W., Michalski, R.S.: Multistrategy task-adaptive learning using dynamically interlaced hierarchies. In: Michalski, R.S., Wnek, J. (eds.) Proceedings of the Third International Workshop on Multistrategy Learning (1996)
4. Althoff, K., Bach, K., Deutch, J., Hanft, A., Manz, J., Muller, T., Newo, R., Reichle, M., Schaaf, M., Weis, K.: Collaborative multi-expert-systems realizing knowledge-lines with case factories and distributed learning systems. In: Proceedings of the 3rd Workshop on Knowledge Engineering and Software Engineering (2007)
5. Boon Toh, L., Ho Chung, L., Ah Hwee, T., Hoon Heng, T.: Connectionist expert system with adaptive learning capability. IEEE Trans. Knowl. Data Eng. **3**(2), 200–207 (1991)
6. Collins, A., Michalski, R.S.: The logic of plausible reasoning: a core theory. Cogn. Sci. **13**, 1–49 (1989)
7. Drezewski, R., Sepielak, J., Filipkowski, W.: The application of social network analysis algorithms in a system supporting money laundering detection. Inf. Sci. **295**, 18–32 (2015)
8. Gabbay, D.M.: LDS - Labeled Deductive Systems. Oxford University Press, Oxford (1991)
9. Hart, P., Nilsson, N.J., Raphael, B.: A formal basis for the heuristic determination of minimum cost path. IEEE Trans. Syst. Sci. Cybern. **4**(2), 100–107 (1968)
10. Legień, G., Śnieżyński, B., Wilk-Kołodziejczyk, D., Kluska-Nawarecka, S., Nawarecki, E., Jaśkowiec, K.: Expert system with web interface based on logic of plausible reasoning. In: Chen, Q., Hameurlain, A., Toumani, F., Wagner, R., Decker, H. (eds.) DEXA 2015. LNCS, vol. 9262, pp. 13–20. Springer, Heidelberg (2015)

11. Ligeza, A.: Logical Foundations for Rule-Based Systems. Springer, Heidelberg (2006)
12. Michalski, R.S.: Inferential theory of learning: developing foundations for multistrategy learning. In: Michalski, R.S. (ed.) Machine Learning: A Multistrategy Approach, vol. IV. Morgan Kaufmann Publishers, Burlington (1994)
13. Michalski, R.S., Larson, J.: AQVAL/1 (AQ7) user's guide and program description. Technical report 731, Department of Computer Science, University of Illinois, Urbana, June 1975
14. Morgan, C.G.: Autologic. Logique et Anal. **28**(110–111), 257–282 (1985)
15. Neapolitan, R.E.: Probabilistic Reasoning In Expert Systems: Theory and Algorithms. CreateSpace Independent Publishing Platform, North Charleston (2012)
16. Quinlan, J.: C4.5: Programs for Machine Learning. Morgan Kaufmann, Burlington (1993)
17. Riley, G.: Clips - an expert system building tool. In: Proceedings of Technology 2001 Conference, San Jose, CA (1991)
18. Sniezynski, B.: Probabilistic label algebra for the logic of plausible reasoning. In: Kłopotek, M., et al. (eds.) Intelligent Information Systems 2002. ASC, vol. 17, pp. 267–277. Springer, Heidelberg (2002)
19. Sniezynski, B.: Proof searching algorithm for the logic of plausible reasoning. In: Kłopotek, M., et al. (eds.) Intelligent Information Processing and Web Mining. ASC, vol. 22, pp. 393–398. Springer, Heidelberg (2003)
20. Szydlo, T., Nawrocki, P., Brzoza-Woch, R., Zielinski, K.: Power aware MOM for telemetry-oriented applications using GPRS-enabled embedded devices - levee monitoring use case. In: Proceedings of the 2014 Federated Conference on Computer Science and Information Systems, vol. 2, pp. 1059–1064. IEEE, September 2014
21. Tran, L.P., Hancock, J.P.: An adaptive-learning expert system for maintenance diagnostics. In: Proceedings of the IEEE 1989 National Aerospace and Electronics Conference, NAECON 1989, vol. 3, pp. 1034–1039, May 1989
22. Wiriyacoonkasem, S., Esterline, A.C.: Adaptive learning expert systems. In: Proceedings of the IEEE Southeastcon 2000, pp. 445–448. IEEE (2000)
23. Zadeh, L.A.: Fuzzy sets. Inf. Control **8**, 338–353 (1965)

A Reverse Nearest Neighbor Based Active Semi-supervised Learning Method for Multivariate Time Series Classification

Yifei Li[1], Guoliang He[1,2(✉)], Xuewen Xia[3], and Yuanxiang Li[1,2]

[1] State Key Laboratory of Software Engineering,
Wuhan University, Wuhan, China
{yfli814,glhe,yxli}@whu.edu.cn
[2] College of Computer Science, Wuhan University, Wuhan, China
[3] School of Software, East China Jiaotong University, Nanchang, China
laughkid@163.com

Abstract. Time series widely exist in many areas. In reality, the number of labeled time series data is often small and there is a huge number of unlabeled data. Manually labeling these unlabeled examples is time-consuming and expensive, and sometimes it is even impossible. To reduce manual cost and obtain high confident labeled training data for multivariate time series classification, in this paper a reverse nearest neighbor based active semi-supervised learning method is proposed. First, based on information entropy and distribution density of the training data, a sampling strategy is introduced to select the most informative examples for manual annotation. Second, in terms of the newly labeled example by experts, a reverse nearest neighbor based semi-supervised learning method is presented to automatically and accurately label some confident examples. We evaluate our work with a comprehensive set of experiments on diverse multivariate time series data. Experimental results show that our approach can obtain a confident labeled training data with less manual cost.

Keywords: Multivariate time series · Active learning · Semi-supervised learning

1 Introduction

Time series widely exist in many areas such as speech recognition, finance and engineering. Time series data mining has been attracting great interest, and time series classification is one of fundamental tasks. Traditional time series classification methods usually require significant amount of labeled training data to learn a high-quality classification model. However, the number of labeled data is often smaller in real-life applications, and there is a huge number of unlabeled data as the quick development of network and information technology. On the other hand, manually labeling examples is time-consuming and expensive, and sometimes it is even impossible. Therefore, how to make better use of these unlabeled data for learning a good classification model is critical.

© Springer International Publishing Switzerland 2016
S. Hartmann and H. Ma (Eds.): DEXA 2016, Part I, LNCS 9827, pp. 272–286, 2016.
DOI: 10.1007/978-3-319-44403-1_17

To make best use of unlabeled data for learning a good classifier, recently some efficient semi-supervised learning methods have been introduced to automatically annotate some of unlabeled data by using only handful of labeled data [1–5]. However, existing semi-supervised learning is not efficient to deal with multiple sub-concepts in a class when the size of labeled examples is extremely smaller [6]. Furthermore, semi-supervised methods are difficult to find confident examples with the limited labeled multivariate time series (MTS) data because comparing with univariate time series, MTS data has multiple variables and data distribution is more complex. Therefore, it is a challenging work to obtain sufficient and reliable labeled MTS data from positive and unlabeled data for classification.

To handle the issue, in this paper we focus on combing active learning and semi-supervised learning for annotating the vast majority of original unlabeled MTS data, which could provide much more information to learn a good classifier. First, we present an efficient way to measure the density of an unlabeled example based on its reverse K-nearest neighbors. Then, a valid method is proposed to evaluate the informativeness of unlabeled examples for manual annotation. Next, based on reverse nearest neighbor technique, we advance a semi-supervised learning method to automatically label some confident unlabeled examples in terms of the newly labeled example. Last, we introduce an efficient active semi-supervised learning framework. We make several contributions.

1. Based on information entropy and density, a sampling strategy is proposed to find the most informative unlabeled MTS examples for manual annotation.
2. A reverse nearest neighbor based semi-supervised learning method is presented to automatically and accurately identify some unlabeled MTS data to further enlarge the scale of the labeled training data.
3. An efficient active semi-supervised learning framework is proposed to accurately annotate the vast majority of original unlabeled MTS data as cheap as possible. And the labeled training data could provide the information to learn a good classifier as much as possible.
4. Experimental results on benchmark datasets show that our proposed method is competitive.

The remainder of this paper is organized as follows. In Sect. 2, we review related work and some basic concepts are defined in Sect. 3. Section 4 introduces sample selection strategy and active semi-supervised learning framework. In Sect. 5, we perform a comprehensive set of experiments on several datasets. Finally, we conclude our work and suggest directions for future work in Section 6.

2   Related Work

2.1   Semi-supervised Learning

In general, traditional classification methods need lots of labeled training examples for learning. In practice, labeled data are often insufficient and labeling examples is time-consuming and expensive. To make better use of unlabeled examples, semi-supervised approaches have been proposed to improve the performance of time series classification. For instance, Wei and Keogh proposed a semi-supervised way to automatically label the nearest example closest to labeled examples iteratively [1]. However, it's hard to find a good stopping criterion to decide when to stop the iteration, which leads to automatically labeled examples being not confident.

To find an accurate boundary between positive and negative samples in unlabeled samples, Nguyen et al. proposed an effective technique called LCLC (Learning from Common Local Clusters) [3]. The unlabeled samples are clustered and all samples within a cluster are assumed to be from same class. Then, a chaining approach was applied to find the boundary between positive and negative clusters in terms of their similarity distances. However, the assumption that all samples within a cluster are from same class is unpractical. To overcome the drawbacks of LCLC, Nguyen et al. proposed a novel ensemble based approach to obtain multiple diverse classifiers [4].

Moreover, Mabel et al. [5] delved into the stopping criterion problem in the self-training context. They proposed a family of parameter-free stopping criteria for k-nearest neighbor in positive unlabeled time series classification, in terms of the minimum distances achieved by the k-NN in each iteration. It used a specialized graphical analysis technique to identify the boundary between classes.

To the best of our knowledge, up to now few work touched the issue of semi-supervised learning on multivariate time series due to the complexity of its data distribution.

2.2   Active Learning

Active learning is an efficient way to find the most valuable instances from lots of unlabeled data for manual labeling. It makes use of user's feedback to enlarge the size of the training data and enhance the confidence of examples, which could improve effectiveness of learning algorithm. In the process of active learning, the key issue is how to select limited subset of most informative examples in the unlabeled data [7, 8].

Among different sample selection strategies, uncertainty based sampling and query by committee are two most popular sampling strategies [7, 8]. Uncertainty sampling is the simplest and most commonly sampling way. In the process of the uncertainty sampling, some instances whose classes are most uncertain with the classifier are selected and further annotated [9, 10]. The assumption is that instances which are harder to identify are more helpful to further enhance the classification performance by updating the classifier. Query by committee strategy selects unlabeled examples whose class is the most disagreement among the committee numbers is chosen for annotation. To implement a Query by Committee algorithm, a committee of hypotheses should be firstly constructed to consistent with the labeled data. Then, a measure is designed to evaluate the disagreements between committee members [11, 12].

However, examples selected by the uncertainty sampling method do not always provide more information than other unlabeled examples for learning a good model. The reason is that some outliers that are most uncertain are also selected during the process of the uncertainty sampling. To avoid outliers being selected by in the process of sampling, He et al. [13] addressed the issue of active learning for multivariate time series classification with positive unlabeled data, and proposed an uncertainty and utility based sampling strategy to find informative unlabeled examples for manual annotation. First, a few of unlabeled examples with the highest uncertainty are selected. Then, among these chosen examples, the best one with the highest utility is manually annotated finally. Although the experimental results show clearly that this method was competitive, we observed that it has drawbacks. It could not consider uncertainty and utility at the same time, which leads to the most informative examples with high local density being not selected due to its little lower uncertainty. However, firstly annotating the most informative one could offer more information for classification and automatically labeling more confident unlabeled examples in the semi-supervised learning process. Therefore, this method results in more examples for manual annotation, which lead to excessive costs. This motivates use to propose a more robust approach to reduce manual cost as much as possible.

3 Definitions and Notations

In this section, we define some basic concepts and the notations used in this paper.

Definition 1. Univariate time series: a univariate time series $s = t_1, t_2,..., t_L$ is an ordered set of L real-valued variables, and L is the length of the time series s (for simplicity, $L = |s|$).

Definition 2. Multivariate time series: a multivariate time series is a vector of sequences $X = (x_1, x_2,..., x_T)$, where each component x_j is a univariate time series.

    This MTS object X has T variables, and the corresponding component of the ith variable is x_i.

    For the similarity distance, there are several methods to measure the similarity between two MTS examples. Here we adopt cosine distance method [14].

Definition 3. Cosine Similarity: For two multivariate time series $B = (b_1, b_2, ..., b_m)$ and $S = (s_1, s_2, ..., s_m)$, supposing $|b_i| \leq |s_i|$, the cosine similarity between B and S is calculated by $Sim(B, S) = \sum_{i=1}^{m} \min_{b'_i \in b_i, len(b'_i) = len(s_i)} \frac{b'_i \cdot s_i}{||b'_i|| \cdot ||s_i||}$.

Definition 4. P is a subset of all positive examples from the training data.

    At the beginning, P contains all positive examples of the training data, which is often smaller. As some unlabeled exampled being annotated, the number of the dataset P increases. At the same time, the size of new negative examples is also augmented.

Definition 5. U is a subset of all unlabeled examples from the training data.

    Originally, the dataset U consists of all the unlabeled examples in the primitive training data. As some unlabeled exampled being annotated as positive or negative, the scale of the dataset U gradually reduces. In this paper my try to minimize the size of the dataset U, which could provide more information to learn a good classifier.

4 Our Proposed Method

In this paper, our goal is to accurately annotate the vast majority of original unlabeled MTS data, which could provide much more information to learn a good classifier. We know that semi-supervised learning is an efficient method to automatically label a large of unlabeled data. However, as we mentioned, it is inefficient to deal with positive unlabeled MTS data. Therefore, in this paper we try to combine active learning and semi-supervised learning to enlarge the number of labeled MTS training data. Because active learning is too costly, we try to improve the quality of the data annotations as well as reducing the amount of manual annotation.

Here we just deal with issue of two-class data. If the data is more than two classes, usually there are two ways to generate a binary class dataset. One is to use a class as positive class and the rest is as negative class. The other way is to select two classes among all classes. In this paper we adopt the later. Moreover, we assure that the data is in general balanced. When the original training data is imbalanced, it is more complex, which is beyond our discussion in this paper.

In this section, we discuss how to combine active learning and semi-supervised learning methods to enlarge the scale of labeled data. First, we present an uncertainty and density based sampling strategy for manual annotation. Then, an active semi-supervised learning framework is proposed.

4.1 Sampling Strategy

As mentioned above, in literature [13] we proposed an uncertainty and utility based sampling strategy. However, the performance of overall this algorithm is less satisfactory than expected. We notice that the cardinality of an example's reverse K-nearest neighbors is more suitable to measure its local density [15, 16]. Therefore, we introduce an efficient sampling strategy to select most informative examples for annotation by considering its uncertainty and local density synchronously.

We are now in a position to present the definitions necessary to describe our idea.

Definition 6. Uncertainty. For an unlabeled MTS sample u, suppose its nearest positive sample is u_p and nearest negative sample is u_n. Based on information entropy, we can calculate the uncertainty of this unlabeled sample u by

$$\text{UCT(u)} = - (P_u * \log P_u + N_u * \log N_u) \tag{1}$$

$$P_u = \frac{Sim(u, u_p)}{Sim(u, u_p) + Sim(u, u_n)}$$

$$N_u = \frac{Sim(u, u_n)}{Sim(u, u_p) + Sim(u, u_n)}$$

where $Sim(u, u_p)$ and $Sim(u, u_n)$ is the cosine similarity between its nearest positive sample u_n and its nearest negative sample u_n, respectively.

Definition 7. Reverse K-nearest Neighbor. Given a data set D, for a MTS example S ∈ D, its reverse K-nearest neighbor, i.e. RkNN (S), is a subset of D that treat S as its K-nearest neighbors. Specifically, $\forall X \in$ RkNN(S), S ∈ kNN(X), where kNN(X) means the K-nearest neighbors of example X.

Definition 8. Density. Given a data set D, for a MTS example S ∈ D, its density denotes as

$$\text{Density}(S, K) = \frac{1}{min_{Y \in S \cup kNN(S)}\{Sim(S, Y)\}} \sum_{X \in S \cup kNN(S)} |RkNN(X)| \qquad (2)$$

where $|RkNN(X)|$ denotes the size of the RkNN(X).

Because in the process of active learning, the unlabeled examples are annotated by an oracle iteratively, each time only an example is annotated in order. Sometimes the average density could not represent the data distribution variation of an example's the local area. To handle this issue, we introduce the weight of density of an example.

Definition 9. Weight. Given a data set D, for a MTS example S ∈ D, the weight of S denotes as

$$\text{Weight}(S, k) = \frac{|RkNN(S)| + 1}{min_{X \in kNN(S)}|RkNN(X)| + 1} \qquad (3)$$

Definition 10. Given a data set D, for a MTS example S ∈ D, based on the density and weight of S, the weighted density of S

$$\text{W\_Density}(S, K) = \text{Weight}(S, k) * \text{Density}(S, k) \qquad (4)$$

Definition 11. Score. Given a data set D, for a MTS example S ∈ D, based on the uncertainty and weighted density, the score of an unlabeled example S denotes as

$$\text{Score}(S) = \text{UCT}(S) * \text{W\_Density}(S, k) \qquad (5)$$

Based on the definition of score, we rank all the unlabeled examples and select the example with the highest score for manual annotation.

According to the definitions in the above, we could easily discover some distinguished and informative unlabeled examples, which to a large extent could provide more information to identify the remaining unlabeled examples.

4.2 Active Semi-supervised Learning for Data Annotations

Based on the sample selection strategy presented in the above, the next work is to select most informative examples for manual annotation. However, the original data is a positive unlabeled dataset, in which it is hard to evaluate the degree of the uncertainty of all unlabeled examples. Therefore, to evaluate the uncertainty degree of each unlabeled example, we first try to find a confident negative example from unlabeled data. In general, the most dissimilar example far from original positive examples is likely to be a negative example. However, this situation is not always true due to the diversity of class and the complexity of the MTS data distribution. Here we continuously select the most dissimilar examples far from positive examples and manually annotate its label until the chosen example is negative. Then, we select the most informative unlabeled sample using sample selection strategy. Next, we adopt semi-supervised learning to confidently identify some of unlabeled data. For clarity, we provide the framework of semi-supervised active learning as following.

Algorithm 1. Active Semi-supervised Learning for MTS data annotation

Input: a training dataset with a few positive samples P and lots of unlabeled samples U

Output: the labeled dataset D

1:    initialize D = P

2:    Do

3:        Selecting the most dissimilar example u in U far from positive examples

4:        D = P + {u}; U= U- {u}

5:    While (u is positive)

6:    Do

7   Computing the score of each unlabeled example in U in terms of the definition 5

8:        Selecting an optimal unlabeled samples x in U in terms of its score

9:        Labeling x manually

10:        using semi-supervised learning method to identify some unlabeled samples Y in U based on x

11:        D = D + {x} + Y; U= U- {x} - Y

12:    While (stopping criterion is not satisfied)

13:    Return D

In the process of semi-supervised learning, to confidently identify some of unlabeled data, we introduce a reverse nearest neighbor based semi-supervised learning method. Specifically, for the newly labeled example x, all unlabeled examples Y in U that treat x as its k-nearest neighbors are automatically labeled as the class of x.

4.3   The Stopping Criterion

A naïve stopping criterion for active semi-supervised learning method is to dynamically evaluate the performance of the trained classifier as the labeled training examples enlarges. When the performance of the classification stops to improve, the process of the active learning would be quit. However, measuring the performance on the labeled training data is costly and time-consuming, in this paper we focus on annotating the vast majority of original unlabeled examples. Therefore, when the percentage of original unlabeled examples having been annotated is above 95, we end the process of active semi-supervised learning. This valid stopping criterion has two advantages: One is that vast majority of original unlabeled examples are annotated. The other is that some abnormal examples are not labeled, which could effectively avoid reducing the performance of classification.

4.4   Multivariate Time Series Classification

As we know, lots of algorithms can be used for MTS classification. Here our aim is to obtain confident labeled training data with less manual cost. To illustrate the efficiency of our proposed method, we adopt a sample and effective classification model, 1-NN, to evaluate the classification performance.

Here we use the F-measure and accuracy to evaluate the performance of these approaches. The F-measure is defined as $F = 2 * p * r / (p + r)$, where p and r means the precision and recall of classification, respectively. F-measure is larger when both of precision and recall are good.

5   Experimental Evaluation

In this section, we empirically study the proposed method for MTS classification on positive unlabeled data. The algorithms are implemented in Matlab, and the parameter $k = 1$. In the beginning, there is only one positive example in P and the rest of examples are treated as U. Therefore, the performance of the trained classifier can be sensitive to the initial labeled example. To mitigate this sensitivity, we repeat the training process ten times by each time starting from a different training example. Finally, we average the F-measure and accuracy of the classifier over all runs.

5.1   Datasets

The experiments are carried out on five real-world datasets, that is, Japanese Vowels, Character Trajectories, Pen-Based Recognition of Handwritten Digits, Wafer, and uWaveGestureLibrary [17–19]. For the Japanese Vowels dataset, there are 9 classes, and here we combine the first 4 classes as positive and the next 4 classes as negative to form a new dataset JV. For the Character Trajectories dataset, we select "b" class as positive and select "d" class as negative to form a new dataset CT. For the Pen-Based Recognition of Handwritten Digits dataset, we select "i" class as positive and select "j" class as negative to form a new dataset PRH. For Wafer dataset, its original dataset is imbalanced. Here we randomly select 127 negative examples and all positive examples to form a new balanced dataset. For uWaveGestureLibrary dataset, we select the second class as positive and select the fourth class as negative to form a new dataset WG.

Table 1 shows the summary of all of the datasets used in the experiments.

Table 1. Summary of datasets used in the experiments

	JV	CT	PRH	Wafer	WG
Num of labels	2	2	2	2	2
Num of variables	12	3	2	6	3
Max length	29	174	8	198	315
Min length	7	174	8	37	315
Num of samples	581	298	2110	254	1120

5.2   Analysis of Sample Selection Strategy

To illustrate the efficiency of our proposed sample selection method in terms of information entropy and reverse-nearest-neighbor based density (IR), in this section we first compare with other three classical sample selection approaches as following.

UU: A sampling method based on uncertainty and utility proposed in literature [13].
UD: A sampling method based on uncertainty and density in literature [14].
DR: A sampling method based on in terms of density based re-ranking in literature [14].

To analyze the function of these four sample selection methods, we compare their F-measure and accuracy when they manually label the same number of unlabeled examples. To show the changes in the performance of classification as the manually labeled examples gradually augmented, we do experiments on different percentage of manually labeled examples.

Figure 1 shows classification results of the four sample selection strategies on five datasets. It is clear that IR produces the best results consistently on different percentage of manually labeled examples. The reason is that at each time IR could select the most informative unlabeled example for manual annotation, which could effectively improve the classification performance when the expert resource is limited. At the same time, we notice that when the number of manually labeled examples is smaller, our proposed method is more competitive. This phenomenon gradually slows down when the percentage of manually labeled examples becomes larger. It is reasonable that as the number of manually labeled examples is larger, informative examples are mostly labeled and added to the labeled training data by all sampling methods for classification.

On the other hand, we also analyze the performance of these four methods when the stopping Criterion is satisfied, which is natural for applications. The comparison of the classification performances are shown in Table 2, where "% (manual annotation)" means the percentage of all original unlabeled examples have been manually annotated when the active semi-supervised learning process ends. We see that when the labeling process ends, classification performance of four active learning methods is similar. Meanwhile, we notice that that the number of manually labeled examples with IR is much smaller than other methods on all datasets. It shows the selected example to be manually annotated with our proposed method each time is a better representative of unlabeled examples, which could further automatically label more confident unlabeled

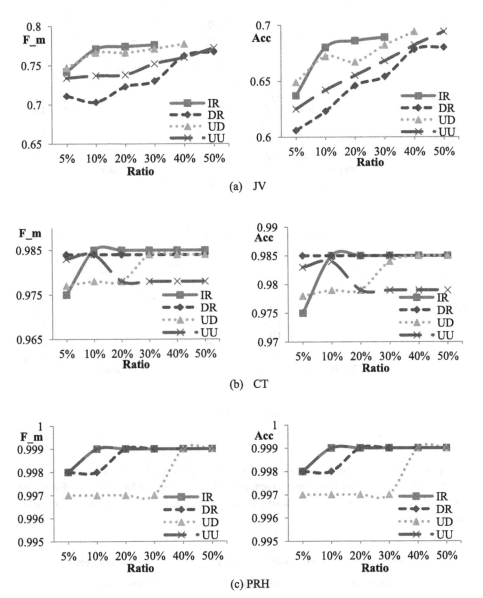

Fig. 1. The classification performances of 4 sampling methods with different ratio of manually labeled examples to original unlabeled examples on five datasets

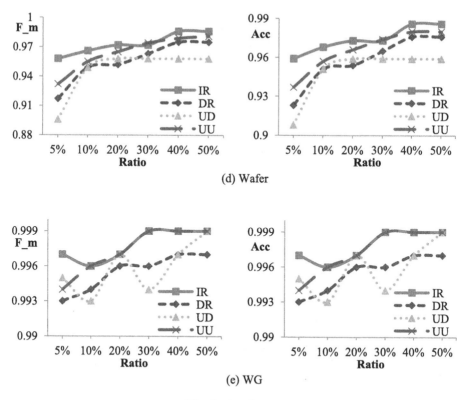

(d) Wafer

(e) WG

Fig. 1. (continued)

examples in the semi-supervised process. More examples need to be labeled by experts, which is more time-consuming and expensive. Therefore, considering the performance and the manual cost, our proposed active learning method is much effective among four sampling methods.

Table 2. Classification results of 4 sampling methods on five datasets

	F-measure				Accuracy				% (manual annotation)			
	IR	UU	ID	DR	IR	UU	ID	DR	IR	UU	ID	DR
JV	0.78	0.77	0.78	0.78	0.69	0.7	0.7	0.69	28.45	92.05	44.35	51.46
CT	0.99	0.98	0.99	0.98	0.99	0.98	0.99	0.98	48.7	81.17	51.3	66.88
PRH	0.99	0.99	0.99	0.99	0.99	0.99	0.99	0.99	45.3	76.06	52.82	64.79
WAFER	0.99	0.99	0.99	0.99	0.99	0.99	0.99	0.99	44.69	69.27	51.4	58.66
WG	0.99	0.99	0.99	0.99	0.99	0.99	0.99	0.99	42.86	80.65	49.31	72.81

5.3 Analysis of Proposed Semi-supervised Learning

In this section, we further analyze our proposed reverse nearest neighbor based semi-supervised learning method. To discuss its performance, we compare with the most commonly method, 1-NN based semi-supervised learning method. Specifically, In the process of the semi-supervised learning, based on the newly labeled example, in

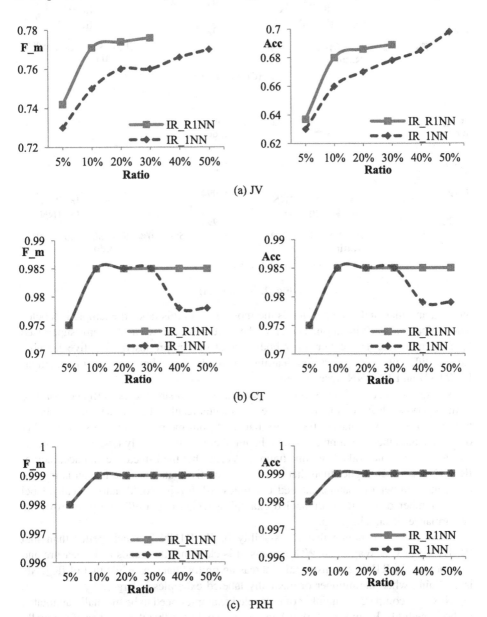

Fig. 2. The classification performances of two semi-supervised learning methods on 5 Datasets

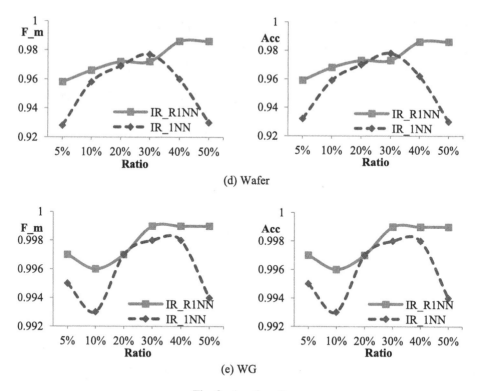

Fig. 2. (continued)

the data an unlabeled example that is the most close to this labeled example is labeled automatically at each iteration. Therefore, here we generate two active semi-supervised learning methods. To be fair, they both consist of our proposed IR active learning except different semi-supervised learning methods. For short, we define them as IR_R1NN and IR_1NN, respectively.

We do some experiments on five datasets to compare the performance of two semi-supervised learning methods. The experimental results of both methods are shown in Fig. 2. Firstly, we analyze the performance of both methods with the same number of the manual labeled examples. When the number of the manually labeled examples is the same in two methods, from this figure it is clear that the F-measure and accuracy of IR_R1NN is much higher than that of IR_1NN on five datasets. The reason is that for the same number of manual labeled examples, IR_R1NN could automatically label larger number of confident unlabeled examples, which could efficiently improve the performance of classification.

Meanwhile, we notice that the stability of IR_R1NN is much better than that IR_1NN. For CT, Wafer and WG datasets, it is clear that the trends of F-measure and accuracy of IR_R1NN is consistently increasing on five datasets while that of IR_1NN is declining when the number of manually labeled examples is larger.

Next, we compare the number of unlabeled examples need to be manually annotated by both methods. From Fig. 2(a), (d) and (e) we could see that the number of manually labeled examples with IR_R1NN is much smaller on JV, Wafer and WG datasets while on CT and PRH datasets the number of manually labeled examples with both methods are similar. It is reasonable because IR_R1NN can automatically label several examples based on a newly labeled example while IR_1NN can only automatically label one. So the IR_R1NN could annotate the vast majority of original unlabeled examples with less number of manually labeled examples, which could reduce manual cost.

Therefore, the performance of reverse nearest neighbor based semi-supervised learning method is more competitive.

5.4 Comparison with State-of-the-Art Semi-supervised Methods

The last experiment analyzes the function for our proposed active semi-supervised learning method IR_R1NN against existing state-of-the–art semi-supervised methods such as WK [5], RK and CBD [1], for MTS classification. The experimental results on five datasets are listed in Table 3.

Table 3. Summary results of various methods on five datasets

	F-measure				Accuracy			
	IR_R1NN	WK	RK	CBD	IR_R1NN	WK	RK	CBD
JV	0.78	0.68	0.65	0.68	0.69	0.59	0.58	0.59
CT	0.99	0.92	0.35	0.56	0.99	0.91	0.64	0.70
PRH	1.00	0.62	0.90	0.98	1.00	0.76	0.95	0.98
Wafer	0.99	0.68	0.40	0.37	0.99	0.79	0.64	0.62
WG	1.00	0.33	0.91	0.73	1.00	0.60	0.95	0.80

It is clear that our proposed methods are more efficient than existing semi-supervised methods. The reason is that in our proposed method, active learning process focus on selecting most informative examples and manually labeling them, and the semi-supervised process just automatically label the confident unlabeled examples which are the reverse nearest neighbors to the newly manually labeled examples. However, existing semi-supervised methods label repeatedly unlabeled examples in terms of the labeled ones, and it does not guarantee that the classes of these labeled examples are all correct because some of them are labeled by the semi-supervised method. Therefore, Comparing with existing state-of-the–art semi-supervised methods, our proposed active semi-supervised method could offer more confidently labeled training data for classification. It could improve the performance of classification.

6 Conclusions

In this paper we focus on combing active learning and semi-supervised learning for annotating the vast majority of original unlabeled examples. First, based on information entropy and density, a sampling strategy is proposed to find the most informative examples for manual annotation. Second, in terms of the newly labeled example by experts, a reverse nearest neighbor based semi-supervised learning method is presented to automatically and accurately label some examples. Last, an efficient active semi-supervised learning framework is proposed to accurately annotate the vast majority of original unlabeled data. The experimental results show clearly that our proposed method is competitive.

In future, we plan to perform research on sampling method on imbalanced unlabeled data to obtain a balanced labeled training data for multivariate time series classification.

References

1. Wei, L., Keogh, E.: Semi-supervised time series classification. In: KDD (2006)
2. Begum, N., Hu, B., Rakthanmanon, T., Keogh, E.: Towards a minimum description length based stopping criterion for semi-supervised time series classification. In: IEEE IRI (2013)
3. Nguyen, M.N., Li, X.-L., Ng, S.-K.: Positive unlabeled learning for time series classification. In: IJCAI (2011)
4. Nguyen, M.N., Li, X.-L., Ng, S.-K.: Ensemble based positive unlabeled learning for time series classification. In: Lee, S.-g., Peng, Z., Zhou, X., Moon, Y.-S., Unland, R., Yoo, J. (eds.) DASFAA 2012, Part I. LNCS, vol. 7238, pp. 243–257. Springer, Heidelberg (2012)
5. Mabel, G., Chrisoph, B., Isaac, T., Taneet, R., Kosé, B.: On the stopping criteria for k-nearest neighbor in positive unlabeled time series classification problems. Inf. Sci. **328**, 42–59 (2016)
6. He, G., Duan, Y., Peng, R., Jing, X., Qian, T., Wang, L.: Early classification on multivariate time series. Neurocomputing **149**, 777–787 (2015)
7. Yifan, F., Zhu, X., Li, B.: A survey on instance selection for active learning. Knowl. Inf. Sys. **35**, 249–283 (2013)
8. Settles, B.: Active learning literature survey. Computer Sciences Technical report, University of Wisconsin–Madison, 26 January 2010
9. Guo, H., Wang, W.: An active learning-based SVM multi-class classification model. Pattern Recogn. **48**(5), 1577–1597 (2015)
10. Huang, S.-J., Jinm, R., Zhou, Z.: Active learning by querying informative and representative examples. IEEE Trans. Pattern Anal. Mach. Intell. **36**(10), 1936–1949 (2014)
11. Hady, M.F.A., Schwenker, F.: Combing committee-based semi-supervised learning and active learning. J. Comput. Sci. Technol. **25**(4), 681–698 (2010)
12. Seung, H., Opper, M., Sompolinsky, H.: Query by committee. In: ACM Workshop on Computational Learning Theory, pp. 287–294 (1992)
13. He, G., Duan, Y., Li, Y., Qian, T., He, J., Jia, X.: Active learning for multivariate time series classification with positive unlabeled data. In: ICTAI (2015)
14. Zhu, J., Wang, H., Tsou, B.K., Ma, M.: Active learning with sampling by uncertainty and density for data annotations. IEEE Trans. Audio Speech Lang. Process. **18**(6), 1323–1331 (2010)
15. Huang, H., He, Q., He, J., Ma, L.: RADRA: rare category detection iva computation of boundary degree. In: PAKDD (2011)
16. Xia, C., Hsu, W., Lee, M.L., Ooi, B.C.: BORDER: efficient computation of boundary points. IEEE Trans. Knowl. Data Eng. **18**(3), 289–303 (2006)
17. http://archive.ics.uci.edu/ml/datasets.html
18. http://www.cs.cmu.edu/~bobski/
19. http://www.cs.ucr.edu/~eamonn/time_series_data/

Leveraging Structural Hierarchy for Scalable Network Comparison

Rakhi Saxena[1], Sharanjit Kaur[2(✉)], Debasis Dash[3], and Vasudha Bhatnagar[4]

[1] Deshbandhu College, University of Delhi, Delhi, India
rsaxena@db.du.ac.in
[2] Acharya Narendra Dev College, University of Delhi, Delhi, India
sharanjitkaur@andc.du.ac.in
[3] CSIR-Institute of Genomics and Integrative Biology, Delhi, India
ddash@igib.res.in
[4] Department of Computer Science, University of Delhi, Delhi, India
vbhatnagar@cs.du.ac.in

Abstract. *K-core* decomposition is a popular method that segments a network revealing the underlying hierarchy. We explore the propensity of this decomposition method for structural discrimination among networks by extracting features from each level of the hierarchy. We propose a novel algorithm for *N*etwork *C*omparison using *k-core D*ecomposition (*NCKD*). The method is effective, efficient and scalable, with computational complexity of $O(|\mathcal{E}|)$, where $\mathcal{E}$ is the set of edges in the network. The low computational complexity of the method makes it attractive for scalable network comparison.

NCKD algorithm decomposes networks and extracts features from the resulting shells. Jensen-Shannon distance between extracted features quantifies structural differences between networks. We establish that probability distributions of coreness and intra/inter-shell edges are capable of characterizing different genres of networks and capturing finer structural differences between networks of the same genre. We experiment with synthetic and real-life networks up to eight million edges on a single PC. Comparison with two recent state-of-the-art network comparison methods affirms that *NCKD* outperforms in terms of effectiveness and scalability.

Keywords: Network comparison · K-core decomposition · Graph analytics · Social networks · Jensen-Shannon distance

1 Introduction

Complex networks have attracted immense attention because of their ability to model social relations, power grids, transportation links, biological processes etc. [7]. One of the challenging tasks in network analytics is to assess and quantify similarity between two networks. Applications of network comparison include construction of phylogenetic trees and function prediction in biological networks, studying evolution in social networks, analysing semantic structure in natural languages, detecting code theft by comparing two executable objects etc. [7,9].

© Springer International Publishing Switzerland 2016
S. Hartmann and H. Ma (Eds.): DEXA 2016, Part I, LNCS 9827, pp. 287–302, 2016.
DOI: 10.1007/978-3-319-44403-1_18

Similarity between two networks is a function of similarities between their orders, sizes, and topological features. While similarities in orders and sizes are trivial to assess, capturing topological and structural similarities is the core challenge in the task of network[1] comparison. Comparison of two networks essentially entails analogizing structural properties such as nature of hierarchy, clustering tendency, neighborhood characterization, correlation between topological attributes etc. Networks may be compared either at a local or global level depending upon the application. For example, construction of a phylogenetic tree using biological networks involves clustering organisms with similar biological evolution. This task demands a global comparison of networks. On the other hand, comparison of two metabolic networks for the purpose of discovering causal factors for functional differences calls for local level comparison.

Extraction of global features like diameter, average clustering coefficient, characteristic path length, betweenness centrality etc. for comparison purpose is unattractive because of high computational complexity even for medium-sized graphs. Computation of local features, on the other hand, involves examining configuration and properties of small subgraphs, conferring scale independence to the comparison method. Hence, it is tempting to adopt local properties for structural comparison of massive networks. Earlier approaches for network comparison pursued this trend and deployed local features including degree, clustering coefficient, degree centrality, triad census, graphlet distribution etc. [10]. Lamentably, these approaches fail to capture underlying structural hierarchy prevailing in real-life networks.

Therefore, it is desirable to devise methods that summarize network structure both locally and globally. Since hierarchical k-core decomposition promotes the local feature *degree* to the global feature *coreness*, we explore k-core decomposition as a tool to quantify the structural similarity between two networks. K-core decomposition has been recognized as an important technique for understanding complex networks by decomposing them in hierarchy [2,12,24]. We posit that hierarchical segmentation using k-core decomposition method has potential to reveal structural differences between networks at all levels of hierarchy.

1.1   Motivation

Motivational factors for using hierarchical k-core decomposition approach for scalable network comparison are listed below.

  i. Real-life networks exhibit structural hierarchy and comparing analogous signals at all levels of hierarchies has potential to reveal the structural disparity between networks.
 ii. The proposed algorithm is particularly appealing for comparing large and sparse graph since k-core decomposition method has computational complexity of $O(|\mathcal{E}|)$, $\mathcal{E}$ being the set of edges [4].
iii. Massive networks that cannot fit in main memory can be decomposed using distributed k-core decomposition [22].

---

[1] We use terms network/graph, node/vertex, and edge/link interchangeably.

1.2 Contributions

In this paper, we propose a novel and scalable method for *Network Comparison* using *k*-core *Decomposition* (*NCKD*). According to Faust [10], network comparison studies are designed to answer two questions. First, does a pair of networks exhibit common structural tendencies?, and second, which structural features distinguish among different relations between nodes? We demonstrate that node distribution in shells of the network is an effective and efficient implement to answer the first question. Augmenting node distribution in shells with edge distribution boosts its power to cogently answer the second question. Research contributions of the paper are listed below:

  i. A novel algorithm (*NCKD*) that uses k-core decomposition to quantify network similarity through network signatures generated using probability distribution of nodes and edges in shells respectively (Sect. 4).
  ii. Comparison of *NCKD* with two state-of-the-art network comparison algorithms (Sect. 5.2).
  iii. Extensive experimentation to demonstrate effectiveness, scalability and robustness of *NCKD* (Sects. 5.3 and 5.4).

2 Related Work

Several decent algorithms for network comparison, that quantify similarity between networks, have been proposed in recent years. A related but different problem is *network alignment*, addressed in bio-informatics, where the goal is to map nodes of one network to the nodes of another. Our focus is on recent representative network comparison algorithms followed by a brief overview of applications of k-core decomposition.

2.1 Network Comparison

Popular approaches for comparing networks include (i) graph isomorphism, (ii) graph edit distance, (iii) iterative methods, and (iv) feature extraction [6].

Graph isomorphism, a theoretically sound approach, has been traditionally employed to establish exact matching between two graphs [15]. Approximate matching is commonly obtained by graph edit distance, which essentially is an error-tolerant method [11]. Iterative methods compute the pairwise similarity between nodes by capturing similarity/dissimilarity of their neighborhoods [21]. These three approaches lead to algorithms with high computational complexity and are hence non-scalable [18]. This deters their applicability to large networks.

Feature extraction approach has recently found favour with the community interested in analyzing massive graphs. The strategy involves constructing features from the compared graphs and computing distance between them to quantify differences. Banerjee [3] used eigenvalues of normalized graph Laplacian spectra to capture global topological properties for computing pairwise networks similarity. Recently, Lu et al. [18] compared complex networks using the heat content estimated by lazy random walk.

Macindoe et al. [19] considered all induced subgraphs of a parametrized radius centered on each vertex and computed three socially relevant structural features, Leadership, Bonding, and Diversity (L,B,D), driven by social theories for each subgraph. Earth mover's distance between LBD distributions of networks quantifies their similarity. Netsimile [6] algorithm composes network signature from moments of distributions of selected local topological properties of the network. The pairwise similarity score of networks is computed using Canberra distance between their signatures. Scale-independent nature of selected properties renders a computational complexity of $O(N)$, where N is the order of the graph.

These algorithms make use of either local or global features, each of which is individually ineffective and non-scalable for network discrimination. NCKD algorithm plugs the gap as it is scalable and exploits local feature while taking into account the global hierarchical structure of the network.

2.2 K-Core Decomposition

Seidman [24] introduced k-core decomposition for characterizing network structure.The k-core of a network is a maximal subgraph in which every node is connected to at least k other nodes. Batagelj et al. [4] present an $O(|\mathcal{E}|)$ algorithm for k-core decomposition of a graph $\mathcal{G}$ with $|\mathcal{E}|$ edges. Analysis of the k-core structure of a graph has been effectively used in identification of social cores and influential nodes in social networks, acceleration of community detection, evaluation of co-operation in communities, and as a visualization tool to highlight the topological and hierarchical structure of graphs [2,12,23]. Recently proposed k-truss decomposition method also presents a hierarchical view of the network yielding the largest subgraph in which every edge is contained in at least (k-2) triangles within it [25]. The method is effective for focusing on smaller and cohesive areas, which are subgraphs of k-core. However higher computational complexity of order $O(|\mathcal{E}|^{1.5})$ for k-truss decomposition is a discouraging factor. Hence, we chose to use k-core decomposition method for network comparison.

To the best of authors' knowledge, NCKD is first-ever application of k-core decomposition for scalable network comparison using single PC.

3 Preliminaries and Notation

We introduce formal notation and definitions used in the paper. Let $\mathcal{G}$ be a simple, undirected graph $\mathcal{G} = (\mathcal{V}, \mathcal{E})$, where $\mathcal{V}$ is the set of vertices and $\mathcal{E}$ is the set of edges. An edge $\mathsf{e}_{ij} \in \mathcal{E}$ if it connects vertices v_i and v_j; $\mathsf{v}_i, \mathsf{v}_j \in \mathcal{V}$. The order of $\mathcal{G}$ is $|\mathcal{V}|$ and its size is $|\mathcal{E}|$. The degree of a vertex v is denoted by $\rho(\mathsf{v})$. The k-core decomposition algorithm iteratively prunes vertices of degree less than k resulting in a hierarchy of nested k-core sub-graphs, within which each node is connected to at least k other nodes. Formal definitions as adapted from [2] follow.

Definition 3.1. *A subgraph,* $\mathcal{G}'_k = (\mathcal{V}'_k, \mathcal{E}'_k)$ *of* $\mathcal{G} = (\mathcal{V}, \mathcal{E})$, *induced by the set* $\mathcal{V}'_k \subseteq \mathcal{V}$ *is a k-core (core of order k) of* $\mathcal{G}$ *if* $\forall v \in \mathcal{V}'_k: \rho(v) \geq k$, *(k $\geq$ 0) and* $\mathcal{G}'_k$ *is a maximal connected subgraph with this property.*   □

Definition 3.2. *Coreness* $\zeta(v)$ *of vertex* v *is k if it belongs to a k-core but not to any (k+1)-core. Coreness of a graph* $\mathcal{G} = (\mathcal{V}, \mathcal{E})$ *is* $max\{\zeta(v) \forall v \in \mathcal{V}\}$.   □

Definition 3.3. *A k-shell ($\mathcal{S}_k$) of* $\mathcal{G} = (\mathcal{V}, \mathcal{E})$ *is the set of all vertices with coreness k, i.e.,* $\mathcal{S}_k = \{v | v \in \mathcal{V} \wedge \zeta(v) = k\}$.   □

The k-core decomposition reflects the structure of a network by faithfully capturing the inherent hierarchy as nested cores. The lower bound on number of nodes in a k-core is (k+1) and a loose upper bound is $|\mathcal{V}|$. The lower bound on the number of edges is $\binom{k+1}{2}$, while a loose upper bound is $|\mathcal{E}|$ [5]. If both endpoints of an edge have the same coreness, the edge is termed as an *intra-shell* edge, otherwise it is an *inter-shell* edge.

Example 3.1. *Figure 1 shows graph* $\mathcal{G}$ *with* $|\mathcal{V}| = 32$ *and* $|\mathcal{E}| = 47$. *Dashed circles, marked* k = i, *demarcate the cores. Nodes within a dashed circle and having same color denote shell* $\mathcal{S}_k$. *Shell* $\mathcal{S}_4$, *the highest order shell induces the 4-core of* $\mathcal{G}$. *Subgraph induced by* $\mathcal{S}_3 \cup \mathcal{S}_4$ *is the 3-core of* $\mathcal{G}$.   □

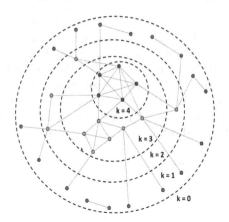

Fig. 1. K-core decomposition of $\mathcal{G}$. Nodes with same color constitute a shell. (Color figure online)

4   Characterizing Networks Using K-Core Decomposition

Adaptation of k-core decomposition for designing a similarity measure is non-trivial because two networks with the same hierarchical structure can have vastly different topology. The challenge is to identify and extract suitable features of the decomposed graph for effective and scalable network discrimination. We hypothesize that differences in the node/edge distribution of shells in the decomposed graph are effective discriminators for the overall structure of underlying networks. We first explain a simple and effective network feature i.e. node distribution followed by the statement of its limitation, and reasoning behind inclusion of edges arrangements.

4.1 Coreness Distribution

Distribution of nodes within shells captures the spread of nodes and reflects the underlying structure [24]. It is synonymous with the distribution of coreness of nodes in the decomposed graph.

Consider a graph $\mathcal{G} = (\mathcal{V}, \mathcal{E})$ with coreness k and shells $\{\mathcal{S}_0, \ldots, \mathcal{S}_k\}$. Let X be a discrete random variable denoting coreness of a node in $\mathcal{G}$ and defined on the sample space $\Lambda = \{0, \ldots, k\}$. We define probability mass function for X as $p(x) = p(X = x) = \frac{|\mathcal{S}_x|}{|\mathcal{V}|}$, where $|\mathcal{S}_x|$ is the cardinality of shell $\mathcal{S}_x$. It is clear that $\sum_{x=0}^{k} |\mathcal{S}_x| = |\mathcal{V}|$. Here, $p(x)$ denotes the probability that a node has coreness value x. Alternatively, $p(x)$ is the probability of an arbitrary node lying in shell $\mathcal{S}_x$. Following example explains computation of probability distribution (p) of nodes (coreness) in the shell.

Example 4.1. *Graph $\mathcal{G}$ in Fig. 1 has 32 nodes and 5 shells. Probability distribution (p) of nodes in $\mathcal{G}$, is given by $p = \langle 0/32,\ 17/32,\ 4/32,\ 6/32,\ 5/32 \rangle$.*

We studied probability distribution of coreness for several synthetic and real-life networks (Table 1) to test its propensity for network comparison. We show the plot of coreness distribution of six metabolic and five co-author networks in Figs. 2a and b. The striking similarity between the coreness probability distributions of graphs belonging to the same genre strongly indicates its utility as a discriminating network feature. Preliminary experimentation (not reported due to space constraint) however quickly revealed the inadequacy of this feature to effectively capture structural differences arising in the real world networks.

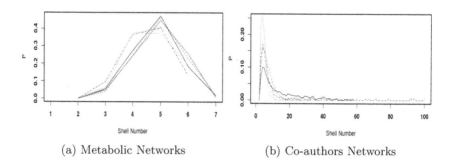

(a) Metabolic Networks   (b) Co-authors Networks

Fig. 2. Plots for probability distribution of coreness for two genres of real-life networks.

This insufficiency arises because the arrangement of edges in a graph, which is the cause of topological variations, is completely ignored by the coreness distribution. Extreme topologies of star and chain with n nodes having identical hierarchical structure and probability distribution of coreness, present a very clear example to substantiate the argument. Since *coreness is inadequate to capture finer structural differences between networks, it is myopic to depend on it as a sole distinguishing feature.*

4.2 Edges Distribution

Theoretically, a graph of order n with coreness k and coreness distribution p is a random sample from the family $\mathcal{F}_{k,p,n}$ of graphs [14]. Graph $\mathcal{G}$ in question is one realization from this family. All graphs in $\mathcal{F}_{k,p,n}$ will have similar coreness distribution, even though they may be topologically different. This is unacceptable in both theory and practice. *Rewiring* and *swapping* lemmas stated in [5] reinforce this argument.

According to the *rewiring* lemma, two adjacent nodes in a shell can disconnect and connect independently to nodes with higher coreness, and vice versa without changing the coreness distribution of the graph. The *swapping* lemma allows non-adjacent nodes in the same shell to swap end-nodes without altering the coreness distribution of the graph. It is reasonable to conclude that *coreness distribution is inadequate to capture finer structural differences between networks.* For better discrimination between the members of $\mathcal{F}_{k,p,n}$ family, we incorporate arrangements of edges influencing the network structure in addition to nodes distribution.

Let $\mathcal{G}$ be a graph with coreness k. Then, E^l denotes the lower triangular matrix representing the arrangement of edges of G. E^l_{ij} is the count of inter-shell links between shells $\mathcal{S}_i$ and $\mathcal{S}_j$. E^l_{ii} is the count of intra-shell links in shell $\mathcal{S}_i$. Clearly $\sum_i \sum_j E^l_{ij} = |\mathcal{E}|$, $(0 \leq i \leq k, 0 \leq j \leq i)$. Example 4.2 clarifies the idea of intra- and inter-shells links using matrix representation used in [5].

Example 4.2. *The count of intra- and inter-shell edges in $\mathcal{G}$ of Fig. 1 is shown below in the lower triangular matrix (E^l). Shell $\mathcal{S}_2$ has one intra-shell link indicated by $E^l_{22} = 1$. It also has six inter-shell links with $\mathcal{S}_1$ indicated by $E^l_{21} = 6$.*

$$E^l = \begin{pmatrix} 0 \\ 0\ 4 \\ 0\ 6\ 1 \\ 0\ 5\ 2\ 9 \\ 0\ 1\ 5\ 4\ 10 \end{pmatrix}$$

We vectorize matrix E^l to a vector V of size $\frac{(k+1)(k+2)}{2}$ such that $V = [E^l_{00}, E^l_{10}, E^l_{11}, E^l_{20}, \ldots, E^l_{kk}]$. Index r in V for E^l_{ij} is obtained by using the following rule.

$$r \leftarrow j + \frac{i * (i+1)}{2} \tag{1}$$

Clearly, the vectorization expresses isomorphism between V and E^l. Let R be a discrete random variable defined on sample space $\Lambda' = \left(0, 1, \ldots, \frac{(k+1)(k+2)}{2} - 1\right)$ denoting linkage count within and between shells in the graph. When $R = r$, it denotes linkage between shells $\mathcal{S}_i$ and $\mathcal{S}_j$, with the mapping defined by Eq. 1.

The probability mass function $u(r)$ of random variable R is defined as $u(r) = p(R = r) = \frac{E_{ij}^l}{|\mathcal{E}|}$. Here $u(r)$ denotes the probability that an arbitrary edge in $\mathcal{G}$ connects a node in $\mathcal{S}_i$ to a node in $\mathcal{S}_j$. It is easy to show that u corresponds to probability distribution of intra-shell and inter-shell links. Following example shows the edge probability distribution u for the lower triangular matrix given in Example 4.2.

Example 4.3. *Given 47 edges in $\mathcal{G}$, probability distribution of edges (u) is computed as:* $\langle$ *0/47, 0/47, 4/47, 0/47, 6/47, 1/47, 0/47, 5/47, 2/47, 9/47, 0/47, 1/47, 5/47, 4/47, 10/47* $\rangle$. $\qquad\square$

If the number of nodes and edges in two graphs with identical coreness distribution are same, structural differences between them arise due to rewiring of edges. To examine the sensitivity of E^l towards rewiring, we focus on the cases that do not alter the coreness of the involved nodes post-rewiring. There are two possibilities for rewiring of a node. It can either connect to a node in the same shell or to a node in a higher shell. Rewiring of a node in a lower shell can be considered as the former situation from the viewpoint of the node in the lower shell. We explain these cases below.

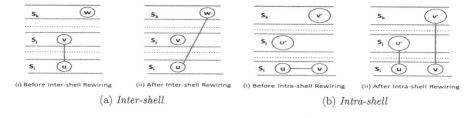

(i) Before Inter-shell Rewiring   (ii) After Inter-shell Rewiring   (i) Before Intra-shell Rewiring   (ii) After Intra-shell Rewiring

(a) *Inter-shell*   (b) *Intra-shell*

Fig. 3. Example for edge rewiring. $\mathcal{S}_i, \mathcal{S}_j, \mathcal{S}_k$ are the shells.

R1. *Rewiring an inter-shell edge*: Consider nodes $u, v, w \in \mathcal{V}$, located in distinct shells $\mathcal{S}_i, \mathcal{S}_j, \mathcal{S}_k$ respectively, s.t. $i \neq j \neq k$ and $i < j, k$. Let edges $(u, v) \in \mathcal{E}$ and $(u, w) \notin \mathcal{E}$ Then R1 leads to

$$\mathcal{E} := \mathcal{E} \setminus (u, v) \cup (u, w) \tag{2}$$

Figure 3(a) exhibits this case. Since node u lies in shell $\mathcal{S}_i$, it has at least i links with nodes in higher shells. After deleting edge (u, v) and adding edge (u, w), link count of u in higher shell remains same. Consequently, coreness of u remains unchanged. Coreness of v and w remains unchanged since links to lower shells do not impact coreness (by Definition 3.1). Consequently, post-rewiring coreness distribution remains unchanged. In the edge distribution matrix, two entries change as follows: E_{ik}^l is incremented and E_{ij}^l is decremented by 1. Hence altered structure of the graph is captured by the edge distribution.

R2. *Rewiring an intra-shell edge*: Consider nodes $u, v, u', v' \in \mathcal{V}$ s.t. $u, v \in \mathcal{S}_i$, $u' \in \mathcal{S}_j$, $v' \in \mathcal{S}_k$ and $i \neq j \neq k$, $i < j, k$. Let edges $(u, v) \in \mathcal{E}$, $(u, u') \notin \mathcal{E}$, $(v, v') \notin \mathcal{E}$. Then R2 leads to

$$\mathcal{E} := \mathcal{E} \setminus (u, v) \cup (u, u') \cup (v, v') \tag{3}$$

Figure 3(b) exhibits this case. Following similar arguments as in R1, intra-shell rewiring does not alter coreness distribution, but is reflected in edge distribution.

4.3 NCKD Algorithm

The proposed algorithm for Network Comparison using k-core Decomposition (NCKD) uses probability distribution of *nodes* as well as intra-shell/inter-shell *edges*. The problem of pair-wise network comparison reduces to computing the statistical distance between probability distributions representing signatures of the networks. We use *Jensen-Shannon Distance (JSD)* as it is a popular metric for comparing probability distributions due to its property of non-negativity, identity, symmetry, and boundedness [17]. Equation 4 gives JSD between two probability distributions p and q, with respective weights w_1 and w_2 ($w_1, w_2 \geq 0$ and $w_1 + w_2 = 1$).

$$JSD(p,q) = [H(w_1 * p + w_2 * q) - w_1 * H(p) - w_2 * H(q)]^{\frac{1}{2}} \tag{4}$$

Here, H is the Shannon entropy function. Equipped with a tool to capture finer distinctions of graph topologies, we quantify the structural difference (distance) between two networks as average of differences (distance) between the (i) distribution of coreness and (ii) distribution of edges.

Let p and q respectively denote the probability distributions of coreness of graphs $\mathcal{G}_1$ and $\mathcal{G}_2$. Further, u and v denote the edge probability distributions of graphs $\mathcal{G}_1$ and $\mathcal{G}_2$. Applying JSD on these distributions and averaging the result gives the net distance between two networks. Equation 5 formally defines distance between networks.

$$\mathcal{D}(\mathcal{G}_1, \mathcal{G}_2) = avg(JSD(p,q), JSD(u,v)) \tag{5}$$

Algorithm 1. Algorithm *NCKD*

Input : Graphs $\mathcal{G}_1$ and $\mathcal{G}_2$
Output: Distance between $\mathcal{G}_1$ and $\mathcal{G}_2$

Begin
    Decompose $\mathcal{G}_1$ and $\mathcal{G}_2$ into cores
    $p \leftarrow$ Prob. dist. of coreness of nodes in $\mathcal{G}_1$
    $q \leftarrow$ Prob. dist. of coreness of nodes in $\mathcal{G}_2$
    $u \leftarrow$ Prob. dist. of intra and inter-shell links in $\mathcal{G}_1$
    $v \leftarrow$ Prob. dist. of intra and inter-shell links in $\mathcal{G}_2$
    $\mathcal{D}(\mathcal{G}_1, \mathcal{G}_2) \leftarrow avg(JSD(p,q), JSD(u,v))$ //*Jensen-Shannon Distance*
End

Algorithm 1 summarizes the steps for *NCKD*. Please note that in all experiments reported in paper, we assign equal weights to the distributions while computing JSD. We are conscious that weights can be constructively manipulated to capture preference for one graph over other during the comparison.

5   Experiments

The experimental study is designed to assess and compare effectiveness, scalability and robustness of *NCKD* algorithm against two recent network comparison algorithms *Netsimile* [6] and *LBD* [19]. We implemented *NCKD* algorithm in *Python* (64 bits, v 2.7.3) and executed on Intel Core i5-3201M CPU @2.50 GHz with 8 GB RAM, running UBUNTU 12.04.

5.1   About Datasets

We performed experiments with both synthetic and real-life datasets (Table 1). Synthetic datasets allow controlled variation of data characteristics and hence enable close scrutiny of algorithmic behaviour. Real-life datasets expose the strengths and weakness of the algorithm in practical scenarios.

Synthetic datasets were generated using *igraph* package of *R*. Erdös-Rényi (E), Forest-Fire (F), Watts-Strogatz (W), and Barabási-Albert (B) models were used for analysis. Order (number of nodes) of the network in thousands (K) is included in nomenclature. Since each network is one probabilistic realization of the model parameters, we generated multiple networks with same parameters. Thus, B10K-n meant n^{th} realization of Barabási-Albert network of order 10K. Three real-life genres include (i) Co-author (CA), (ii) Autonomous Systems (A), and (iii) Metabolic (M) networks. Large datasets used for scalability experiment are described in Sect. 5.4.

5.2   Effectiveness of *NCKD*

We compute effectiveness of *NCKD* by comparing $\mathcal{D}(\mathcal{G}_1, \mathcal{G}_2)$ (Eq. 5) with distance measures defined in two state-of-the-art algorithms *LBD* and *Netsimile*. We compute pairwise distances for networks given in Table 1, using distance measures used in three algorithms. *LBD* algorithm was unable to generate network signatures for large graphs even after running for more than 24 h. We, therefore, restrict experiment to 14/32 graphs that *LBD* algorithm was able to process in reasonable time (<4 h) and cluster the networks using hierarchical agglomerative clustering[2]. We compute purity, precision, recall, accuracy, and Normalized Mutual Information (NMI) measures [20] to assess the quality of clustering (Table 2).

It is clear from Table 2 that resultant clustering of 14 small networks by *NCKD* is better than those delivered by *Netsimile* and *LBD* algorithms. Timings (averaged over 3 runs) for generating network signatures (Table 3) for *small* networks show that *NCKD* is also faster. We dropped *LBD* algorithm for further experimentation since it was patently non-scalable and the clustering, as evidenced by NMI, was also poorer in quality.

---

[2] *hclust* and *cutree* functions of *stats* package in *R* were used for agglomerative clustering and to cut dendrogram by specifying known number of classes.

Table 1. Characteristics of synthetic and real-life networks used; D: Diameter, C: Connected components, GCC: Global clustering coefficient, α: Parameter for power law distribution

Networks		Nodes	Edges	D	C	GCC	α	Remarks
Synthetic networks using generative models								
Erdos Reyni [8]	E10K-1	9827	19823	15	7	0.0004	11.47	Generator G(n,m = 2n)
	E10K-2	9807	19772	14	9	0.0005	11.58	m: number of edges
	E1K-1	983	2010	10	3	0.0030	13.568	n: number of nodes
	E1K-2	984	2022	10	3	0.0037	10.61	
Forest Fire [16]	F10K-1	10000	58901	6	1	0.0598	3.06	4 ambassador vertices
	F10K-2	10000	58823	6	1	0.0588	3.105	20 % backward burning probability
	F1K-1	1000	5873	5	1	0.0894	3.05	30 % forward burning probability
	F1K-2	1000	5717	5	1	0.0899	3.111	
Watts Strogatz [26]	W20K-1	20000	80000	8	1	0.0714	8.30	Lattice dimension = 1
	W20K-2	20000	80000	8	1	0.0694	8.25	Degree = 4
	W2K-1	2000	8000	7	1	0.0757	8.19	Rewiring probability = 0.3
	W2K-2	2000	8000	7	1	0.0731	8.22	
Barabasi Albert [1]	B10K-1	10000	9999	2	1	0	1.33	Non-assortative version
	B10K-2	10000	9999	3	1	0	1.33	added 4 edges/iteration
	B1K-1	1000	999	3	1	0	2	
	B1K-2	1000	999	2	1	0	2	
Real-life networks								
Co-author [16]	CA-1	18772	396159	14	290	0.3180	1.71	Papers submitted to arXiv during
	CA-2	23133	186935	15	567	0.2643	2.21	period January 1993 to April 2003
	CA-3	5242	28979	17	355	0.6298	2.23	Astro Physics, Condensed Matter
	CA-4	12008	237009	13	278	0.6595	1.74	General Relativity, High Energy
	CA-5	9877	51970	18	429	0.2840	2.36	Physics (HEP) and HEP Theory
Autonomous [16]	A-1	10670	22002	10	1	0.0093	2.17	Oregon route-views for period
	A-2	10729	21999	12	1	0.0085	2.19	March 31 to May 26, 2001
	A-3	10790	22469	10	1	0.0094	2.20	
	A-4	10859	22747	10	1	0.0097	2.206	
	A-5	10886	22493	10	1	0.0089	2.19	
Metabolic [13]	M-1	1268	3011	14	1	0	2.17	Three types of Organisms
	M-2	490	1163	11	1	0	2.18	Archaea (M-1, M-2),
	M-3	993	2368	12	2	0	2.21	Bacteria (M-3, M-4)
	M-4	409	880	9	7	0	2.35	and Eukaryotes (M-5, M-6)
	M-5	665	1514	14	3	0	2.25	
	M-6	1511	3833	14	1	0	2.37	

Next, we execute *Netsimile* and *NCKD* on all networks mentioned in Table 1. Figures 4a and b show the dendrograms generated from pairwise distances computed by two algorithms. It is evident that *NCKD* algorithm performs better grouping than *Netsimile*. Cluster quality metrics of 32 networks (Table 2) for two algorithms vindicate the visual observation. Comparison of execution time of two algorithms reveals that *NCKD* is several orders faster than *Netsimile* (See *Large* synthetic and real-life networks in Table 3). The swift execution of *NCKD* indicates its scalability.

Table 2. Quality metrics for hierarchical clustering of 14 small networks and all 32 networks.

Datasets	Algorithm	Purity	Precision	Recall	Accuracy	NMI
Small networks (14)	NCKD	**1.0**	**1.0**	**1.0**	**1.0**	**1.0**
	LBD	0.8571	0.8182	0.9474	0.9451	0.8921
	NetSimile	0.8947	0.625	0.5263	0.8352	0.8213
All networks (32)	NCKD	**0.875**	**0.688**	**0.8983**	**0.9395**	**0.9161**
	Netsimile	0.656	0.382	0.5763	0.8387	0.6885

Table 3. Signature generation time (in seconds) for *NCKD*, *Netsimile* and *LBD* on selected networks from Table 1. A - indicates that the algorithm did not complete even after running for 24 h.

Algorithm → Networks ↓		NCKD	Netsimile	LBD
Small	M-2	**0.015**	0.783	120.615
	F1K-1	**0.018**	0.899	1825.11
	CA-5	**0.076**	6.711	6583.153
Large synthetic	B10K-1	**0.0202**	111.534	–
	E10K-1	**0.0503**	4.817	–
	W20K-1	**0.097**	25.353	–
	F10K-1	**0.077**	13.643	–
Large real-life	M-6	**0.013**	0.821	–
	A-5	**0.065**	30.676	–
	CA-1	**0.245**	198.789	–

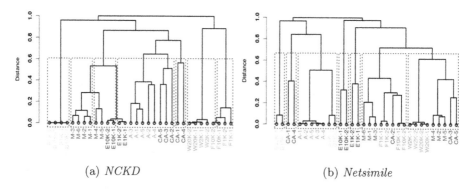

(a) *NCKD*   (b) *Netsimile*

Fig. 4. Dendrogram for networks described in Table 4. Networks belonging to same genre have same color. (Color figure online)

In order to capture finer distinctions between networks of the same genre, we selected eleven metabolic networks in two sub-categories (six Archaea (A) and five Eukaryotes (E)) whose order ranges from 490 to 1511 and size ranges from 1148 to 3807 [13]. We performed hierarchical agglomerative clustering of these networks from distance matrices generated by *NCKD* and *NetSimile* (Fig. 5). Algorithm *NCKD* is able to identify one pure group of Eukaryotes, which *Netsimile* missed. The clustering quality metrics for metabolic networks shown in Table 4 confirm the effectiveness of *NCKD* over *Netsimile*.

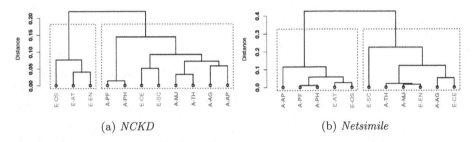

(a) *NCKD*   (b) *Netsimile*

Fig. 5. Metabolic networks in two sub-categories - A: Archaea, and E: Eukaryote.

Table 4. Quality metrics for dendrograms shown in Fig. 5

N/w type →	Metabolic networks				
Algorithm ↓	Purity	Precision	Recall	Accuracy	NMI
NCKD	**0.8182**	**0.6129**	**0.76**	**0.6727**	**0.4393**
Netsimile	0.5455	0.40	0.40	0.4545	0.0073

5.3 Handling of Missing Data

We compare effectiveness of *NCKD* and *Netsimile* towards missing data. For this experiment, we compared networks with themselves after applying random edge deletion systematically. For network G, we created $G'_{x_1}, G'_{x_2} \cdots G'_{x_k}$ variations by deleting $x_i\%$ of edges from it. Intuitively, both algorithms should yield similarity score (SS) of 1 while comparing G with G'_0, with the score falling as deleted edges increase. Fall in SS is expected to be different for different graphs due to structural differences. In order to beat the effect of randomness, reported results are averaged over three runs.

Three real-life networks (A-1, DC-1, CA-1) and one synthetic network (E10K-1) were perturbed by deleting edges from 0 % to 20 % (in steps of 2) and compared using two algorithms. Similarity scores obtained by *NCKD* and *Netsimile* are plotted in Figs. 6a and b. *Netsimile* registered a fall of maximum 10 % for the real-life datasets even after deleting 20 % edges while *NCKD* revealed significant differences in networks. This observation indicates superior ability of *NCKD* to suitably react to missing data.

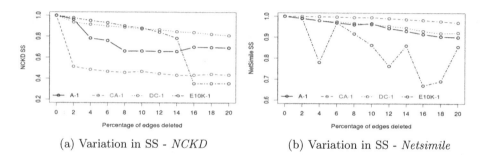

(a) Variation in SS - *NCKD*   (b) Variation in SS - *Netsimile*

Fig. 6. Comparison of robustness towards missing data - *NCKD* vs. *Netsimile*.

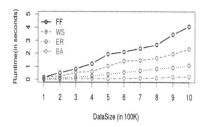

Large networks	Nodes (10^6)	Edges (10^6)	Runtime (seconds)
Amazon product	0.33	0.93	0.799
Road n/w of Texas	1.38	1.92	1.528
Road n/w of California	1.96	2.77	3.4824
Youtube OSN	1.13	2.99	3.532
Web graph of Berkeley and Stanford	0.69	7.796	10.889

Fig. 7. Feature generation time of *NCKD* for synthetic networks

Fig. 8. Feature generation time of *NCKD* for *massive* real-life networks. n/w: Network

5.4 Scalability w.r.t Large Datasets

Networks generated from different models allow convenient variations in the order of graphs to examine scalability of *NCKD*. We generated 10 graphs for each generative model (description in Table 1) with varying number of nodes 100K to 1000K in steps of 100K, and edges proportionally depending on the model. *Netsimile* was unable to process graphs of order >100K even after running for more than 24 h. Hence, it was dropped for scalability analysis. *NCKD* was executed five times for each graph to average out the timing observations. Figure 7 shows approximately sublinear growth in timings for each model. The increase in timings for the models varies with the number of edges in the corresponding networks. Edges increase fastest in FF model and slowest in BA model, which is faithfully reflected by the timings for two models.

Figure 8 shows execution timings of *NCKD* for five real-life large datasets downloaded from SNAP[3], which strengthens the claim of scalability. Since k-core decomposition algorithm is $O(\mathcal{E})$, time increases linearly with edges.

6 Conclusion

Each large-scale network is unique at the microscopic level. However, at different levels of resolutions, commonalities emerge among different pairs of graphs.

---

[3] http://snap.stanford.edu/data.

Discovery of these commonalities and their quantification is the goal of the proposed algorithm *NCKD* (Network Comparison using k-core Decomposition), which is intuitive, effective and scalable. The algorithm decomposes the graph into cores, analyses shells and constructs node and edge related probability distributions, which serve as network signatures. Jensen-Shannon distance is applied on these signatures to find distance between networks. We establish that node and edge distributions adequately discriminate networks.

Extensive experimentation and comparison of *NCKD* with *Netsimile* and *LBD* algorithms establish its superiority in terms of effectiveness and scalability. Execution timings for large synthetic and real-life networks affirm its scalability. We also demonstrate that *NCKD* is sensitive to the underlying topological structure of the graph, but needs to be improved to take cognizance of size and order of the network. The agenda for future is to overcome its deficiency to clearly segregate networks of the same genre by including more shell features.

References

1. Albert, R., lászló Barabsi, A.: Statistical mechanics of complex networks. Rev. Mod. Phys. **74**, 47 (2002)
2. Alvarez-Hamelin, J.I., Barrat, A., Vespignani, A.: Large scale networks fingerprinting and visualization using the k-core decomposition. Adv. Neural Inf. Process. Syst. **18**, 41–50 (2006)
3. Banerjee, A.: Structural distance and evolutionary relationship of networks. Biosystems **107**(3), 186–196 (2012)
4. Batagelj, V., Zaversnik, M.: An O(m) algorithm for cores decomposition of networks. CoRR cs.DS/0310049 (2003)
5. Baur, M., Gaertler, M., Grke, R., Krug, M.: Generating graphs with predefined k-core structure. Technical report, DELIS - Dynamically Evolving, Large-Scale Information Systems (2007)
6. Berlingerio, M., Koutra, D., Eliassi-Rad, T., Faloutsos, C.: Network similarity via multiple social theories. In: Proceedings of International Conference on ASONAM, pp. 1439–1440. IEEE (2013)
7. Dorogovtsev, S., Goltsev, A., Mendes, J.: Critical phenomena in complex networks. Rev. Mod. Phys. **80**, 1275 (2008)
8. Erdös, P., Rényi, A.: On random graphs I. Publicationes Math. **6**, 290–297 (1959). Debrecen
9. Faloutsos, C., Koutra, D., Vogelstein, J.T.: DELTACON: a principled massive-graph similarity function. In: Proceedings of the 13th SIAM International Conference on Data Mining, pp. 162–170 (2013)
10. Faust, K.: Comparing social networks: size, density and local structure. Adv. Methodol. Stat. **3**(2), 185–216 (2006)
11. Gao, X., Xiao, B., Tao, D., Li, X.: A survey of graph edit distance. Pattern Anal. Appl. **13**(1), 113–129 (2010)
12. Giatsidis, C., Thilikos, D.M., Vazirgiannis, M.: Evaluating cooperation in communities with the k-core structure. In: Proceedings of International Conference on ASONAM, pp. 87–93. IEEE (2011)
13. Jeong, H., Tombor, B., Albert, R., Oltvai, Z.N., Barabasi, A.L.: The large-scale organization of metabolic networks. Nature **407**(6804), 651–654 (2000)

14. Karwa, V., Pelsmajer, M.J., Petrovic, S., Stasi, D., Wilburne, D.: Statistical models for cores decomposition of an undirected random graph. CoRR abs/1410.7357 (2014)
15. Kollias, G., Mohammadi, S., Grama, A.: Network similarity decomposition (NSD): a fast and scalable approach to network alignment. Technical report. Purdue University (2011)
16. Leskovec, J., Kleinberg, J., Faloutsos, C.: Graph evolution: densification and shrinking diameters. ACM Trans. Knowl. Discov. Data 1(1), 2 (2007)
17. Lin, J.: Divergence measures based on the shannon entropy. IEEE Trans. Inf. Theory 37(1), 145–151 (1991)
18. Lu, S., Kang, J., Gong, W., Towsley, D.: Complex network comparison using random walks. In: Proceedings of 23rd International WWW Conference, pp. 727–730 (2014)
19. Macindoe, O., Richards, W.: Graph comparison using fine structure analysis. In: Proceedings of the 2nd IEEE International Conference on Social Computing, pp. 193–200 (2010)
20. Manning, C.D., Raghavan, P., Schütze, H.: Introduction to Information Retrieval. Cambridge University Press, New York (2008)
21. Melnik, S., Garcia-Molina, H., Rahm, E.: Similarity flooding: a versatile graph matching algorithm and its application to schema matching. In: Proceedings of the 18th ICDE, pp. 117–128 (2002)
22. Montresor, A., Pellegrini, F.D., Miorandi, D.: Distributed k-core decomposition. IEEE Trans. Parallel Distrib. Syst. 24(2), 288–300 (2013)
23. Peng, C., Kolda, T.G., Pinar, A.: Accelerating community detection by using k-core subgraphs. CoRR abs/1403.2226 (2014)
24. Seidman, S.B.: Network structure and minimum degree. Soc. Netw. 5, 269–287 (1983)
25. Wang, J., Cheng, J.: Truss decomposition in massive networks. Proc. VLDB Endowment 5(9), 812–823 (2012)
26. Watts, D.J., Strogatz, S.H.: Collective dynamics of small-world networks. Nature 393(6684), 440–442 (1998)

Data Streams

Incremental Stream Processing
of Nested-Relational Queries

Leonidas Fegaras[(✉)]

University of Texas at Arlington, Arlington, USA
fegaras@cse.uta.edu

Abstract. Current work on stream processing is focused on approximation techniques that calculate approximate answers to simple queries by focusing on a fixed or sliding window that contains the most recent tuples from an input stream and by using condensed synopses to summarize the state. It is widely believed that without using approximation techniques, most interesting queries would be blocking (i.e., they would have to wait for the end of stream to release their results) or unbounded (i.e., their memory requirements would grow proportionally to the stream size, which may be infinite). The goal of this paper is to convert nested-relational queries to incremental stream processing programs automatically. In contrast to most current stream processing systems that calculate approximate answers, our system derives incremental programs that return accurate results. This is accomplished by retaining a state during the query evaluation lifetime and by using incremental evaluation techniques to return an accurate snapshot answer at each time interval that depends on the current state and the data in the current fixed window. Our methods can handle most forms of declarative queries on nested data collections, including arbitrarily nested queries, group-by with aggregation, and equi-joins. We report on a prototype system implementation and we show some preliminary results on evaluating queries on a small computer cluster running Spark.

1 Introduction

New frameworks in Big Data analytics have become indispensable tools for large-scale data mining and scientific data analysis. Currently, the most popular framework for Big Data processing is Map-Reduce [12], which has emerged as a powerful, generic, and scalable solution for a wide range of data analysis applications. Unfortunately, to simplify fault tolerance and recovery, the Map-Reduce model does not preserve data in memory between consecutive jobs, which inflicts a high overhead on complex workflows and repetitive algorithms, such as PageRank and data clustering. Although the Map-Reduce framework was originally designed for batch processing, there are several recent systems that have extended Map-Reduce with on-line processing capabilities, such as MapReduce Online [11], Incoop [7], and i^2MapReduce [37]. In addition, many distributed stream processing engines (DSPEs) have emerged recently, such as

© Springer International Publishing Switzerland 2016
S. Hartmann and H. Ma (Eds.): DEXA 2016, Part I, LNCS 9827, pp. 305–320, 2016.
DOI: 10.1007/978-3-319-44403-1_19

Apache Storm [28], Spark's D-Streams [36], and Flink Streaming [16]. Most of these systems use data stream processing techniques based on fixed or sliding windows and incremental operators [4], which have already been used successfully in relational stream processing systems, such as Aurora [1] and Telegraph [10].

Furthermore, there is a recent interest in incremental data analysis, where data are analyzed in incremental fashion, so that existing results on current data are reused and merged with the results of processing the new data. In many cases, incremental data processing can achieve better performance and may require less memory than batch processing for many common data analysis tasks. Incremental processing can also be used for analyzing large amounts of data incrementally, in small batches that can fit in memory, thus enabling to process more data with less hardware. It can also be very valuable to stream-based applications that need to process continuous streams of data in real-time with low latency, which is not possible with existing batch analysis tools.

We are presenting a novel incremental stream processing framework for large-scale data analysis queries that run on a distributed stream processing engine (DSPE). Our design objective is to be able to convert any batch data analysis query to an incremental distributed stream processing program automatically, without requiring the user to rewrite the query. Furthermore, in contrast to most current stream processing systems that calculate approximate answers, we want our system to derive incremental programs that return accurate results. Such a task requires a query analysis to separate the query parts that can be used to process the incremental batches of data from the query parts that merge the current results with the new results of processing the incremental batches of data. Such analysis is more tractable if it is performed on declarative queries than on algorithmic programming languages, such as Java. We have developed our framework on Apache MRQL [26], because it is the only Big Data query language powerful enough to express complex data analysis tasks, such as PageRank, data clustering, and matrix factorization. We have developed general methods to transform batch MRQL queries to incremental queries. The derived incremental queries retain a state during their evaluation lifetime and use incremental evaluation techniques to return an accurate snapshot answer at each time interval that depends on the current state and the latest batches of data.

The first step in our approach is to transform a query so that it propagates the join and group-by keys to the query output. This is known as lineage tracking [5,6]. That way, the values in the query output are grouped by a key combination that corresponds the join/group-by keys used in deriving these values during query evaluation. If we also group the new data in the same way, then computations on existing data can be combined with the computations on the new data by joining the data on these keys. Our approach requires that we can combine computations on data that have the same lineage to derive incremental results. In our framework, this is accomplished by transforming a query to a *monoid homomorphism* by removing the non-homomorphic parts of the query, using algebraic transformation rules, and by combining them to form an answer function. The remaining query, which is now a monoid homomorphism, is used to derive the state transformation that merges the existing state with the new results to form a new state.

We have implemented our incremental processing framework on MRQL [26] on top of Spark Streaming [36], which is an in-memory distributed stream processing platform. Our system is called *Incremental MRQL*. MRQL is currently the best choice for implementing our framework because other query languages for data-intensive, distributed computations provide limited syntax for operating on data collections, in the form of simple relational joins and group-bys, and cannot express complex data analysis tasks, such as PageRank, data clustering, and matrix factorization, using SQL-like syntax exclusively. Our framework, though, can be easily adapted to apply to other query languages, such as Hive, PigLatin, SQL, XQuery, and Jaql.

The contribution of this work can be summarized as follows:

- We present a general framework for translating nested-relational data analysis queries to incremental stream processing programs that can run on a distributed stream processing platform.
- This framework can handle many forms of queries over nested data, including deeply nested queries, group-by with aggregation, and equi-joins.
- We report on a prototype system implementation and we show some preliminary results on evaluating three queries, groupBy, join-groupBy, and PageRank step, on a small computer cluster running Spark.

2   MRQL Overview

Apache MRQL [26] is a query processing and optimization system for large-scale, distributed data analysis. MRQL was originally developed by the author [13,14], but is now an Apache incubating project with many users and developers. The MRQL language is an SQL-like query language for large-scale data analysis on clusters of commodity hardware. The MRQL query processing system can evaluate MRQL queries in four modes: in Map-Reduce mode using Apache Hadoop, in BSP mode (Bulk Synchronous Parallel model) using Apache Hama, in Spark mode using Apache Spark [31], and in Flink mode using Apache Flink. The MRQL query language is sufficiently powerful to express many data analysis tasks over many different kinds of raw data, such as XML documents, JSON documents, binary files, and text documents in CSV format. The design of MRQL has been influenced by XQuery and OQL. In fact, when restricted to XML data, MRQL is as powerful as XQuery. MRQL is more powerful than other current high-level Map-Reduce languages, such as Hive and PigLatin, since it can operate on more complex data and supports more powerful query constructs, thus eliminating the need for using explicit procedural code. With MRQL, users are able to express complex data analysis tasks, such as PageRank, k-means clustering, matrix factorization, etc., using SQL-like queries exclusively, while the MRQL query processing system is able to compile these queries to efficient Java code that can run on various distributed processing platforms. The MRQL system stack is shown in Fig. 1.

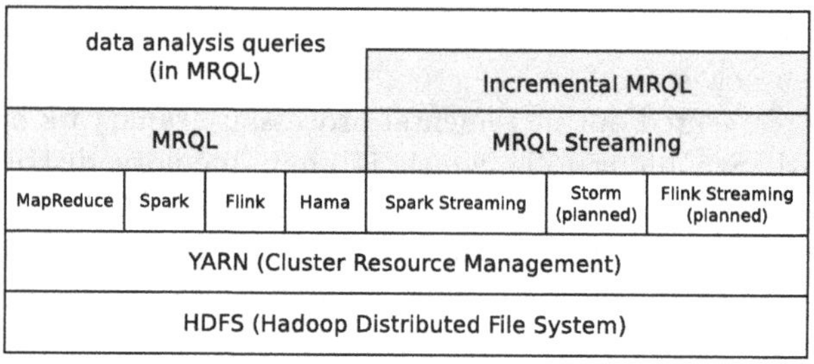

data analysis queries (in MRQL)				Incremental MRQL		
MRQL				MRQL Streaming		
MapReduce	Spark	Flink	Hama	Spark Streaming	Storm (planned)	Flink Streaming (planned)
YARN (Cluster Resource Management)						
HDFS (Hadoop Distributed File System)						

Fig. 1. The MRQL system stack

A recent extension to MRQL, called *MRQL Streaming*, supports the processing of continuous MRQL queries on streams of batch data (that is, data that come in continuous large batches). Before our incremental MRQL work presented in this paper, MRQL Streaming supported traditional window-based streaming based on a sliding window during a specified time interval. Currently, MRQL Streaming works on Spark Streaming only but we are currently adding support for Storm and Flink Streaming. The work reported here, called *Incremental MRQL*, extends the current MRQL Streaming engine with incremental stream processing. Incremental MRQL is now available in the latest official MRQL release (MRQL-0.9.6).

3   Incremental Query Processing

The MRQL data model consists of collections types, such as lists (sequences), bags (multisets), and key-value maps. The difference between a list and a bag is that a list supports order-based operations, such as indexing and subsequence. In addition, MRQL supports records, tuples, algebraic data types (union types), and basic types, such as integers and booleans. These types can be freely nested, thus supporting nested relations and hierarchical data. For example, XML data can be represented as a recursive algebraic data type with two value constructors, Node and CData:

```
data XML = Node: < tag: String, attributes: bag( (String, String) ),
                   children :  list (XML) >
         | CData: String
```

Non-streaming MRQL queries work on datasets, which are stored in the distributed file system (HDFS). A dataset is a bag (multiset) that consists of arbitrarily complex values. Datasets are stored as text files, such as XML, JSON, and CSV, or sequence (binary) files. Streaming MRQL queries work on 'batch' data streams, where stream data arrive in batches (bags) of new data. For example, the query:

```
select  (x,avg(y)) from (x,y) in  stream(binary ," data/points") group by x
```

groups a stream of points (x, y) by x and returns the average y values in each group. The MRQL Streaming engine will first process all the existing sequence files in the directory data/points and then will check this directory periodically for new files. When a new file is inserted in the directory, MRQL will process the new batch of data using distributed processing. In addition to directory of files, MRQL Streaming supports a special socket input format for listening to TCP sockets for text input, based on one of the current supported MRQL Parsed Input Formats (XML, JSON, CSV). A query may work on multiple stream sources and multiple batch dataset sources. If there is at least one stream source in the query, the query becomes continuous, that is, it never stops. The output of a continuous query is dumped into a directory HDFS as a sequence of files, so that each file in the sequence contains the results of processing a single batch of streaming data.

Our incremental query processing framework can handle continuous queries over a number of streaming data sources, $S_1, \ldots, S_n$, denoted by $\overline{S}$. A data stream S_i in our framework consists of an initial dataset, followed by a continuous stream of incremental batches ΔS_i, which arrive at regular time intervals Δt. In MRQL Streaming, these are batch data streams, where stream data arrive in batches of new data in the form of new files created inside some pre-specified directories. Then, a streaming query can be expressed as $q(\overline{S})$, where an $S_i \in \overline{S}$ is a streaming data source. Incremental stream processing is attainable if we can derive the query results at time $t + \Delta t$ by combining the query results at time t with the results of processing ΔS_i, rather than processing the query over the entire streams $S_i \uplus \Delta S_i$, where $\uplus$ is bag union. This is possible if $q(\overline{S \uplus \Delta S})$ can be expressed in terms of the current query result, $q(\overline{S})$, and the incremental query result, $q(\overline{\Delta S})$, that is, when $q(\overline{S})$ is a monoid homomorphism over $\overline{S}$. In abstract algebra, a monoid $\otimes$ is an associative function with a zero element $\otimes_z$, such that $x \otimes \otimes_z = \otimes_z \otimes x = x$. In addition, h is a monoid homomorphism if $h(x \oplus y) = h(x) \otimes h(y)$, for two monoids $\oplus$ and $\otimes$.

Unfortunately, some queries, such as counting the number of distinct elements in a stream or calculating average values after a group-by, are not (monoid) homomorphisms. The first query is not a homomorphism because when we count the distinct elements in a bag X we loose the information about which elements are contained in X, and therefore, counting the distinct elements of $X \uplus Y$ cannot be derived by combining the counts of the distinct elements in X and Y separately, since X and Y may have common elements. The average in the second query prevents the query from becoming a homomorphism because the average value is the sum divided by the count of values, and both these values are lost after we derive the query result. To handle a non-homomorphic query $q(\overline{S})$, we break q into two functions a and h, so that h is a homomorphism with $q(\overline{S}) = a(h(\overline{S}))$. Recall that function h is a homomorphism if $h(\overline{S \uplus \Delta S}) = h(\overline{S}) \otimes h(\overline{\Delta S})$ for some monoid $\otimes$. For example, the first query that counts the number of distinct elements can be broken into the query h that returns the list of distinct elements, which is a homomorphism, and the answer query a that counts the derived distinct elements. For this approach to be effective, most of the computations in q must be done in h, possibly leaving some computationally inexpensive data mappings to the answer function a. After we split the query q into an answer function a and a homomorphism h, we can calculate the results of h incrementally by storing its results into a state, which is maintained across the stream processing, and then combine the current state with the new data to calculate the next state instance. More specifically, at each time interval Δt, the query answer $h(\overline{S \uplus \Delta S})$ is calculated from the new state, state $\leftarrow$ state $\otimes h(\overline{\Delta S})$, and is equal to $a(\text{state})$, which is the snapshot of the query answer at time $t + \Delta t$.

Our framework translates incremental MRQL queries into incremental query plans as follows. First, it pulls all non-homomorphic parts of a query q out from the query using algebraic transformations, leaving an algebraic homomorphic term h, such that $q(\overline{S}) = a(h(\overline{S}))$. Then, it combines these non-homomorphic parts into an answer function a. Finally, from the homomorphic algebraic term $h(\overline{S})$, our framework synthesizes a merge function $\otimes$, such that $h(\overline{S \uplus \Delta S}) = h(\overline{S}) \otimes h(\overline{\Delta S})$. All these tasks are performed using algebraic transformations on the MRQL query algebraic terms [13,14]. Although all algebraic operations used in MRQL are homomorphic, their composition may not be. We have developed transformation rules to derive homomorphisms from compositions of homomorphisms, and for pulling non-homomorphic parts outside a query. Our methods can handle most forms of queries on nested data sets, including complex nested queries with any form of nesting and any number of nesting levels, complex group-bys with aggregations, and general one-to-one and one-to-many equi-joins. Our methods cannot handle non-equi-joins and many-to-many equi-joins, as they are very difficult to implement efficiently in a streaming or an incremental computing environment.

For example, consider the following MRQL query $q(S_1, S_2)$:

```
select (x, avg(z))
  from (x,y) in S₁, (y,z) in S₂
  group by x
```

where S_1 and S_2 are stream data sources. Here, the streams S_1 and S_2 are joined so that the second column of S_1 is equal to the first column of S_2, then the join result is grouped by the first column of S_1, and finally the average value of all z values in the group is returned. This query is not a homomorphism over both S_1 and S_2, that is $q(S_1 \uplus \Delta S_1, S_2 \uplus \Delta S_2)$ cannot be expressed in terms of $q(S_1, S_2)$ and $q(\Delta S_1, \Delta S_2)$, because the query result does not contain any information on how the avg(z) value is related to the x value. That is, there is no lineage in the query output that links a pair in the query result to the join key that contributed to this pair. Hence, it is impossible to tell how the new data ΔS_1 and ΔS_2 will contribute to the previous query results if we do not know how these results are related to the previous inputs S_1 and S_2. Our approach is to establish links between the query results and the parts of the data sources that were used to form their values. This is called *lineage tracking* and has been used for propagating annotations in relational queries [6]. In our case, this lineage tracking can be done by propagating all keys used in joins and group-bys along with the values associated with these keys, so that, for each combination of keys, we return one group of result values. For our query, this is done by including the join key y in the group-by keys. That is, the query is transformed to $h(S_1, S_2)$:

```
select ((x,y), (sum(z), count(z)))
   from (x,y) in S₁ , (y,z) in S₂
group by (x,y)
```

Hence, the join key y is propagated to the output values so that the avg components, sum and count, are aggregations over groups associated with unique combinations of x and y. This query is a homomorphism over S_1 and S_2, provided that the join is not on a many-to-many relationship. In general, a query with N join/group-by/order-by steps is transformed to a query that injects the join/group-by/order-by keys to the output so that each output value is associated with a unique combination of N keys. The answer query a that returns the final result in $q(S_1, S_2) = a(h(S_1, S_2))$ is:

```
select (x, sum(s)/sum(c))
   from ((x,y),(s,c)) in State
group by x
```

where State is the current state, equal to $h(S_1, S_2)$. This query removes the lineage y (the join key) from the State but also groups the result by the group-by key again, since there may be duplicate x values, and returns the final average value. (Note that sum(s) adds the partial sums while sum(c) adds the partial counts.) The merge function H1 ⊗ H2 of the homomorphism h, which combines tuples with the same lineage key, is a full outer-join on the lineage key that aggregates the matches:

```
select (k,(s1+s2,c1+c2))
   from (k,(s1,c1)) in H1,
        (k,(s2,c2)) in H2
union select (k,(s2,c2)) from (k,(s2,c2)) in H2 where k not in π₁(H1)
union select (k,(s1,c1)) from (k,(s1,c1)) in H1 where k not in π₁(H2)
```

where π_1 is a bag projection that retrieves the keys. The first select is an equi-join between H1 and H2 over the lineage key k, equal to the pair (x,y) that contains the group-by key x and the join key y. The first union returns all H2 pair that are not joined with H1, while the second union returns all H1 pair that are not joined with H2.

MRQL Streaming has been implemented on Apache Spark [31] running on an Apache Yarn cluster. More specifically, the incremental state is cached in memory as a Distributed DataSet (an RDD [35]), which is distributed across the worker nodes, while the streaming data sources are implemented as Discretized Streams (D-Streams [36]), which are also distributed across the worker nodes. The full outer-join used in the homomorphism h, which merges the query result on the new data with the current state, is implemented efficiently as a distributed hash-partitioned join (a Spark's coGroup operation), by keeping the state partitioned on the lineage keys and shuffling only the new results to worker nodes to be combined locally with the state using merging. That is, the results of processing the new data, which are typically substantially smaller than the state, are shuffled across the worker nodes before coGroup. Our approach is to keep the state partitioned on the lineage keys by simply leaving the partitions of the newly created state by coGroup at the place they were generated, since hash-partitioning generates data partitioned by the same join key. That way, only the results from the new data would have to be partitioned and shuffled across the working nodes to be combined with the current state.

4   The Translation Framework

The MRQL algebra consists of a small number of higher-order homomorphic operators [14], which are defined using structural recursion based on the union representation of bags [15]. The most important operation is cMap (also known as flatten-map in functional programming), which generalizes the select, project, join, and unnest operators of the nested relational algebra. Given two arbitrary types α and β, the operation $\mathrm{cMap}(f, X)$ maps a bag X of type $\{\alpha\}$ to a bag of type $\{\beta\}$ by applying the function f of type $\alpha \to \{\beta\}$ to each element of X, yielding one bag for each element, and then by merging these bags to form a single bag of type $\{\beta\}$. Using a set former notation on bags, $\mathrm{cMap}(f, X)$ can be expressed as $\{ z \mid x \in X, z \in f(x) \}$. Using structural recursion, it can also be defined as a homomorphism:

$$\mathrm{cMap}(f, X \uplus Y) = \mathrm{cMap}(f, X) \uplus \mathrm{cMap}(f, Y)$$
$$\mathrm{cMap}(f, \{a\}) = f(a)$$
$$\mathrm{cMap}(f, \{\,\}) = \{\,\}$$

The second in importance operator is groupBy, which groups a bag of pairs by their first value. Given the arbitrary types κ and α, and a bag X of type $\{(\kappa, \alpha)\}$, the operation $\mathrm{groupBy}(X)$ groups the elements of the bag X by their first component and returns a bag of type $\{(\kappa, \{\alpha\})\}$. For example, $\mathrm{groupBy}(\{(1,10),(2,20),(1,30),(1,40)\})$ returns $\{(1,\{10,30,40\}),(2,\{20\})\}$. The groupBy operation can be defined using structural recursion:

$$\mathrm{groupBy}(X \uplus Y) = \mathrm{groupBy}(X) \Uparrow_{\uplus} \mathrm{groupBy}(Y)$$
$$\mathrm{groupBy}(\{(k, a)\}) = \{(k, \{a\})\}$$
$$\mathrm{groupBy}(\{\,\}) = \{\,\}$$

where the parametric monoid $\Uparrow_{\oplus}$ is a full outer-join that merges groups associated with the same key using the monoid $\oplus$ (equal to $\uplus$ for groupBy). It is expressed using a set-former notation for bags:

$$
\begin{aligned}
X \Uparrow_{\oplus} Y = & \{ (k, a \oplus b) \mid (k, a) \in X, (k', b) \in Y, k = k' \} & \text{(join between } X \text{ and } Y) \\
& \uplus \{ (k, a) \mid (k, a) \in X, \forall (k', b) \in Y : k' \neq k \} & (\subseteq X \text{ not joined with } Y) \\
& \uplus \{ (k, b) \mid (k, b) \in Y, \forall (k', b) \in X : k' \neq k \} & (\subseteq Y \text{ not joined with } X)
\end{aligned}
$$

In other words, the monoid $\Uparrow_{\oplus}$ constructs a set of pairs whose unique key is the first pair element. In fact, any bag X can be converted to a set:

$$\mathrm{distinct}(X) = \mathrm{cMap}(\lambda(k, s).\{k\}, \mathrm{groupBy}(\mathrm{cMap}(\lambda \mathrm{x}. \{(x, x)\}, X)))$$

Equi-joins and outer-joins between a bag X of type $\{(\kappa, \alpha)\}$ and a bag Y of type $\{(\kappa, \beta)\}$ over their first component of a type κ, are captured by coGroup(X, Y), which returns a bag of type $\{(\kappa, (\{\alpha\}, \{\beta\}))\}$:

$$\text{coGroup}(X_1 \uplus X_2, Y_1 \uplus Y_2) = \text{coGroup}(X_1, Y_1) \Uparrow_{\uplus \times \uplus} \text{coGroup}(X_2, Y_2)$$

(plus more equations for cases with singleton and empty bags). Here, the product of two monoids, $\oplus \times \otimes$ is a monoid that, when applied to the two pairs (x_1, x_2) and (y_1, y_2), returns $(x_1 \oplus y_1, x_2 \otimes y_2)$. That is, the monoid $\Uparrow_{\uplus \times \uplus}$ merges two bags of type $\{(\kappa, (\{\alpha\}, \{\beta\}))\}$ by unioning together their $\{\alpha\}$ and $\{\beta\}$ values that correspond to the same key κ. For example, coGroup$(\{(1,10),(2,20),(1,30)\}, \{(1,100),(2,200),(3,300)\})$ returns $\{(1,(\{10,30\}, \{100\})),(2,(\{20\},\{200\})),(3,(\{\},\{300\})))\}$. Finally, aggregations are captured by reduce$(\oplus, X)$, which aggregates a bag X using a commutative monoid $\oplus$. For example, reduce$(+, \{2, 1, 1\}) = 4$.

Algebraic terms can be normalized using the following rule:

$$\text{cMap}(f, \text{cMap}(g, S)) \rightarrow \text{cMap}(\lambda x.\, \text{cMap}(f, g(x)), S)$$

which fuses two cascaded cMaps into a nested cMap, thus avoiding the construction of the intermediate bag. If we apply this transformation repeatedly, any algebraic term can be normalized to a tree of groupBy/coGroup operations connected via cMaps, while the root of the tree may be either a cMap or a reduce operation, if the query is a total aggregation.

A query q in our framework is transformed in such a way that it propagates the lineage to the query output, starting with the empty lineage () at the sources and extended with the join and group-by keys. Each value v returned by the transformed query is annotated with a lineage θ, in the form of a pair (θ, v). The lineage θ of the query result v is a tree of groupBy and coGroup keys that are used in deriving the result v. The lineage tree θ has the same shape as the groupBy/coGroup tree of the algebraic term of the query.

The transformation of an algebraic term to a term that propagates the join and group-by keys is done using rewrite rules, which make use of the fact that normalized algebraic terms are trees of groupBy/coGroup operations connected via cMaps. The first rule is:

$$\text{cMap}(f, \text{groupBy}(X))$$
$$\rightarrow \{ (k', ((k, \theta), v)) \mid ((k, \theta), s) \in \text{groupBy}(\{ ((k, \theta), x) \mid (k, (\theta, x)) \in X \}),$$
$$(k', v) \in f(k, s) \}$$

Here, X at the left-hand side is a bag of type $\{(k, x)\}$ that contains the groupBy key k, while the cMap result is a bag of type $\{(k', v)\}$ so that the new key k' is used for the subsequent groupBy or coGroup (the parent operation) in the algebraic tree. In the transformed term, X is lifted to a type $\{(k, (\theta, x))\}$, which includes the incoming lineage θ, while the cMap result is lifted to

$\{(k', ((k, \theta), v))\}$, which extends the lineage with the groupBy key k. The second rule handles joins:

$$\text{cMap}(f, \text{coGroup}(X, Y))$$
$$\rightarrow \{ (k', ((k, (\theta_x, \theta_y)), v)) \mid (k, (s_1, s_2)) \in \text{coGroup}(X, Y),$$
$$(\theta_x, xs) \in \text{groupBy}(s_1), \ (\theta_y, ys) \in \text{groupBy}(s_2),$$
$$(k', v) \in f(k, (xs, ys)) \}$$

Here, both inputs X and Y have been lifted to $\{(k, (\theta_x, x))\}$ and $\{(k, (\theta_y, y))\}$, respectively, with possibly different lineages. Hence, these two lineages θ_x and θ_y must be paired with k to form the new lineage. Finally, if there is a reduce$(\oplus, X)$ at the tree root, it is lifted to reduce$(\Uparrow_{\oplus}, X)$, which aggregates data with the same lineage.

Transforming an algebraic term to propagate the group-by and join keys alone does not guarantee that the resulting term would be a homomorphism, but it is a required step. Although all algebraic operators used in MRQL are homomorphisms, their composition may not be. For instance, cMap$(f, \text{groupBy}(X))$ is not a homomorphism for certain functions f, because, in general, cMap does not distribute over $\Uparrow_{\uplus}$. Since both groupBy and coGroup are $\Uparrow_{\oplus}$ homomorphisms, our goal is to make sure that cMap is a $\Uparrow_{\oplus}$ homomorphism too, so that the entire query would be a $\Uparrow_{\oplus}$ homomorphism. More specifically, it can be proved by induction that cMap(f, X) is a $\Uparrow_{\otimes}$ homomorphism if X is a $\Uparrow_{\oplus}$ homomorphism and $f(k, s)$ is a $\Uparrow_{\otimes}$ homomorphism, if s is a $\oplus$ homomorphism. That is, a cMap is a homomorphism if its functional argument is a homomorphism too. This rule is part of a monoid inference system that we have developed, inspired by type inference systems used in programming languages, which infers the monoid of any algebraic term, if exists.

If a cMap term is not a $\Uparrow_{\otimes}$ homomorphism, our approach is to split cMap into two cMaps, one homomorphic and one not, and pull out and fuse all non-homomorphic cMaps at the root of the algebraic tree, thus splitting the query into two parts: the answer query and a homomorphism. Consider the term cMap$(\lambda v. e, X)$. In our framework, we find the largest subterms in the algebraic term e, namely $e_1, \ldots, e_n$, that are homomorphisms. This is accomplished by traversing the tree that represents the term e, starting from the root, and by checking if the node can be inferred to be a homomorphism. If it is, the node is replaced with a new variable. Thus, e is mapped to a term $f(e_1, \ldots, e_n)$, for some term f, and the terms $e_1, \ldots, e_n$ are replaced with variables when f is pulled outwards. That is, cMap$(\lambda v. e, X)$ is split into two cMaps:

$$\text{cMap}(\lambda(v, v_1, \ldots, v_n). f(v_1, \ldots, v_n), \text{cMap}(\lambda v. \{(v, e_1, \ldots, e_n)\}, X))$$

5 Performance Evaluation

We have implemented our incremental processing framework using Apache MRQL [26] on top of Apache Spark Streaming [36]. The Spark streaming engine monitors the file directories used as stream sources in an MRQL query, and when a new file is inserted in one of these directories or the modification time of a file changes, it triggers the MRQL query processor to process the new files, based on the state derived from the previous step, and creates a new state. The platform used for our evaluations is a small cluster of 9 nodes, built on the Chameleon cloud computing infrastructure, www.chameleoncloud.org. This cluster consists of nine m1.medium instances running Linux, each one with 4 GB RAM and 2 VCPUs at 2.3 GHz. For our experiments, we used Hadoop 2.6.0 (Yarn) and MRQL 0.9.6. The cluster front-end was used exclusively as a NameNode/ResourceManager, while the rest 8 compute nodes were used as DataNodes/NodeManagers. For our experiments, we used all the available 16 VCPUs of the compute nodes for Map-Reduce tasks.

We have experimentally validated the effectiveness of our methods using three queries: groupBy, join-groupBy, and a PageRank step. The groupBy and join-groupBy queries are expressed in MRQL as follows:

```
select (x,avg(y))
  from (x,y) in stream(binary,"S")
group by x;
```

```
select (x,avg(z))
  from (x,y) in stream(binary,"S1"),
       (y,z) in stream(binary,"S2")
group by x;
```

The PageRank algorithm computes the importance of the web pages in a web graph based exclusively on the topology of the graph. For a graph with vertices V and edges E, the PageRank P_i of a vertex $v_i \in V$ is calculated from the PageRank P_j of its incoming neighbors $v_j \in V$ with $(v_j, v_i) \in E$ using the rule $P_i = \sum_{(v_j,v_i)\in E} \frac{P_j}{|\{v_k \mid (v_j,v_k)\in E\}|}$. We represent the web graph as a bag of edges (i, j) from node i to node j. The following MRQL query computes one step of the PageRank algorithm:

```
select < id: a, rank: (1−factor)/graph_size +factor*sum(in_rank) >
  from n in ( select < id: src, rank: 0.1, adjacent: dst >
                from (src,dst) in stream(binary,"S")
              group by src )
       a in n.adjacent,
       in_rank = n.rank/count(n.adjacent)
group by a;
```

where factor=0.85 is the dumping factor and graph_size is the number of graph nodes. The inner select-query converts the bag of edges from the input stream to a nested bag of type {< id: int, rank: double, adjacent: {int} >}, that is, to a bag of nodes where each node is associated with a unique id, a current PageRank value rank, and a bag of its outgoing neighbors adjacent. Then, the outer-select query binds the in_rank variable to one incoming PageRank contribution from node n to node a (= in MRQL is for binding a variable to a value, not ranging through a bag), and all these contributions are added in the sum to form the new rank of a. This query is over nested relations (it uses the node adjacent list) and is also a nested query.

The data streams used by the first two queries (groupBy and join-groupBy) consist of a large set of initial data, which is used to initialize the state, followed by a sequence of 9 equal-size batches of data (the increments). The groupBy initial

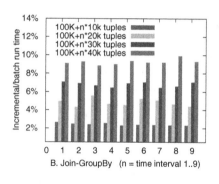

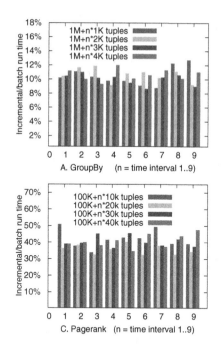

Fig. 2. Incremental query evaluation of groupBy, join-groupBy, and PageRank step

dataset had size 1M tuples, while the two join-groupBy inputs had sizes 100K tuples. The experiments were repeated for increments of size 1K–4K tuples for groupBy and 10K–40K tuples for join-groupBy, always starting with a fresh state (constructed from the initial data only). For the PageRank step query, we generated random graphs using the R-MAT algorithm using the Kronecker graph generator parameters: $a = 0.30$, $b = 0.25$, $c = 0.25$, and $d = 0.20$. The initial graph used in our evaluations had size 10K nodes with 10K edges and the four different increments have sizes 100, 200, 300, and 400 edges, respectively. The performance results are shown in Fig. 2. The x-axis represents the time points Δt when we get new batches of data in the stream. The 9 increments arrive at the time points $1\Delta t$ through $9\Delta t$. The y-axis is the incremental execution time for each batch of data divided by the total processing time of the initial data (percentage).

We can see from Fig. 2 that incremental processing can give an order of magnitude speed-up compared to batch processing that processes all the data every time the data changes. More importantly, the time to process each new batch of data remains nearly constant through time, even though the state grows with new data each time. The reason that the incremental processing time does not substantially increase through time is that we have carefully implemented state merging using Spark's coGroup operation (a hash-partitioned join), which is configured to partitioned the new batch of data only, but not the existing state, since the state has already been partitioned during the previous incremental step. That is, the new state created by coGroup is cached as a Spark RDD and is already partitioned by the join key, so that next time, when the state is merged with the new batch of data, it does not need to be partitioned again. Repartitioning data is very expensive because it requires to suffle the data across the worker nodes. That way, the state merging is done very fast, because the new batch of data is expected to be substantially smaller than the state.

6 Related Work

Recently, there is a large number of Big Data stream processing systems, also known as distributed stream processing engines (DSPEs), that have emerged. The most popular one is Twitter's Storm [28], which is now part of the Apache ecosystem for Big Data analytics. It provides primitives for transforming streams based on a user-defined topology, consisting of spouts (stream sources) and bolts (which consume input streams and may emit new streams). Other popular DSPE platforms include Spark's D-Streams [36], Flink Streaming [16], Apache S4, and Apache Samza. Many of these systems build on the well-established research on incremental stream processing of relational data, based on sliding windows and incremental operators [4], which includes systems such as Aurora [1] and Telegraph [10]. In addition, there are several recent systems that have extended Map-Reduce with online processing capabilities, since Map-Reduce was originally designed for batch processing only. MapReduce Online [11] maintains state in memory for a chain of Map-Reduce jobs and reacts efficiently to additional input records. It also provides a memoization-aware scheduler to reduce communication across a cluster. Incoop [7] is a Hadoop-based incremental processing system with an incremental storage system that identifies the similarities between the input data of consecutive job runs and splits the input based on the similarity and file content. i^2MapReduce [37] implements incremental iterative Map-Reduce jobs using a store, MRB-Store, that maps input values to the reduce output values. This store is used for detecting delta changes and propagating these changes to the output. Google's Percolator [28] is a system based on BigTable for incrementally processing updates to a large data set. It updates an index incrementally as new documents are crawled. Microsoft Naiad [25] is a distributed framework for cyclic dataflow programs that facilitates iterative and incremental computations. It is based on differential dataflow computations, which allow incremental computations to have nested iterations. CBP [20] is a continuous bulk processing system on Hadoop that provides a stateful group-wise operator that allows users to easily store and retrieve state during the reduce stage as new data inputs arrive. Their incremental computing PageRank implementation is able to cut running time in half. REX [24] handles iterative computations in which changes in the form of deltas are propagated across iterations and state is updated efficiently. In contrast to our automated approach, REX requires the programmer to explicitly specify how to process deltas, which are handled as first class objects. Trill [9] is a high throughput, low latency streaming query processor for temporal relational data, developed at Microsoft Research. The Reactive Aggregator [33], developed at IBM Research, is a new sliding-window streaming engine that performs many forms of sliding-window aggregation incrementally. Furthermore, the incremental query processing is related to the problem of incremental view maintenance, which has been extensively studied in the context of relational views (see [17] for a literature survey).

In programming languages, self-adjusting computation [2] refers to a technique for compiling batch programs into programs that can automatically respond to changes to their data. It requires the construction of a dependence graph at run-time so that when the computation data changes, the output can be updated by re-evaluating only the affected parts of the computation. In contrast to our work, which requires only the state to reside in memory, self-adjusting computation expects both the input and the output of a computation to reside in memory, which makes it inappropriate for unbounded data in a continuous stream. Furthermore, such dynamic methods impose a run-time storage and computation overhead by maintaining the dependence graph. The main idea in [2] is to manually annotate the parts of the input type that is changeable, and the system will derive an incremental program automatically based on these annotations. Each changeable value is wrapped by a mutator that includes a list of reader closures that need to be evaluated when the value changes. A read operation on a mutator inserts a new closure, while the write operation triggers the evaluation of the closures, which may cause writes to other mutators, etc., resulting to a cascade of closure execution triggered by changed data only. This technique has been extended to handle incremental list insertions (like our work), but it requires the rewriting of all list operations to work on incremental lists. Recently, there is a proof-of-concept implementation of this technique on mapreduce [3], but it was tested on a serial machine. It is doubtful that such dynamic techniques can be efficiently applied to a distributed environment, where a write in one compute node may cause a read in another node. Finally, there is recent work on static incrementalization based on derivatives [8]. In contrast to our work, it assumes that the merge function that combines the previous result with the result on the delta changes uses exactly the same delta changes, a restriction that excludes aggregations and group-bys.

7   Conclusion

We have presented a general framework for incremental distributed stream processing that translates SQL-like data analysis queries to incremental distributed stream processing programs. In contrast to other stream processing approaches, our framework derives incremental programs that return accurate results, rather than approximate answers. Our framework can also be used on a batch distributed system to process data larger than the total available memory in the cluster, by processing these data incrementally, in batches that can fit in memory. As a future work, we are planning to improve our state transformations by using a map join, where the results of processing the new batch of data are broadcast to all workers and joined with their state partitions locally. We are also planning to store each state partition as a local key-value map at each worker to implement state updates more efficiently.

Acknowledgments. This work is supported in part by the National Science Foundation under the grant CCF-1117369. Our performance evaluations were performed at the Chameleon cloud computing infrastructure, www.chameleoncloud.org, supported by NSF.

References

1. Abadi, D.J., Carney, D., Cetintemel, U., et al.: Aurora: a new model and architecture for data stream management. VLDB J. **12**(2), 120–139 (2003)
2. Acar, U.A., Blelloch, G.E., Blume, M., Harper, R., Tangwongsan, K.: An experimental analysis of self-adjusting computation. ACM Trans. Program. Lang. Syst. **32**(1), 3:1–3:53 (2009)
3. Acar, U.A., Chen, Y.: Streaming big data with self-adjusting computation. In: Workshop on Data Driven Functional Programming (DDFP) (2013)
4. Babcock, B., Babu, S., Datar, M., Motwani, R., Widom, J.: Models and issues in data stream systems. In: Symposium on Principles of Database Systems (PODS), pp. 1–16 (2002)
5. Benjelloun, O., Sarma, A.D., Halevy, A., Widom, J.: ULDBs: databases with uncertainty and lineage. In: International Conference on Very Large Data Bases (VLDB), pp. 953–964 (2006)
6. Bhagwat, D., Chiticariu, L., Tan, W.C., Vijayvargiya, G.: An annotation management system for relational databases. In: International Conference on Very Large Data Bases (VLDB), pp. 900–911 (2004)
7. Bhatotia, P., Wieder, A., Rodrigues, R., Acar, U.A., Pasquin, R.: Incoop: MapReduce for incremental computations. In: ACM Symposium on Cloud Computing (SoCC) (2011)
8. Cai, Y., Giarrusso, P.G., Rendel, T., Ostermann, K.: A theory of changes for higher-order languages. Incrementalizing λ-calculi by static differentiation. In: ACM SIGPLAN Conference on Programming Language Design and Implementation (PLDI), pp. 145–155 (2014)
9. Chandramouli, B., Goldstein, J., Barnett, M., DeLine, R., Fisher, D., Platt, J.C., Terwilliger, J.F., Wernsing, J.: Trill: a high-performance incremental query processor for diverse analytics. In: International Conference on Very Large Data Bases (VLDB), pp. 401–412 (2014)
10. Chandrasekaran, S., Cooper, O., Deshpande, A., Franklin, M.J., Hellerstein, J.M., Hong, W., Krishnamurthy, S., Madden, S., Raman, V., Reiss, F., Shah, M.: TelegraphCQ: continuous data flow processing for an uncertain world. In: Conference on Innovative Data System Research (CIDR) (2003)
11. Condie, T., Conway, N., Alvaro, P., Hellerstein, J.M., Elmeleegy, K., Sears, R.: MapReduce online. In: USENIX Symposium on Networked Systems Design and Implementation (NSDI), vol. 10, no. (4) (2010)
12. Dean, J., Ghemawat, S.: MapReduce: simplified data processing on large clusters. In: Symposium on Operating System Design and Implementation (OSDI) (2004)
13. Fegaras, L., Li, C., Gupta, U., Philip, J.J.: XML query optimization in MapReduce. In: International Workshop on the Web and Databases (WebDB) (2011)
14. Fegaras, L., Li, C., Gupta, U.: An optimization framework for Map-Reduce queries. In: International Conference on Extending Database Technology (EDBT), pp. 26–37 (2012)
15. Fegaras, L., Maier, D.: Optimizing object queries using an effective calculus. ACM Trans. Database Syst. (TODS) **25**(4), 457–516 (2000)

16. Apache Flink. http://flink.apache.org/
17. Gupta, A., Mumick, I.S.: Maintenance of materialized views: problems, techniques, and applications. IEEE Bull. Data Eng. **18**(2), 145–157 (1995)
18. Apache Hadoop. http://hadoop.apache.org/
19. Apache Hive. http://hive.apache.org/
20. Logothetis, D., Olston, C., Reed, B., Webb, K.C., Yocum, K.: Stateful bulk processing for incremental analytics. In: ACM Symposium on Cloud Computing (SoCC) (2010)
21. Low, Y., Gonzalez, J., Kyrola, A., Bickson, D., Guestrin, C., Hellerstein, J.M.: Distributed GraphLab: a framework for machine learning and data mining in the cloud. Proc. VLDB Endow. **5**(8), 716–727 (2012)
22. Malewicz, G., Austern, M.H., Bik, A.J.C., Dehnert, J.C., Horn, I., Leiser, N., Czajkowski, G.: Pregel: a System for large-scale graph processing. In: ACM Symposium on Principles of Distributed Computing (PODC) (2009)
23. McSherry, F., Murray, D.G., Isaacs, R., Isard, M.: Differential dataflow. In: Conference on Innovative Data System Research (CIDR) (2013)
24. Mihaylov, S.R., Ives, Z.G., Guha, S.: REX: recursive, delta-based data-centric computation. Proc. VLDB Endow. **5**(11), 1280–1291 (2012)
25. Murray, D.G., McSherry, F., Isaacs, R., Isard, M., Barham, P., Abadi, M.: Naiad: a timely dataflow system. In: ACM Symposium on Operating Systems Principles (SOSP) (2013)
26. Apache MRQL (incubating). http://mrql.incubator.apache.org/
27. Olston, C., Reed, B., Srivastava, U., Kumar, R., Tomkins, A.: Pig Latin: a not-so-foreign language for data processing. In: ACM SIGMOD International Conference on Management of Data, pp. 1099–1110 (2008)
28. Peng, D., Dabek, F.: Large-scale incremental processing using distributed transactions and notifications. In: Symposium on Operating System Design and Implementation (OSDI) (2010)
29. Power, R., Li, J.: Piccolo: building fast, distributed programs with partitioned tables. In: Symposium on Operating System Design and Implementation (OSDI) (2010)
30. Shinnar, A., Cunningham, D., Herta, B., Saraswat, V.: M3R: increased performance for in-memory Hadoop jobs. Proc. VLDB Endow. **5**(12), 1736–1747 (2012)
31. Apache Spark. http://spark.apache.org/
32. Apache Storm: A System for Processing Streaming Data in Real Time. http://hortonworks.com/hadoop/storm/
33. Tangwongsan, K., Hirzel, M., Schneider, S., Wu, K.-L.: General incremental sliding-window aggregation. Proc. VLDB Endow. **8**(7), 702–713 (2015)
34. Valiant, L.G.: A bridging model for parallel computation. Commun. ACM (CACM) **33**(8), 103–111 (1990)
35. Zaharia, M., Chowdhury, M., Das, T., Dave, A., Ma, J., McCauley, M., Franklin, M.J., Shenker, S., Stoica, I.: Resilient distributed datasets: a fault-tolerant abstraction for in-memory cluster computing. In: USENIX Symposium on Networked Systems Design and Implementation (NSDI) (2012)
36. Zaharia, M., Das, T., Li, H., Hunter, T., Shenker, S., Stoica, I.: Discretized streams: fault-tolerant streaming computation at scale. In: Symposium on Operating Systems Principles (SOSP) (2013)
37. Zhang, Y., Chen, S., Wang, Q., Yu, G.: i^2 MapReduce: incremental MapReduce for mining evolving big data. IEEE Trans. Knowl. Data Eng. (TKDE) **27**(7), 1906–1919 (2015)

Incremental Continuous Query Processing over Streams and Relations with Isolation Guarantees

Salman Ahmed Shaikh[1]([✉]), Dong Chao[2], Kazuya Nishimura[2], and Hiroyuki Kitagawa[1]

[1] Center for Computational Sciences, University of Tsukuba, Tsukuba, Japan
salman@kde.cs.tsukuba.ac.jp, kitagawa@cs.tsukuba.ac.jp
[2] Graduate School of Systems and Information Engineering, University of Tsukuba, Tsukuba, Japan
{dongchaotina,sigkan311}@kde.cs.tsukuba.ac.jp

Abstract. Stream processing has become an important research issue with the increase in stream data sources. Many stream processing systems need to reference non-streaming resources such as database relations to answer real world queries. Since database relation is a shared entity, it may be updated during the continuous query (CQ) execution by other database clients resulting in inconsistent query results (partly using the relation before update and partly after update). For this problem, an isolation model is needed to define the way in which these updates are reflected in the output of the stream-relation join. In this work we propose an incremental CQ processing approach with isolation guarantees which makes use of a monitor operator to transform the relational updates into stream tuples. Since database relations tend to be large, an in-memory T*-Tree index is used to increase the stream-relation join efficiency. Experiments are performed to prove that guaranteeing isolation solves the inconsistency problem and to show that the incremental computation and indexing improves the query throughput significantly.

Keywords: Continuous query processing · Stream processing · Isolation guarantees · Stream-relation join · Inconsistency problem · Incremental computation · T*-Tree index

1 Introduction

Stream processing has gained a lot of attention recently with the rise of real-time processing and analysis requirements. To fulfil such requirements, many stream processing engines (SPEs) have been developed [2–9]. Most of the SPEs run in main memory as their main goal is to produce quick results for real-time analysis. While real-time analysis is an essential requirement, a lot of stream processing applications require access to stored data such as database relations (hereafter called database) along with data streams to answer real world queries [19–26]. Since a database is a shared entity, it may be updated by other database clients

© Springer International Publishing Switzerland 2016
S. Hartmann and H. Ma (Eds.): DEXA 2016, Part I, LNCS 9827, pp. 321–335, 2016.
DOI: 10.1007/978-3-319-44403-1_20

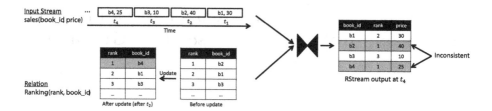

Fig. 1. Inconsistency example

during the query execution resulting in inconsistent query results (partly using the database before update and partly after update). Example 1 discusses this problem in detail.

Example 1. Consider a data stream *Sales* of books selling transactions. Let a query on this data stream needs to know if the book being sold is in the top-3 popular books list, which is available in a database relation *Ranking*. This query can be modelled as a join between the stream *Sales* and the relation *Ranking*, which we call a *mixed join*[1] in this work. We assume that the CQ in this example uses Rstream[2] (discussed in Sect. 2) to produce output. Figure 1 shows the inputs and output at timestamp t_4. Assuming that the relation *Ranking* is updated after the arrival of stream tuple <b2,40> at t_2, the Rstream operator at t_4 produces the output as shown in Fig. 1, which is inconsistent. Although the query is expected to output top-3 books, however due to the relation update four tuples are generated with book ids $b2$ and $b4$ both ranked 1.

To solve the inconsistency problem discussed in Example 1, an isolation model is required to determine the way in which the database updates must be reflected in the output of the mixed-join. Hence we propose an incremental CQ processing approach for mixed join with isolation guarantees. The proposed approach introduces a monitor operator to transform the relational updates into stream tuples. The monitor operator employs an in-memory mapping table. The mapping table exploits the cheap and large memory available in the machines these days, to map the primary keys of the database tuples to the unique IDs for the identification of tuples within the SPE. Since database relations tend to be large, a T*-Tree index [13] is used to increase the mixed join efficiency. Although there exist a few other stream processing frameworks dealing with the inconsistency problem (discussed in Sect. 5), none of them focus on the incremental computation which is an important framework and adopted by many state-of-the-art SPEs [4,8]. Our main contributions can be summarized as follows.

– A monitor operator employing an in-memory mapping table to transform the database updates into stream tuples.
– Use of an index structure for the synopsis handling the database tuples to increase the mixed join efficiency.

---

[1] The term *mixed join* is used by Neil Conway in [10] for the stream-relation join.
[2] The Rstream operator is commonly used to stream the mixed-join results.

– Experimental study to prove that the proposed approach solves the inconsistency problem by guaranteeing isolation and to show that the incremental computation and indexing improves the query throughput significantly.

The rest of the paper is organized as follows. Section 2 presents preliminaries and assumptions. In Sect. 3, the proposed incremental CQ processing with isolation guarantees is presented. In Sect. 4 we present an experimental study to prove that the proposed approach solves the inconsistency problem and to show that the incremental computation along with the indexing improves the query throughput significantly. We discuss related work in Sects. 5 and 6 finally concludes this paper and discusses some of the future directions.

2 Preliminaries and Assumptions

To fully understand the proposed incremental CQ processing for mixed join, understanding of the basic incremental computation is needed. For completeness, we summarize the incremental computation given by Arasu et al. for their STREAM SPE [4] and the related concepts in this section.

To process and query continuously evolving data streams, many SPEs have been developed. STREAM [4], Borealis [5], Aurora [2] and Storm [3] are a few examples of the well-known and commonly used SPEs. When a user registers a CQ on an SPE, it is executed continuously on the incoming stream tuples and continuous output is generated. CQL [1] is an SQL-based declarative language for CQs. It is a state of the art continuous query language, initially designed for STREAM, the prototype Data Stream Management System [4]. CQL is more general than many other continuous query languages and is therefore adopted by many SPEs. We summarize the CQL abstract semantics, its query plan and its incremental computation from [1,4] in the following subsections.

2.1 CQL Abstract Semantics

The CQL abstract semantics is based on two data types, streams and relations. Let Γ be discrete, ordered time domain then a *stream* is an unbounded multiset of pairs $\langle e, t \rangle$, where e is a tuple and $t \in \Gamma$ is the timestamp that denotes the arrival time of tuple e on stream S. Similarly, a *relation* is a time-varying multiset of tuples. The multiset at time $t \in \Gamma$ is denoted by $R(t)$, where $R(t)$ is an instantaneous relation.

The abstract semantics uses three classes of operators over streams and relations. (1) *relation-to-relation* operator takes one or more relations as input and produces a relation as output. (2) *stream-to-relation* operator takes a stream as input and produces a relation as output. (3) *relation-to-stream* operator takes a relation as input and produces a stream as output. A CQ Q is a tree of operators belonging to the three classes. Q uses leaf operators to receive inputs which could be streams and relations, and a root operator to produce output which could be either a stream or a relation. At time t, an operator of Q produces new outputs corresponding to t which depends on its inputs up to t.

CQL is defined by instantiating the operators of the abstract semantics. For the *relation-to-relation* operators, CQL uses existing SQL constructs. The *stream-to-relation* operators in CQL are based on sliding window over a stream, which are specified using window specification language derived from SQL-99. A window at any point of time holds a historical snapshot of a finite portion of the stream. Although various window operators have been given by SPEs and discussed by researchers, here we summarize two of the most commonly used. (1) *tuple-based window* operator on a stream S is specified using an integer n. At any time t it returns a relation R of n most recent tuples from stream S. (2) *time-based window* takes a parameter τ and at any time t returns a relation R containing tuples with timestamps between $t - \tau$ and t from stream S. CQL has three *relation-to-stream* operators which are also adopted in our framework: Istream, Dstream and Rstream. At time t, the Istream (insert stream) applied to relation R results in a stream element $\langle e, t \rangle$ whenever tuple e is in $R(t) - R(t-1)^3$. The Dstream (delete stream) returns a stream element $\langle e, t \rangle$ from R whenever tuple e is in $R(t-1) - R(t)$. The Rstream (relation stream) applied to R, results in a stream element $\langle e, t \rangle$ whenever tuple e is in R.

An example of a query written in CQL is shown in Query 1.1, which performs continuous binary-join with respect to common integer attribute A of streams S1 and S2.

```
Select S1.B, S2.C
From S1[Range τ], S2[Rows n]
Where S1.A = S2.A
```

Query 1.1. A simple CQL query

2.2   CQL Query Plan

A CQL query is translated into a query plan and is executed continuously. Query plans are composed of operators, queues and synopsis. *Operators* perform actual processing on data streams. The data arrive at an operator as a sequence of timestamped tuples, where each tuple is additionally flagged as either an insertion ($+$) or deletion ($-$) as explained later. These tuple-timestamp-flag triples are referred as elements. Each operator reads from one or more input queues, processes the input and writes any output to the output queue. *Queues* buffer elements as they move between operators. *Synopsis* is a buffer which belongs to a specific operator. It stores an operator's state that may be required for future evaluation of that operator.

A query plan for Query 1.1 is shown in Fig. 2. The query plan in Fig. 2 consists of seven operators: a root, an Rstream, a binary join, two instances of window operators and two instances of leaf operators. Note that the projection is performed as part of the binary join, so no separate projection operator or synopsis is employed. Queues q_1 and q_2 hold the input stream elements read by their respective leaf operators. Queues q_3 and q_4 hold elements representing the

---

[3] For simplicity, we assume that a new tuple arrives at every time instant t.

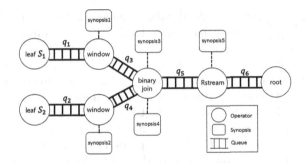

Fig. 2. Query plan for Query 1.1

relations $S1$[Range τ] and $S2$[Rows n], respectively. Queue q_5 holds elements for the result of joining relations $S1$[Range τ] and $S2$[Rows n]. Queue q_6 holds the elements coming out of the Rstream operator, which may lead to output or input to other query. The query plan has five synopsis, *synopsis1* ~ *synopsis5*. Each window operator has a synopsis so that it can hold the current window elements and generate '−' elements when elements expire from the sliding window. The binary-join operator has two synopses, one for each input, to materialize each of its relational inputs. The Rstream operator has a synopsis to convert its relational input to stream output.

2.3 Incremental Computation

A CQL query logically outputs elements based on $R(t)$ and $R(t-1)$, but computations required for $R(t)$ and $R(t-1)$ often have a lot of overlap. To eliminate redundant computation, the incremental computation is used.

Considering the query plan shown in Fig. 2, the window operator on $S1$, on being executed reads element $\langle e, t, + \rangle$. It inserts element e in *synopsis1*, and if an old element e' expires, it removes that element from the synopsis. The window then outputs elements $\langle e, t, + \rangle$ and $\langle e', t, - \rangle$ to q_3 to reflect the addition and deletion of elements e and e' respectively. The other window operator executes in the similar fashion. When the binary-join operator is executed, it reads the newly arrived element from one of its two input queues, i.e., q_3 or q_4. If it reads an element $\langle e, t, + \rangle$ from q_3, then it inserts e into *synopsis3* and joins e with the contents of *synopsis4*, generating output elements $\langle e.f, t, + \rangle$ for each matching element f in *synopsis4*. Similarly, if the binary-join operator reads an element $\langle e', t, - \rangle$ from q_3, it generates $\langle e'.f, t, - \rangle$ for each matching element f in *synopsis4*. The same process is done for the elements read from q_4. The output elements from the binary join are enqueued to q_5. The Rstream operator reads the data from q_5, inserts it into *synopsis5*, converts the relational input to stream output and enqueues it to q_6 which is then output by the root operator.

Now consider this incremental computation for the case of mixed join as shown in Fig. 1, which can be expressed as Query 1.2 in CQL. The input stream *sales* behaves similarly as discussed above. However, due to the absence of well-defined semantics to handle database tuples incrementally in the basic mixed join computation, a trigger is used to update the database related query operators' synopses (*DB synopses* for short). That is, the trigger is fired at the start of the query to populate the DB synopses and during the execution of the query it is fired as a result of the database update to update the DB synopses. Let us assume that there are three tuples in the database relation *Ranking* as shown in Fig. 1. The trigger populates the DB synopses in the start of the query. When the *Ranking* is updated after t_2, the trigger is fired to update the DB synopses, i.e., the tuple with Rank 1 is updated from Book-id $b2$ to $b4$. However this update is not communicated to the query downstream operators in an incremental fashion, resulting in multiple book ids with Rank 1 in the Rstream operators' synopsis. This causes inconsistent query output as shown in the Fig. 1.

```
Select S1.book-id, T1.rank, S1.price
From Sales[Rows n] as S1, Ranking as T1
Where S1.book-id = T1.book-id
```

Query 1.2. Stream-relation join query

3   Isolation Guaranteed Incremental CQ Processing

In order to solve the inconsistency problem discussed in Sects. 1 and 2, we propose an incremental CQ approach for mixed join with isolation guarantees. To guarantee isolation or to define the way in which the database updates are reflected in the mixed join output without resulting in inconsistency, a monitor operator is proposed in this work. For the details on the different isolation models, please refer [10]. In this section, we first present our proposed *monitor* operator employing the mapping table and then discuss the use of indexing to increase the mixed join efficiency.

3.1   Monitor Operator

The monitor operator is placed exactly after the leaf operator responsible for receiving database tuples as shown in Fig. 3. (For brevity leaf and root operators are not shown in Fig. 3.) Its main task is to transform database updates into stream tuples. The monitor operator receives all the database tuples at the start of a CQ involving mixed join. Since a tuple within the SPE (we assume the use of the JsSpinner SPE [9]) is identified by a unique ID, generated by the SPE, the monitor operator assigns a unique ID to each incoming database tuple. Additionally it maps the assigned IDs to the database tuples' primary keys in a one-to-one fashion and stores them in a mapping table. The mapping table is an in-memory data structure maintained in the monitor operator's synopsis.

In order to keep track of the database modifications (insert, update or delete), a trigger is associated with it. When the database is modified the trigger is fired, as a result modifications are sent to the monitor operator of the SPE. The monitor operator updates the mapping table and also sends the modifications to the downstream operators in the query plan. The modifications are sent as '+' and '−' tuples, i.e., for the insert a '+' tuple is sent, for the delete a '−' tuple is sent and for the update both the '+' and '−' tuples are sent to the downstream operators as shown in Fig. 3 and discussed in the Example 2. Hereafter, the term *database update* is used for the modifications in database.

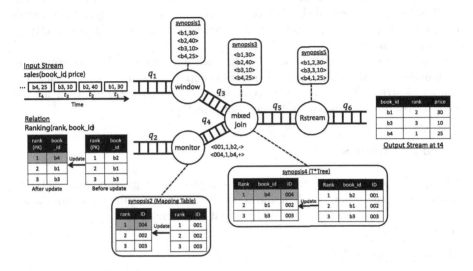

Fig. 3. Isolation guaranteed incremental CQ processing

Example 2. Once again consider the data stream *Sales* of books selling transactions and a database relation *Ranking* containing the top-3 popular books, as shown in Fig. 3. For the same query as of Example 1, we assume that the database is updated after timestamp t_2 and the figure shows the synopses snapshot at timestamp t_4. The monitor operator receives the database update and updates its mapping table by assigning a new ID (004) to rank 1. The database updates are also sent to the downstream operators in an incremental fashion, i.e., a '+' tuple $< 004, 1, b4, + >$ is generated for the updated tuple and a '−' tuple $< 001, 1, b2, −>$ is generated for its corresponding old tuple and are sent to the downstream operators. The downstream operators on receiving the '−' tuple delete its corresponding '+' tuple from their synopses while append the '+' tuple into their synopses. As a result of the synopses update, the mixed join produces consistent results which is then output by the Rstream operator.

3.2   Indexed Synopsis

Joining is computationally expensive operation, especially when one of the joining relation is big, as in our case. To increase the join efficiency between database relation and stream tuples, we use T*-Tree [13] index, which is an extension of T-Tree [12].

T-Tree [12] is a binary tree with many elements in a node. It is evolved from AVL Tree [27], which is an effective main memory data structure, and B-Tree [28], which has good update and storage characteristics. Therefore, T-Tree is an effective main memory data structure that supports binary search. It is known to be one of the best index structures for ordered data in main memory. T*-Tree, which is an extension of T-Tree provides better support for the range queries in addition to the features of the T-Tree. T*-Tree structure is exactly the same as that of the T-Tree, except that it has a successor pointer pointing directly to the successor node.

The T*-Tree index is used only for the database tuples and is maintained in the monitor and the mixed join operators synopses containing them. Since the number of stream tuples is relatively small, no index structure is used for the synopses containing stream tuples, as shown in Fig. 3.

4 Experiments

This section presents experimental study to prove that guaranteeing isolation solves the inconsistency problem and to show that the incremental computation and indexing improves the query throughput significantly.

4.1 Experimental Setup

For the sake of experiments a prototype SPE, JsSpinner [9], which enables users to register CQL style Jaql queries is used. The JsSpinner is being developed at the KDE Lab, University of Tsukuba. The experiments are performed on Dell Precision T3610 with Intel Xeon 3.7 GHz processor and 16 GB RAM running Ubuntu 14.10 OS.

For the relational database, we choose Oracle Express 11.2.0 on the same machine running the SPE. To connect the oracle to the JsSpinner, Oracle C++ Call Interface (OCCI) [29] is used. It is a high-performance and comprehensive API to access the Oracle database.

To simulate situations where database referenced by the continuous queries are updated by other systems, we register a simple CQ to the SPE similar to the one shown in Query 1.2. While the SPE processes the query continuously, the relational table referenced by the join operation is updated periodically.

Table 1. Default parameter values

Parameter	Meaning	Default value
N_T	Relation size	10,000 tuples
U_T	Database update frequency	300 tuples/second
R_S	Input stream arrival rate	10,000 tuples/second
W_S	Window size	100 tuples

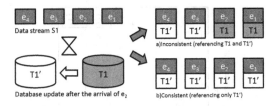

Fig. 4. An example of mixed join output

The maximum and minimum T*-Tree node sizes used for the experiments are 16 and 12 respectively. Unless stated otherwise the parameter values listed in the Table 1 are used in the experiments.

Stream and Relational Data: For the experiments, a synthetic data stream S1 and a database relation T1 is used. The S1 contains a string attribute A and an integer attribute B, while the T1 contains three integer attributes PK, A and B. The synthetic data stream is generated at different rates using random strings for the string attribute A. For the T1 attributes PK and A, consecutive integer values are used. Join attribute B of S1 and T1 is generated in a way to keep the mixed join selectivity equals to 0.01.

4.2 Experimental Evaluation

The experimental evaluation in this section is divided into inconsistency, efficiency and memory usage.

Inconsistency: Here we perform experiments to prove that not guaranteeing isolation results in the inconsistent results. To show this, we execute the Query 1.2 using the basic mixed join computation (referred as *Basic computation* in Fig. 5) as discussed in Sect. 2.3 on the stream S1 and database relation T1 with the default parameter values. To measure the inconsistency in the mixed join output, we update 100 T1 tuples every 5 s.[4] The inconsistency percentage in the Fig. 5 is measured as follows.

$$\text{Inconsistency (\%)} = \frac{\text{no. of inconsistent tuples in the output}}{\text{total number of tuples in the output}}$$

Mixed join output can be inconsistent in the following ways.

- Join computed using tuples referencing the old version (before update) of the database as shown in Fig. 4.
- Presence of tuples in the output which must had been deleted but exist due to the absence of the incremental approach.
- Absence of tuples in the output which have been inserted in the database.

---

[4] For the sake of experiments, the inconsistent tuples are checked manually.

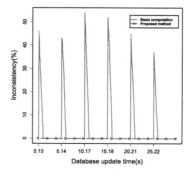

Fig. 5. Inconsistency in query results ($N_T = 10K$ tuples, $U_T = 300$ tuples/s, $R_S = 10K$ tuples/s, $W_S = 100$ tuples)

In Fig. 5, soon after the database update, the inconsistency reaches to its local maximum because in the absence of the proposed incremental approach the tuples in the query plan operators' synopses are not updated, resulting in the output referencing multiple versions of the database. As time proceeds, arrival of new stream tuples causes the old and inconsistent tuples to expire, resulting in the reduction in inconsistency as can be observed from Fig. 5. The inconsistency reaches zero once all the inconsistent tuples expire. Similarly, the next update causes the rise of inconsistency to local maximum once again which reduces to zero with the expiration of the inconsistent tuples and so on. On the other hand, the proposed approach continues to be consistent throughout the experiment.

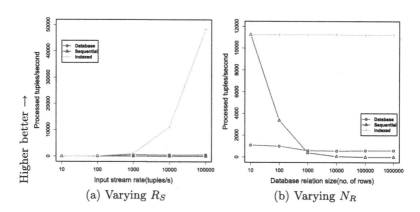

(a) Varying R_S (b) Varying N_R

Fig. 6. Default: $N_T = 10K$ tuples, $U_T = 300$ tuples/s, $R_S = 10K$ tuples/s, $W_S = 100$ tuples

Efficiency: Next we perform experiments to show that the incremental computation and indexing improves the query throughput significantly. For this we compare the following three approaches, all of which guarantees the consistent

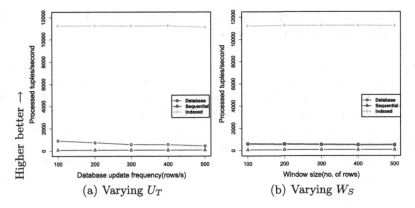

Fig. 7. Default: $N_T = 10\text{K}$ tuples, $U_T = 300$ tuples/s, $R_S = 10\text{K}$ tuples/s, $W_S = 100$ tuples

query results: (1) **Database:** Using this approach, the mixed join is performed by joining the stream tuples directly to the database tuples. The proposed monitor operator is still being used but it only maps the database tuples' primary keys (PKs) to the SPE generated IDs for the sake of incremental computation. (2) **Sequential:** This approach uses the monitor operator for mapping the PKs to the IDs and sending the updates to the downstream operator, however for the mixed join processing no index structure is used. Instead, for each stream tuple, all the database tuples in its respective synopsis are scanned sequentially to process the join. (3) **Indexed:** This approach is similar to that of the sequential approach, except that it uses an index structure for the synopsis handling the database tuples, i.e., monitor operator and join operator synopses, to increase the mixed join efficiency.

Firstly, experiments are performed by varying the input stream arrival rate R_S from 10 tuples/second to 100K tuples/second. From Fig. 6(a) one can observe that, all the approaches can process lower R_S, however with the increase in R_S the database and the sequential approaches cannot handle all the stream tuples. This is because in the database approach, with the increase in R_S the number of required IO increases, making it difficult for the higher R_S to get processed. In the sequential approach, absence of index requires the sequential scan of all the database tuples in the synopsis for each stream tuple join, making the join processing expensive. On the other hand, the indexing in the indexed approach improves the stream processing rate significantly.

Next, the experiments are performed by varying the database relation size N_T. Figure 6(b) shows that both the sequential and the indexed approaches can process equal number of stream tuples at $N_T = 10$. However as N_T increases, the stream processing rate decreases significantly. This is due to the absence of index in the sequential approach making the join processing expensive. The database approach can process much smaller number of tuples though the N_T does not affect the performance a lot. That is again due to the requirement to fetch the database tuples from the secondary storage, resulting in a lot of IO, thus reducing the overall stream processing rate.

We also performed experiments by varying the parameters U_T and W_S as shown in Figs. 7(a) and (b), respectively. Both the figures prove the advantage of using the proposed indexed incremental approach. The number of tuples processed by the indexed approach is several times higher than that of the database and the sequential approaches. Additionally, one can observe from Fig. 7(a) that with the increase in U_T, the number of processed tuples by all the approaches decreases slightly. This is due to the increase in the number of update tuples that need to be processed by the system with the increase in U_T. Furthermore, from Figs. 7(a) and (b) we can observe that the database app-roach performs slightly better than the sequential approach. This is because in the database approach, for the join of each stream tuple we use an SQL query to search the matching tuples in the database. Since the database search is backed by indexes, reduction in the search time for the matching tuples for the mixed join processing results in an improved performance compared to the sequential approach which uses the sequential scan for the same.

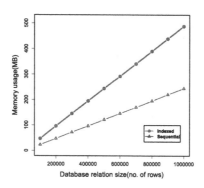

Fig. 8. Memory consumption ($U_T = 300$ tuples/s, $R_S = 10$K tuples/s, $W_S = 100$ tuples)

Memory Usage: In this experiment, we compared the memory usage of the indexed and the sequential approaches by varying the database relation size N_T. Figure 8 shows that the indexed approach consumes a little more memory than the sequential. This is due to the utilization of the T*-Tree index structure by the indexed approach in the synopses for the database tuples. This is the cost that need to be paid for the better system throughput, however with the availability of the cheap and large memories these days, this cost is bearable.

5   Related Work

Golab et al. [14] are among the pioneers to point out the difference between rela-tions and streams. Their work discusses the role of database relations in contin-uous queries and proposes to model relations as look-up time-varying relations.

For mixed join, they define an order for events and updates according to which an update on the relation at time t should only affect the stream events that arrive after t.

The first comprehensive work on transactional stream processing supporting queries on both streams and stored data sources was given by Botan et al. [15]. They extended the page model [18] used by DBMS for streams and CQs. Stream sources in [15] are treated as stored data sources with read/write operations and CQs as a series of one-time queries activated by stream data arrivals. Hence, each one time query is equivalent to a sequence of read and/or write operations. Operations from one or more sequences are grouped into transactions. Conflict serializability and event arrival ordering are ensured when executing these transactions. Hence in their work, isolation is achieved by grouping and correct ordering of read/write operations in a transaction.

Oyamada et al. [11] also studied the inconsistency problem in the join results when the referenced non-streaming source in the mixed join is updated during the query execution. To tackle this problem the authors introduced CQ-derived transactions, a concept that derives read-only transactions from continuous queries. Based on this they showed that the above-mentioned inconsistency problem can be solved by ensuring serializable schedules. Furthermore, to ensure serializable schedules they proposed three continuous query processing strategies, i.e., two-phase lock strategy, snapshot strategy and optimistic strategy.

S-Store [16] is another transactional processing system that supports mixing of streaming transactions with OLTP transactions. It is built on top of H-Store [17], which is an in-memory OLTP system to offer low-latency query response and provide consistency guarantees. In order to guarantee the isolation of a transaction's operations from others', S-Store puts constraints on the ordering of stored procedure in the dataflow graph (DAG in which nodes represent streaming transactions and edges represent an execution ordering) in addition to the ordering of stream tuples.

In contrast to above works, this work presents an incremental approach to reflect the database updates in the output of the mixed join or to guarantee the isolation and consistency of results, which is an important requirement of many application.

Neil Conway in his work [10], presented an isolation model to guarantee the consistency of mixed join results at a window level. According to this model, single window's worth of tuples must be joined to a single consistent snapshot of all the relations in the database. However, as mentioned by the author this model is not suitable for queries containing multiple, differing window clauses. For example, consider a query with two mixed joins, the outputs of which are then joined. If the two mixed joins access the same database relation, while the window sizes of two streams are different, the join joining their outputs will receive stream input derived from two different snapshots of the database relation, resulting in inconsistent query results. The proposed isolation model avoids such a problem by sending the incremental updates to the query downstream operators immediately as they occur.

6   Conclusion and Future Work

In this work, we addressed the inconsistency problem that arrive when the non-streaming data sources referenced by the mixed join are updated during the continuous query execution. To solve this problem, we proposed an isolation guaranteed incremental computation of mixed join. The proposed solution employs a monitor operator to transform the database updates into stream tuples with the help of a mapping table. Furthermore, we used a T*-Tree index structure for the monitor and join operators' synopses handling the database tuples to increase the mixed join efficiency. Experimental study proved that the proposed approach effectively solves the inconsistency problem of the mixed join. Additionally, the incremental computation and the use of index improve the query throughput significantly.

As part of the future work, we are working on the extension of the proposed incremental model to incorporate other features of transactional stream processing. Additionally we have plans to perform intensive experiments on real data streams and relational data.

Acknowledgements. This research was partly supported by the program "Research and Development on Real World Big Data Integration and Analysis" of MEXT, Japan.

References

1. Arasu, A., Babu, S., Widom, J.: The CQL continuous query language: semantics foundations and query execution. VLDB J. **15**(2), 121–142 (2006)
2. Abadi, D.J., Carney, D., Cetintemel, U., Cherniack, M., Convey, C., Lee, S., Stonebraker, M., Tatbul, N., Zdonik, S.: Aurora: a new model and architecture for data stream management. VLDB J. **12**(2), 120–139 (2003)
3. Storm project. https://storm.apache.org/. Accessed 16 Feb 2016
4. Arasu, A., Babcock, B., Babu, S., Cieslewicz, J., Datar, M., Ito, K., Motwani, R., Srivastava, U., Widom, J.: STREAM: the stanford data stream management system. Technical report, Stanford InfoLab (2003). IEEE Data Eng. Bull. **26**(1)
5. Abadi, D.J., Ahmad, Y., Balazinska, M., Cherniack, M., Hyon Hwang, J., Lindner, W., Maskey, A.S., Rasin, E., Ryvkina, E., Tatbul, N., Xing, Y., Zdonik, S.: The design of the borealis stream processing engine. In: Proceedings of the CIDR, pp. 277–289 (2005)
6. Wu, Y., Tan, K.: ChronoStream: elastic stateful stream computation in the cloud. In: Proceedings of the ICDE, pp. 723–734 (2015)
7. Cetintemel, U., Du, J., Kraska, T., Madden, S., Maier, D., Meehan, J., Pavlo, A., Stonebraker, M., Sutherland, E., Tatbul, N., Tufte, K., Wang, H., Zdonik, S.B.: S-Store: a streaming NewSQL system for big velocity applications. In: Proceedings of VLDB, pp. 1633–1636 (2014)
8. Chandramouli, B., Goldstein, J., Barnett, M., DeLine, R., Fisher, D., Platt, J.C., Terwilliger, J.F., Wernsing, J.: Trill: a high-performance incremental query processor for diverse analytics. In: Proceedings of VLDB, pp. 401–412 (2014)
9. Shaikh, S.A., Watanabe, Y., Wang, Y., Kitagawa, H.: Smart query execution for event-driven stream processing. In: Proceedings of IEEE BigMM (2016, to appear)
10. Conway, N.: Transactions and data stream processing, pp. 1–28 (2008)

11. Oyamada, M., Kawashima, H., Kitagawa, H.: Continuous query processing with concurrency control: reading updatable resources consistently. In: Proceedings of ACM SAC, pp. 788–794 (2013)
12. Lehman, T.J., Carey, M.J.: A study of index structures for main memory database management systems. In: Proceedings of VLDB, pp. 294–303 (1986)
13. Choi, K.-R., Kim, K.-C.: T*-tree: a main memory database index structure for real time applications. In: Proceedings of International Workshop on Real-Time Computing Systems and Applications, pp. 81–88 (1996)
14. Golab, L., Tamer Özsu, M.: Update-pattern-aware modeling and processing of continuous queries. In: Proceedings of ACM SIGMOD, pp. 658–669 (2005)
15. Botan, I., Fischer, P.M., Kossmann, D., Tatbul, N.: Transactional stream processing. In: Proceedings of EDBT, pp. 204–215 (2012)
16. Meehan, J., Tatbul, N., Zdonik, S., Aslantas, C., Cetintemel, U., Jiang, D., Kraska, T., Madden, S., Maier, D., Pavlo, A., Stonebraker, M., Tufte, K., Wang, H.: S-Store: streaming meets transaction processing. Proc. VLDB **8**(13), 2134–2145 (2015)
17. Kallman, R., Kimura, H., Natkins, J., Pavlo, A., Rasin, A., Zdonik, S., Jones, E.P.C., Madden, S., Stonebraker, M., Zhang, Y., Hugg, J., Abadi, D.J.: H-Store: a high-performance, distributed main memory transaction processing system. Proc. VLDB **1**(2), 1496–1499 (2008)
18. Weikum, G., Vossen, G.: Transactional Information Systems: Theory, Algorithms, and the Practice of Concurrency Control and Recovery. Morgan Kaufmann Publishers, San Francisco (2001)
19. Chakraborty, A., Singh, A.: A partition-based approach to support streaming updates over persistent data in an active DW. In: Proceedings of IPDPS, pp. 1–11 (2009)
20. Chandramouli, B., Goldstein, J., Duan, S.: Temporal analytics on big data for web advertising. In: Proceedings of ICDE, pp. 90–101 (2012)
21. Golab, L., Johnson, T.: Consistency in a stream warehouse. In: Proceedings of CIDR, pp. 114–122 (2011)
22. Golab, L., Johnson, T., Seidel, J.S., Shkapenyuk, V.: Stream warehousing with DataDepot. In: Proceedings of ACM SIGMOD, pp. 847–854 (2009)
23. Jubatus. http://jubat.us/. Accessed 16 Feb 2016
24. Naeem, M.A., Dobbie, G., Weber, G., Alam, S.: R-MESHJOIN for near-real-time data warehousing. In: Proceedings of International Workshop on DOLAP, pp. 53–60 (2010)
25. Polyzotis, N., Skiadopoulos, S., Vassiliadis, P., Simitsis, A., Frantzell, N.E.: Supporting streaming updates in an active DW. In: Proceedings of ICDE, pp. 476–485 (2007)
26. Polyzotis, N., Skiadopoulos, S., Vassiliadis, P., Simitsis, A., Frantzell, N.: Meshing streaming updates with persistent data in an active data warehouse. IEEE TKDE **20**(7), 976–991 (2008)
27. Aho, A., Hopcroft, J., Ullman, J.D.: The Design and Analysis of Computer Algorithms. Addison-Wesley Publishing Company, Boston (1974)
28. Comer, D.: Ubiquitous B-tree. ACM Comp. Surv. **11**(2), 121–137 (1979)
29. OCCI. http://www.oracle.com/. Accessed 18 Dec 2015

An Improved Method of Keyword Search over Relational Data Streams by Aggressive Candidate Network Consolidation

Savong Bou[1(✉)], Toshiyuki Amagasa[2], and Hiroyuki Kitagawa[2]

[1] Graduate School of Systems and Information Engineering,
University of Tsukuba, Tsukuba, Japan
`savong.bou@kde.cs.tsukuba.ac.jp`
[2] Center for Computational Sciences, University of Tsukuba, Tsukuba, Japan
`{amagasa,kitagawa}@cs.tsukuba.ac.jp`

Abstract. Keyword search over relational streams is useful when allowing users to query on streams without understanding the details about the streams and query language as well. There have been several research works on this direction, and the state-of-the-art approaches exploit Candidate Networks (CNs), which are schema-level descriptions of possible joining networks of tuples, and generate query plans based on CNs. However, in fact, the performance of these approaches seriously degrades in particular when the maximum size of CNs (T_{max}) and/or the number of query keywords are large due to the explosive increase in number of CNs. To cope with this problem, we propose a novel query plan called *MX-structure* to consolidate all CNs as much as possible. We suppress explosive blowup of nodes in query plans by consolidating all common edges among CNs. The experimental results prove that the proposed algorithm performs much better than the state-of-the-art approaches.

Keywords: Keyword search · Relational streams · Candidate network

1 Introduction

With the recent trends of Cyber Physical Systems [7,13], Internet of Things [6,18], etc., the number of real-time information sources has been explosively increasing. Besides, it has become common to extract information from various social medias, such as Twitter and Facebook, in real-time for making analysis of diverse social activities. Such stream data sources can typically be modeled as *relational streams*, where structured records (relational tuples) are transmitted. Therefore, the importance of query processing over relational streams has been increasing.

When querying relational streams, *keyword search* is considered to be an attractive and practical approach due to several reasons. One of the major reasons is that users do not need to learn neither (potentially) complicated query language, like CQL [3], nor the schemas of streams being queried, which is also very complicated in many real applications. Instead, what they only need to do is

S. Hartmann and H. Ma (Eds.): DEXA 2016, Part I, LNCS 9827, pp. 336–351, 2016.
DOI: 10.1007/978-3-319-44403-1_21

to give several query keywords. So far, keyword search over permanently-stored relational data [2,9,10,12,14,17] has been extensively studied, but only a few works have addressed keyword search over relational streams [11,15].

In the works [11,15], they employ candidate network-based approach for improving query performance. Specifically, for a given set of query keywords and a parameter that defines the maximum size of resulting networks of tuples (T_{max}), they first enumerate all candidate networks (CNs) that represent all possible combinations of keyword occurrences on join paths, and the generated CNs are merged to generate a query plan. Then, the actual streams are processed according to the query plan. More precisely, in S-KWS [11], a set of CNs are merged only if they share at least one leaf node called *root*, and possible subtrees are merged to remove redundant processing as much as possible. In SS-KWS [15], common partial networks are merged more aggressively from every leaf node of CNs, thereby generating more compact query plans.

However, it should be noted that the performance of S-KWS and SS-KWS considerably degrades when the number of query keywords and/or network size (T_{max}) are increased. The increase of these two parameters causes rapid increase in the number of CNs, which results in a lot of common partial networks remain unintegrated. To exemplify the problem, let us take TPC-H dataset [1] as an example. When the number of keywords and T_{max} are increased from four to five, the number of CNs increases from 3,600 to 85,803 [15]. Likewise, the total number of edges in the query execution plans exponentially increases from 4,276 to 73,596 in S-KWS and from 7,486 to 222,040 in SS-KWS. (More detailed discussion can be found in Sect. 5.1.) Thus the performance of S-KWS and SS-KWS would deteriorate in particular when dealing with a lot of query keywords and/or large relational streams consisting of many relations. As reported in [8], the average query length to the search engines has been increasing. For example, the ratio of queries containing more than five words has increased by 10 % over the years, while that of single keyword queries has decreased by 3 %.

How can we cope with such exponential blow up of CNs and the complication of query plans? If we consider the edges in CNs, each of them can be associated to one of the primary/foreign-key relationship between two tables, whose number is in general small. In other words, we can consider that the edges in CNs are intensively duplicated from the primary/foreign-key relationships in the schema. With the same example above when the number of keywords and T_{max} are increased from four to five, the total number of unique edges in all CNs grows linearly from 1,088 to 3,536. Under this observation, to cope with the problem of CN's exponential blow up, it is possible to consolidate the edges sharing the same primary/foreign-key relationship into one edge when generating a query plan, which leads to great performance improvement.

This paper proposes a novel approach to processing keyword search over relational streams by taking into account the above idea. Specifically, an *MX-structure* is proposed to consolidate common edges in different CNs as much as possible. The experimental results reveal that the proposed approach greatly outperforms comparative methods in both CPU running time and memory usage.

2 Problem Statement

In this section, we shall first introduce keyword search on relational databases. As a common basis, graph representation of relational database is used to define the semantics of keyword search [16]. In a data graph, each node represents a tuple, and an edge represents a primary/foreign-key reference between two tuples. Now, let us assume a relational schema and a database that conforms to the schema. Given a set of user-specified query keywords, $\{k_1, k_2, \ldots, k_n\}$, keyword search on the database is to find all minimal total joining networks of tuples (MTJNT) [10]. More precisely, *total* means that all keywords are contained in each joining networks of tuples, and *minimal* means that removing any tuple from a network of tuples leads to loss of eligibility for query results. Figures 1(a) and (c) show an example of MTJNT. Notice that the maximum size of data graphs is bounded by parameter T_{max}.

In contrast to conventional relational data, *relational streams* [3] can be modeled as possibly unbounded sequences of relational tuples that conform to relational schemas. In other words, each tuple in a stream can be represented by a pair of (1) a relational tuple and (2) a time instant of a discrete and ordered time domain, e.g., integer. Thus tuples are regarded that they are arrived according to their timestamps. Figure 1 illustrates a sample schema and its instances.

When dealing with (relational) streams, we often use *sliding windows* to convert an infinite stream of tuples to a relation of finite tuples. In such window semantics, two tuples can be joined only if both tuples are in the sliding window.

Having defined relational streams and sliding windows, keyword search over relational streams can be defined as follows: given a set of query keywords $\{k_1, k_2, \ldots, k_n\}$, a maximum network size T_{max}, and a window specification W, it continuously reports (1) new MTJNTs when new tuples are delivered and (2) invalidation of existing MTJNTs due to deletion or aging of tuples.

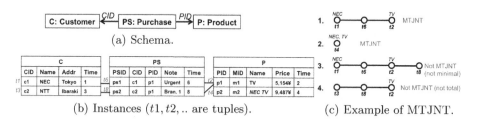

(a) Schema.

(b) Instances ($t1, t2, ..$ are tuples).

(c) Example of MTJNT.

Fig. 1. An example of keyword search "NEC, TV" on relational streams in 1(b). The joining networks of tuples (JNTs) 1 and 2 are MTJNT; while, 3 and 4 are not as shown in 1(c). Therefore, search results (MTJNTs) are JNTs 1 and 2.

3 Existing Works

As mentioned in Sect. 1, S-KWS [11] and SS-KWS [15] are the predecessors of this work. In this section, we briefly overview these works.

3.1 Overview

In S-KWS and SS-KWS the process of keyword search on relational streams comprises two main steps: *preprocessing* and *filtering* steps as shown in Fig. 4.

Preprocessing Step. Given a schema, a set of query keywords, and T_{max}, all *Candidate Networks* (CNs) [11,15] are generated. A CN is a tree, where (1) each node represents a relation and (2) each edge represents a relational *join* operation. Notice that all CNs must conform to the concept of MTJNT [11]. Figure 2 shows three examples of linear CNs from the schema in Fig. 1(a). Then a query plan is generated from all CNs.

Filtering Step. In this step, the query plan is evaluated over relational streams. When new MTJNTs are detected due to arrivals of new tuples, they are reported. On the other hand, expired tuples are removed by using either eager or lazy approaches [11].

3.2 S-KWS

S-KWS [11] is one of the pioneering works for this framework. In this work, for each CN, the *root* node is defined as a node containing one chosen query keyword. Then, left-deep operator tree is created for each CN.

To improve performance, they propose a query plan, called operator mesh, by grouping all left-deep operator trees that share the same root into a cluster so that all common join operators can be consolidated, resulting in improved performance by sharing common operations on the same data. Figure 3 shows two clusters of operator mesh created from the three CNs in Fig. 2.

When processing relational streams, all partial results are cached in each operator's buffer for efficient retrieval of matched results. However, caching all partial results causes a performance bottleneck due to its high memory cost.

3.3 SS-KWS

SS-KWS [15] is a successor of S-KWS and can be regarded as the state-of-the-art approach. The novel idea of SS-KWS is to aggressively merge more sub-networks

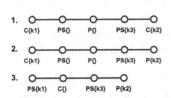

Fig. 2. Some CNs created from schema in Fig. 1(a) for query $\{k_1, k_2, k_3\}$.

(a) Cluster #1. (Combine the first two CNs.)

(b) Cluster #2. (Only CN 3 is included.)

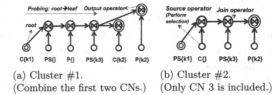

Fig. 3. Operator mesh for CNs in Fig. 2.

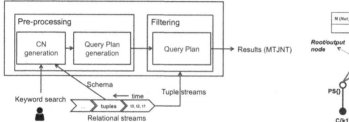

Fig. 4. General framework.

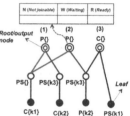

Fig. 5. Lattice for CNs in Fig. 2. (Color figure online)

in CNs not only at single leaf, but also at all leaves. Unlike S-KWS, the root is the center node of the CNs. Besides, instead of operator mesh, a query plan, called lattice, is created by combining all CNs so that the query processing cost is reduced by sharing common subtrees except for the root nodes in CNs as much as possible. For example, the lattice structure for the three CNs in Fig. 2 is shown in Fig. 5. In this example, nodes marked with double lines are root nodes; black colored nodes are leaf nodes; and the rests are other non-leaf nodes.

To fully reduce all partial results, SS-KWS proposes selection/semi-join approach by dividing buffer of each node into three sub-buffers: N (not joinable), W (waiting), and R (ready). It adopts a bottom-up probing sequence. If the tuple is joinable with other tuples, it is stored in sub-buffer W; otherwise, in N. If MTJNT of any CN is detected, all related tuples are stored in sub-buffer R. Thus SS-KWS successfully reduces memory usage compared to S-KWS. Figure 5 shows the lattice structure and all sub-buffers of node P.

3.4   Scalability Issues in Existing Approaches

We discuss in detail the scalability issues of these approaches. As a common problem, the number of CNs grows exponentially as the number of keywords and/or T_{max} increase. This gives a significant impact on both time and space.

In S-KWS, partial results are maintained in the buffers in an operator mesh. Due to the low sharing rate of common subtrees in CNs; i.e., in an operator mesh, we can find a lot of edges connecting the same relations but are not consolidated, because they are either in different clusters or do not have same root node (e.g., common edge PS{k3}-P{k2} in CNs 2 and 3 in Fig. 3). Consequently, in query processing, a lot of partial results have to be duplicated in buffers and need also to be processed independently.

In SS-KWS, the problem of the low sharing rate of common subtrees is mitigated by sharing common subtrees in all possible subtrees. However, there still exists a restriction that it is impossible to consolidate common paths in internal nodes, because (1) sharing is only allowed for common subtrees; and (2) root nodes are not allowed to be shared (e.g., common edges PS{}-$P{} and

P{}-PS{k3}, marked in red and green, are not consolidated in lattice in Fig. 5). Therefore, the number of unconsolidated paths grows rapidly as the number of CNs grows. For the same reason discussed above, such duplicated paths cause high memory consumption in internal buffers and also cause high computational cost for possibly useless processing of (duplicated) intermediate results.

4 Proposed Approach

4.1 Overview

We propose a novel query plan representation, called *MX-structure* (maximal-sharing structure), that combines all CNs by consolidating common edges. By using MX-structure, we can avoid redundant nodes and edges to be expanded. To make it possible to evaluate queries using MX-structure, we introduce fine-grained node buffers and branch maps for managing existing partial/full query results. To deal with expiration of tuples, we employ *lazy approach* [11] where expired tuples are removed when node buffers are probed. Due to the page limitation, we do not give detailed explanation. Interested readers may refer to the original paper [11].

4.2 MX-Structure

First, we introduce the proposed MX-structure. In each CN, the root (and the output node as well) is determined as the center of the CN (the node which its path to all leaf nodes is minimal). Then, all CNs are merged in such a way that all edges are unique; i.e., edges in MX-structure are created only for different combinations of nodes regardless of the node's position (root or leaf). Such information needs to be maintained as well. In the sequel discussion, we denote by () a leaf node and by [] a root node. Notice that, in MX-structure, a source node and an edge represent selection operation and join operation between two connected nodes, respectively. When all CNs are generated, each of them is labeled by a unique ID, which is used detect the matched MTJNTs.

Due to the space limitation, we cannot show the algorithm, but it can be constructed in the following way. Basically, all CNs are processed and added to an MX-structure one by one. When adding a new CN, we take each edge, and check its existence; we add one only if it has not been added yet. Next, the ID of CN is added to each of its edges in MX-structure. The information about each CN's root and leaf nodes is also maintained.

Figure 6 illustrates an example of MX-structure for the CNs in Fig. 2. Nodes marked with double lines show root nodes, and black nodes are leaf nodes. The label on each edge represents the set of corresponding CNs in term of IDs. The numbers in () and [] are the IDs of CNs of leaf and root nodes, respectively.

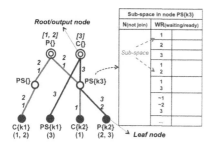
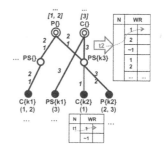

Fig. 6. MX-structure for CNs in Fig. 2. (Color figure online)   **Fig. 7.** Example of probing sequence.

4.3   Query Evaluation in MX-Structure

To evaluate queries over relational streams using MX-structure, we need to track the matching status of each tuple in the respective CNs. For example, look at Fig. 6. If all joins between all edges in C{k1}-PS{}-P{}-PS{k3}-C{k2} (marked in red) of CN 1 are detected, tuples that contribute to form MTJNT of CN 1 need to be output as a query result. This is allowed by the fine-grained status management of (existing) tuples using *node buffers*. More precisely, for each incoming tuple, its joinability is checked according to the *probing sequence*, and is stored in an appropriate sub-space in a sub-buffer w.r.t. the corresponding CN, which is allocated dynamically when necessary. Thus the proposed scheme achieves better performance while consuming less memory space.

4.3.1   Node Buffers

In an MX-structure, each node buffer is divided into two sub-buffers, N and WR. Sub-buffer N is for storing tuples that are *not joinable*, while WR is for storing tuples that are *joinable* with other tuples. Moreover, sub-buffer WR is divided into sub-spaces according to the CNs it belongs to. A sub-space indicates the joint status of each joinable tuple to its matched CNs. In the following discussion we denote by $\sim n$ the sub-space for tuples that are *fully matched* (as part of the complete query results) w.r.t. CN n, whereas by n the sub-space for tuples that are *partially matched* (not part of the complete query results) w.r.t. CN n. The table in Fig. 6 shows the buffer of node PS{k3} in the MX-structure. As can be seen, node PS{k3} appears in CNs 1, 2, and 3. For this reason, some sub-spaces are created in sub-buffer WR; e.g., $\{\sim 1, \sim 2, 3\}$ is for those tuples that fully match in CNs 1 and 2 and partially match in CN 3. Notice that we dynamically create sub-spaces when necessary to avoid allocation of unnecessary (unpopulated) sub-spaces.

4.3.2   Probing Sequence

To systematically evaluate queries, for each incoming tuple, we check its joinability with other existing tuples in the node buffers in other child and/or parent nodes, and such probes are performed in the bottom-to-root direction; if a new tuple arrives at a leaf node, then we probe its parent nodes; otherwise, we first probe the child nodes, then probe the parent nodes. More precisely, when probing child nodes, we probe existing tuples in both sub-buffers N and WR if the nodes being probed are at the leaf level, but do so only in WR if the nodes are at non-leaf levels. If it turns out that the incoming tuple is not joinable with any other tuples in the node buffers in the child nodes, then current probing is finished, and the tuple is stored in sub-buffer N (not joinable); otherwise, it is stored in a sub-space in sub-buffer WR that corresponds to the CN(s) to which the incoming tuple contributes to form resulting MTJNT(s).

Note here that we call the CN(s) that the incoming tuple contributes to form MTJNT(s) *active CN(s)*. The set of active CNs are defined as follows:

$$cn_{active} = cn_{edge} \cap (cn_{leaf} \cup cn_{ecsubspace}) \tag{1}$$

where cn_{edge} is the set of IDs of CNs assigned to a connected edge being traversed, cn_{leaf} is the set of IDs of CNs assigned to a leaf node if the probed child node is s leaf, and $cn_{ecsubspace}$ is the set of IDs of CNs of non-empty sub-spaces in the child node(s). Notice that, if the probed child node is a non-leaf node, cn_{leaf} is empty. Similarly, in sub-buffer N, $cn_{ecsubspace}$ is also empty. Determining active CNs is beneficial to avoid unnecessary probings due to the fact that inactive CNs in child nodes can never be active in parent nodes. Thus, once active CNs are determined by probing child nodes, only the parent nodes that are connected via edges of active CNs are probed, thereby avoiding unnecessary probings in upper levels.

Look at Fig. 7 as an example. Notice that node buffers that store tuples are shown for simplicity. Let us assume that tuples t1 and t2, which are (1) of relations C and PS, resp., (2) contain keywords k2 and k3, resp., and (3) joinable with each other, arrive in this order. When t1 arrives, we immediately probe the parent node PS{k3}, because C{k2} is at the leaf level. As a result, it turns out that t1 is not joinable because of empty node buffer in PS{k3}, and is stored in the sub-buffer N in C{k2}. Afterwards, when t2 arrives, we probe the child nodes, C{k2} and P{k2}. When probing C{k2}, t2 turns out to be joinable with t1 w.r.t. CN 1 by applying the formula explained above[1], and (1) t1 is moved to the sub-space {1} in sub-buffer WR and (2) t2 is stored in sub-space {1} in sub-buffer WR in the respective nodes. For subsequent probings of parents, only active CN (CN 1) is taken into consideration. Consequently, only P{} is probed while C{} is not, because P{} is connected by CN 1 edge.

4.3.3   Branch Map

In an MX-structure, in many cases, root/output nodes are internal (non-leaf) nodes in one or more CNs and probing proceeds in the leaf-to-root direction. For

---

[1] We have $cn_{edge} = \{1\}$ (edge C{k2}-PS{k3} belongs to CN 1), $cn_{leaf} = \{1\}$ (node C{k2} is leaf node of CN 1), and $cn_{ecsubspace} = \{\}$ (t1 is currently in sub-buffer N). As a result, we get $cn_{active} = \{1\}$.

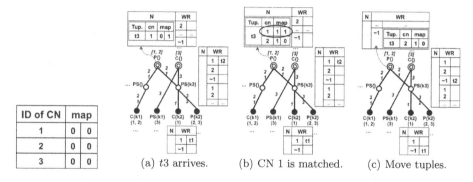

ID of CN	map	
1	0	0
2	0	0
3	0	0

(a) t3 arrives.   (b) CN 1 is matched.   (c) Move tuples.

Fig. 8. Branch maps for CNs in Fig. 2.   **Fig. 9.** Example of MX-structure evaluation. (Color figure online)

this reason, we need to maintain for each tuple in the root its matching status so that we can output new MTJNTs as soon as they are detected. To this end, we use a map called *branch map* to track whether all nodes from the leaves up to the root/output node are matched. More precisely, a branch map is attached to each joinable tuple in the root/output nodes. A branch map is comprises several bits corresponding to the branches from the leaf (or leaves). When all bits are set to one, the MTJNT that contains the root tuple is output as a result. Figure 8 is the branch maps of all CNs in Fig. 2. Since each CN has two leaf nodes, each map has two bits which are initialized by zero.

Continued from the example in Fig. 7. Suppose tuple t3 in P{} has arrived (Fig. 9(a)), and is joinable with t2 w.r.t. active CN 1^2. Since node P{} is the root node, a branch map for active CN (CN 1) is created, and the second bit corresponding to branch C{k2}-PS{k3}-P{} (colored in red) is set to one.

When another branch C{k1}-PS{}-P{}, belonging to CNs 1 and 2, is detected as matched (Fig. 9(b)) due to the arrival of additional tuples, another branch map for CN 2 is created, and the first bits in the branch maps for CNs 1 and 2 in t3 are set to one. Now, CN 1 is detected as matched, and the MTJNT is returned as a query result. Then, all matched tuples are moved to appropriate sub-spaces of their fully matched CN 1 as shown in Fig. 9(c) for subsequent processing[3].

4.3.4 Dynamic Generation of Sub-spaces

As explained earlier, we dynamically populate sub-spaces when necessary, because (1) generating all possible sub-spaces requires huge memory spaces and (2) only a few sub-spaces are used in query processing. To this end, we populate

---

[2] We have $cn_{edge} = \{1, 2\}$ (edge PS{k3}-P{} belongs to CNs 1 and 2), cn_{leaf} is empty (node PS{k3} is not leaf node), and $cn_{ecsubspace} = \{1\}$ (t2 is currently in sub-space $\{1\}$). As a result, we get $cn_{active} = \{1\}$.

[3] Notice that buffers of nodes C{k1} and PS{} are not shown here for simplicity.

a new sub-space according to the following formula:

$$cn_{newsubspace} = cn_{oldsubspace} \cup cn_{active} \qquad (2)$$

where $cn_{newsubspace}$ and $cn_{oldsubspace}$ are respectively the new sub-space and the existing sub-space marked by IDs of CNs for each joinable tuple. Notice that, if tuple just arrives or is currently in sub-buffer N, its $cn_{oldsubspace}$ is empty. The algorithm is omitted due to the space limitation, but its algorithm is straightforward.

4.4 Algorithm Details

The proposed algorithm is shown in Algorithm 1. This algorithm works as follows. If the incoming tuple, t_0, belongs to a non-leaf node, it probes child nodes by calling function Probe_child_nodes (Line 3). This function returns $joinable\_to\_child$ = true if there are joinable tuples in child nodes with the incoming tuple. Otherwise, it returns $joinable\_to\_child$ = false, which results in finishing the current probing, and t_0 is stored in sub-buffer N (Line 4).

This function Probe_child_nodes works as follow. For each sub-space of sub-buffer WR in each child node (and sub-buffer N if child node is leaf node), cn_{active} is computed by Eq. (1). If cn_{active} is not empty, it checks each tuples in that sub-space (Line 2–5). If there are tuples joinable with the incoming tuple, $joinable\_to\_child$ is set to true (Line 6), and function Match_CN is called to check if any partially matched CNs in cn_{active} are fully matched (Line 7). This function returns $joinable\_to\_child$ (Line 12).

In function Match_CN, each CN in cn_{active} is checked if there are fully matched CNs. First, appropriate sub-space, $cn_{newsubspace}$, is computed by Eq. (2) (Line 1). Then, for each partially matched CN, $branch\_map$ is updated (Line 2). There are fully matched CNs if the parent node is root node and all bits in the $branch\_map$ are set (Line 3–4). If any fully matched CN is found, its MTJNT is returned as result, and sub-space, $cn_{newsubspace}$, are updated according to the fully matched CN (Line 5–6). Finally, all matched tuples are moved to appropriate sub-space $cn_{newsubspace}$ (Line 8).

Back to the main algorithm, if the incoming tuple is from leaf nodes or $joinable\_to\_child$ is true, subsequent parent nodes are probed until no parent nodes have joinable tuples (Line 8–18) by calling function Probe_parent_nodes (Line 10) following similar procedure above.

4.5 Discussion

In this section we elaborate the reason why the proposed scheme is advantageous to the existing approaches, S-KWS and SS-KWS. As we observed, the number of CNs exponentially increases as query keywords and/or T_{max} grows. Consequently, even though S-KWS and SS-KWS try to merge the CNs by finding common sub-networks, the size of query plans rapidly grows, which means a large number of CNs cannot share processing and need to be evaluated independently.

Algorithm 1. MX-structure Evaluation

Input: Tuple t_0 just from streams, MX-structure MX

1: $joinable\_to\_child$ = false
2: **if** t_0 from non-leaf nodes **then**
3: | $joinable\_to\_child$ = Probe_child_nodes (t_0, MX)
4: | Put t_0 in sub-buffer N if $joinable\_to\_child$ = false
5: **end if**
6: **if** t_0 from leaf nodes or $joinable\_to\_child$ = true **then**
7: | put t_0 in set_joint_tuples
8: | **while** 1 **do**
9: | | **while** each t in set_joint_tuples **do**
10: | | | $sjtp$ = Probe_parent_nodes $(t, sjtp, MX)$
11: | | **end while**
12: | | **if** $sjtp$ is empty **then**
13: | | | break;
14: | | **else**
15: | | | set_joint_tuples = $sjtp$
16: | | | clear $sjtp$
17: | | **end if**
18: | **end while**
19: **end if**

Function: Probe_child_nodes (t, MX)

1: $joinable\_to\_child$ = false
2: **while** Each child nodes **do**
3: | **while** Each sub-space, sp, in WR (and N if child node is leaf node) **do**
4: | | **if** cn_{active} not empty **then**
5: | | | **while** Each tuple t_1 in sp joinable with t **do**
6: | | | | $joinable\_to\_child$ = true
7: | | | | Matched_CN (cn_{active}, MX)
8: | | | **end while**
9: | | **end if**
10: | **end while**
11: **end while**
12: Return $joinable\_to\_child$

Function: Probe_parent_nodes $(t, sjtp, MX)$

1: **while** Each parent node, pn **do**
2: | **if** cn_{active} not empty **then**
3: | | **while** Each tuple t_1 in pn joinable with t **do**
4: | | | Matched_CN (cn_{active}, MX)
5: | | | put t_1 in $sjtp$
6: | | **end while**
7: | **end if**
8: **end while**
9: Return $sjtp$

Function: Matched_CN (cn_{active}, MX)

1: Compute $cn_{newsubspace}$
2: **while** Update $branch\_map$ of each CN in cn_{active} **do**
3: | **if** All bits in $branch\_map$ set to 1 **then**
4: | | **if** Parent node is root node **then**
5: | | | Return all matched tuples (MTJNT) as result
6: | | | Update $cn_{newsubspace}$ to fully match to $\sim CN$.
7: | | **end if**
8: | | Move all matched tuples into appropriate sub-space $cn_{newsubspace}$
9: | **end if**
10: **end while**

Such redundant evaluation is very costly because it requires to scan all related tuples and check if they are joinable. This leads to very poor performance.

In MX-structure, we combine all CNs by consolidating all common edges without any restriction of node position. Thus, we can avoid the exponential blow up in query plans, which means all CNs having overlapping edges can share processing. We enable MX-structure by keeping track of matching status using sub-space in each node buffer. It is true that the management of the complicated sub-buffers is not cost-free; however, that cost is very trivial comparing to that of independent evaluation of all unconsolidated CNs (very costly operation as explained above) in the query plans of S-KWS and SS-KWS. This leads to much better performance. We confirm this in the following experimental evaluation.

5 Experiments

In this section we compare the performance of the proposed approach with full mesh (FM) and partial mesh (PM) of S-KWS, and SS-KWS. All approaches were implemented using C++ language. All experiments were performed using 2.93 GHz Intel Core i7 CPU with 31.4 GiB memory running Ubuntu 13.10.

We used both synthetic and real datasets. Due to lack of space, we only reported the result of synthetic dataset, TPC-H [1], which deals with ad-hoc decision support system in business environment. In this dataset, there are eight tables with 61 attributes. To generate relational streams, we read datasets from the disk, and fed them in the implement systems. Parameters used in the experiments are shown in Table 1. The default parameters are written in bold.

5.1 Comparison of Query Plans' Size

We first make a comparison of query plans' size (in terms of number of edges) of all approaches because they have great impact on the performance.

For this experiment, we only use two parameters, number of query keywords and T_{max}, because other parameters do not have an impact on the size of query plan. The result is shown in Fig. 10. As can be seen, when the number of query keywords and T_{max} increased, the total number of edges in S-KWS and SS-KWS was exponentially increased, which was caused by the explosion of number of CNs whose edges could not be consolidated in their query plans. Such explosive increase in size of query plans indicates that the performance of S-KWS and SS-KWS will greatly degrade when the number of query keywords and

Table 1. Parameters used in the experiments.

Parameter	Range and default
Window size (mn)	10, 20, **30**, 40, 50
Keyword frequency (%)	0.003, **0.007**, 0.01, 0.013
# of keywords	2, **3**, 4, 5
T_{max}	2, 3, **4**, 5

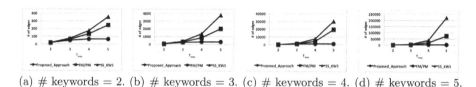

(a) # keywords = 2. (b) # keywords = 3. (c) # keywords = 4. (d) # keywords = 5.

Fig. 10. Comparison of number of edges.

T_{max} increase. However, the growth rate of the proposed scheme was linearly increased because it consolidated unique edges into one, and the total number of unique edges, which were the primary/foreign-key relationships between two tables in the schema (which is usually comparatively small), in all CNs was slightly increased as the number of CNs increased. This proves that the proposed scheme can scale well with the increase in number of query keywords and T_{max}.

5.2   Performance Comparison

We compared CPU running time, memory usage, and total number of probings. Notice that this dataset is specially prepared to favor SS-KWS to S-KWS.

First, we measured the CPU running time and the memory usage when varying the number of keywords (Figs. 11(a) and (b), respectively). As can be seen, CPU running time and the memory usage in FM/PM and SS-KWS were increased exponentially, whereas the proposed scheme was not. As an evidence, the number of probings was also exponentially increased in FM/PM and SS-KWS as shown in Fig. 11(c). This is due to the explosive in size of the query plans of FM/PM and SS-KWS as explained in the above experiment. Similar tendency can be observed when varying T_{max} from two to five (Fig. 12).

Next, we increased the size of window from 10 min, 20 min, 30 min, 40 min, and 50 min. As expected, when the size of window was increased, the CPU running time, memory usage, and number of probings of all approaches also increased as shown in Fig. 13. This was because fewer tuples in the buffers of all approaches were expired and deleted as a result of the increase in size of window. Figure 14 shows the impact on the performance of all approaches when varying keyword frequency. When keyword frequency was increased, there were more tuples containing the keywords of the query. As a result, there were more tuples that need to be joint. Therefore, the CPU running time, memory usage, and number of probings of all approaches also increased. Nevertheless, the total number of CNs did not increase when increasing window size and keyword frequency. Therefore, there is no change in size of query plans of all approaches, which caused little impact on the performance.

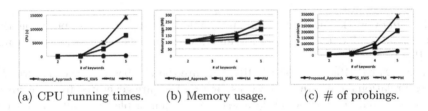

(a) CPU running times.　　(b) Memory usage.　　(c) # of probings.

Fig. 11. Varying # of keywords.

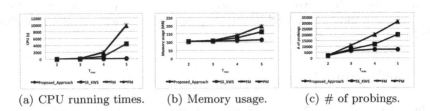

(a) CPU running times.　　(b) Memory usage.　　(c) # of probings.

Fig. 12. Varying T_{max}.

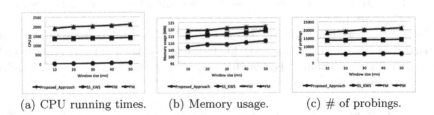

(a) CPU running times.　　(b) Memory usage.　　(c) # of probings.

Fig. 13. Varying window size.

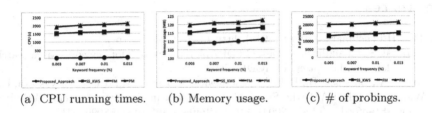

(a) CPU running times.　　(b) Memory usage.　　(c) # of probings.

Fig. 14. Varying keyword frequency.

6 Related Work

So far, many works have been done to enable keyword search on permanently-stored-relational data [2,4,5,10,14] and few proposals on relational streams [11,15].

DISCOVER [10] and DBXPLORER [2] are the CN-based keyword search on (static) relational data. In these works, first all CNs are generated from the given keyword search and relational data's schema. Then, a plan is built for efficient evaluation. Since the total number of CNs can be very big, and evaluation of all CNs is costly, [14] proposes an algorithm to rank all CNs, and only top-k CNs are chosen to evaluate against relational data.

S-KWS [11] is the first work to enable keyword search over relational streams. It is also a candidate network based approach. S-KWS [11] proposes a query plan, called operator mesh. Later, SS-KWS [15] proposes a more compact query plan, called lattice, for better query evaluation.

7 Conclusion

In this paper we have proposed an improved keyword search method over relational streams. In the proposed scheme candidate networks are merged into a novel data structure called MX-structure, and keyword search is efficiently processed based on the proposed algorithms with the help of MX-structure. The experimental results have shown that the proposed scheme significantly outperforms the state-of-the-art methods even when the number of query keywords and/or T_{max} are increased.

In this work we have noticed that CN-based approach has some limitations. In particular some CNs are not used due to the biased keyword distribution with respect to different schemas. For the future work we plan to exploit such locality to enhance the performance.

Acknowledgments. This research was partly supported by the Grant-in-Aid for Scientific Research (B) (#26280037) by JSPS and the program *Research and Development on Real World Big Data Integration and Analysis* of the Ministry of Education, Culture, Sports, Science and Technology, Japan.

References

1. TPC-H benchmark dataset (2015). http://www.tpc.org/tpch/
2. Agrawal, S., Chaudhuri, S., Das, G.: DBXplorer: a system for keyword-based search over relational databases. In: ICDE (2002)
3. Arasu, A., Babcock, B., Babu, S., Cieslewicz, J., Datar, M., Ito, K., Srivastava, U., Widom, J.: STREAM: the Stanford data stream management system. Technical report, Stanford InfoLab (2004). http://ilpubs.stanford.edu:8090/641/
4. Arasu, A., Babu, S., Widom, J.: CQL: a language for continuous queries over streams and relations. In: Lausen, G., Suciu, D. (eds.) DBPL 2003. LNCS, vol. 2921, pp. 1–19. Springer, Heidelberg (2004)

5. Chaudhuri, S., Das, G., Hristidis, V., Weikum, G.: Probabilistic ranking of database query results. In: VLDB, Toronto, Canada (2004)
6. Dyk, M., Najgebauer, A., Pierzchała, D.: Agent-based M&S of smart sensors for knowledge acquisition inside the Internet of Things and sensor networks. In: Nguyen, N.T., Trawiński, B., Kosala, R. (eds.) ACIIDS 2015. LNCS, vol. 9012, pp. 224–234. Springer, Heidelberg (2015)
7. Edward, L.: Cyber physical systems: design challenges. Technical report no. UCB/EECS-2008-8, University of California, Berkeley (2008). Accessed 07 June 2008
8. Hogan, K.: Interpreting hitwise statistics on longer queries. Technical report, Ask.com (2009)
9. Hristidis, V., Gravano, L., Papakonstantinou, Y.: Efficient IR-style keyword search over relational databases. In: VLDB (2003)
10. Hristidis, V., Papakonstantinou, Y.: Discover: keyword search in relational databases. In: VLDB, Hong Kong, China (2002)
11. Markowetz, A., Yang, Y., Papadias, D.: Keyword search on relational data streams. In: SIGMOD, Beijing, China (2007)
12. Mehdi, K., Aijun, A., Nick, C., Parke, G., Jaroslaw, S., Xiaohui, Y.: Meaningful keyword search in relational databases with large and complex schema. In: ICDE, Seoul, Korea (2015)
13. Niggermann, O., Lohweg, V.: On the diagnosis of cyber-physical production systems. In: AAAI, Austin, Texas, USA (2015)
14. Pericles, O., Altigran, S., Edleno, M.: Ranking candidate networks of relations to improve keyword search over relational databases. In: ICDE, Seoul, Korea (2015)
15. Qin, L., Yu, J.X., Chang, L.: Scalable keyword search on large data streams. VLDB J. **20**, 35–57 (2011)
16. Shaul, D., Gadi, E., Shai, G., Eran, P.: DTL's DataSpot: database exploration using plain language. In: VLDB, San Francisco, CA, USA (1998)
17. Xu, Y., Guan, J., Ishikawa, Y.: Scalable top-k keyword search in relational databases. In: Lee, S., Peng, Z., Zhou, X., Moon, Y.-S., Unland, R., Yoo, J. (eds.) DASFAA 2012, Part II. LNCS, vol. 7239, pp. 65–80. Springer, Heidelberg (2012)
18. Zhang, H., Sanin, C., Szczerbicki, E.: Experience-oriented enhancement of smartness for Internet of Things. In: Nguyen, N.T., Trawiński, B., Kosala, R. (eds.) ACIIDS 2015. LNCS, vol. 9012, pp. 506–515. Springer, Heidelberg (2015)

Data Integration, and Interoperability

Evolutionary Database Design: Enhancing Data Abstraction Through Database Modularization to Achieve Graceful Schema Evolution

Gustavo Bartz Guedes[1,2(✉)], Gisele Busichia Baioco[2],
and Regina Lúcia de Oliveira Moraes[2]

[1] Federal Institute of São Paulo, Hortolândia, SP 13183-250, Brazil
gubartz@ifsp.edu.br
http://hto.ifsp.edu.br
[2] University of Campinas, Limeira, SP 13484-332, Brazil
{gisele,regina}@ft.unicamp.br
http://www.ft.unicamp.br/

Abstract. Software systems are not immutable through time, especially in modern development methods such as agile ones. Therefore, a software system is constantly evolving. Besides coding, the database schema design also plays a major role. Changes in requirements will probably affect the database schema, which will have to be modified to accommodate them. In a software system, changes to the database schema are costly, due to application's perspective, where data semantics needs to be maintained. This paper presents a process to conduct database schema evolution by extending the database modularization to work in an evolutionary manner. The evolutionary database modularization process is executed during conceptual design, improving the abstraction capacity of generated data schema and results in loosely coupled database elements, organized in database modules. Finally, we present the process execution in an agile project.

Keywords: Evolutionary database design · Schema evolution · Database evolution · Agile methods

1 Introduction

Software systems manipulate data through a database management system, which provides a set of resources that allow the transparent creation, definition and manipulation of databases. An important process in the database design is its definition, made as a schema that describes the metadata about how information is represented and stored.

During an evolution process, new requirements demand the evaluation of the necessary changes and their impact along all levels that compose the software system. Therefore, it is most likely that the database schema also needs to be updated in order to support new functionalities.

© Springer International Publishing Switzerland 2016
S. Hartmann and H. Ma (Eds.): DEXA 2016, Part I, LNCS 9827, pp. 355–369, 2016.
DOI: 10.1007/978-3-319-44403-1_22

Nowadays, the database evolution is mainly performed with the refactoring technique, which consists in small changes to the database schema in order to support new functionalities [1]. However, these changes are costly and may result on inconsistencies and semantic lost, regarding the original database schema. Therefore, it is desirable to produce a high quality data schema at first, rich in abstraction, to facilitate further evolution.

In this way, the choice of the database model has a great impact on schema evolution. Schemas built upon low expressiveness data models make further evolution harder, because they have a poor representation of reality. Therefore, conceptual data models are preferable, since they provide a high-level abstract representation. According to Batini, Ceri and Navathe [2] "the quality of the resulting schemas depends not only on the skill of the database designers, but also on the qualities of the selected data model".

In respect to the software development, Sommerville [3] states that modularity is one of the attributes towards software quality, and if the modularization process is well conducted it results in improved software quality. Thus, modularity assists in software evolution, since each module contains a set of loosely coupled functionalities. In addition, the emergence of the agile methods introduced a new standard in software development, allowing the delivery of independent functional modules in an incremental manner [4–6]. In this way, each increment must be integrated with the already operational software system delivered in previous iterations in order to provide new functionalities. Therefore, modularity can be used at database design process level, in order to produce cohesive and loosely coupled database elements assisting in the schema evolution.

The work of Ferreira and Busichia [7] presents the database modularization process, an extension of the classical database design process. It decomposes a database schema into database modules, according to the application's transactions. In addition, it provides a representation for a database module, which enhances the overall expressiveness of the database design, by representing the conceptual data schema of a module along with its related functionalities.

Database modularization provides a standardized way to conduct the database design process, resulting in loosely coupled and autonomous schemas, providing an evolutionary design approach. It narrows down the scope when assessing the changes and impacts caused by evolution, improving traceability from database and software perspective. A preliminary work towards evolutionary database modularization process was introduced in Guedes, Busichia and Moraes [8].

This paper presents the extension of the database modularization process to support an evolutionary design. The result is a graceful schema evolution amplifying the schema's abstraction to database designers, assisting future changes. Also, we propose a metadata schema to hold the database modules definitions providing the application with a transparent view of the database modules.

The remainder of this paper is organized as follows: related works are presented in Sect. 2; Sect. 3 presents an overview of the database modularization process; the evolutionary database modularization process, including new phases, is described in Sect. 4; Sect. 5 presents our approach for the evolutionary database modules integration; an application of our approach is presented in Sect. 6; finally Sect. 7 concludes the paper and indicates future works.

2 Related Works

Database schema evolution is not a recent topic and some works date back to the seventies [9,10]. Whereas the software-developing paradigm has changed, from procedural to object oriented, database technologies did not have such a dramatic change and are mostly supported by the relational model [11].

Most works in schema evolution includes database refactoring, co-evolution of the conceptual and physical schemas, impact and prediction analysis of schema changes and schema mapping and versioning. The tight coupling between application and database schema is a common topic of these works.

A database refactoring [1,12] consists in a small change to the schema in order to support new functionalities while it must preserve the behavioral and informational semantics. Thus, after a refactoring, the information, from the application's point of view, must be preserved.

According to Ambler [12], a database refactoring comprises: (i) Changes in the database schema, such as modification in relations, views, stored procedures, triggers; (ii) Data migration to the new schema. This may require data transformation, such as type conversion and update in referential integrity constraints. When multiple applications use the database, a transition period may exist, in which the old and new schemas coexist; (iii) Changes to the application's source code to reflect the schema's modifications.

Cleve et al. [13,14] approach explores the co-evolution of the conceptual and physical layer, by automating generation of the relational schema from its corresponding conceptual schema.

The Hecataeus is a "what-if" analysis tool that simulates the propagation of events caused by schema changes [15]. Hence, focused on the coupling between database elements. Hecataeus maps a relational database schema as well as queries and views into a directed graph, where each node represents a dependency. A set of metrics is provided to evaluate coupling.

The work of Curino et al. [16] presents a set of operators, called Schema Modification Operators (SMO) that modify the physical database schema while preserving previous versions and establishing a mapping among them. This allows different applications to query multiple versions of the database. In that work a prototype tool, the PRISM workbench, is presented in order to support the SMOs.

This paper presents a standardized method to design the database schema in evolutionary and modular way, resulting in a set of cohesive and independent database modules, improving the schema abstraction and reducing the need for database changes. In addition, it provides interoperability between the software and the database, through a catalog that stores the database modules metadata definition.

3   Background

This work is based on the extension of the database modularization design process proposed in Ferreira and Busichia [7]. Therefore, in this section we present the background regarding the phases involved in this process.

3.1   Overview of Database Modularization Design

In this section, we describe the four stages of the "Modularization Design" phase as it is in Ferreira and Busichia [7].

Stage 1 - Partitioning of the conceptual global data schema. The input is the global application schema, which is partitioned in one subschema per subsystem. A subsystem is defined as a group of functionalities.

Stage 2 - Treatment of information sharing. Previous stage may result in overlaps, when a schema element belongs to more than one subschema. In these situations each element is classified according to read or write operations executed by each subsystem. Three classifications are defined as follows:

1. Non-shared: the element belongs to a single subschema and a single subsystem carries out both reading and writing operations on the element.
2. Unidirectional sharing: the element belongs to two or more subschemas and a single subsystem performs the writing operations on the element, while the others carry out only reading operations.
3. Multidirectional sharing: the element belongs to two or more subschemas and two or more subsystems perform writing operations on the element.

Stage 3 - Definition of the database modules. Here a group G_{Si} is created for each subsystem Si containing the elements that it maintains, which are those that perform writing operations. An intersection with all groups is made, in order to handle multidirectional sharing and to define the database modules.

Groups, which intersection with all others is empty, originates a database module per group.

$G_{S1} \cap G_{S2} \cap ...G_{Sn} = \emptyset$, a Module M_n is created for each G_{Sn}.

Groups, which intersection with all others is non-empty, can originate a database module with all elements of the groups.

$G_{S1} \cap G_{S2} \cap ...G_{Sn} \neq \emptyset$, a Module M_n is created with $G_{S1} \cup G_{S2} \cup ...G_{Sn}$.

Another possibility is to create a database module based on the difference between the elements of the intersection result and each subsystem elements group.

$G_{S1} \cap G_{S2} = G_x$, generates module M_x.

G_{S1}-G_x, generates Module M_n.

G_{S2}-G_x, generates Module M_{n+1}.

Stage 4 - Definition of the interface of the database modules. This stage encapsulates subschemas in database modules with public and private procedures, read access and write access respectively. Figure 1 shows a graphical representation of a database module, enhancing abstraction by including both the conceptual schema of the database module as well as the associated procedures.

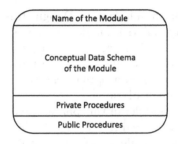

Fig. 1. A module representation [7].

4 Evolutionary Database Modularization Design Process

Our approach towards database evolution is an extension of the database modularization design process along with new data modeling concepts.

The original database modularization process was designed as a top-down database project, which means that a global conceptual schema was upfront designed and partitioned (modularized). On the other hand, our scenario is an incremental database schema design, where subsequent development cycles start from an already implemented schema. For example, this is the case on agile projects.

The evolutionary database modularization design process is conducted with both top-down and bottom-up design strategies. The top-down is applied at each cycle to define new subsystems or associate new requirements and functionalities to the already existing ones. Concurrently, it is necessary to consider the existence of database modules created in previous cycles that can contemplate fully or partially new functionalities, thus characterizing a bottom-up design.

Six phases comprises the evolutionary database modularization design process as showed in Fig. 2. Gray rectangles represent our extension of the original design process.

1. Requirements Collection and Analysis: it consists in identifying and documenting both data and functional requirements.
2. Evolutionary Analysis of Modularization Requirements: in an evolutionary approach, it becomes an iterative process, where all phases are executed at each cycle, resulting in a new increment to the database. Therefore, this phase aims to analyze new requirements taking into account the already implemented databases modules from previous iterations. The output is the Modularization Data Requirements, which indicates how the iteration will affect the subsystems definitions, according to the functional requirements of the previous stage, and the impact on the database modules, resulting in either extending or creating new modules.
3. Iteration's Conceptual Design: the requirements are analyzed in order to produce the iteration's conceptual schema through the use of the entity-relationship model.

4. Evolutionary Database Modularization Design: this phase was extended to perform the database modularization design in an evolutionary manner, considering the stages presented in Sect. 3.1. Thus, we added the "Evolutionary definition of the database modules" in respect to the original modularization process. In addition, the conceptual schema of each database module is adapted to support future evolutions.
5. Logical Design to each Module: in this phase, the conceptual schema of each database module is transformed into a logical schema.
6. Physical Design to each Module: lastly the logical schema of each module is deployed physically into the database, using a specific database management system.

From the second iteration on, we consider the deployed database modules, represented by the dotted arrow in Fig. 2. The next subsections describe, in detail, the new phase called "Evolutionary Analysis of Modularization Requirements" and the extended one called "Evolutionary Database Modularization Design".

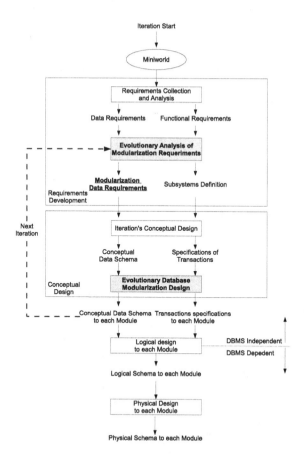

Fig. 2. Evolutionary database modularization design process.

4.1   Evolutionary Analysis of Modularization Requirements

This phase assess the completeness of the database modules in respect to an iteration's functionalities. The activity diagram of Fig. 3 shows the requirements analysis according to the "Evolutionary Analysis of Modularization Requirements" phase as well as the possible outputs.

1. The already existing database modules support the new requirements; therefore, no change is required and the functionalities are implemented into the correspondent database modules.
2. There is no support, from the existing database modules, to the new requirements; thus, new database modules will be created with the correspondent elements and functionalities.
3. If the existing database modules contemplate partially the new requirements, new database modules can be created as well as the extension of the existing ones, in order to support the new requirements.

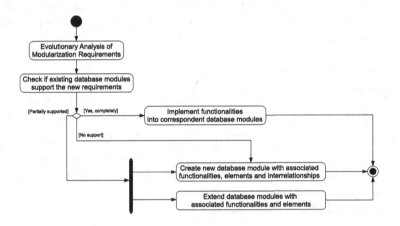

Fig. 3. Activity diagram: evolutionary approach to modularization design.

It is important to note that there are no database modules to assess at the first iteration. Therefore, it results in the creation of the early database modules, which means that the output is the "Create new database module with associated functionalities, elements and interrelationships" of Fig. 3. From the second iteration, this phase receives also the conceptual data schema of database modules generated in previous iterations, represented by the dotted arrow of Fig. 2.

4.2   Evolutionary Database Modularization Design

Our work presents interrelationship and intergeneralization as new semantic data modeling concepts that arise from modules definitions. In order to produce cohesive database modules to support further evolution, we need to consider coupling that in modularization occurs in interrelationships and generalization hierarchies among modules, a semantic data modeling concept that we named intergeneralization.

An intergeneralization occurs when a future requirement needs to specialize an entity type that belongs to an existing database module. In this scenario, the specific entity type is created in a new module. This avoid unnecessary coupling between the specific and generic entity types and it conforms to the database modularization definition where "a highly detailed degree of data abstraction is required to make a partition of the conceptual data schema" [7]. The dotted generalization hierarchy of Fig. 4, represents an intergeneralization, where the generic entity type E1 and the specific E1' belong to two distinct database modules. A specific entity type (EX) is added to Module X and shares the $A_{E1'}$ attributes domain. Therefore, intergeneralization provides independent evolution of each entity type reducing coupling among the elements. As of interrelationships, they represent the point where two database modules share data. In an evolutionary design there is a high probability of multiplicity change in a future iteration, which would require refactoring. Therefore, interrelationships are multiplicity change safe, since they are mapped as many-to-many relationships, as it is the case of R relationship type in Fig. 4.

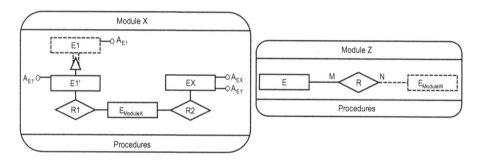

Fig. 4. Intergeneralization representation (left) and interrelationship mapped as many-to-many (right).

With the concepts of interrelationship and intergeneralization presented previously, now we discuss the "Evolutionary Database Modularization Design" stages, that include adaptations and extensions of the original "Modularization Design" stages presented in Sect. 3.1.

Stage 1 - Partitioning of the conceptual database schema. The conceptual schema is still being partitioned in one subschema per subsystem. However, new subsystems and modifications on the already existing ones can occur on future iterations, which can result in a different set of subschemas in respect to the previous iterations.

Stage 2 - Treatment of information sharing. The overlaps of the elements between the subschemas of the stage 1 can change from one iteration to the other.

The introduction of new subsystems and the modification of the already existing ones can change the classification of an element:

1. From Non-shared to Unidirectional Sharing: if a subsystem starts to read data from an element maintained by another subsystem;
2. From Unidirectional to Multidirectional Sharing: if a subsystem starts to write data on an element maintained by another subsystem.

Stage 3 - Evolutionary definition of the database modules. In this stage, the elements may have had their classification changed according to the stage 2, which means intersection operations can generate new interrelationships. Hence, this stage's role was extended and we have added the "Evolutionary" to the original's stage name "Definition of the database modules" (see Sect. 3.1).

According to the database modularization process, there are two options to handle multidirectional sharing. One is to create a database module with the intersected elements; the other is to perform a union among all these elements. Our approach considers only the first, since it will result in loosely coupled modules. First, from one iteration to another the multiplicity of the relationship types can change, however when this occurs in an interrelationship there is no need to refactoring because it is already a many-to-many relationship. The same happens with newly introduced generalization among modules, because they will be mapped as intergeneralizations, hence the specialized entity type is created in the correspondent database module.

This stage defines the database modules following the database modularization requirements, which is the output of the "Evolutionary Analysis of Modularization Requirements" phase, resulting in:

1. Creation of new database modules when:
   (a) a set of elements that were non-shared or unidirectional shared become multidirectional shared;
   (b) new elements are not associated to any previously existing database module.
2. A database module is extended when the newly introduced set of elements were associated to a previously existing database module and have only non-shared or unidirectional sharing.

Stage 4 - Definition of the interface of the database modules. The public and private procedures are associated with each defined database module according to stage 3.

5   Evolutionary Database Modules Integration

In order to support the evolutionary design, we propose an Integration Object Catalog that holds the modularization's process metadata at each iteration, such as subsystems definitions, entity types, relationships types, attributes and the database modules definitions. Figure 5 shows the conceptual schema of catalog, which is updated to reflect database modules changes. The gray highlighted subschema represents the extension proposed by this work, while the remaining subschema was introduced in Busichia and Ferreira [17,18].

The conceptual schema holds interrelationships metadata in the "Belongs" and "Participates" relationship types. The first relates each database module's elements to its module, while the second holds the relationship types of the application's conceptual schema. Hence, a relationship between two entity types of two distinct database modules is a interrelationship. The former multiplicity of an interrelationship is store to enforce integrity rules, since physically they are many-to-many. Similarly, "Belongs" and "Generalizes" hold intergeneralizations, where the latest indicates if an entity type participates in a generalization hierarchy. If the generic and specific entity types belong to two distinct database modules, then it is characterized as an intergeneralization.

The catalog has a major role in the process and it is used as the input of the "Evolutionary Analysis of Modularization Requirements" phase (see Fig. 2), since it maintains the metadata of the database modularization process. Moreover, it can be used to automate the generation of database modules at each iteration, by verifying the changes in the metadata, such as the conceptual schema, subsystems and the access type to each element. In this way, it is necessary to keep at least the previous instance of catalog data, in order to compare changes and apply them to the database schema, according to the definitions of the evolutionary database modularization design process.

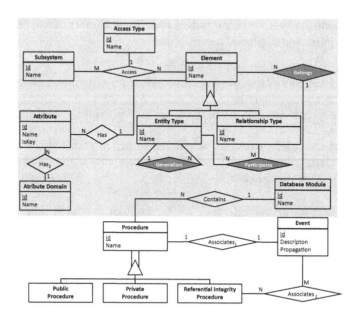

Fig. 5. Conceptual schema of the Integration Object Catalog.

6   Case Study: Applying the Evolutionary Database Modularization in an Agile Project

In this section we use the software specification contained in Ambler [19], a simple karate school system at first, to demonstrate the Evolutionary Database Modularization Design. Table 1 presents a brief description of the requirements.

Table 1. Brief description of the requirements.

Iteration	Description
1	Maintain student contact information
	Enroll student
	Drop student
	Record payment
2	Promote student to higher belt
	Invite student to grading
	Email membership to student
	Print membership for student
3	Schedule grading
	Print certificate
	Put membership on hold
4	Enroll child student
	Offer family membership plan
	Support child belt system
5	Enroll student in Tai Chi
	Support Tai Chi belt system
	Enroll student in cardio kick boxing
	Support the belt order for each style
6	Maintain product information
	Sell product

The first iteration results in two subsystems, one to handle the student functionalities (S1) and one to control payment (S2). These subsystems originate database modules M1 and M2 respectively. Also, we create the Student entity type, instead of creating a generic Person entity type upfront.

Second iteration introduces the belt control system, whose functionalities are grouped in S3 subsystem. When applying the database modularization process it indicates that the existing database modules attend the data requirements partially, since module M1 maintain student data. Next, "treatment of information sharing" indicates that there is only unidirectional sharing between new elements; therefore, new M3 module is created to support the belt system. Iterations 3 and 4 also have similar situations where either a database module is created or an existing one is extended. Figure 6 shows the generated database modules until the fourth iteration.

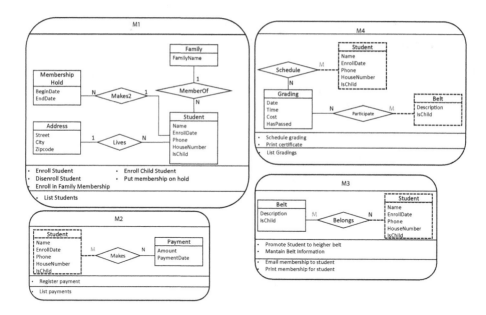

Fig. 6. Generated modules until fourth iteration.

6.1 Fifth Iteration

Initially, the system was design to support only one type of training (karate). However, fifth iteration requirements introduce support for other types of coaching. First, training management functionalities are grouped in a new subsystem, S5. The "Evolutionary Analysis of Modularization Requirements" phase indicates that M1 and M3 modules partially support multi-training functionality. The result of "Evolutionary Database Modularization Design" phase is summarized on Fig. 7.

6.2 Sixth Iteration

The sixth iteration requirements introduce a store functionalities that are grouped into a new subsystem (S6). An intergeneralization is introduced to support both student and general customers buys (Fig. 8). Note that Credit-Card attribute are added to both "StudentCustomer" and "Customer" entity types. On the other hand, DiscountIndex attribute is only present in "Student-Customer" entity type, since it only makes sense to students byers. In this way, "StudentCustomer" and "Customer" entity types evolve separately in further iterations.

Module	Iteration	Element	Element Type	Type of Operation (R- Read / W - Write) S1	S2	S3	S4	S5	S6	Type of Sharing	Maintained by S1	S2	S3	S4	S5	S6
M1	1	Student	Entity Type	R/W	R	R	R	R	R	Unidirectional						
	6	Lives	Interrelationship	R/W						Non-Shared						
	3	MembershiHold	Entity Type	R/W						Non-Shared						
		Makes2	Relationship Type	R/W						Non-Shared						
	4	Family	Entity Type	R/W						Non-Shared						
		MemberOf	Relationship Type	R/W						Non-Shared						
	5	Enroll	Interrelationship	R/W				R		Unidirectional						
M2	1	Payment	Entity Type		R/W					Non-Shared						
		Makes2	Interrelationship		R/W					Non-Shared						
M3	2	Belt	Entity Type			R/W	R	R		Unidirectional						
		Belongs	Interrelationship			R/W				Non-Shared						
M4	3	Grading	Entity Type				R/W			Non-Shared						
		Schedule	Interrelationship				R/W			Non-Shared						
		Participate	Interrelationship				R/W			Non-Shared						
M5	5	Training	Entity Type					R/W		Non-Shared						
		Has	Interrelationship					R/W		Non-Shared						
M6	6	StudentCustomer	Intergeneralization	R					R/W	Unidirectional						
		Deliver	Interrelationship						R/W	Non-Shared						
		Deliver2	Interrelationship						R/W	Non-Shared						
		Makes3	Relationship Type						R/W	Non-Shared						
		Makes4	Relationship Type						R/W	Non-Shared						
		Customer	Entity Type						R/W	Non-Shared						
		Order	Entity Type						R/W	Non-Shared						
		Item	Entity Type						R/W	Non-Shared						
M7	6	Address	Entity Type	R/W					R/W	Multidirectional						

Fig. 7. Evolutionary database modularization design phase until sixth iteration.

Figure 8 shows the extension of M1 module with "Lives" relationship type as an interrelationship. Also, a new database module M7 accommodates "Address" entity type, with a multidirectional sharing, since both S1 and S6 subsystems perform writing operations. Intergeneralization avoids over modeling, like it would have been the case if a "Person" entity type was upfront created at the first iteration. It allows independent evolution of entity types that would participate in a same generalization hierarchy. This is achieved by representing most detailed level of data abstraction, creating the specific entity types only.

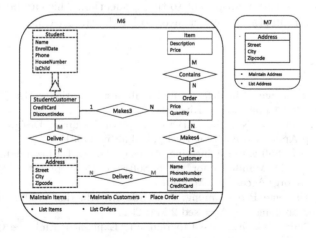

Fig. 8. M6 and M7 modules.

7 Conclusion

Preserve software system integrity through an evolutionary process demands a great deal of effort, especially due to two different paradigms: coding and database, whose schema must evolve to support new functionalities. Therefore, in a constantly changing environment it is ideal to have an evolutionary database design process. Thus, this paper presented the evolutionary database modularization design process.

At first our design increases the abstraction level of the general schema. First, because the process is conducted at conceptual design phase; secondly by introducing two new data modeling concepts, interrelationship and intergeneralization.

Interrelationships, can stand further multiplicity changes without refactoring. That is also the case of intergeneralizations, that evolve independently in each module.

The scope of change assessment is narrowed down within each database module. In addition it improves traceability, since a database module representation includes both the conceptual schema and its related transactions.

We presented an Integration Object Catalog to store database modules definition. Therefore, it is possible to use database modules metadata to assist the evolution process of database modules.

In short, a standardized evolutionary database design method is provided facilitating further schema evolution. Agile projects can benefit from our approach due to its incremental and iterative characteristics.

Future work includes the extension of evolutionary database modularization design process to intra-modules changes, when they occur inside a database module. Further, the creation of a database module definition language. Another issue concerns the coexistence of multiple database schemas for a certain period. In this situation, public and private procedures could encapsulate the transactions, making schema changes transparent to the application. This can be achieved by implementing an integration object layer to handle transactions.

References

1. Ambler, S., Sadalage, P.J.: Refactoring Databases: Evolutionary Database Design. Addison Wesley, Upper Saddle River (2006)
2. Batini, C., Ceri, S., Navathe, S.B.: Conceptual Database Design: An Entity-Relationship Approach. Addison Wesley, California (1991)
3. Sommerville, I.: Software Engineering. Addison-Wesley, Harlow (2007)
4. Beck, K., et al.: Manifesto for Agile Software Development (2001). http://agilemanifesto.org. Accessed 2 Mar 2015
5. Wells, D.: Extreme Programming: A Gentle Introduction (1999). http://www.extremeprogramming.org. Accessed 2 Mar 2015
6. Beck, K., Andres, C.: Extreme Programming Explained: Embrace Change, 2nd edn. Addison-Wesley, Boston (2004)

7. Ferreira, J.E., Busichia; G.: Database modularization design for the construction of flexible information systems. In: Proceedings of the 1999 International Symposium on Database Engineering & Applications (IDEAS 1999), pp. 415–422. IEEE Computer Society, Montreal (1999)

8. Guedes, G.B., Busichia, G., Moraes, R.L.O.: Database modularization applied to system evolution. In: Sixth Latin American Symposium on Dependable Computing (LADC 2013), Rio de Janeiro, Brazil, pp. 81–82 (2013)

9. Navathe, S.B., Fry, J.P.: Restructuring for large databases: three levels of abstraction. ACM Trans. Database Syst. 1, 138–158 (1976)

10. Sockut, G.H., Goldberg, R.P.: Database reorganization-principles and practice. ACM Comput. Surv. 11(4), 371–395 (1979)

11. Codd, E.F.: A relational model of data for large shared data banks. Commun. ACM 13(6), 377–387 (1970)

12. Ambler, S.: Agile Database Techniques: Effective Strategies for the Agile Software Developer. Wiley, New York (2003)

13. Cleve, A., Hainaut, J.-L.: Co-transformations in database applications evolution. In: Lämmel, R., Saraiva, J., Visser, J. (eds.) GTTSE 2005. LNCS, vol. 4143, pp. 409–421. Springer, Heidelberg (2006)

14. Cleve, A., Brogneaux, A.-F., Hainaut, J.-L.: A conceptual approach to database applications evolution. In: Parsons, J., Saeki, M., Shoval, P., Woo, C., Wand, Y. (eds.) ER 2010. LNCS, vol. 6412, pp. 132–145. Springer, Heidelberg (2010)

15. Papastefanatos, G., Anagnostou, F., Vassiliou, Y., Vassiliadis, P.: Hecataeus: a what-if analysis tool for database schema evolution. In: Proceedings of the 2008 12th European Conference on Software Maintenance and Reengineering (CSMR 2008), pp. 326–328. IEEE Computer Society, Washington (2008)

16. Curino, C.A., Moon, H.J., Ham, M., Zaniolo, C.: The PRISM workwench: database schema evolution without tears. In: Proceedings of the 2009 IEEE International Conference on Data Engineering (ICDE 2009), pp. 1523–1526. IEEE Computer Society, Washington (2009)

17. Busichia, G., Ferreira, J.E.: Sharing of heterogeneous database modules by integration objects. In: Eder, J., Rozman, I., Welzer, T. (eds.) Proceedings of the Third East European Conference on Advances in Databases and Information Systems (ADBIS 1999), pp. 1–8. Institute of Informatics, Faculty of Electrical Engineering and Computer Science, Smetanova 17, IS-2000 Maribor, Slovenia (1999)

18. Busichia, G., Ferreira, J.E.: Compartilhamento de Módulos de Bases de Dados Heterogêneas através de Objetos Integradores. In: Simpósio Brasileiro de Banco de Dados (SBBD), pp. 395–409. SBC, Florianpolis (1999)

19. Ambler, S.: Agile/Evolutionary Data Modeling: From Domain Modeling to Physical Modeling (2004). http://www.agiledata.org/essays/agileDataModeling.html. Accessed 2 Mar 2015

Summary Generation for Temporal Extractions

Yafang Wang[1](✉), Zhaochun Ren[2], Martin Theobald[3], Maximilian Dylla[4],
and Gerard de Melo[5]

[1] Shandong University, Jinan, China
yafang.wang@sdu.edu.cn
[2] University of Amsterdam, Amsterdam, The Netherlands
z.ren@uva.nl
[3] University of Ulm, Ulm, Germany
martin.theobald@uni-ulm.de
[4] Max Planck Institute of Informatics, Saarbrücken, Germany
mdylla@mpi-inf.mpg.de
[5] Tsinghua University, Beijing, China
gdm@demelo.org

Abstract. Recent advances in knowledge harvesting have enabled us to collect large amounts of facts about entities from Web sources. A good portion of these facts have a temporal scope that, for example, allows us to concisely capture a person's biography. However, raw sets of facts are not well suited for presentation to human end users. This paper develops a novel abstraction-based method to summarize a set of facts into natural-language sentences. Our method distills temporal knowledge from Web documents and generates a concise summary according to a particular user's interest, such as, for example, a soccer player's career. Our experiments are conducted on biography-style Wikipedia pages, and the results demonstrate the good performance of our system in comparison to existing text-summarization methods.

Keywords: Temporal information extraction · Knowledge harvesting · Summarization

1 Introduction

In recent years, we have seen a number of major advances in large-scale text mining and information extraction (IE). Amongst others, such efforts have led to the emergence of large knowledge graphs, which are collected by companies like Google, Microsoft, and Facebook, as well as open efforts such as DBpedia [2] and YAGO [22]. Given a piece of input text, numerous open-domain tools, such as NELL [4], ReVerb [7] or PRAVDA [31], are readily available for extracting subject-predicate-object triples from the text. However, while the extracted triples concisely capture the essential information conveyed by the original text also with respect to their temporal scope, they are usually not directly suitable for presentation to human end users. In this paper, we thus present a novel method to automatically generate natural-language summaries of such extractions.

© Springer International Publishing Switzerland 2016
S. Hartmann and H. Ma (Eds.): DEXA 2016, Part I, LNCS 9827, pp. 370–386, 2016.
DOI: 10.1007/978-3-319-44403-1_23

There are a number of challenges to be addressed. Existing summarization methods for natural-language text mostly just return existing sentences from the text, instead of summarizing the content at a more "abstract" level. A sentence may be long and include both key facts but also large amounts of less essential information. Additionally, key facts may need to be ranked and aggregated. What, for instance, should a short summary of, say, 100 words for a soccer player's career include? For famous players, with long careers, it may well be quite impossible to list all their clubs and games. An ideal summary might thus focus on the most important clubs and honors. Also, when extracting facts, we often just obtain observations about a series of individual time points that needs to be aggregated to obtain a larger picture, e.g., that a person not only played for Arsenal in 2013 and 2015 but over a longer period of time. Thus, it is important to move towards systems that attempt to go beyond selecting pre-existing sentences and are able to produce more concise summaries.

Contributions. We propose a method that, unlike previous approaches, attempts to identify the key facts in a document, much like a human would, and then generates concise summaries from them. We propose a new method that (1) summarizes facts extracted from multiple documents, (2) deals with temporal reordering and aggregation of potentially noisy pieces of evidence, and (3) produces a coherent abstractive text summary. Our experiments on Wikipedia biographies demonstrate the strength of this method.

Overview. Figure 1 provides an overview of our approach. Our approach is designed to follow the way a human would summarize information, by first digesting it and then capturing, aggregating, and rearranging the essential pieces of knowledge. To harvest information from both semi-structured and textual data sources, we rely on a number of extraction rules to mine a set of seed facts for our relations of interest from the semi-structured parts of the input documents (e.g., tables and Wikipedia infoboxes). These seed facts are then used to identify

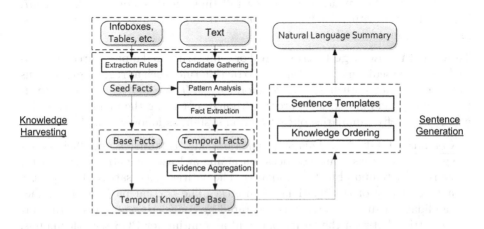

Fig. 1. Overall workflow

characteristic patterns to harvest more facts from the textual data sources. For this purpose, we rely on the general architecture of the PRAVDA system [31] to extract such facts (including temporal ones) from free-text sources. Multiple occurrences of temporal facts are reconciled via a form of evidence aggregation, which serves to condense the extracted knowledge, and in particular to extract high-confidence time intervals at which these new facts are found to be valid. Finally, for better readability and coherence of the final summary, these facts along with their time intervals are ordered chronologically and presented as natural-language sentences by mapping the temporally aligned facts onto a set of handcrafted sentence templates.

2   Knowledge Harvesting

Model. We are given a set of input sources $\mathcal{D} = \{D_1, D_2, \ldots, D_n\}$ (e.g., documents) and a set of binary target relations $\mathcal{R} = \{R_1, R_2, \ldots, R_m\}$ of interest. Then, the knowledge harvesting step aims at extracting instances of these relations from the input sources. Each relation R has an associated type signature (T_R^s, T_R^o), providing valid entity types for the subjects and objects of this relation. We distinguish between base facts and temporal facts. A base fact (*b-fact*, for short) is of the form $R(e_1, e_2)$, where the entity e_1 is of type T_R^s and e_2 has type T_R^o.

Temporal Facts. A temporal fact includes an additional temporal marker, which we denote as $R(e_1, e_2)@t$. This indicates that the relationship holds (i.e., is valid) at time t, which may refer to either a time point or a time interval. In our system, we define the finest granularity to be days, and all coarser granularities are converted to time intervals (e.g., January 2012 to *[1-Jan-2012, 31-Jan-2012]*). For example, *playsForClub(David_Beckham, Real_Madrid)@2005* is consistent with *playsForClub*'s type signature (*Person, Club*) and indicates that *David_Beckham* played for *Real_Madrid* in 2005. This may reflect just one statement in a document, while the overall time interval would be 2003–2007. We thus further distinguish between *event* and *state* relations, which we describe in Sect. 3 in more detail.

Temporal Knowledge Extraction. For semi-structured input sources, such as tables or Wikipedia infoboxes, simple extraction rules such as regular expressions suffice to extract both base and temporal facts. These are then used as seeds to find more facts in textual sources. Although not being the main focus of this paper, we briefly summarize our extraction system as follows:

1. Candidate Gathering: This step generates fact candidates and their corresponding patterns from sentences containing at least two entities (and a time marker for temporal fact candidates). The entities must satisfy the type signature of any of the relations of interest. The textual pattern of the fact candidate in such a sentence is generated by considering n-grams of the surface string between the entity pair and accounting for POS tags (for nouns, verbal phrases, prepositions, etc.).

2. Pattern Analysis: We compute the initial weight of each pattern based on the seed facts and the output of the previous step. The weight depends on the number of co-occurrences between the seed facts and the textual patterns. Patterns with co-occurrence weights above a threshold are initialized with this weight for the algorithm we apply in the next step, while the initial value for other patterns is zero.

3. Fact Extraction: In our final step, a graph is built from the fact candidates and patterns. Edges between fact candidates and patterns are added if they co-occur within a sentence. Similar patterns are also connected this way. Then, a form of label propagation [25] is utilized to determine the most likely relation for each of the fact candidates. Once a fact candidate is labeled with a particular relation R, it is called a *valid observation* and added to the set of event facts that are returned as result of the knowledge harvesting phase.

3   Evidence Aggregation Model

A main challenge in extracting and mining temporal knowledge is the proper distinction between *event* and *state* relations. For an event relation, a t-fact is valid only at a single time point. For example, *visits(François_Hollande, Berlin)* is valid on 24-Aug-2015. Actually, President Hollande visits Germany frequently, so there could be multiple such facts, each with different time points. State relations hold for an extended time interval during which a fact is valid at any time point within a given interval. For example, *playsForClub(Diego_Maradona, FC_Barcelona)* is valid in the entire interval *[1-July-1982, 30-June-1984]*. Multiple non-contiguous time spans are represented by several such state facts. The extraction of time periods for state facts is challenging, because there are typically only few occurrences of facts in input sentences with explicit time intervals. Ideally, we would encounter sentences like "Maradona had a contract with FC Barcelona from July 1982 to June 1984". However, such explicit sentences are rare in both news and web sources. Instead, we can find cues that refer to the *begin*, *end*, or some time point *during* the desired interval. For example, news articles would often mention sentences such as "Maradona did not play well in the match against Arsenal London" with a publication date of 15-May-1983 (a time point presumably contained within the corresponding state fact's interval). Thus, having extracted specific time points, we need to aggregate these into intervals for state-oriented t-facts. To address this, our method (1) aggregates individual time points into time histograms, and (2) computes a high-confidence time interval from these histograms.

We aim to aggregate individual *begin*, *end*, and *during* observations of a fact into a concise time histogram. So even if we are aiming at state-oriented t-facts, we first collect and aggregate event-style cues. Ideally, these point-wise observations would then form a compact time interval that captures the validity of the fact. However, a general problem of such an approach is the inherent ambiguity when individual events are mapped to an initially unknown amount of time intervals. This gets even more difficult due to frequent extraction errors, overlapping occurrences of *begin* and *end* events, or other inconsistencies. However, the observations are often noisy and require non-trivial reconciliation for each base fact. First, we construct histograms for each of the begin, end, and during events. After that, the histograms are combined into a single state-oriented histogram, which is distilled into a single high-confidence interval that represents the fact's temporal validity. Finally, an algorithm computes the confidence interval of the histogram.

3.1   Aggregating Events into State Histograms

Among all observations of event facts with matching entities found in the input sentences, we first determine the time range $[t_b, t_e]$ of the largest possible validity interval of a corresponding state fact by selecting the earliest time point t_b and the latest time point t_e encountered, respectively. According to the relation an event fact has been labeled with, we classify the individual facts as *begin*, *end*, and *during* observations that mark either the possible begin or end time point, or a time point during which the corresponding state fact may be valid.

Next, all observations of *begin*, *end*, and *during* facts are aggregated into three initial histograms, each ranging over $[t_b, t_e]$. This yields one frequency value $f[t_i]$ per time point t_i. Initially, the i-th bin's frequency value $f[t_i]$ refers to the plain number of observations corresponding to this time point, for each of the three types of event facts. Subsequent time points with equal frequencies are coalesced into a single histogram bin. In each of the histograms, the bins' frequencies are then normalized to 1. For combining the three event-oriented histograms into a single histogram of the corresponding state fact, we apply the following assumptions:

1. A *during* observation at time point t_j should increase the confidence in the state fact being correct at t_j (for all time points captured by the interval of the *during* observation).
2. A *begin* observation at time point t_j should increase the confidence in the state fact for all time points ranging from t_j to t_e.
3. An *end* observation at time point t_j should decrease the confidence in the state fact for all time points t_j to t_e.

Our approach produces a multi-modal histogram if *end* facts interleave with *begin* and *during* events at different time points, which we can exploit to extract multiple validity intervals for the state fact (there are two time intervals in Fig. 2). In case none of the different event types interleave (i.e., all *begin* events occur before all *end* events, and all *during* indeed occur between all *begin* and *end* events), we obtain a uni-modal histogram from which we can extract just a single validity interval for the resulting state fact.

Algorithm 1 describes how we combine the *begin*, *end* and *during* histograms. We first merge the two *begin* and *end* histograms, before we merge the resulting *begin-end* histogram with the *during* histogram as follows (using De Morgan's law):

$$P = P_{\text{during}} \cup P_{\text{begin,end}} = \overline{\overline{P_{\text{during}}} \bigcap \overline{P_{\text{begin,end}}}} \tag{1}$$
$$= 1 - (1 - P_{\text{during}}) \cdot (1 - P_{\text{begin,end}})$$

Here, P denotes the final frequency obtained after all aggregation steps, P_{during} denotes the frequency of the *during* event, and $P_{\text{begin,end}}$ is the output (i.e., the $f[t_i]$ after the inner *for* loop in Algorithm 1) of aggregating the *begin* and *end* histograms. For all the non-empty bins in the *during* histogram, we use Eq. 1 to compute the new frequency value P. P_{during} refers to the probability of a time point indicating *during* given the observations from *during* events. The new frequency is thus the union of the probability of a time point by considering all types of events. Finally, all consecutive bins with the same frequency values are merged, and the bins are once more normalized to 1 (cf. Algorithm 1 and Fig. 2).

Algorithm 1. Aggregating events into state histograms.

Require: Event histograms with frequencies f_{begin}, f_{end}, f_{during} over the time range $[t_b, t_e]$
For all $t_i \in [t_b, t_e]$ **do**
　　$f[t_i] \leftarrow 0$
　　For all $t_j \in [t_i, t_e]$ **do**　　　　　　　　　$\triangleright$ Aggregate *begin* and *end* histograms
　　　　$f[t_j] \leftarrow f[t_j] + f_{\text{begin}}[t_i]$　　　　　　　　　　　$\triangleright$ aggregate *begin*
　　　　$f[t_j] \leftarrow \max(0, f[t_j] - f_{\text{end}}[t_i])$　　　　　　　　　$\triangleright$ reduce *end*
　　End
　　$f[t_i] \leftarrow (1 - (1 - f_{\text{during}}[t_i]) \cdot (1 - f[t_i]))$
　　　　　　　　　　　　　　　$\triangleright$ Combine *begin,end* histogram with *during* one
End
Reorganize the bins and normalize their frequencies to 1
Return: State histogram with frequencies f

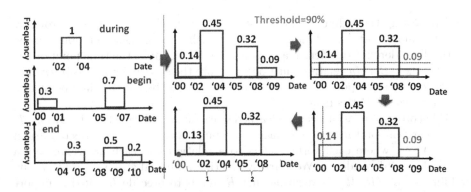

Fig. 2. Aggregating events into state histograms.

3.2   Extracting High-Confidence Intervals

The combined state histogram reflects the confidence distribution for a fact's validity over time. The value of a bin can be interpreted as the probability of the fact being valid during this bin's interval. For our temporal summarization, we next simplify the possibly very fine-grained histogram by discarding bins with a low confidence. Assuming, for example, we are interested in a final histogram that captures at least 90 % of the confidence mass of the original histogram, we discard all low-confidence bins whose cumulative frequencies sum up to at most 10 %.

Since the original histogram's bins form a discrete confidence distribution, we pursue an iterative strategy. Starting from the lowest-frequency bin, we first sort all bins by their frequency values and then check for the remaining confidence mass when cutting off these bins horizontally. Let τ be the expected threshold of the confidence interval (e.g., 90 %). Our algorithm stops as soon as we have cut off more than a threshold of $1 - \tau$ (e.g., 10 %) of the overall confidence mass. We then pick the previous solution, which must still be above τ. This procedure is further refined by a final vertical trimming step of the remaining bins. To this end, we assume a uniform confidence distribution within each bin, and we adjust the frequency $f[i]$ of the trimmed bin proportionally to its cut-off width (cf. Fig. 2) until we reach τ.

4   Sentence Generation and Reordering

When summarizing multiple sources, possibly containing randomly ordered facts, it is usually not a-priori clear in which order to present these facts to the user. For short texts, we conjecture that a chronological order is appropriate in many cases.

4.1   Knowledge Ordering

Before sorting the individual facts about an entity of interest, we first roughly sort the more abstract relations associated with t-facts. Some relations can be naturally ordered. Considering a person's life, for example, the time point of a t-fact for the *isBornIn* relation must occur before the start point of a *isMarriedTo* t-fact for the same person, which in turn must occur before the time point of a *diedIn* t-fact of that person.

This order of relations can be learned statistically. Given a set of relations $\mathcal{R}$ and their temporal instances (t-facts), we build a time-ordered directed graph $G = (V, A)$, where each vertex refers to a relation and each arc represents a chronological dependency. We start by creating an initial graph $G' = (\mathcal{R}, E)$ by adding an arc (R_i, R_j) (indicating that R_i tends to precede R_j) if the support s_{ij} of R_i occurring before R_j is much greater than the inverse s_{ji}. s_{ij} is calculated by counting the instances of R_i and R_j having the same subject, i.e. $(a, b)@t_1 \in R_i$ and $(a, c)@t_2 \in R_j$, satisfying that t_1 precedes t_2. The final graph

G is then obtained from G' by adding two extra vertices representing the *start* and *end* states to G' and by removing all transitive dependencies from G'. For example, *isBornIn* may have an edge with many relations, such as *graduatedFromHighSchool*, *graduatedFromUniversity* and *diedIn*. These edges are removed according to the transitive dependencies among these relations, and only a path from *isBornIn* through *graduatedFromHighSchool*, *graduatedFromUniversity* to *diedIn* is kept. If the graph G contains a cycle, we remove the cycle by dropping the edge with the lowest support within the cycle. Figure 3 illustrates an example for transforming a set of relations into G, while Algorithm 2 shows details about how to determine the chronological order of both t-facts and b-facts according to G. For a state fact, which is valid during an entire time interval, only the start time point is taken into consideration. For example, suppose we captured that *David_Beckham* played for *Manchester_United* from 1991 to 2003, *Real_Madrid* from 2003 to 2007 and got married on 4-July-1999. The three temporal facts are ordered as {*playsForClub*(*David_Beckham*, *Manchester_United*), *getsMarriedTo*(*David_Beckham*, *Victoria_Beckham*, *playsForClub*(*David_Beckham*, *Real_Madrid*)}, according to the time points {1-January-1991, 4-July-1999, 1-January-2003}. Base facts (b-facts), which generally cannot be ordered explicitly by time, are inserted into G after the temporal facts (t-facts) in the same relation according to the topological order (Line 14).

4.2 Natural Language Generation

Similar to many other abstractive summarization methods [18], we rely on templates for natural language generation. For each relation, we manually define a number of sentence templates to construct the summary sentences. After the knowledge ordering, t-facts of the same relation are ordered next to each other due to the topological order in the relation graph. For each relation, we randomly

---

Algorithm 2. Knowledge Ordering

Require: Graph G; the base and temporal facts F_b and F_t.
  1: $S = \emptyset$ ▷ Empty list that will contain the sorted facts.
  2: $L \Longleftarrow$ Set of all vertices with no incoming edges.
  3: **while** L is non-empty **do**
  4:     remove a vertex n from L
  5:     **if** n has not been visited yet **then**
  6:         insert all t-facts (in temporal order) of relation n into S
  7:         insert all b-facts of relation n into S
  8:     **for** each node m with an edge e from n to m **do**
  9:         remove edge e from the graph G
 10:         **if** m has no other incoming edges **then**
 11:             insert m into L and mark m as visited
 12:             sort all the t-facts of relation m by time
 13:             insert sorted t-facts into S
 14:             insert b-facts of relation m into S
 15: **return** S ▷ Facts sorted by topological order of relations in G.

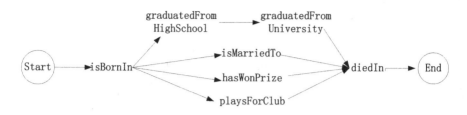

Fig. 3. Relation graph.

choose among the templates for a given subject to improve the diversity of the output. For a given subject, sentences representing the same relation are likely to contain a lot of redundancy. Thus we enable merging of arguments. For example, *"David Beckham played for Manchester United from 1993 to 2003"* and *"David Beckham played for Real Madrid from 2003 to 2007"* are merged into *"David Beckham played for Manchester United (1993–2003) and Real Madrid (2003–2007)"*. For his honors, we similarly obtain the merged sentence *"David Beckham won the Premier League (1996), the FA Cup (1999), the UEFA Champions League (1999), the Intercontinental Cup (1999), and the La Liga (2007), etc."* There are many honors, so we resort to only show the first ones.

In case there are too many facts holding the same relation, the method chooses among omitting unimportant facts, reporting the total number, or choosing only some examples for the summary sentences. Similarly, repeated occurrences of the main subject name (e.g., "David Beckham") are replaced by the corresponding pronoun ("he"), as heuristically determined by the most frequent pronoun in the source text, if available. Hence, the final summary is compressed into *"David Beckham has played for about eight clubs. He joined Manchester United in 1993. During his career in Manchester United, he won about fifteen honors including the Premier League (1996), the FA Cup (1999), etc."*. The initially redundant sentences were thus condensed into just three sentences with the key facts about David Beckham.

5   Experiments

5.1   Experimental Setup

We evaluate our method on Wikipedia articles from two domains: soccer players and movie stars. The corpora include Wikipedia articles for soccer players from the "FIFA 100 list"[1], and movie stars from the "Top 100 movies stars"[2]. For extraction, we preprocessed the corpora by replacing the most frequent pronoun by the title of the Wikipedia article, and all the entity mentions were disambiguated against the YAGO [22] knowledge base using the AIDA [10] framework for named entity disambiguation.

---

[1] http://en.wikipedia.org/wiki/FIFA_100/.

[2] http://articles.cnn.com/2003-05-06/entertainment/movie.poll.100_1_star-movies-godfather?_s= PM:SHOWBIZ/.

Table 1. Example sentence templates for relations.

Relation	Templates
isBornIn	ARG1 was born in ARG2
worksForClub	ARG1 served for ARG2; ARG1 worked for ARG2
actedIn	ARG1 acted in ARG2; ARG1 appeared in ARG2
hasWonHonor	ARG1 has won ARG2; ARG1 received ARG2

Relations. We list some example templates in Table 1. These are used for base facts. For temporal facts, an additional time point or time interval placeholder is added. For example, the template for temporal facts of *isBornIn* is "ARG1 was born in ARG2 on TIME". The template for temporal facts of *worksFor-Club* with a single time interval is "ARG1 served for ARG2 from begin_TIME to end_TIME", and for multiple time intervals we use "ARG1 served for ARG2 (begin_TIME*1*-end_TIME*1*, begin_TIME*2*-end_TIME*2*,...,begin_TIME*n*-end_TIME*n*)". For both domains, we query the system for summaries about facts associated with the birth and death dates of the respective persons, their family life (including marriage and children), honors they won, and their career (including the relations *worksForClub* for soccer players or *actedIn* for movie stars, as well as playing positions for soccer players).

Baseline Systems. We compare our system to four alternative approaches. For existing extraction-based multi-document summarization methods, we choose **NIST-Wiki** as the baseline in our experiments. NIST-Wiki extracts the first n sentences from a Wikipedia article. Since the top paragraphs in a Wikipedia article usually contain a short biography of the subject of the article, this is a very strong baseline. **LDA** here refers to an latent dirichlet allocation-based summarization method [1], which uses probabilistic topic distributions to calculate the salience for each input sentence. Additionally, as a representative model for recent abstractive summarization methods, we use Opinosis as another baseline. **Opinosis** [9] is a graph-based abstractive summarization framework. It constructs a graph from a set of input sentences set by considering redundancy and generates an optimal path from the graph. Finally, we also add a **Random** baseline to our comparison. The Random baseline just randomly selects n sentences from the data source. Since these baseline systems only support textual input data, the semi-structured sources (such as infoboxes) are translated to natural-language sentences via the sentence templates, yielding, e.g. "David Beckham has won FIFA 100".

Evaluation Procedures. We conduct two experiments. (1) We generate the summary with all the facts about a person, and (2) we generate a summary with only the most important facts and aggregated statistics. Since the first summary is longer than the second one, the corresponding baselines generate more sentences for the first experiment. We call the results from each experiment a *long summary* and a *short summary*, respectively. For the short summaries, we limit the number of words to at most 100, for long summaries 200. Since Opinosis is limited based on the number of sentences, a short summary is limited to 10 sentences, while a long summary contains 20 sentences. We evaluate the summary for *informativeness, diversity, coherence,* and *precision* [13, 16] by performing a user study. We randomly sample thirty summaries for each domain. For each of the above metrics, two human judges rate the summary on a Likert scale from one to five, where *one* means { "least informative", "least diverse", "very incoherent", "very imprecise"}, depending on the measure; while a rating of *five* means { "very informative", "very diverse", "very coherent", "very precise"}, respectively. The final score of each metric then is the average of all thirty summaries. The score of each metric in Table 4 is the average of all sixty summaries on both domains. The overall score then is the average over all metrics for each system.

5.2   Experimental Results

Table 2 provides examples of generated summaries, while the experimental results are given in Tables 3 and 4. In terms of the average over all metrics, our system outperforms all baseline systems (see Table 5). More specifically, it outperforms all the others on diversity and informativeness. On precision and coherence, NIST-Wiki is slightly better, because it just picks the first n sentences from each Wikipedia article, which are essentially human-written summaries. The precision of NIST-Wiki in the soccer domain is not 100 % correct, because in some cases it misinterpreted URL links as sentences. Since there is no 100 % perfect extraction methodology, incorrect extractions obviously affect the precision of our method. Furthermore, extraction recall can affect short summaries, since we report an aggregate number, as in *"David Beckham won about one honor"*, which is incorrect. To reduce such errors, we might consider including also vague statements like "at least". Opinosis compresses the text by considering the sentence redundancy, so the newly generated sentences may change the semantics of the original sentences. This holds for semi-structured contents, which is presented as natural language, e.g. "Rui Costa has won Toulon Tournament in 1992. Rui Costa has won FIFA U-20 World Cup in 1991." Opinosis is able to compress them into one meaningful sentence "Rui Costa has won Toulon Tournament in 1992 and FIFA U-20 World Cup in 1991." While for other sentences in the Wikipedia article, most generated sentences are meaningless and often incorrect, as evident in Table 2. Opinosis generates the sentence *"Beckham's marriage in 2007--/:."*, but Beckham actually married Victoria on July 4, 1999. NIST-Wiki produces perfect coherence, as it just returns contiguous n input sentences. Other extraction-based methods, such as LDA and Random, introduce incoherence. They also introduce imprecision when the extracted sentence contains indicative pronouns, such as "after this", and temporal phrases, such as "one year later", when the prior sentences were not chosen. As for the abstractive method, some sentences generated by Opinosis are meaningless, increasing the difficulty of reading the summary. On the contrary, our system exploits simple templates that are easy to understand. Only when too many facts hold for the same relation, the generated sentence feels non-fluent. For example, *"David Beckham won*

Table 2. Example summaries.

Our System (long): *David Beckham was born in London in 1975/05/02. He played as Midfielder. He served in Manchester United F.C. (1991-2003), Real Madrid C.F. (2003-2007), Los Angeles Galaxy (2007-). He has won FA Youth Cup (1992), FA Community Shield (1993,1994,1996,1997), Premier League (1996,1997,1999,2000,2001,2003), FA Cup (1996,1999), UEFA Club Football Awards (1999), MLS Cup (2011), FIFA 100, etc.*

Our System (short): *David Beckham was born in London in 1975/05/02. He played as Midfielder. He has played for about 3 clubs and won about 45 honors. In 1991 he joined Manchester United F.C. and served for 13 years. During this time period, he was awarded FA Youth Cup (1992), FA Cup (1996), FA Cup (1999), Intercontinental Cup (1999), UEFA Club Football Awards (1999), etc.*

NIST-Wiki: *David Robert Joseph Beckham, Order of the British Empire (born 1975-05-02) is an England association footballer who plays for Los Angeles Galaxy. David Beckham has played for Manchester United F.C., Preston North End F.C., Real Madrid C.F., A.C. Milan, and the England national football team for which David Beckham holds the appearance record for a Outfield#Association football. David Beckham's career began when David Beckham signed a professional contract with Manchester United, making his first-team debut in 1992 aged 17.*

LDA: *Beckham scored the equaliser and United went on to win the match and the league. Beckham scored 9 goals that season, all in the Premier League. The income from his new contract, and his many endorsement deals, made Beckham the highest-paid player in the world at the time. In the first nine matches David Beckham started, Real lost 7. David Beckham returned to play in the final home match of the season. Beckham is Officers of the Order of the British Empire. Beckham is England expatriates in the United States.*

Opinosis: *David Beckham enjoyed tremendous following. Beckham's right midfield position. Beckham's contract became public knowledge. Beckham's maternal grandfather was Jewish. Beckham's best season as united player and united. Beckham is England under-21 international footballers. Beckham England people of Jewish descent. Beckham's marriage in 2007- -/:. Beckham crumpled hard to the ground. Beckham of the most recognisable athletes throughout the world, not concentrating on the tournament and England 's next match.*

Manual: *David Beckham, born in 2 May, 1975, is a midfielder. Beckham began his career with Manchester United in 1991. During his 13 years career there, he won several honors. He received Premier League 10 Seasons Awards for his contribution from the 1992-93 to 2001-02 seasons. He also played for Real Madrid, LA Galaxy, etc. To honor his contribution, he was named FIFA 100. On 4 July 1999, David married Victoria. They have four children: sons Brooklyn Joseph, Romeo James, and Cruz David; and daughter Harper Seven.*

---

the Primier League (1996,1997,1999,2000,2001,2003), FA Cup (1996), La Liga (2007), MLS Cup (2011) ...". Notice also that the informativeness and diversity are affected by recall. Our system managed to find the key information. The sentences from the semi-structured input contents facilitate LDA and Opinosis to find this key information. Specifically for LDA, those sentences get higher

Table 3. Long summaries.

System	Diversity	Informativeness	Coherence	Precision	
Ours	3.93	4.73	4.33	4.57	*Soccer*
NIST-Wiki	3.13	3.73	4.97	4.97	
LDA	3.10	4.10	3.47	4.73	
Opinosis	1.97	3.87	1.87	3.10	
Random	1.63	2.27	1.63	4.53	
Ours	3.40	4.83	4.10	4.70	*Movie Star*
NIST-Wiki	2.23	3.63	4.47	5.00	
LDA	1.87	3.63	1.97	4.77	
Opinosis	1.20	3.20	1.77	3.37	
Random	1.60	2.47	1.87	4.83	

Table 4. Short summaries.

System	Diversity	Informativeness	Coherence	Precision	
Ours	3.73	4.23	4.40	4.17	*Soccer*
NIST-Wiki	2.73	2.93	4.93	4.97	
LDA	2.40	3.63	3.23	4.73	
Opinosis	1.80	3.07	1.77	3.07	
Random	1.27	1.50	1.63	4.80	
Ours	3.37	4.53	4.47	4.03	*Movie Star*
NIST-Wiki	1.90	3.27	4.53	5.00	
LDA	1.33	3.10	2.03	4.83	
Opinosis	1.10	2.80	1.63	3.33	
Random	1.13	1.70	2.17	4.80	

topic saliency than other sentences from the free text contents in the article for each topic. Thus, LDA could extract more information from those structured sentences into the final summary. Because of this, the score of LDA and Opinosis for informativeness is better than or close to NIST-Wiki (the natural biography), according to Table 5. The diversity is not very good for all systems. No system managed to extract all information of interest. Looking at the last parts of the examples in Table 2, no system extracted summaries about Beckham's marriage and children. Considering the honors, even if our system extracted all the honors for Beckham, it is difficult to decide which ones are the most important ones to be shown in the summary, since it takes expert knowledge to judge which are the most significant. Since the LDA-based summarization strategy calculates the saliency in multiple topics, it could get different sentences focusing on different sub-topics for each article. Therefore, as shown in the results, for diversity, the LDA-based method could obtain scores close to those of NIST-Wiki.

Table 5. Overall score.

System	Diversity	Informativeness	Coherence	Precision	Overall
Ours	**3.61**	**4.58**	4.33	4.37	**4.22**
NIST-Wiki	2.50	3.39	**4.72**	**4.98**	3.90
LDA	2.17	3.61	2.68	4.77	3.31
Opinosis	1.52	3.24	1.76	3.22	2.44
Random	1.41	1.98	1.83	4.74	2.49

6   Related Work

Summarization strategies for text can broadly be categorized as either extractive or abstractive. Extractive frameworks produce a summary by selecting existing sentences from the input text and concatenating them. For example, MEAD [19] relies on a centroid clustering-based strategy to score the saliency of input sentences, while others use random walks [28] and coverage maximization with bigram concepts [20]. For the supervised methods, HMMs [6], CRFs [21] and system combinations [11] have proven effective for extractive document summarization. However, all of these approaches merely pick sentences from input documents, without attempting to identify the key facts expressed in them.

Abstractive document summarization methods seek to produce novel sentences summarizing the contents at a more abstract level. Some methods apply sentence compression techniques to remove less important parts of existing sentences [8,12]. Opinosis [9,15] generate a summary from redundant data sources by building a graph-based representation. [3] constructs new sentences by selecting and merging informative phrases. Still, all of these works aim at summarizing text, which is different from our goal of summarizing key facts extracted from both semi-structured and unstructured sources while aggregating temporal evidence.

There are also some works that aim to summarize factual information from a knowledge base. [32] introduced the notion of RDF sentences and to summarize an ontology by ranking in the ontology graph. [27] retrieves the salient type properties for a certain entity. [23] presented a diversity-aware algorithm for graphical entity summarization. [5] generates a ranked list of textual summaries for the two-length entity chains. These works only consider existing knowledge bases as input, and the summary is merely given in the form of a subgraph or list of properties, while our work automatically harvests knowledge from heterogeneous data sources, aggregates temporal and other evidence (which is much noisier and incomplete in automatic extractions than in knowledge graphs), and produces a textual summary.

There has been some previous work on temporal extraction. For instance, [29] use a combination of statistical aggregation, label propagation, and integer linear programming to extract fact. [14,26] connect time events in documents by using unimodal time histograms, whereas our aggregation approach also supports multimodal histograms. [17,24] study the properties of relations, e.g. whether a relation is time-dependent and unique. However, all of these works aim at temporal information extraction-related tasks and do not address the issue of summarization.

An approach that handles queries over uncertain temporal facts has been presented by [30]. However it was mainly about probabilistic reasoning with rules and lineage and histograms played only an auxiliary role. CoTS [26] applies a classifier to publication dates to determine the *begin* and *end* dates of temporal facts, but does not make use of temporal expressions in text. Most importantly, both methods are limited to coping with unimodal distributions. So they cannot express that a football player was with the same club during two non-contiguous time-spans. In contrast, aggregation in this work can handle multimodal distributions.

7   Conclusion

Given the wealth of new knowledge graphs and knowledge harvesting efforts, we have proposed the novel task of summarizing temporal extractions. Our system achieves this by aggregating information in a temporally aware manner, supporting both semi-structured and textual sources. This leads to abstractive multi-document summaries beyond the capabilities of current summarization tools for text, opening up important new avenues of research on how to exploit extraction techniques in information retrieval and information management.

Acknowledgments. We thank the anonymous reviewers for their valuable comments. This project was sponsored by National Natural Science Foundation of China (No. 61503217), Shandong Provincial Natural Science Foundation of China (No. ZR2014FP002), and The Fundamental Research Funds of Shandong University (Nos. 2014TB005, 2014JC001).

References

1. Arora, R., Ravindran, B.: Latent Dirichlet allocation based multi-document summarization. In: Second Workshop on Analytics for Noisy Unstructured Text Data (AND), pp. 91–97. ACM (2008)
2. Auer, S., Bizer, C., Kobilarov, G., Lehmann, J., Cyganiak, R., Ives, Z.G.: DBpedia: a nucleus for a web of open data. In: Aberer, K., et al. (eds.) ASWC 2007 and ISWC 2007. LNCS, vol. 4825, pp. 722–735. Springer, Heidelberg (2007)
3. Bing, L., Li, P., Liao, Y., Lam, W., Guo, W., Passonneau, R.J.: Abstractive multi-document summarization via phrase selection and merging. In: ACL, pp. 1587–1597 (2015)
4. Carlson, A., Betteridge, J., Wang, R.C., Hruschka Jr., E.R., Mitchell, T.M.: Coupled semi-supervised learning for information extraction. In: WSDM (2010)
5. Chhabra, S., Bedathur, S.: Towards generating text summaries for entity chains. In: de Rijke, M., Kenter, T., de Vries, A.P., Zhai, C.X., de Jong, F., Radinsky, K., Hofmann, K. (eds.) ECIR 2014. LNCS, vol. 8416, pp. 136–147. Springer, Heidelberg (2014)

6. Conroy, J., O'leary, D.: Text summarization via hidden Markov models. In: SIGIR, pp. 406–407. ACM (2001)
7. Fader, A., Soderland, S., Etzioni, O.: Identifying relations for open information extraction. In: EMNLP, Edinburgh, Scotland, UK, pp. 1535–1545, 27–31 July 2011
8. Filippova, K.: Multi-sentence compression: finding shortest paths in word graphs. In: ACL, pp. 322–330 (2010)
9. Ganesan, K., Zhai, C., Han, J.: Opinosis: a graph-based approach to abstractive summarization of highly redundant opinions. In: ACL, pp. 340–348 (2010)
10. Hoffart, J., Yosef, M.A., Bordino, I., Fürstenau, H., Pinkal, M., Spaniol, M., Thater, S., Weikum, G.: Robust disambiguation of named entities in text. In: EMNLP, pp. 782–792 (2011)
11. Hong, K., Marcus, M., Nenkova, A.: System combination for multi-document summarization. In: EMNLP, pp. 107–117 (2015)
12. Knight, K., Marcu, D.: Summarization beyond sentence extraction: a probabilistic approach to sentence compression. Artif. Intell. **139**(1), 91–107 (2002)
13. Li, L., Zhou, K., Xue, G., Zha, H., Yu, Y.: Enhancing diversity, coverage and balance for summarization through structure learning. In: WWW, pp. 71–80. ACM (2009)
14. Ling, X., Weld, D.S.: Temporal information extraction. In: AAAI, pp. 1385–1390, 11–15 July 2010
15. Liu, F., Flanigan, J., Thomson, S., Sadeh, N.M., Smith, N.A.: Toward abstractive summarization using semantic representations. In: NAACL, pp. 1077–1086 (2015)
16. Mani, I.: Summarization evaluation: an overview (2001)
17. McClosky, D., Manning, C.D.: Learning constraints for consistent timeline extraction. In: EMNLP-CoNLL, pp. 873–882 (2012)
18. McDonald, D., Pustejovsky, J.: Natural language generation. In: IJCAI. Citeseer (1986)
19. Radev, D., Allison, T., Blair-Goldensohn, S., Blitzer, J., Celebi, A., Dimitrov, S., Drabek, E., Hakim, A., Lam, W., Liu, D., et al.: MEAD-a platform for multidocument multilingual text summarization. In: LREC, vol. 2004 (2004)
20. Schluter, N., Søgaard, A.: Unsupervised extractive summarization via coverage maximization with syntactic and semantic concepts. In: ACL, pp. 840–844 (2015)
21. Shen, D., Sun, J., Li, H., Yang, Q., Chen, Z.: Document summarization using conditional random fields. IJCAI **7**, 2862–2867 (2007)
22. Suchanek, F.M., Kasneci, G., Weikum, G.: YAGO: a core of semantic knowledge. In: WWW, pp. 697–706. ACM, New York (2007)
23. Sydow, M., Pikula, M., Schenkel, R.: The notion of diversity in graphical entity summarisation on semantic knowledge graphs. J. Intell. Inf. Syst. **41**(2), 109–149 (2013)
24. Takaku, Y., Kaji, N., Yoshinaga, N., Toyoda, M.: Identifying constant and unique relations by using time-series text. In: EMNLP-CoNLL, pp. 883–892 (2012)
25. Talukdar, P.P., Crammer, K.: New regularized algorithms for transductive learning. In: Buntine, W., Grobelnik, M., Mladenić, D., Shawe-Taylor, J. (eds.) ECML PKDD 2009, Part II. LNCS, vol. 5782, pp. 442–457. Springer, Heidelberg (2009)
26. Talukdar, P.P., Wijaya, D., Mitchell, T.: Coupled temporal scoping of relational facts. In: WSDM. Association for Computing Machinery, Seattle, February 2012
27. Tylenda, T., Sozio, M., Weikum, G.: Einstein: physicist or vegetarian? Summarizing semantic type graphs for knowledge discovery. In: WWW (Companion Volume), pp. 273–276 (2011)
28. Wan, X., Yang, J.: Multi-document summarization using cluster-based link analysis. In: SIGIR, pp. 299–306. ACM (2008)

29. Wang, Y., Dylla, M., Spaniol, M., Weikum, G.: Coupling label propagation and constraints for temporal fact extraction. In: ACL, vol. 2, pp. 233–237 (2012)
30. Wang, Y., Yahya, M., Theobald, M.: Time-aware reasoning in uncertain knowledge bases. In: MUD, pp. 51–65 (2010)
31. Wang, Y., Yang, B., Qu, L., Spaniol, M., Weikum, G.: Harvesting facts from textual web sources by constrained label propagation. In: CIKM, pp. 837–846 (2011)
32. Zhang, X., Cheng, G., Qu, Y.: Ontology summarization based on RDF sentence graph. In: WWW, pp. 707–716 (2007)

SuMGra: Querying Multigraphs via Efficient Indexing

Vijay Ingalalli[1,2(✉)], Dino Ienco[2], and Pascal Poncelet[1]

[1] Université de Montpellier, LIRMM, Montpellier, France
{vijay,pascal.poncelet}@lirmm.fr
[2] IRSTEA Montpellier, UMR TETIS, F-34093 Montpellier, France
dino.ienco@irstea.fr

Abstract. Many real world datasets can be represented by a network with a set of nodes interconnected with each other by multiple relations. Such a rich graph is called a multigraph. Unfortunately, all the existing algorithms for subgraph query matching are not able to adequately leverage multiple relationships that exist between the nodes. In this paper we propose an efficient indexing schema for querying single large multigraphs, where the indexing schema aptly captures the neighbourhood structure in the data graph. Our proposal SuMGra couples this novel indexing schema with a subgraph search algorithm to quickly traverse though the solution space to enumerate all the matchings. Extensive experiments conducted on real benchmarks prove the time efficiency as well as the scalability of SuMGra.

1 Introduction

Many real world datasets can be represented by a network with a set of nodes interconnected with each other by multiple relations. Such a rich graph is called multigraph and it allows different types of edges in order to represent different types of relations between vertices [1,2]. Example of multigraphs are: social networks spanning over the same set of people, but with different life aspects (e.g. social relationships such as Facebook, Twitter, LinkedIn, etc.); protein-protein interaction multigraphs created considering the pairs of proteins that have direct interaction/physical association or they are co-localised [15]; gene multigraphs, where genes are connected by considering the different pathway interactions belonging to different pathways; RDF knowledge graph where the same subject/object node pair is connected by different predicates [10].

One of the difficult operation in graph data management is subgraph querying [6]. Although subgraph querying is an NP-complete [6] problem, practically, we can find embeddings in real graph data by employing a good matching order and intelligent pruning rules. In literature, different families of subgraph matching algorithms exist. A first group of techniques employ *Feature based indexing* followed by a filtering and verification framework. During filtering, some graph patterns (subtrees or paths) are chosen as indexing features to minimize the number of candidate graphs. Then the verification step checks for the subgraph

© Springer International Publishing Switzerland 2016
S. Hartmann and H. Ma (Eds.): DEXA 2016, Part I, LNCS 9827, pp. 387–401, 2016.
DOI: 10.1007/978-3-319-44403-1_24

isomorphism using the selected candidates [4,11,14,16]. All these methods are developed for transactional graphs, i.e. the database is composed of a collection of graphs and each graph can be seen as a transaction of such database, and they cannot be trivially extended on the single multigraph scenario. A second family of approaches avoids indexing and it uses *Backtracking algorithms* to find embeddings by growing the partial solutions. In the beginning, they obtain a potential set of candidate vertices for every vertex in the query graph. Then a recursive subroutine called SUBGRAPHSEARCH is invoked to find all the possible embeddings of the query graph in the data graph [5,7,13]. All these approaches are able to manage graphs with only a single label on the vertex. Although index based approaches focus on transactional database graphs, some backtracking algorithms address the large single graph setting [9]. All these methods are not conceived to manage and query multigraphs and their extension to manage multiple relations between nodes cannot be trivial. A third and recent family of techniques defines *equivalence classes* at query and/or database level, by exploiting vertex relationships. Once the data vertices are grouped into equivalence classes, the search space is reduced and the whole process is speeded up [6,12].

Adapting these methods to multigraph is not straightforward since, the different types of relationships between vertices can exponentially increase the number of equivalent classes (for both query and data graph) thereby drastically reducing the efficiency of these strategies. Among the vast literature on subgraph isomorphism, [3] is the unique approach, which is a backtracking approach, that is able to directly manage graph with (multiple) labels on the edges. It proposes an approach called RI that uses light pruning rules in order to avoid visiting useless candidates.

Due to the availability of multigraph data and the importance of performing query on multigraph data, in this paper, we propose a novel method SUMGRA that supports subgraph matching in a multigraph via efficient indexing. In particular, we capture the multigraph properties in order to build the index structures, and we show that by exploiting multigraph properties, we are able to perform subgraph matching very efficiently. As observed in Table 1, the proposed SUMGRA is almost one order of magnitude better than the benchmark approach *RI*, for the DBPEDIA dataset and for the query sizes from 3 to 11.

Deviating from all the previous proposed approaches, we conceive an indexing schema to summarize information contained in a single large multigraph. SUMGRA involves two main phases: (i) an off-line phase that builds efficient indexes for the information contained in the multigraph; (ii) an on-line phase, where a search procedure exploits the indexing schema previously built.

Table 1. Time (msec) taken by *RI* and SUMGRA for DBPEDIA dataset

Approach	3	5	7	9	11
SUMGRA	160.5	254.1	545.7	971.5	1610.9
RI	7045.7	7940.8	7772.7	6872.6	8502.5

The rest of the paper is organized as follows. Background and problem definition are provided in Sect. 2. An overview of the proposed approach is presented in Sect. 3, while Sects. 4 and 5 describe the indexing schema and the query subgraph search algorithm, respectively. Section 6 presents experimental results. Conclusions are drawn in Sect. 7.

2   Preliminaries and Problem Definition

Formally, we can define a multigraph G as a tuple of four elements (V, E, L_E, D) where V is the set of vertices and D is the set of dimensions, $E \subseteq V \times V$ is the set of undirected edges and $L_E : V \times V \rightarrow 2^D$ is a labelling function that assigns the subset of dimensions to each edge it belongs to. In this paper, we address the sub-graph isomorphism problem for undirected multigraphs.

Definition 1. Subgraph isomorphism for undirected multigraph. *Given a multi-graph* $Q = (V^q, E^q, L_E^q, D^q)$ *and a multigraph* $G = (V, E, L_E, D)$, *the subgraph isomorphism from* Q *to* G *is an injective function* $\psi : V^q \rightarrow V$ *such that:*

$$\forall (u_m, u_n) \in E^q, \exists (\psi(u_m), \psi(u_n)) \in E \text{ and } L_E^q(u_m, u_n) \subseteq L_E(\psi(u_m), \psi(u_n)).$$

Problem Definition. Given a query multigraph Q and a data multigraph G, the subgraph query problem is to enumerate all the embeddings of Q in G.

For the ease of representation, in the rest of the paper, we simply refer to a data multigraph G as a *graph*, and a query multigraph Q as a *subgraph*. We also enumerate (for unique identification) the set of query vertices by U and the set of data vertices by V.

In Fig. 1, we introduce a query multigraph Q and a data multigraph G. The two valid embeddings for the subgraph Q are marked by the thick lines in the graph G and are enumerated as follows: $R_1 := \{[u_1, v_4], [u_2, v_5], [u_3, v_3], [u_4, v_1]\}$; $R_2 := \{[u_1, v_4], [u_2, v_3], [u_3, v_5], [u_4, v_6]\}$; where, each query vertex u_i is matched to a distinct data vertex v_j, written as $[u_i, v_j]$.

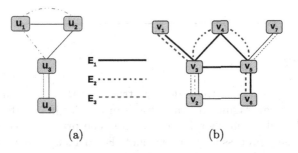

(a)   (b)

Fig. 1. A sample (a) query multigraph Q and (b) data multigraph G

3   An Overview of SuMGra

In this section, we sketch the main idea behind our proposal. The entire procedure can be divided into two parts: (i) an indexing schema for the graph G that exploits edge dimensions and the vertex neighbourhood structure (Sect. 4) (ii) a subgraph search algorithm, that integrates recent advances in the graph data management field, to enumerate the embeddings of the subgraph (Sect. 5).

The overall idea of SuMGra is depicted in Algorithm 1. Initially, we order the set of query vertices U using a heuristic proposed in Sect. 5.1. With an ordered set of query vertices U^o, we use the indexing schema to find a list of possible candidate matches only for the initial query vertex u_{init} by calling SELECT-CAND (Line 5), as described in Sect. 5.2. Then, for each possible candidate of the initial query vertex, we call the recursive subroutine SUBGRAPHSEARCH, that performs the subgraph isomorphism test.

The SUBGRAPHSEARCH procedure (Sect. 5.3), finds the embeddings starting with the possible matches for the initial query vertex u_{init} (Lines 7–11). Since u_{init} has $|C_{u_{init}}|$ possible matches, SUBGRAPHSEARCH iterates through $|C_{u_{init}}|$ solution trees in a depth first manner until an embedding is found. That is, SUBGRAPHSEARCH is recursively called to find the matchings that correspond to all ordered query vertices U^o. The partial embedding is stored in $M = [M_q, M_g]$ - a pair that contains the already matched query vertices M_q and the already matched data vertices M_g. Once the partial embedding grows to become a complete embedding, the repository of embeddings R is updated.

Algorithm 1. SuMGra

---
1 INPUT: subgraph Q, graph G, indexes $\mathcal{S}$, $\mathcal{N}$
2 OUTPUT: R: all the embeddings of Q in G
3 $U^o = $ ORDERQUERYVERTICES(Q, G)
4 $u_{init} = u | u \in U^o$
5 $C_{u_{init}} = $ SELECTCAND$(u_{init}, \mathcal{S})$
6 $R = \emptyset$ /* Embeddings of Q in G */
7 **for** each $v_{init} \in C_{u_{init}}$ **do**
8      $M_q = u_{init};$ /* Matched initial query vertex */
9      $M_d = v_{init};$ /* Matched possible data vertex */
10     $M = [M_q, M_g]$ /* Partial matching of Q in G */
11   UPDATE: $R := $ SUBGRAPHSSEARCH$(R, M, \mathcal{N}, Q, G, U^o)$

12 **return** R

---

4   Indexing

In this section, we propose the indexing structures that are built on the data multigraph G, by leveraging the multigraph properties in specific; this index is used during the subgraph querying procedure. The primary goal of indexing is to make the query processing time efficient. For a lucid understanding of our indexing schema, we introduce a few definitions.

Definition 2. Vertex signature. *For a vertex v, the vertex signature $\sigma(v)$ is a multiset containing all the multiedges that are incident on v, where a multiedge* *between v and a neighbouring vertex v' is represented by a set that corresponds to edge dimensions. Formally, $\sigma(v) = \bigcup_{v' \in N(v)} L_E(v, v')$ where $N(v)$ is the set of neighbourhood vertices of v, and $\cup$ is the union operator for multiset.*

For instance, in Fig. 1(b), $\sigma(v_6) = \{\{E_1, E_3\}, \{E_1\}\}$. The vertex signature is an intermediary representation that is exploited by our indexing schema.

The goal of constructing indexing structures is to find the *possible candidate set* for the set of query vertices u, thereby reducing the search space for the SUBGRAPHSEARCH procedure, making SuMGra time efficient.

Definition 3. Candidate set. *For a query vertex u, the candidate set $C(u)$ is defined as $C(u) = \{v \in g | \sigma(u) \subseteq \sigma(v)\}$.*

In this light, we propose two indexing structures that are built offline: (i) given the vertex signature of all the vertices of graph G, we construct a vertex signature index $\mathcal{S}$ by exploring a set of features f of the signature $\sigma(v)$ (ii) we build a vertex neighbourhood index $\mathcal{N}$ for every vertex in the graph G. The index $\mathcal{S}$ is used to select possible candidates for the initial query vertex in the SELECT-CAND procedure while the index $\mathcal{N}$ is used to choose the possible candidates for the rest of the query vertices during the SUBGRAPHSEARCH procedure.

4.1   Vertex Signature Index $\mathcal{S}$

This index is constructed to enumerate the possible candidate set only for the initial query vertex. Since we cannot exploit any structural information for the initial query vertex, $\mathcal{S}$ captures the edge dimension information from the data vertices, so that the non suitable candidates can be pruned away.

We construct the index $\mathcal{S}$ by organizing the information supplied by the vertex signature of the graph; i.e., observing the vertex signature of data vertices, we intend to extract some interesting features. For example, the vertex signature of v_6, $\sigma(v_6) = \{\{E_1, E_3\}, \{E_1\}\}$ has two sets of dimensions in it and hence v_6 is eligible to be matched with query vertices that have at most two sets of items in their signature. Also, $\sigma(v_2) = \{\{E_2, E_3, E_1\}, \{E_1\}\}$ has the edge dimension set of maximum size 3 and hence a query vertex must have the edge dimension set size of at most 3. More such features (e.g., the number of unique dimensions, the total number of occurrences of dimensions, etc.) can be proposed to filter out irrelevant candidate vertices. In particular, for each vertex v, we propose to extract a set of characteristics summarizing useful features of the neighbourhood of a vertex. Those features constitute a *synopses* representation (surrogate) of the original vertex signature.

In this light, we propose six $|f| = 6$ features, that leverage the multigraph properties; the features will be illustrated with the help of the vertex signature $\sigma(v_3) = \{\{E_1, E_2, E_3\}, \{E_1, E_3\}, \{E_1, E_2\}, \{E_1\}\}$:

f_1 Cardinality of vertex signature, $(f_1(v_3) = 4)$
f_2 The number of unique dimensions in the vertex signature, $(f_2(v_3) = 3)$
f_3 The number of all occurrences of the dimensions (repetition allowed), $(f_3(v_3) = 8)$

f_4 Minimum index of the lexicographically ordered edge dimensions, $(f_4(v_3) = 1)$
f_5 Maximum index of the lexicographically ordered edge dimensions, $(f_5(v_3) = 3)$
f_6 Maximum cardinality of the vertex sub-signature, $(f_6(v_3) = 3)$

By exploiting the aforementioned features, we build the synopses to represent the vertices in an efficient manner that will help us to select the eligible candidates during query processing.

Once the synopsis representation for each data vertex is computed, we store the synopses in an efficient data structure. Since each vertex is represented by a synopsis of several fields, a data structure that helps in efficiently performing range search for multiple elements would be an ideal choice. For this reason, we build a $|f|$-dimensional R-tree, whose nodes are the synopses having $|f|$ fields.

The general idea of using an R-tree structure is as follows: A synopses $F = \{f_1, \ldots, f_{|f|}\}$ of a data vertex spans an axes-parallel rectangle in an f-dimensional space, where the maximum co-ordinates of the rectangle are the values of the synopses fields $(f_1, \ldots, f_{|f|})$, and the minimum co-ordinates are the origin of the rectangle (filled with zero values). For example, a data vertex represented by the synopses with two features $F_v = (2, 3)$ spans a rectangle in a 2-dimensional space in the interval range $([0, 2], [0, 3])$. Now if we consider synopses of two query vertices, $F_{u_1} = (1, 3)$ and $F_{u_2} = (1, 4)$, we observe that the rectangle spanned by F_{u_1} is wholly contained in the rectangle spanned by F_v but F_{u_2} is not wholly contained in F_v. Formally, the possible candidates for vertex u can be written as $\mathcal{P}(u) = \{v | \forall_{i \in [1, \ldots, f]} F_{u(i)} \leq F_{v(i)}\}$, where the constraints are met for all the $|f|$-dimensions. Since we apply the same inequality constraint to all the fields, we need to pre-process few synopses fields; e.g., the field f_4 contains the minimum value of the index, and hence we negate f_4 so that the rectangular containment problem still holds good. Thus, we keep on inserting the synopses representations of each data vertex v into the R-tree and build the index $\mathcal{S}$, where each synopses is treated as an $|f|$-dimensional node of the R-tree.

4.2   Vertex Neighbourhood Index $\mathcal{N}$

The aim of this indexing structure is to find the possible candidates for the rest of the query vertices.

Since the previous indexing schema enables us to select the possible candidate set for the initial query vertex, we propose an index structure to obtain the possible candidate set for the subsequent query vertices. The index $\mathcal{N}$ will help us to find the possible candidate set for a query vertex u during the SUBGRAPH-SEARCH procedure by retaining the structural connectivity with the previously matched candidate vertices, while discovering the embeddings of the subgraph Q in the graph G.

The index $\mathcal{N}$ comprises of neighbourhood trees built for each of the data vertex v. To understand the index structure, let us consider the data vertex v_3 from Fig. 1(b), shown separately in Fig. 2(a). For this vertex v_3, we collect all the neighbourhood information (vertices and multiedges), and represent this information by a tree structure. Thus, the tree representation of a vertex v contains the neighbourhood vertices and their corresponding multiedges, as shown in Fig. 2(b), where the nodes of the tree structure are represented by the edge dimensions.

In order to construct an efficient tree structure, we propose the structure - Ordered Trie with Inverted List (OTIL). Consider a data vertex v_i, with a set of n neighbourhood vertices $N(v_i)$. Now, for every pair $(v_i, N^j(v_i))$, where $j \in \{1, \ldots, n\}$, there exists a multiedge (set of edge dimensions) $\{E_1, \ldots, E_d\}$, which is inserted into the OTIL structure. Each multiedge is ordered (with the increasing edge dimensions), before inserting into OTIL structure, and the order is universally maintained for both query and data vertices. Further, for every edge dimension E_i that is inserted into the OTIL, we maintain an *inverted list* that contains all the neighbourhood vertices $N(v_i)$, that have the edge dimension E_i incident on them. For example, as shown in Fig. 2(b), the edge E_2 will contain the list $\{v_2, v_4\}$, since E_2 forms an edge between v_3 and both v_2 and v_4.

To construct the OTIL index as shown in Fig. 2(b), we insert each ordered multiedge that is incident on v at the root of the trie structure. To make index querying more time efficient, the OTIL nodes with identical edge dimension (e.g., E_3) are internally connected and thus form a linked list of data vertices. For example, if we want to query the index in Fig. 2(b) with a vertex having edges $\{E_1, E_3\}$, we do not need to traverse the entire OTIL. Instead, we perform a pre-ordered search, and as soon as we find the first set of matches, which is $\{V_2\}$, we will be redirected to the OTIL node, where we can fetch the matched vertices much faster (in this case $\{V_1\}$), thereby outputting the set of matches as $\{V_2, V_1\}$.

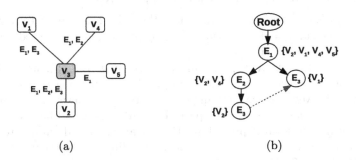

(a)   (b)

Fig. 2. (a) Neighbourhood structure of v_3 and (b) Neighbourhood index for vertex v_3

5   Subgraph Query Processing

We now proceed with the subgraph query processing. In order to find the embeddings of a subgraph, we not only need to find the valid candidates for each query vertex, but also retain the structure of the subgraph to be matched.

5.1   Query Vertex Ordering

Before performing query processing, we order the set of query vertices U into an ordered set of query vertices U^o. It is argued that an effective ordering of the query vertices improves the efficiency of subgraph querying [9]. In order to achieve this, we propose a heuristic that employs two scoring functions.

The first scoring function relies on the number of multiedges of a query vertex. For each query vertex u_i, the number of multiedges incident on it is assigned as a score; i.e., $r_1(u_i) = \sum_{j=1}^{m} |\sigma(u_i^j)|$, where u_i has m multiedges, $|\sigma(u_i^j)|$ captures the number of edge dimensions in the jth multiedge. Query vertices are ordered in ascending order considering the scoring function r_1, and thus $u_{init} = \text{argmax}(r_1(u_i))$. For example, in Fig. 1(a), vertex u_3 has the maximum number of edges incident on it, which is 4, and hence is chosen as an initial vertex.

The second scoring function depends on the structure of the subgraph. We maintain an ordered set of query vertices U^o and keep adding the next *eligible* query vertex. In the beginning, only the initial query vertex u_{init} is in U^o. The set of next eligible query vertices U^o_{nbr} are the vertices that are in the 1-neighbourhood of U^o. For each of the next eligible query vertex $u_n \in U^o_{nbr}$, we assign a score depending on a second scoring function defined as $r_2(u_n) = |\{U^o \cap adj(u_n)\}|$. It considers the number of the adjacent vertices of u_n that are present in the already ordered query vertices U^o.

Then, among the set of next eligible query vertices U^o_{nbr} for the already ordered U^o, we give first priority to function r_2 and the second priority to function r_1. Thus, in case of any tie ups, w.r.t. r_2, the score of r_1 will be considered. When both r_2 and r_1 leave us in a tie up situation, we break such tie at random.

5.2   Select Candidates for Initial Query Vertex

For the initial query vertex u_{init}, we exploit the index structure $\mathcal{S}$ to retrieve the set of possible candidate data vertices, thereby pruning the unwanted candidates for the reduction of search space.

During the SELECTCAND procedure (Algorithm 1, Line 5), we retrieve the possible candidate vertices from the data graph by exploiting the vertex signature index $\mathcal{S}$. However, since querying $\mathcal{S}$ would not prune away all the unwanted vertices for u_{init}, the corresponding partial embeddings would be discarded during the SUBGRAPHSEARCH procedure. For instance, to find candidate vertices for $u_{init} = u_3$, we build the synopses for u_3 and find the matchable vertices in G using the index $\mathcal{S}$. As we recall, synopses representation of each data vertex spans a rectangle in the d-dimensional space. Thus, it remains to check, if the rectangle spanned by u_3 is contained in any of rectangles spanned by the synopses of the data vertices, with the help of R-tree built on data vertices, which results in the candidate set $\{v_3, v_5\}$.

5.3   Subgraph Searching

The SUBGRAPHSEARCH recursive procedure is described in Algorithm 2. Once an initial query vertex u_{init} and its possible data vertex $v_{init} \in C_{u_{init}}$, that could be a potential match, is chosen from the set of select candidates, we have the partial

Algorithm 2. SUBGRAPHSEARCH($R, M, \mathcal{N}, Q, G, U^o$)

```
1  FETCH u_nxt ∈ U^o                              /* Fetch query vertex to be matched */
2  M_C = FINDJOINABLE(M_q, M_g, N, u_nxt)         /* Matchable candidate vertices */
3  if |M_C| ≠ ∅ then
4  |   for each v_nxt ∈ M_C do
5  |   |   M_q = M_q ∪ u_nxt;
6  |   |   M_g = M_g ∪ v_nxt;
7  |   |   M = [M_q, M_g]                          /* Partial matching grows */
8  |   |   SUBGRAPHSEARCH(R, M, N, Q, G, U^o)
9  |   |   if (|M| == |U^o|) then
10 |   |   |   R = R ∪ M                           /* Embedding found */

11 return R
```

solution pair $M = [M_q, M_g]$ of the subgraph query pattern we want to grow. If v_{init} is a right match for u_{init}, and we succeed in finding the subsequent valid matches for U^o, we will obtain an embedding; else, the recursion would revert back and move on to next possible data vertex to look for the embeddings.

In the beginning of SUBGRAPHSEARCH procedure, we fetch the next query vertex u_{nxt} from the set of ordered query vertices U^o, that is to be matched (Line 1). Then FINDJOINABLE procedure finds all the valid data vertices that can be matched with the next query vertex u_{nxt} (Line 2). The main task of subgraph matching is done by the FINDJOINABLE procedure, depicted in Algorithm 3. Once all the valid matches for u_{nxt} are obtained, we update the solution pair $M = [M_q, M_g]$ (Line 5–7). Then we recursively call SUBGRAPHSEARCH procedure until all the vertices in U^o have been matched (Line 8). If we succeed in finding matches for the entire set of query vertices U^o, then we update the repository of embeddings (Line 9–10); else, we keep on looking for matches recursively in the search space, until there are no possible candidates to be matched for u_{nxt} (Line 3).

Algorithm 3. FINDJOINABLE($M_q, M_g, \mathcal{N}, u_{nxt}$)

```
1  A_q := M_q ∩ adj(u_nxt)                         /* Matched query neighbours */
2  A_g := {v|v ∈ M_g}                              /* Corresponding matched data neighbours */
3  INTIALIZE: M_C^temp = 0, M_C = 0
4  M_C^temp = ∩_{i=1}^{|A_q|} NEIGHINDEXQUERY(N, A_g^i, (A_q^i, u_nxt))
5  for each v_c ∈ M_C^temp do
6  |   if σ(v_c) ⊇ σ(u_nxt) then
7  |   |   add v_c to M_C                          /* A valid matchable vertex */

8  return M_C
```

The FINDJOINABLE procedure guarantees the structural connectivity of the embeddings that are outputted. Referring to Fig. 1, let us assume that the already matched query vertices $M_q = \{u_2, u_3\}$ and the corresponding matched data vertices $M_g = \{v_3, v_5\}$, and the next query vertex to be matched $u_{nxt} = u_1$. Initially, in the FINDJOINABLE procedure, for the next query vertex u_{nxt}, we collect all the neighbourhood vertices that have been already matched, and store them in A_q; formally, $A_q := M_q \cap adj(u_{nxt})$ and also collect the corresponding matched data vertices A_g (Line 1–2). For instance, for the next query vertex u_1, $A_q = \{u_2, u_3\}$ and correspondingly, $A_g = \{v_3, v_5\}$.

Now we exploit the neighbourhood index $\mathcal{N}$ in order to find the valid matches for the next query vertex u_{nxt}. With the help of vertex $\mathcal{N}$, we find the possible candidate vertices M_C^{temp} for each of the matched query neighbours A_q^i and the corresponding matched data neighbour A_g^i.

To perform querying on the index structure $\mathcal{N}$, we fetch the multiedge that connects the next matchable query vertex u_{nxt} and the ith previously matched query vertex A_q^i. We now take the multiedge (A_q^i, u_{nxt}) and query the index structure $\mathcal{N}$ of the correspondingly matched data vertex A_g^i (Line 4). For instance, with $A_q^i = u_2$, and $u_{nxt} = u_1$ we have a multiedge $\{E_1, E_2\}$. As we can recall, each data vertex v_j has its neighbourhood index structure $\mathcal{N}(v_j)$, represented by an OTIL structure. The elements that are added to OTIL are nothing but the multiedges that are incident on the vertex v_j, and hence the nodes in the tree are nothing but the edge dimensions. Further, each of these edge dimensions (nodes) maintain a list of neighbourhood (adjacent) data vertices of v_j that contain the particular edge dimension as depicted in Fig. 2(b). Now, when we look up for the multiedge (A_q^i, u_{nxt}), which is nothing but a set of edge dimensions, in the OTIL structure $\mathcal{N}(A_g^i)$, two possibilities exist. (1) The multiedge (A_q^i, u_{nxt}) has no matches in $\mathcal{N}(A_g^i)$ and hence, there are no matchable data vertices for the next query vertex u_{nxt}. (2) The multiedge (A_q^i, u_{nxt}) has matches in $\mathcal{N}(A_g^i)$ and hence, NEIGHINDEXQUERY returns a set of possible candidate vertices M_C^{temp}. The set of vertices M_C^{temp}, present in the OTIL structure as a linked list, are the possible data vertices since, these are the neighbourhood vertices of the already matched data vertex A_g^i, and hence the structure is maintained. For instance, multiedge $\{E_1, E_2\}$ has a set of matched vertices $\{v_2, v_4\}$ as we can observe in Fig. 2(a).

Further, we check if the next possible data vertices are maintaining the structural connectivity with all the matched data neighbours A_g, that correspond to matched query vertices A_q, and hence we collect only those possible candidate vertices M_C^{temp}, that are common to all the matched data neighbours with the help of intersection operation $\cap$. Thus we repeat the process for all the matched query vertices A_q and the corresponding matched data vertices A_g to ensure structural connectivity (Line 4). For instance, with $A_q^1 = u_2$ and corresponding $A_g^1 = v_3$, we have $M_C^{temp1} = \{v_2, v_4\}$; with $A_q^2 = u_3$ and corresponding $A_g^2 = v_5$, we have $M_C^{temp2} = \{v_4\}$, since the multiedge between (A_q^i, u_{nxt}) is $\{E_2\}$. Thus, the common vertex v_4 is the one that maintains the structural connectivity, and hence belongs to the set of matchable candidate vertices $M_C^{temp} = v_4$.

The set of matchable candidates M_C^{temp} are the valid candidates for u_{nxt} both in terms of edge dimension matching and the structural connectivity with the already matched partial solution. However, at this point, we propose a strategy that predicts whether the further growth of the partial matching is possible, w.r.t. to the neighbourhood of already matched data vertices, thereby pruning the search space. We can do this by checking the condition whether the vertex signature $\sigma(u_{nxt})$ is contained in the vertex signature of $v \in M_C^{temp}$ (Line 11–13). This is possible since, the vertex signature σ contains the multiedge information about the unmatched query vertices that are in the neighbourhood of already matched data vertices. For instance, v_4 can be qualified as M_C since $\sigma(v_4)$

$\supseteq \sigma(u_1)$. That is, considering the fact that we have found a match for u_1, which is v_4, and that the next possible query vertex is u_4, the superset containment check will assure us the connectivity (in terms of edge dimensions) with the next possible query vertex u_4. Suppose a possible candidate data vertex fails this superset containment test, it means that, the data vertex will be discarded by FINDJOINABLE procedure in the next iteration, and we are avoiding this useless step in advance, thereby making the search more time efficient.

In order to efficiently address the superset containment problem between the vertex signatures $\sigma(v_c)$ and $\sigma(u_{nxt})$, we model this task as a maximum matching problem on a bipartite graph [8]. Basically, we build a bipartite graph whose nodes are the sub-signatures of $\sigma(v_c)$ and $\sigma(u_{nxt})$; and an edge exists between a pair of nodes only if the corresponding sub-signatures do not belong to the same signature, and the ith sub-signature of v_c is a superset of jth sub-signature of u_{nxt}. This construction ensures to obtain at the end a bipartite graph. Once the bipartite graph is built we run a maximum matching algorithm to find a maximum match between the two signatures. If the size of the maximum match found is equal to the size of $\sigma(u_{nxt})$, the superset operation returns true otherwise $\sigma(u_{nxt})$ is not contained in the signature $\sigma(v_c)$. To solve the maximum matching problem on the bipartite graph, we employ the *Hopcroft-Karp* [8] algorithm.

6 Experimental Evaluation

In this section, we evaluate the performance of SUMGRA on real multigraphs.

We evaluate the performance of SUMGRA by comparing it with two baseline approaches (its own variants) and a competitor *RI* [3]. The two baseline approaches are: (i) SUMGRA-No-SC that does not consider the vertex signature index S and it initializes the candidate set of the initial vertex $C(u_{init})$ with the whole set of data nodes; (ii) SUMGRA-Rand-Order that consider all the indexing structure but it employs a random ordering of the query vertices preserving connectivity. The *RI* approach is able to manage graphs with multi-edges, and we obtain the implementation from the original authors. For the purpose of evaluation. we consider three real world multigraphs: *DBLP* data set built by following the procedure adopted in [1]; *FLICKR*[1] crawled from Flickr, which is an image and video hosting website, web services suite, and an online community; *DBPEDIA*[2] that is the well-known knowledge base built by the Semantic Web Community. For *DBLP*, vertices correspond to different authors and each dimensions represent one of the top 50 Computer Science conferences. Two authors are connected over a dimension if they co-authored at least one paper together in that conference. In *FLICKR*, users are represented by nodes, and blogger's friends are represented using edges. Multiple edges exist between two users if they have common multiple memberships. The RDF format in which *DBPEDIA* is stored can naturally be modeled as a multigraph where vertices are subjects and objects of the RDF triplets and edges represent the predicates between them. Benchmark characteristics are reported in Table 2.

---

[1] http://socialcomputing.asu.edu/pages/datasets.

[2] http://dbpedia.org/.

Table 2. Benchmark statistics.

Dataset	Nodes	Edges	Dim	Density
DBLP	83 901	141 471	50	4.0e-5
FLICKR	80 513	5 899 882	195	1.8e-3
DBPEDIA	4 495 642	14 721 395	676	1.4e-6

Table 3. Index construction time (secs.).

Dataset	S	$\mathcal{N}$
DBLP	1.15	0.37
FLICKR	1.55	8.89
DBPEDIA	64.51	66.59

To test the behavior of the different approaches, we generate *random* queries [7,13] varying their size (in terms of vertices) from 3 to 11 in steps of 2. All the generated queries contain one (or more) edge with at least two dimensions. In order to generate queries that can have at least one embedding, we sample them from the corresponding multigraph. For each dataset and query size we obtain 1 000 samples. Following the methodology previously proposed [6,11], we report the average time values considering the first 1 000 embeddings for each query. It should be noted that the queries returning no answers were not counted in the statistics (the same statistical strategy has been used by [7,11]).

All the experiments were run on a server, with 64-bit Intel 6 processors @ 2.60 GHz, and 250 GB RAM, running on a Linux OS - Ubuntu. Our methods have been implemented using C++.

6.1   Performance of SuMGra

Table 3 reports the index construction time of SuMGRA for each of the employed dataset. As we can observe for the bigger datasets like *FLICKR*, and *DBPEDIA*, construction of the index $\mathcal{N}$ takes more time when compared to the construction of S. This happens due to either huge number of edges, or nodes or both in these two datasets. For *DBLP* we can observe the opposite phenomenon. This can be explained by the small number of edges and dimensions present in this dataset. Among all the datasets, *DBPEDIA* is the most expensive dataset in terms of indices construction but it always remains reasonable as time consumption for the off-line step is around one minute for each index.

Query Processing Time. Figures 3, 4 and 5 summarize time results. All the times we report are in milliseconds; the Y-axis (logarithmic in scale) represents the query matching time; the X-axis represents the increasing query sizes.

We also analyse the time performance of SuMGRA by varying the number of edge dimensions in the subgraph. In particular, we perform experiments for query multigraphs with two different edge dimensions: $d = 2$ and $d = 4$. That is, a query with $d = 2$ has at least one edge that exists in at least 2 dimensions. The same analogy applies to the queries with $d = 4$.

For *DBLP* dataset, we observe in Fig. 3 that SuMGRA performs the best in all the situations, and in fact it outperforms the other approaches by a huge margin. This happens thanks to both: a rigorous pruning of candidate vertices

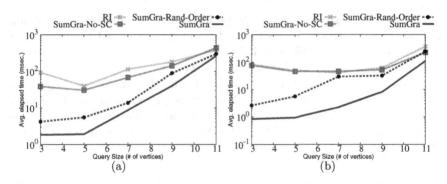

Fig. 3. Query time on *DBLP* dataset for (a) with $d = 2$ (b) with $d = 4$

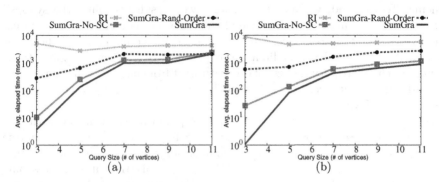

Fig. 4. Query time on *FLICKR* dataset for (a) with $d = 2$ (b) with $d = 4$

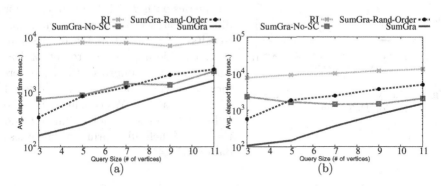

Fig. 5. Query time on *DBPEDIA* dataset (a) for $d = 2$ (b) $d = 4$

for initial query vertex as underlined by the gain w.r.t. SUMGRA-No-SC and an efficient query vertex ordering strategy as highlighted by the difference w.r.t. SUMGRA-Rand-Order. For the *FLICKR* dataset (Fig. 4) SUMGRA, SUMGRA-No-SC and SUMGRA-Rand-Order outperform *RI*. For many query instances, especially for *FLICKR*, SUMGRA-No-SC obtains better performance than RI while SuMGra still outperforms competitors. We can observe that random query ordering drastically affects the performance pointing out the importance of this step. Moving to *DBPEDIA* dataset in Fig. 5, we observe a significant deviation between *RI* and SuMGra, with SuMGra winning by a huge margin.

To conclude, we note that SuMGra outperforms the considered competitors, for all the employed benchmarks for all query size. Its performance is reported as best for multigraphs having many edge dimensions - *FLICKR* and high sparsity - *DBPEDIA*. Thus, we highlight that SuMGra is robust in terms of time performance varying both the query size and dimensions.

Assessing the Set of Synopses Features. In this experiment we assess the quality of the features composing the synopses representation for our indexing schema. To this end, we vary the features we consider to build the synopsis representation to understand if some of them can be redundant and/or do not improve the final performance. Since visualizing the combination of the whole set of features will be hard, we limit this experiment to a subset of combinations. Hence, we choose to vary the size of the feature set from one to six, by considering the order defined in Sect. 4.1. Using all the six features results in the proposed approach SuMGra.

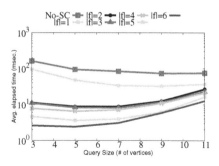

Fig. 6. Query time with varying synopses fields for *DBLP* with $d = 4$

We perform experiments with different configurations that have varying number of synopses features; for instance $|f| = 3|$ means that it considers only first three features to build synopses. Although we report plots only for *DBLP* for queries with $d = 4$, the behaviour for different datasets has similar behaviour. Results are reported in Fig. 6. We note that, considering the entire set of features drastically improves the time performance, when compared to a subset of these six features. We conclude that the different features are not redundant and they are all helpful in pruning the useless data vertices.

7 Conclusion

We proposed an efficient strategy to support Subgraph Matching in a Multigraph via efficient indexing. The proposed indexing schema leverages the rich structure available in the multigraph. The different indexes are exploited by a subgraph search procedure that works on multigraphs. The experimental section highlights the efficiency, versatility and scalability of our approach over different real datasets. The comparison with a state of the art approach points out the necessity to develop specific techniques to manage multigraphs.

As a future work, we are interesting in testing new synopses features as well as try novel vertex ordering strategies more rigorously. Further, we will be addressing dynamic multigraphs where nodes and multiedges are being added or removed over time.

Acknowledgments. This work has been funded by Labex NUMEV (NUMEV, ANR-10-LABX-20).

References

1. Boden, B., Günnemann, S., Hoffmann, H., Seidl, T.: Mining coherent subgraphs in multi-layer graphs with edge labels. In: KDD, pp. 1258–1266 (2012)
2. Bonchi, F., Gionis, A., Gullo, F., Ukkonen, A.: Distance oracles in edge-labeled graphs. In: EDBT, pp. 547–558 (2014)
3. Bonnici, V., Giugno, R., Pulvirenti, A., Shasha, D., Ferro, A.: A subgraph isomorphism algorithm and its application to biochemical data. BMC Bioinform. **14**(S–7), S13 (2013)
4. Cheng, J., Ke, Y., Ng, W., Lu, A.: Fg-index: towards verification-free query processing on graph databases. In: SIGMOD, pp. 857–872. ACM (2007)
5. Cordella, L.P., Foggia, P., Sansone, C., Vento, M.: A (sub) graph isomorphism algorithm for matching large graphs. IEEE TPAMI **26**(10), 1367–1372 (2004)
6. Han, W.-S., Lee, J., Lee, J.-H.: Turbo ISO: towards ultrafast and robust subgraph isomorphism search in large graph databases. In: SIGMOD, pp. 337–348. ACM (2013)
7. He, H., Singh, A.K.: Graphs-at-a-time: query language and access methods for graph databases. In: SIGMOD, pp. 405–418. ACM (2008)
8. Hopcroft, J.E., Karp, R.M.: An $n^{5/2}$ algorithm for maximum matchings in bipartite graphs. SIAM J. Comput. **2**(4), 225–231 (1973)
9. Lee, J., Han, W.-S., Kasperovics, R., Lee, J.-H.: An in-depth comparison of subgraph isomorphism algorithms in graph databases. In: PVLDB, pp. 133–144 (2012)
10. Libkin, L., Reutter, J., Vrgoč, D.: Trial for RDF: adapting graph query languages for RDF data. In: PODS, pp. 201–212. ACM (2013)
11. Lin, Z., Bei, Y.: Graph indexing for large networks: a neighborhood tree-based approach. Knowl. Based Syst. **72**, 48–59 (2014)
12. Ren, X., Wang, J.: Exploiting vertex relationships in speeding up subgraph isomorphism over large graphs. PVLDB **8**(5), 617–628 (2015)
13. Shang, H., Zhang, Y., Lin, X., Yu, J.X.: Taming verification hardness: an efficient algorithm for testing subgraph isomorphism. PVLDB **1**(1), 364–375 (2008)
14. Yan, X., Yu, P.S., Han, J.: Graph indexing: a frequent structure-based approach. In: SIGMOD, pp. 335–346. ACM (2004)
15. Zhang, A.: Protein Interaction Networks: Computational Analysis. Cambridge University Press, Cambridge (2009)
16. Zhao, X., Xiao, C., Lin, X., Wang, W., Ishikawa, Y.: Efficient processing of graph similarity queries with edit distance constraints. VLDB J. **22**(6), 727–752 (2013)

Semantic Web, and Data Semantics

Re-constructing Hidden Semantic Data Models by Querying SPARQL Endpoints

María Jesús García-Godoy[✉], Esteban López-Camacho,
Ismael Navas-Delgado, and José F. Aldana-Montes

Departamento de Lenguaje y Ciencias de la Computación, Universidad de Málaga,
Andalucía Tech, Ada Byron Research Building, 29071 Málaga, Spain
{mjgarciag,jfam}@lcc.uma.es
http://khaos.uma.es/

Abstract. Linked Open Data community is constantly producing new repositories that store information from different domains. The data included in these repositories follow the rules proposed by the W3C community, based on standards such as Resource Description Framework (RDF) and the SPARQL query language. The main advantage of this approach is the possibility of external developers accessing the data from their applications. This advantage is also one of the main challenges of this new technology due to the cost of exploring how the data is structured in a given repository in order to construct SPARQL queries to retrieve useful information. According to the reviewed literature, there are no applications to reconstruct the underlying semantic data models from an SPARQL endpoint. In this paper, we present an application for the reconstruction of the data model as an OWL (Ontology Web Language) ontology. This application, available as Open Source at http://github.com/estebanpua/ontology-endpoint-extraction uses a set of SPARQL queries to discover the classes and the (object and data) properties for a given RDF database. A web application interface has also been implemented for users to browse through classes, properties of the ontology generated from the data structure (http://khaos.uma. es/oee). The ontologies generated by this application can help users to understand how the information is semantically organized, making easier the design of SPARQL queries.

Keywords: Semantic Web · Ontology · OWL · Linked Open Data · Endpoint

1 Introduction

Linked (Open) Data has become a set of repositories that stores information from different domains and interlinks related data that has not been previously linked [6]. This approach makes possible to integrate or solve queries accessing data repositories retrieving the requested information. Linked Data is based on the Semantic Web principles proposed by the W3C community of using standards

© Springer International Publishing Switzerland 2016
S. Hartmann and H. Ma (Eds.): DEXA 2016, Part I, LNCS 9827, pp. 405–415, 2016.
DOI: 10.1007/978-3-319-44403-1_25

such as RDF to formally describe the data [7] and SPARQL as query language
[10]. Following these principles, data is published in RDF and accessible through
SPARQL queries.

The building of an RDF database is guided by a semantic model that provides
all the elements to represent the RDF graphs. The semantic model defines the
Classes and Properties that can be used in the description of the data. Thus,
SPARQL queries use these elements to build the query graph.

VoID[1] is an RDF Schema vocabulary for expressing metadata about RDF
datasets. This vocabulary aims at connecting the publishers and users of RDF
data. This standard provides two mechanisms to provide documentation on the
SPARQL Endpoint semantics:

- To publish the documentation in a document with the address .well-
  known/void.
- To include the link to the documentation into a triplet of the dataset, using
  the property *void:inDataset*, for example: <document.rdf> void:inDataset
  <void.ttl#MyDataset>. So, this information could be obtained with an
  SPARQL query:

```
PREFIX void:<http://rdfs.org/ns/void#>
SELECT *
FROM <http://sparql.uniprot.org/taxonomy/>
WHERE
{
?s void:Dataset ?o .
}
```

However, the development of RDF databases does not always use this approach to make explicit the underlying semantic model. Database developers have
in mind the conceptual design of the database when developing queries on applications using databases. However, the access to Linked Data sources are usually
made by developers that have not participated in the database design. Thus,
the absence of an explicit semantic model makes either the design of SPARQL
queries or the RDF navigation a complex issue that requires an effort for understanding the explicit semantic model.

Aemoo [9] is an application that supports exploratory search through exploiting Semantics and explicit links. This application is based on encyclopaedic
knowledge patterns (EKPs) which are only applicable to RDF schemas such as
DBpedia. This tool does not provide the hidden OWL semantic data model, so
it is not comparable to the proposed approach. Another approach to mention is
Linked Data summaries [4]. The objective of this is the evaluation of conjunctive
queries that are answered using a structure index which summarizes the content
of sources. The purpose of this approach is different than the presented in this
paper, which tries to make easier the task of linked data exploration to end-users
by providing the endpoint semantic model in OWL.

---

[1] https://www.w3.org/TR/void/.

According to the literature reviewed, there are no proposals for reconstructing the underlying semantic data models in OWL from an RDF database. In this paper, we present an application for discovering part of the explicit semantic model of an RDF database. This approach is based on a set of simple queries for exploring the RDF graph. As a result of the exploration procedure the application returns an OWL ontology.

This paper is organized as follows: Sect. 2 presents the methodology that has been followed to develop the application; Sect. 3 presents some cases in which the proposed application has been used for the discovery of the explicit semantic model and how it can be used to build SPARQL queries; Finally, Sect. 4 shows the main results of this application and how it can be extended.

2   Methods

The approach for extracting the hidden semantic data model is based on a set of SPARQL queries that are executed to infer the implicit RDF structure of any Linked Data endpoint. These queries are combined to produce a single OWL ontology. The query patterns are described below:

1. The first query pattern aims at returning all the classes describing the objects (instances) stored in the database. This pattern requires that the RDF graph contains this information as triplets defining the type of each data node. The use of blank nodes ([]) allows the SPARQL engine to avoid linking the class URIs to a variable, and so reducing the evaluation cost.

```
SELECT DISTINCT ?class
WHERE {
  [] a ?class .
}
```

2. The second query pattern returns all the properties that link all the data instances between them.

```
SELECT DISTINCT ?property
WHERE {
  [] ?property [] .
}
```

3. After having obtained the set of classes and properties (using patterns 1 and 2), the relationship between them has to be specified by knowing which classes are part of the domain or range of the properties. This query pattern will be executed for each property discovered by pattern 2. The pattern shown below uses as example the property with URI http://www.example.org/property_1.

```
SELECT DISTINCT ?domain
WHERE {
  ?s <http://www.example.org/property_1> [] .
  ?s a ?domain .
}
```

4. This query pattern aims at discovering the range of object properties. Thus, this pattern will discover entities used as the object in a property, and later to know if this is an object or a literal. In fact, this is a combination of patterns 1 and 2.

```
SELECT DISTINCT ?range
WHERE {
  [] <http://www.example.org/property_1> ?o .
  ?o a ?range .
}
```

5. Data properties connect entities with literals. In this case, we aim at discovering the data type of such literals (such as integer or string).

```
SELECT DISTINCT (datatype(?o) as ?datatype)
WHERE {
  [] <http://www.example.org/property_1> ?o .
}
```

According to the range of the property, it is possible to divide the obtained list of properties into two subsets: object properties (when its range is composed by classes) and data properties (when its range is composed by data types).

The application developed is a Java program that orchestrates the execution of all these SPARQL queries and use the query results to build the OWL ontology. This ontology can be explored in dedicated programs such as Protégé[2]. The code of the program is publicly shared in GitHub[3] to allow other users and developers to contribute to new features and improve those already included. The application uses the Apache Jena open framework[4] to build and execute the SPARQL queries and the OWL API[5] for building the ontology from the extracted data results. The following pseudo-code shows how this application manages the use of the query patterns to reconstruct the semantic data models:

This pseudo-code is based in functions using the described query patterns:

- *query_classes()* makes use of the query pattern 1.
- *query_properties()* makes use of the query pattern 2.
- *query_domain(property)* makes use of the query pattern 3.
- *query_range(property)* makes use of the query pattern 4.
- *query_datatype(property)* makes use of the query pattern 5.

When executing the program on large remote databases such as DBpedia[6], queries or server time-out may occur. This problem is avoided using some parameters that can be configured in the developed application:

Algorithm 1.

```
 1: procedure CONSTRUCT ONTOLOGY FROM ENDPOINT
 2:     classes = query_classes();
 3:     properties = query_properties();
 4:     for property in properties do
 5:         domains += query_domain(property);
 6:         ranges += query_range(property);
 7:         datatypes += query_datatype(property);
 8:     end for
 9:     object_properties = get_object_properties(properties);
10:     data_properties = get_data_properties(properties);
11:     build_owl_file(classes, object_properties, data_properties, domains, ranges,
            datatypes);
12: end procedure
```

- **Limit of Results:** Instead of trying to get all classes or properties in our single query, a limit parameter can be set to retrieve the results in batches. This approach is useful when databases have a large number of distinct elements, so it is easier to get results with several smaller queries than with a single all-inclusive query. Default value: 1000
- **Number of Retries:** Sometimes a remote database can fail to return some data because of having to respond to multiple requests simultaneously. In that case, a number of retries of the same query can be established. Default value: 5
- **Cool-down Time:** Time in milliseconds between different query retries after a remote error. Default value: 10000

Each database endpoint has a better parameter set-up, but the user can modify them as they prefer and try different configurations. In the case of the remote database to continue failing, the program will build the ontology with the data that has been retrieved successfully instead of failing to do so.

3   Use Cases

This application has been tested using a collection of endpoints to manually test its accuracy. Some use cases obtained from the experimentation process are presented below, including two SPARQL Endpoints developed by us (Sects. 3.1 and 3.2), a well-know SPARQL Endpoint in the Life Sciences domain (Sect. 3.3) and a well-known SPARQL Endpoint in the Linked Data community (Sect. 3.4).

3.1   Kpath

Kpath [8] is a database that integrates information related to metabolic pathways from different sources such as Bio2Rdf Kegg's, NCBI Taxonomy and Protein data from SwissProt. This Endpoint contains classes that refer to concepts related to

Table 1. Model obtained from the Kpath endpoint. Full URI paths are not shown for simplicity, but the complete OWL ontology is available at http://khaos.uma.es/oee/examples/kpath.owl.

CLASSES		
Compound, Enzyme, Gene, Glycan, Organism, Pathway, Protein, Reaction		

OBJECT PROPERTIES		
Property	Domain	Range
enzymaticActivity	Protein	Enzyme
enzyme	Reaction	Enzyme
gene	Enzyme	Gene
left	Reaction	Glycan, Compound
organism	Protein, Pathway	Organism
reaction	Pathway	Reaction
relatedPathway	Pathway	Pathway
right	Reaction	Glycan, Compound
sameAs	-	Organism, Glycan, Compound, Reaction

DATA PROPERTIES		
Property	Domain	Range
comment	Reaction, Compound, Organism, Pathway, Protein	string
ecNumber	Enzyme	string
formula	Compound, Glycan	string
geneName	Protein	string
id	Gene	string
keggCode	Organism	string
keyword	Protein	string
mass	Compound, Glycan	string
metabolism	Pathway	string
name	Enzyme, Glycan, Protein, Pathway, Organism, Reaction, Compound	string
synonim	Enzyme, Glycan, Compound, Organism, Protein	string
uniprotID	Protein	string
seeAlso	Gene	string

metabolism (such as pathway, reaction, organism, enzyme, etc.), object and data properties with domain and range. This SPARQL endpoint is available at http://sparql.kpath.khaos.uma.es/.

Table 1 summarizes all classes, object and data properties obtained from the resulting semantic model. As we have mentioned previously, the knowledge of the inferred semantic model can help users to design queries. For example, in this case, it can be observed that the right and left properties have the *Reaction* class as domain and either the *Compound* or *Glycan* classes as range. With this information about the data structure, a user can make an SPARQL query that retrieves the chemical reactions whose right participant elements are glycans:

```
PREFIX rdf: <http://www.w3.org/1999/02/22-rdf-syntax-ns#>
select DISTINCT *
WHERE {
  ?reaction rdf:type <http://khaos.uma.es/pathways/Reaction>.
  ?reaction <http://khaos.uma.es/pathways/right> ?right .
  ?right rdf:type <http://khaos.uma.es/pathways/Glycan>.
}
```

Or another that retrieves the chemical reactions whose left participant elements are chemical compounds:

```
PREFIX rdf: <http://www.w3.org/1999/02/22-rdf-syntax-ns#>
select DISTINCT *
WHERE {
   ?reaction rdf:type <http://khaos.uma.es/pathways/Reaction>.
   ?reaction <http://khaos.uma.es/pathways/left> ?left .
   ?left rdf:type <http://khaos.uma.es/pathways/Compound>.
}
```

3.2   ReprOlive

ReprOlive [1] is an easy-to-use olive tree (*Olea europaea L.*) database containing its reproductive transcriptome obtained from pollen and stigma (both together and separately). This database has been migrated to RDF and provides an SPARQL Endpoint at http://150.214.214.6/sparql inside the graph http://khaos.uma.es/olivedb. Table 2 shows all classes, object and data properties of the semantic model extracted from the ReprOlive endpoint.

With the semantic model of the ReprOlive Endpoint, as example in this case, some queries can be designed. For example, the object property *owl:has_annotation* has the *Protein* class as domain and the *InterPro, description, annotation, GO, orthologue, EC* and *Gene* classes as range. The object property *owl:has_pathway* has the *Gene* as domain and *Annotation* classes and the *Annotation_keggs* class as range. With this information, an SPARQL query can retrieve all those proteins that have an annotation which is a gene and this gene has a role in a pathway that is annotated as a Kegg annotation:

```
PREFIX rdf: <http://www.w3.org/1999/02/22-rdf-syntax-ns#>
select DISTINCT *
WHERE {
   ?s rdf:type <http://khaos.uma.es/vocab/protein> .
   ?s <http://khaos.uma.es/vocab/has_annotation> ?annotation .
   ?annotation rdf:type <http://khaos.uma.es/vocab/gene> .
   ?annotation <http://khaos.uma.es/vocab/has_pathway> ?annotation_kegg .
}
```

3.3   Biomodels

Biomodels [2] is a repository that stores computational models of biological processes. This database has been migrated to RDF standard and is part of the EBI (European Bioinformatics Institute) RDF platform [5]. The SPARQL endpoint is available at https://www.ebi.ac.uk/rdf/services/biomodels/sparql. As the number of classes, object and data properties are very large to summarize in a table. Table 3 includes the number of ontology elements. In the case of Biomodels ontology, the number of classes, object and data properties are 41, 63 and 107, respectively. A column with the time (in ms unit) that each process took has also been added.

Table 2. Model obtained from the ReprOlive endpoint. Full URI paths are not shown for simplicity, but the complete OWL ontology is available at http://khaos.uma.es/oee/examples/repr_olive.owl.

CLASSES
EC, GO, InterPro, annotation, annotation_keggs, assembly, description, expressions, gene, orthologue, pathway, pollen, protein, ssr, stigma, transcript, vegetative, Property, Class

OBJECT PROPERTIES		
Property	Domain	Range
has_annotation	protein	InterPro, description, annotation, GO, orthologue, EC, gene
has_pathway	gene, annotation	annotation_keggs
has_ssr	transcript, ssr	transcript, ssr
is_produced_by_assembly	transcript	assembly
produces_protein	transcript	protein
produces_transcript	assembly	transcript
type	EC, GO, InterPro, annotation, annotation_keggs, assembly, description, expressions, gene, orthologue, pathway, pollen, protein, ssr, stigma, transcript, vegetative, Property, Class	Class

DATA PROPERTIES		
Property	Domain	Range
ace_name, assembly_description, assembly_name, file, organism, owner_group, tissue	assembly	string
alignment, database, fln_status, nucleotides, transcript_sequence, transcript_subject	transcript	string
annotated, reversed, transcript_length, transcript_orf_end, transcript_orf_start	transcript	integer
assemblies_mean_contig_length, assemblies_total_nt, assembly_id, order, sub_assembly, version	assembly	integer
description_name	annotation, description	string
ec_code	annotation, EC	string
end_position, ssr_id, ssr_length, start_position	ssr	integer
expressions_id, expressions_raw, expressions_unigen_id	expressions, pollen, stigma, vegetative	integer
expressions_rpkm	expressions, pollen, stigma, vegetative	double
gene_id	annotation, gene	string
go_id, go_subtype	annotation, GO	string
interpro_id	annotation, InterPro	string
kegg_id, pathway_name	pathway	string
message, protein_sequence	protein	string
orthologue_name, orthologue_subtype	annotation, orthologue	string
protein_name, transcript_name	protein, transcript	string
repeat_motif, ssr_sequence	ssr	string
test_code	transcript	decimal
label	GO, vegetative, gene, transcript, pollen, InterPro, ssr, annotation, expressions, Property, stigma, orthologue, EC, protein, Class, pathway, assembly, description	string

Table 3. Statistics of analysed SPARQL endpoints.

Endpoint	Classes	Object properties	Data properties	Time taken (ms)
kPath[a]	8	8	10	308, 089
ReprOlive[b]	19	7	50	104, 124
Biomodels[c]	41	63	107	1, 008, 788
LinkedGeoData[d]	1, 164	125	1, 175	9, 553, 433

[a] Endpoint: http://sparql.kpath.khaos.uma.es/.
 OWL file: http://khaos.uma.es/oee/examples/kpath.owl
[b] Endpoint: http://150.214.214.6/sparql.
 Graph: http://khaos.uma.es/olivedb.
 OWL file: http://khaos.uma.es/oee/examples/repr_olive.owl
[c] Endpoint: https://www.ebi.ac.uk/rdf/services/biomodels/sparql.
 OWL file: http://khaos.uma.es/oee/examples/biomodels.owl
[d] Endpoint: http://linkedgeodata.org/sparql.
 OWL file: http://khaos.uma.es/oee/examples/geodata.owl

With the semantic model extracted from Biomodels endpoint, some queries can be designed to end-users with a biological background. For example, the model shows that the data property *owl:initialConcentration* has the *SBMLUnit, LocalParameter, Parameter, FunctionDef, Reaction, SpeciesReference, Compartment, Species, SBMLUnitdef, KineticLaw, ModifiedSpeciesReference* classes as domain and double integer as range. As example, according to the model, an SPARQL query can retrieve all biological models whose initial reactive concentration is equal or greater than 30.5 M.

```
PREFIX sbmlrdf: <http://identifiers.org/biomodels.vocabulary#>
SELECT *
WHERE {
  ?s rsbmlrdf:initialConcentration ?o .
  FILTER (?o >= 30.5)
}
```

3.4 LinkedGeoData

LinkedGeoData [11] was included in this study because of its large size. This repository is focused to adding a spatial dimension to the Semantic Web. It uses the information collected by the OpenStreetMap project[7] and makes it available as RDF. They have available two public endpoints: one that is updated frequently and an static one with data from a certain date which is the one that has been used. This endpoint is available at http://linkedgeodata.org/sparql.

Table 3 includes the number of classes, object and data properties of the inferred LinkedGeoData semantic model and the time that took to generate the model from the endpoint.

---

[7] https://www.openstreetmap.org/.

The semantic model extracted from the endpoint shows that, for example, the data property *owl:altitude* has several classes such as *HistoricThing*, *DrinkingWater*, *Shop*, *Toilets* as domain and a string as range. An SPARQL query example could retrieve all places which have a historical element at an altitude above 2,000 ft and that these places also have drinking water at the same altitude as optional. The corresponding SPARQL query is the following:

```
Prefix lgdo:<http://linkedgeodata.org/ontology/>
SELECT distinct *
WHERE {
  ?s rdf:type lgdo:HistoricThing .
  ?s lgdo:ALTITUDE ?o FILTER (?o > 2000) .
  OPTIONAL{
  ?s rdf:type lgdo:DrinkingWater
  ?s lgdo:ALTITUDE ?o FILTER (?o > 2000)}
}
```

These four examples are different according to its data size and complexity. As mentioned in each subsection, Table 3 included a comparison between their number of classes and properties and the time taken to extract their related data model. After each execution, all this data was used to create an OWL file with the extracted ontology.

4   Conclusions and Future Work

In this paper we have presented an approach to reconstruct the semantic data model behind an RDF database. This technological approach is based on a set of simple SPARQL queries to explore the structure of the RDF graph. As a result, the application is able to partially discover the underlying data model to help users in designing new SPARQL queries. The resulting tools have been published as Open Source at http://github.com/estebanpua/ontology-endpoint-extraction. This Open Source project is open to other developers to improve the tool with new SPARQL patterns to extract other aspects that the current approach could have missed. This, we aim at continuously improving this solution to enhance the semantic models generated.

The experiments carried out for testing the application show that the reconstructed model does not usually include some specific parts of the underlying semantic model. For example, the subsumption relationship between classes is not obtained. This hierarchical relationship cannot be directly extracted if it is not stored as part of the RDF graph. The implicit use of owl:subClass, rdfs:domain and rdfs:range axioms can be used to obtain these relations, but they are nor usually available in most of the endpoints. We are currently developing an extension for using these axioms, in case they are used in the analysed Endpoints. Otherwise, different approaches should be used to discover such relationships. A possible approach is to align these automatically extracted ontologies with existing ones to infer these relationships from how the classes are organized in similar ontologies. Another different approach we are currently testing is the extraction of the instance set for each discovered class in the Endpoint, and calculating if any of them have a subsumption relationship. This method would provide false subclass results, but they could be manually curated.

Additionally, a Web interface has been developed to help users to directly generate the data models given an SPARQL Endpoint. This interface and the application itself have been tested in a number of SPARQL Endpoints, showing that the results are useful for the design of SPARQL queries. This application is being used to help Bioqueries [3] (http://bioqueries.uma.es) in designing queries on SPARQL Endpoints accessing biological data.

Acknowledgements. This work was partially supported by Grants TIN2014-58304-R (Ministerio de Ciencia e Innovación) and P11-TIC-7529 and P12-TIC-1519 (Plan Andaluz de Investigación, Desarrollo e Innovación).

References

1. Carmona, R.M., Zafra, A., Seoane, P., Castro, A.J., Guerrero-Fernndez, D., Castillo-Castillo, T., Medina-García, A., Cánovas, F.M., Aldana-Montes, J., Navas-Delgado, I., Alché, J.D.D., Claros, M.G.: ReprOlive: a database with linked data for the olive tree (Olea europaea L.) reproductive transcriptome. Front. Plant Sci. **6**(625) (2015)
2. Chelliah, V., Juty, N., Ajmera, I., Ali, R., Dumousseau, M., Glont, M., Hucka, M., Jalowicki, G., Keating, S., Knight-Schrijver, V., Lloret-Villas, A., Natarajan, K.N., Pettit, J.B., Rodriguez, N., Schubert, M., Wimalaratne, S.M., Zhao, Y., Hermjakob, H., Le Novre, N., Laibe, C.: BioModels: ten-year anniversary. Nucleic Acids Res. **43**(D1), D542–D548 (2015)
3. García-Godoy, M.J., López-Camacho, E., Navas-Delgado, I., Aldana-Montes, J.F.: Sharing and executing linked data queries in a collaborative environment. Bioinformatics **29**(13), 1663–1670 (2013)
4. Harth, A., Hose, K., Karnstedt, M., Polleres, A., Sattler, K.U., Umbrich, J.: Data summaries for on-demand queries over linked data, pp. 411–420 (2010)
5. Jupp, S., Malone, J., Bolleman, J., Brandizi, M., Davies, M., Garcia, L., Gaulton, A., Gehant, S., Laibe, C., Redaschi, N., Wimalaratne, S.M., Martin, M., Le Novère, N., Parkinson, H., Birney, E., Jenkinson, A.M.: The EBI RDF platform: linked open data for the life sciences. Bioinformatics (Oxf., Engl.) **30**(9), 1338–1339 (2014). http://dx.doi.org/10.1093/bioinformatics/btt765
6. LOD: Open linked data. http://linkeddata.org/
7. Manola, F., Miller, E.: RDF Primer. World Wide Web Consortium, February 2004
8. Navas-Delgado, I., García-Godoy, M.J., López-Camacho, E., Rybinski, M., Reyes-Palomares, A., Medina, M., Aldana-Montes, J.F.: Kpath: integration of metabolic pathway linked data. Database **2015**, bav053 (2015)
9. Nuzzolese, A.G., Presutti, V., Gangemi, A., Musetti, A., Ciancarini, P.: Aemoo: exploring knowledge on the web. In: Proceedings of the 5th Annual ACM Web Science Conference, WebSci 2013, pp. 272–275 (2013)
10. Prudhommeaux, E., Seaborne, A.: SPARQL query language for RDF, W3C recommendation. http://www.w3.org/TR/rdf-sparql-query/
11. Stadler, C., Lehmann, J., Höffner, K., Auer, S.: LinkedGeoData: a core for a web of spatial open data. Semant. Web J. **3**(4), 333–354 (2012). http://jens-lehmann.org/files/2012/linkedgeodata2.pdf

A New Formal Approach to Semantic Parsing of Instructions and to File Manager Design

Alexander A. Razorenov and Vladimir A. Fomichov$^{(\boxtimes)}$

Faculty of Business and Management, School of Business Informatics,
Department of Innovations and Business in the Sphere of Informational
Technologies, National Research University Higher School of Economics,
Kirpichnaya Street 33, 105187 Moscow, Russia
{arazorenov,vfomichov}@hse.ru

Abstract. During roughly the last seven years, an increase of interest in semantic parsing of instructions in natural language (NL) could be observed. The principal applications of developed algorithms are NL-interfaces for interaction with robots and the personages of videogames, navigation in virtual space, and for developing programs by means of NL. However, the known algorithms are able to process only simple instructions, including one verb with dependent words. This paper has the following theoretical objectives: (a) to formally define in a new way the semantic-syntactic component of a linguistic database and semantic-syntactic structure of a NL-text; (b) to develop a new algorithm of semantic parsing of instructions satisfying the following conditions: (i) being able to process complex NL-instructions including several verbs; (ii) having a relatively compact form due to a high-level angle of look; (iii) being easy to implement and to expand; (iv) finding semantic-syntactic relationships in the input text without constructing a pure syntactic representation of the input text; (v) being convenient for processing texts not only from English but also from Russian, German, French and many other languages. The practical objective of the study was to develop a useful NL-interface to a file manager. A file manager with a NL-interface NLC-2 (Natural Language Commander - Version Two) has been developed. This study is underpinned by the theory of K-representations (knowledge representations) developed by the second author.

Keywords: Natural language processing · Semantic parsing · Theory of K-representations · SK-language · Morphological basis · Dictionary of lexical frames · Linguistic database model · Graph-like semantic-syntactic representation of a text · Software management · File system natural language management · Natural language commander · Haskell

1 Introduction

The field of designing natural language (NL) processing systems has been quickly progressed during last fifteen years. During last five-seven years, one has been able to observe a considerable growth of interest in NL-interfaces being able to fulfill semantic parsing of instructions. The major part of the publications consider principal aspects of interaction with robots [1, 2, 4, 17, 19], governing the personages of videogames,

© Springer International Publishing Switzerland 2016
S. Hartmann and H. Ma (Eds.): DEXA 2016, Part I, LNCS 9827, pp. 416–430, 2016.
DOI: 10.1007/978-3-319-44403-1_26

navigation in virtual two- and three-dimensional space [5, 18], and developing programs by means of NL [6, 16].

The problem of instructions' semantic parsing shifts the accent from statistical processing of NL-texts to developing the methods of extracting meanings from texts. The formal methods for designing such NL-interfaces have been still insufficiently developed. The principal reason is that the most popular formal means for representing texts' meanings are based either on first order logic (FOL) or on lambda-calculus. Both approaches are oriented at considering assertions, the sets of their well-constructed expressions (formulas) don't include formal analogues of imperatives.

That is why the listed above NL-interfaces for dealing with instructions realized the idea of recognizing inputs, proceeding from preliminary prepared templates. The common shortcomings of the approaches described in the mentioned papers are as follows: (a) orientation at simple systems of instructions; (b) the necessity to foresee the full spectrum of the supported instructions.

The analysis of publications on formal semantics of NL shows that nowadays there is at least one theory providing expressive mechanisms for describing structured meanings (or semantic structure) of arbitrary complex instructions in NL. It is the theory of K-representations (knowledge representations), or TKR; it was called the theory of K-calculuses and K-languages during the first stage of its development. It is an original theory of designing semantic parsers of NL-texts with the broad use of formal means for representing input, intermediary, and output data. This theory is set forth in numerous publications in English [7–9, 11–14] and Russian (see [10]).

This paper continues the line of the works [15, 20] describing the methodology of designing the file manager with a NL-interface NLC-1 (Natural Language Commander – Version 1) under the framework of TKR.

The first objective of this paper is to develop a new formal approach to designing the algorithms of complex instructions' semantic parsing. The constructed semantic representations (SR) of instructions belong to the class of K-representations, because they are the expressions of SK-languages (standard knowledge languages). The second objective is to propose a new approach to designing file managers with NL-interface. The principles of designing the NL-interface NLC-2 are set forth.

The structure of this paper is as follows. Section 2 describes related approaches to semantic parsing of instructions; in particular, it includes a short introduction to TKR. The Sects. 3–8 contain a new mathematical model of a linguistic database, a new formal way of describing semantic-syntactic structure of NL-texts, and a new algorithm of instructions' semantic parsing. Section 9 describes an implementation of the proposed algorithm – a file manger with a NL-interface NLC-2 (Natural Language Commander - Version 2). Section 10 contains the conclusions.

2 The Main Approaches to Representing Structured Meanings of Instructions

The design decisions in the process of developing the semantic parsing algorithms considerably depend on the used approaches to representing structured meanings (SMs) of instructions in NL. In other words, one often speaks about the approaches to constructing semantic representations (SRs) of instructions. The analysis of the scientific literature shows that the main approaches to this problem used in practice are Abstract Meaning Representation (AMR) [3, 4], lambda-calculus meaning representation (LCMR) [1], and the theory of K-representations [7–14, 20].

For instance, the instruction "Take the book on the table" is represented by means of AMR in [4] as follows: *(t/take - Taking: Theme (b/book); Source (t1/table); location (o/n; traject(b).* In [1], the instruction "Move to the chair in the third intersection" is associated with the following SR being an LCMR:

$$\lambda a.move(a) \wedge to(a, \iota x.sofa(x)) \wedge intersect(order(\lambda y.junction(y), frontdist, 3), x).$$

One of the shortcomings of using SRs of such kinds for reflecting SMs of NL-texts is the loss of expressiveness: if a text is relatively large, its SR provides no possibility of reconstructing this text (e.g., for the control of this process by a user).

The SRs of instructions belonging to the classes AMR and LCMR usually look different. However, their significant common feature is the possibility of constructing designations of various objects mentioned in instructions.

The approaches AMR and LCMR are rather convenient for representing SMs of simple instructions: with one verb and dependent words, containing no connective OR. But the instructions emerging in real applications may be much more complex: include actions joined by the connective AND or OR, indicate the order of actions, time distance between actions, mention compound designations of objects' groups as the operands of actions, include the modal words "necessary", "should", etc.

The analysis shows that expressive power of AMR and LCMR is insufficient for effectively dealing with such complex instructions. However, since the middle of the 1990s, an approach has been available being free from the listed restrictions. It is the theory of K-representations (knowledge representations), or TKR, developed by Fomichov [7–14]. An early version of TKR (the theory of restricted K-calculuses and restricted standard K-languages) [7, 8] introduced a class of formal languages with the expressive power considerably exceeding the expressive power of AMR seventeen years before the description of AMR in [3]. A part of advantages of restricted standard K-languages and SK-languages is indicated in the end of this section.

The part 1 of TKR is a mathematical model (Model 1) of a system of primary units of conceptual level used by an applied intelligent system. This model determines, in particular, a new class of complex formal objects called *conceptual bases*. To construct an arbitrary conceptual basis (c.b.) B is equivalent to defining a certain finite sequence of formal objects *Tuple(B)*. The interpretation of its distinguished components *St, X, V, F, tp* is as follows. *St* is a finite set of symbols called sorts and interpreted as designations of most general notions used in the considered application domains: *physical object, intelligent system, organization, distance value, price value*, etc. The countable set V contains the variables. The countable set X includes the subset *St* of sorts and contains the symbols interpreted as the designations of primary informational (or conceptual) units. The set X is called the primary informational universe of the c.b. B. The finite subset F of X contains the designations of functions.

The component tp of the sequence *Tuple(B)* is a function from the union of X and V into a countable set of strings *Types (B)*, it includes *St*. The elements of this set are called *types* and are interpreted as structured characteristics (labels) of the entities denoted by the elements of X. The mapping *tp* gives us a much more fine-grained structuring of application domains than first order logic.

Example. A c.b. B may satisfy the following conditions: (a) *St* includes the elements (sorts) *dyn.phys.ob* (dynamic physical object), *ints* (intelligent system), *org* (organization), *inf.ob* (informational object); (a) X includes the elements *Leo-Tolstoy, War-and-Piece, person, tourist-group, Suppliers, Authorship,* and

> $tp(person) =\uparrow ints * dyn.phys.ob, tp(Leo-Tolstoy) = ints * dyn.phys.ob,$
> $tp(War-and-Piece) = inf.ob, tp(Authorship) = \{(ints, inf.ob)\},$
> $tp(tourist-group) = =\uparrow \{ints * dyn.phys.ob\}, tp(Suppliers) = \{(org, \{org\})\}.$

Here the symbol ↑indicates a type of a notion; *Suppliers* is the name of the function associating an enterprise with the set of all its suppliers.

A partial order |- is defined on the set of types *Types (B)*, it is called the *concretization relation* (here the symbol |- is used in a non-standard way, i.e., not as in mathematical logic). For instance, the following relationships may take place:

$$phys.ob|-dyn.phys.ob, phys.ob|-ints * dyn.phys.ob, ints|-ints * dyn.phys.ob,$$
$$ints|-ints * dyn.phys.ob, \{phys.ob\}|-\{ints * dyn.phys.ob\}.$$

The part 2 of TKR determines a mathematical model (Model 2) of a system consisting of ten partial operations on conceptual structures. The Model 2 defines, in particular, a new class of formal languages called SK-languages (standard knowledge languages). There are weighty reasons to conjecture that SK-languages is a convenient formal tool for building SRs of arbitrarily complex NL-texts (sentences and discourses) pertaining to mass spheres of professional activity (engineering, medicine, business, sport, etc.). The term "a K-representation" (KR) is used for denoting SRs of NL-texts being the expressions of SK-languages.

The expressions of SK-languages are built from primary semantic units given by a conceptual basis (c.b.) and several service symbols by means of inductive application of some original rules P[0], P[1], …, P[10]. The language corresponding to an arbitrary c.b. B is designated by *Ls(B)*.

The rule P[0] says that the elements of the primary informational universe $X(B)$ and the variables from $V(B)$ belong to *Ls(B)*; in other words, they are K-strings. E.g., the unit *file1* is a K-string. The rules P[1]–P[10] jointly define a system consisting of ten partial operations on conceptual structures [10, 11].

The rule P[1] allows to connect intensional quantifiers and designations (simple or compound) of notions, in particular, to construct the formulas *certain file1, certain file1 * (Extension, "doc"), all file1 * (Extension, "doc")*. The rule P[2] is used for constructing the expressions of the form $f(t_1,..., t_n)$, where f is the name of a function (example: *Creation-date(certain file1)*). The rule P[3] enables us to build the expressions of the form $(a \equiv b)$. Example: *(document $\equiv$ file1 * (Extension, "doc"))*.

One uses the rule P[4] for building the expressions of the form $r(t_1, ..., t_n)$, where r is the name of a relation with n attributes (example: *Earlier (Creation-date(certain file1), #yeasrterday#))*. The rule P[5] provides the possibility to mark a formula or its part by means of a variable. Example: *all file1 * (Extension, "doc") : S1*.

The rule P[6] allows us to join the negation connective ¬ to a formula (example: *¬file1*). The rule P[7] governs the use of the logical connectives ∧ (and) and ∨ (or). Example: *file1 * (Extension, ("doc" ∨ "docx"))*. Using the rule P[8] at the last step of an inference, it is possible to construct compound designations of notions. Example:

*file1 * (Extension, ("doc" ∨ "docx"))(Location, certain desktop)*.

The rule P[9] allows us to use the universal quantifier and existential quantifier (∀ и ∃) in formulas. The rule P[10] enables us to construct the SRs of finite sequences as the strings of the form $<c_1, ..., c_n>$, where $c_1, ..., c_n$ are the elements of a sequence.

Example 1. The instruction "Move to the chair in the third intersection" may have a KR *Semrepr1* of the form *IsAction (#now#, movement1 *(Object, certain chair) (Destination, certain Intersection*(Number, 3)))*.

Example 2. The instruction "Archive documents in folder "Project" and send to somebody@example.org" may be associated with the KR *Semrepr2* of the form

*(IsAction (#now#, archiving1 * (Object1, certain set * (Qualitative-composition, document1 * (Location, certain folder 1 * (Name1, "Project") : S1))(Result, x1), e1 ∧ IsAction (#now#, sending1 * (Object1, x1)(Email-address, "somebody@example. org"), e2) ∧ Immediately-after(e2, e1))*.

SK-languages enable us to construct SRs both of simple and complex instructions, preserving the expressiveness of NL and without the loss of intuitively-understandable connection with the initial text of instruction. Their expressive means exceed the means of other considered above approaches to representing structured meanings of instructions. Besides, it is necessary to underline such additional possibilities of SK-languages as (a) the advantages of using one format both for representing semantics of instructions and for representing knowledge pieces from ontologies; (b) the possibility to represent goals and to connect compound goals of active systems by means of logical connectives; (c) to construct formal semantic representations of compound descriptions of sets and of notions; (d) to reflect the time and other connections between separate goals being the components of a compound goal.

3   The Notion of Morphological Basis

Let's start to describe the principal ideas of transforming NL-texts into K-representations (KR). The main thing is to find semantic-syntactic links between lexical units of the processed text and, on this basis, to form one or several KR. With this aim, we'll define several new formal objects: a morphological basis, a set of K-templates (Sect. 4), a dictionary of lexical frames and a linguistic database (Sect. 5), and a graph-like semantic-syntactic representation of an input text (Sect. 6).

Consider the main ideas of defining a new class of formal objects called morphological bases. Let W be a finite set of elementary textual units (words of word combinations) of an input language, and W be the union of the non-intersecting finite sets of symbols *Wd, Ws, Wu,* and *Wp,* where (a) *Wd* is the set of main lexical units (words or word combinations) associated with meanings, (b) *Ws* is the set of markers (point, comma, etc.); (c) *Wu* is the set of the connectives 'and" "or", and others; (d) *Wp* is the set of prepositions and, possibly, another lexical units used for linking the words.

If Z is an arbitrary finite set, let the power set *P(Z)* be the designation of the set consisting of all subsets of Z, including the empty set Ø and the set Z.

Let *Mprop* be a finite set of symbols interpreted as the values of morphological properties associated with the lexical units of input language: the designations of parts of speech, of grammatical cases, etc. Let M be a certain subset of *Mprop* (it is not excluded that M coincides with the power set *P(Mprop))*, and <=: be a partial order on M (i.e., <=: is binary relation on M being reflexive, transitive and antisymmetric).

Example. Let *Mprop = (verb, noun, past.simple, present.simple, active.voice, passive. voice, singular, plural, artif.name}.* Here *artif.name* is the value of the property "class of textual unit" introduced in the described study; the value *artif.name* have the strings in quotes or apostrophes. Then M may contain the subsets of *Mprop {verb}, {verb, past.simple}, {verb, present.simple}, {noun}, {noun, singular}, {noun, plural},* empty set Ø. Then the following relationships may take place: {*verb*} <=: {*verb, past.simple}, {verb}* <=: {*verb, present.simple}, {noun}* <=: {*noun, singular}, {noun}* <=: {*noun, plural},* Ø <=: Y, where Y is an arbitrary element of M.

Let B be an arbitrary conceptual basis (see an explanation in Sect. 2 and the definition in [11]). Let also the following mappings be given:

- *rm: W -> P(M)* – a mapping associating a unit of an input language with a set of the possible collections of its morphological properties (the empty set is not excluded);
- *rc: W -> W* – a mapping associating an elementary text unit with its basic morphological form;
- *ru: W -> P(X(B))* – a mapping associating an elementary text unit with a finite subset of elements from the primary informational universe X(B); besides, for each d from Wu, *ru(d)* is a subset of the set {∧, ∨ }, i.e., of the set consisting of the connectives conjunction and disjunction.

For instance, for a certain subset of English describing the operations with the file systems, *rm(deleted) = {verb, verb.transitive, past.simple, active.voice}.* The functions *rc* and *ru* may be described in such a way that *rc(deleted) = delete, ru(and) = {∧}, ru (or) = {∨}.* The described ideas can help to grasp the meaning of the definition of a morphological basis. It will be used below as one of the tools for finding the links between lexical units of a text.

Definition 1. Let B be an arbitrary conceptual basis (c.b.). Then a morphological basis co-ordinated with the c.b. B is an arbitrary ordered 10-tuple *Morph* of the form

$$(Wd, Ws, Wu, Wp, Mprop, M, < =:, rm, rc, ru),$$

where

- *Wd, Ws, Wu, Wp, Mprop* are non-intersecting finite sets of symbols;
- *M* is a non-empty subset of the power set *P(Mprop)*, and <=: is a partial order on the set *M;*
- *rm: W -> P(M)* is a mapping from *W* into the power set *P(M); rc: W -> W* is a mapping from *W* into *W; ru: W -> P(X(B))* is a mapping from W into *P(X(B))*, where X(B) is the primary informational universe of the c.b. B.

4 Templates for Building K-Representations

Let's introduce the notion of a formula's template for constructing K-representations (KR). E.g., a template can have the form $x1 * (Object, x2)$, where $x1, x2$ are the variables of the type *inf.object.* Let's make a try to use this string for processing the phrase "Delete the folder "Documents". It is possible to replace the variables $x1$ and $x2$ by KRs describing an action and an object being the operand of this action. E.g., $x1$ may be replaced by the informational unit *deleting1*, and $x2$ – by the KR *certain folder1 * (Name1, "Documents")*. Then the result of replacement will be the KR *deleting1 * (Object, certain folder1 * (Name1, "Documents"))*.

Definition 2. Let *B* be an arbitrary conceptual basis (c.b.). Then *a template of a K-representation* is an ordered triple *(Frame, x_1, x_2)*, where x_1 and x_2 are the variables of the c.b. *B*, and *Frame* is a string that can be transformed into an expression (an l-formula) of the SK-language *Ls(B)* by means of replacing the occurrences of the symbols x_1 and x_2 by certain expressions (l-formulas) of the language *Ls(B)*. The set of all templates of a KR corresponding to the c.b. *B* is denoted by *K-templates (B)*.

5 A Dictionary of Lexical Frames and a Model of Linguistic Database

A dictionary of lexical frames describes the rules used for establishing the semantic-syntactic relationships between the lexical units of the processed text. It formulates the requirements to the lexical units to be connected: the collections of their morphological properties and additional semantic restrictions.

For describing semantic restrictions, the mechanism of structuring thematic domains in the theory of K-representations (TKR) by means of types will be used. This mechanism will prevent us, in particular, from building the semantic structures of the form *certain paint * (Colour, sweet)*, because semantic units corresponding to the colours and tastes will have different types.

Since TKR defines the concretization relation |- on the set of types *Types (B)*, we do have the possibility (if necessary) to indicate most general semantic restrictions and to verify not the coincidence of two types but the fact that one type *t1* is a concretization of the other type *t2* (in particular, the types *t1* and *t2* may coincide). E.g., the type of the notion "a person" may be the string *ints * dyn.phys.ob*, where *ints* is the sort "intelligent system", and *dyn.phys.ob* is the sort "dynamic physical object". That is we are able to consider a concrete person as dynamic physical object, while finding his/her space location, and as an intelligent system, while reading the news about the development of a new technology, etc.

Definition 3. *B* be an arbitrary conceptual basis (c.b.), *Morph* be an arbitrary morphological basis co-ordinated with the c.b. *B*. Then *a dictionary of lexical frames* co-ordinated with *B* and *Morph* is an arbitrary finite set *LexFrames* consisting of the ordered 7-tuples of the form

$$(type_1, morph_1, type_2, morph_2, prep, template, dir),$$

where the following conditions are satisfied:

- *type$_1$* and *type$_2$* are the types from the set *Types(B);*
- *morph$_1$* and *morph$_2$* are the elements of the set *M*, i.e. they are the collections of morphological properties' values;
- *prep* is either an element of the set *Wp* or the empty preposition *nil*;
- *template* is an element of the set *K-templates(B)*;
- *dir* is a number from the set $\{-1, 0, 1\}$.

Example. A certain dictionary of lexical frames to be used below may include the 7-tuples

*(action, {verb}, [object], {noun}, nil, (x1 * (Object1, x2), x1, x2), 0),*
*(action, {verb}, space.object, {noun}, to, (x1 * (Object1, x2), x1, x2), 0),*
*([object],{noun}, literal,{artif.name},(x2 * (Name1, x4), x2, x4), 1).*

A natural partial order on the set *M* being a subset of the power set *P(Mprop)* may be the relation "To be a subset of a set", e.g. {noun} <=: {noun, possessive.case}.

Definition 4. A *linguistic database* is the ordered triple *LingDb* of the form *(B, Morph, LexFrames)*, where *B* is an arbitrary conceptual basis (c.b.), *Morph* is an arbitrary morphological basis (m.b.) co-ordinated with the c.b. *B*, and *LexFrames* is an arbitrary dictionary of lexical frames co-ordinated with the c.b. *B* and m.b. *Morph*.

6   A Graph-like Semantic-Syntactic Structure of a Text

Let's introduce a notion playing the crucial role in the description of the algorithm SemSyntRA in Sect. 7.

Definition 5. A *graph-like semantic-syntactic structure (GSSR)* of the input text $T = t_1...t_n$ for the linguistic database *LingDb = (B, Morph, LexFrames)* is an ordered pair *SemGraph = (V, E)* such that $V = \{1,..., n \mid t_i$ belongs to *Wd* $\}$, and *E* is the set of all possible ordered triples of the form *(i, j, (sem$_1$,sem$_2$,k,f))* such that *i, j* belong to *V* and for a certain element *(type$_1$, morph$_1$, type$_2$, morph$_2$, prep, template, dir)* of the dictionary of lexical frames *LexFrames* the following conditions are satisfied:

(1) the set of semantic units $ru(t_1)$ associated with the lexical unit t_1 includes such unit sem_1 that the type $tp(sem_1)$ is a concretization of the element $type_1$ in the partially ordered set of types $Types(B)$;

(2) the set of the collections of morphological properties' values $rm(t_1)$ associated with the lexical unit t_1 includes a collection x such that $morph_1 <=: x$;

(3) the set of semantic units $ru(t_2)$ associated with the lexical unit t_2 includes such unit sem_2 that the type $tp(sem_2)$ is a concretization of the element $type_2$

(4) the set of the collections of morphological properties' values $rm(t_2)$ associated with the lexical unit t_2 includes a collection y such that $morph_2 <=: y$;

(5) the element $prep$ either is the empty preposition nil or $rc(t_k) = prep$ for a certain k, where $0 < k < j$; besides, if $i < j$, then $i < k < j$.

(6) f is an element of the set K-templates(B);

(7) if $dir = -1$, then $i > j$; if $dir = 1$, then $i < j$; if $dir = 0$, then i is not equal to j.

The indicated conditions are interpreted as follows. The conditions 1–4 formulate the restrictions for the pair of words to be connected by a semantic relationship; later it will be described by a K-representation constructed with the help of the string $template$ (the condition 6). The condition 1 restricts the type of semantic unit corresponding to the principal lexical unit, and the condition 2 restricts the values of morphological properties of the unit. The conditions 3 and 4 set forth similar restrictions for the dependent lexical unit. The condition 5 takes into account the participation of prepositions in establishing semantic relations between the words. Besides, it says that a preposition is located before the dependent lexical unit. The condition 7 restricts or not the order of principal and dependent lexical units.

It should be mentioned that the vertices of a graph are not the words but their indexes. The explanation is that the same word may have several occurrences in a text having different connections with another words.

Example. Using the set of the values of morphological properties and the dictionary of lexical frames described above, it is possible to construct a GSSR of the instruction *Move (1) file (2) "a.txt"(3) to (4) the folder (5) "Documents"(6)*.

The GSSR will be as follows (we indicate only the set of edges, because the set of vertices, obviously, is *{1,2,3,5,6}*):

$$\{(1, 2, (movement1, file1, nil, (x1 * (Object1, x2), x1, x2))),$$
$$(1, 5, ((movement1, file1, folder1, 4, (x1 * (Destination1, x3), x2, x3))),$$
$$(2, 3, (file1, ''a.txt'', nil, (x2 * (Name1, x4), x2, x4))),$$
$$(5, 6, (file1, ''Documents'', nil, (x2 * (Name1, x4), x2, x4))),$$
$$(2, 6, (file1, ''Documents'', nil, (x2 * (Name1, x4), x2, x4)))\}.$$

7   The Algorithm of Semantic Parsing SemSyntRA

The ideas stated above underpin a new algorithm of semantic parsing described below. We combined the methods of descending and ascending design.

Description of the algorithm SemSyntRA

Purpose: transform NL-text into its possible K-representations.

Input: T – text; LingDb – linguistic database.

Output: Result – a list of K-representations.

External specifications of auxiliary algorithms

Specification of procedure Tokenize

Input: T – text.

Output: WT – array of lexical units.

Remark: Algorithm depends on input natural language and could be different, for example, for Russian and English. So, it isn't described in this paper.

Specification of procedure BuildSpanningTrees

Input: GR – a graph-like semantic-syntactic representation (GSSR)

Output: Ltr – list of spanning trees

Remark: Any algorithm of spanning trees search in directed graphs can be used.

Specification of procedure BuildGSSR

Purpose: build a graph-like semantic-syntactic representation (GSSR).

Input: WT = $wt_1,...,wt_n$ – array of lexical units from Wd; LingDb – linguistic database.

Output: GSSR of the input text.

Algorithm BuildGSSR

```
begin
        E := {i | i = 1,...,n : wt_i ∈ Wd(Morph(LingDb))};
        V := empty list;
        loop for i from 1 to n
                loop for j from 1 to n
                        if there exist the elements of
                        LexFrames(LingDb) for the pair
                        (wt_i, wt_j) which meet the
                        conditions of definition 3
                                then add to list V corre-
                                spondent tuples (i, j,
                                (sem1, sem2, k, templ));
                        end if;
                end loop for j;
        end loop for i;
        GR := (E,V);
end.
```

Specification of procedure BuildRepresentations

Purpose: build possible K-representations of user input using graph of semantic and syntactic links.

Input: Ltr – list of spanning trees

Output: Result – list of K-representations

Algorithm BuildRepresentations

```
begin
        Result := empty list;
        Loop for current list item Tree from Ltr
                Try to build a K-representation for Tree
                {see Section 8};
                if it is successfully built
                        then add the built K-
                        representation to Result list;
                end if;
        end loop;
end.
```

Algorithm SemSyntRA

```
begin
        Tokenize (T, WT);
        BuildGSSR (WT, LingDb, GR);
        BuildSpanningTrees (GR, Ltr);
        BuildRepresentations (Ltr, Result)
end.
```

8   The Examples of Processing Instructions by the Algorithm SemSyntRA

The stage of building GSSR was illustrated above by means of examples. For building spanning trees, we used the algorithm from [21]. It was modified for dealing with directed graphs. This algorithm gets all possible subgraphs with all vertices of GSSR and checks that the subgraph is a tree and all vertices of GSSR are reachable from a certain root vertex.

K-representations for every spanning tree are built by tree folding from the deepest nodes to the root of the tree. On every step we reduce the subtrees with depth 1. For every edge of the subtree, the algorithm builds a KR based on a K-representation template. Several K-representations are mapped to single K-representation in accordance with certain rules, in particular:

- K-representations of the form $c^*(r_1, z_1),\ldots,c^*(r_n, z_n)$ are mapped to K-representation $c^*(r_1, z_1)\ldots(r_n, z_n)$;
- K-representations of the form int c, where int is intensional quantifier, together with K-representations of the form $c^*(r_1, z_1),\ldots, c^*(r_n, z_n)$ are mapped to K-representation int $c^*(r_1, z_1)\ldots(r_n, z_n)$;
- K-representation a and $(b_1\ \lambda\ldots\lambda\ b_n)$, where $\lambda \in \{\wedge,\vee\}$, are mapped to K-representations $(a\lambda b_1\lambda\ldots\lambda b_n)$, if $tp(a) = tp(b_1\lambda\ldots\lambda b_n)$.

This K-representation is used on next step of tree folding or (for root node) as the result. Consider the graph-like semantic-syntactic representation (GSSR) from the example above. Its edges: *(2, 3, (certain x2*(Name1, x4), x2, x4))* and *(5, 6, (certain x2*(Name1, x4), x2, x4))* will be mapped to K-representations *certain file1*(Name1, "a.txt")* and *certain folder1*(Name1, "Documents")* respectively. The edges *(1, 2, (x1*(Object1, x2), x1, x2))* and *(1, 5, (x1*(Destination1, x3), x1, x3))* will be mapped to K-representations *movement1*(Object1, certain file1*(Name1, "a.txt"))* and *movement1*(Destination1, folder1*(Name1, "Documents"))* respectively. These K-representations will be mapped to single K-representation:

$$movement1 * (Object1, certain\,file1 * (Name1,''a.txt''))(Destination1, certain$$
$$folder1 * (Name1,''Documents'')).$$

This K-representation is the result of spanning tree folding. Not every spanning tree could be mapped to correct KR. Also spanning trees could be filtered by some rules, for example: all words between the proposition and the noun should be connected to the noun.

Let's consider an example with the conjunctions "and" or "or":
Move (1) files(2) "a.txt"(3) and (4) "b.txt"(5) to(6) the folder(7) "Documents"(8).
The GSSR for this text is the follows: the set of vertices: *{1,2,3,5,6}*; the edges:

$$\{$$

$$(1,2,(Movement1, file1, nil, (certain\, x1 * (Object1, x2), x1, x2))),$$

$$(1,7,(movement1, folder1, \ll to \gg, (certain\, x1 * (Destination1, x3), x1, x3))),$$

$$(2,3,(file1, ''\,a.txt'', nil, (certain\, x2 * (Name1, x4), x2, x4))),$$

$$(2,5,(file1, ''\,b.txt'', nil, (certain\, x2 * (Name1, x4), x2, x4))),$$

$$(7,8,(folder1, ''\,Documents'', nil, (certain\, x2 * (Name1, x4), x2, x4))),$$

$$(2,8,(file1, ''\,Documents'', nil, (certain\, x2 * (Name1, x4), x2, x4)))$$

$$\}.$$

Consider spanning tree without the edge $(2,8, \ldots)$. Let's describe the folding of node 2 subtree. Edges of subtree will be mapped to K-representations *certain file1* (Name1, "a.txt")* и *certain file1*(Name1, "b.txt")*. So, these K-representations should be mapped to KR *certain file1*(Name1, ("a.txt" λ "b.txt"))*, where $\lambda \in \{\wedge, \vee\}$. So we should define binary link between literals *"a.txt" λ "b.txt"*. The vertices of subtree are between 2 and 5. There is the conjunction "and" between 2nd and 5th positions on position 4, so we should use $\lambda = ru(\text{«and»}) = \wedge$. And subtree of the vertex 2 will be mapped to K-representation *certain File1*(Name1, ("a.txt" $\wedge$ "b.txt"))*. The final K-representation will be:

$$movement1 * (Object1, certain\, file1 * (Name1, (''a.txt'' \wedge'' b.txt'')))(Destination1, certain$$
$$folder1 * (Name1, ''\,Documents'')).$$

We should also note that every spanning tree could be mapped to at most one K-representation, because variations are not allowed by the algorithm and edge marking.

9   Application of the Algorithm in the File Manager NLC-2

The model of linguistic database, the notion of GSSR, and the described algorithm underpinned the design of file management system NLC-2 (Natural Language Commander – Version 2). This program is the next generation of NLC-1, which was developed for the studies and experiments in the field of NL-interfaces to action-based applications [15, 20]. NLC-2 processes natural language instructions in accordance with the following scheme on Fig. 1:

Example. Let's look how NLC-2 processed the user instruction from Example 2 of Sect. 2: "Archive documents in folder "Project" and send to "somebody@example. org"". This instruction is transformed by the algorithm described above into the primary K-representation *Semrepr2* described in Sect. 2.

Now if the knowledge base of NLC-2 contains the transformation rule *Document1* $\vdash file1 * (Extention1, (''doc'' \vee ''docx'' \vee ''odt''))$ then the system NLC-2 transforms the constructed primary K-representation of the user instruction into its secondary KR

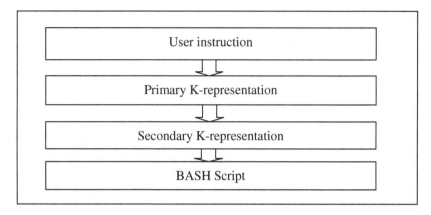

Fig. 1. The scheme of processing instructions in NLC-2

$(IsAction(\#now\#, archiving1 * (Object1, certain\,set * (Qualitative-$
$composition, certain\,file1 * (Extention1, (''doc'' \vee'' docx'' \vee'' odt''))(Location,$
$certain\,folder1 * (Name1,'' Project'') : S1))(Result, x1), e1 \wedge IsAction$
$(\#now\#, sending1 * (Object1, x1)(Email-address,$
$'' somebody@example.org''), e2) \wedge Immediately-after(e2, e1)).$

Then the result shell script for Bourne-Again Shell (BASH) is as follows:
zip/tmp/z000129.zip "Project/.doc" "Project/*.docx" "Project/*.odt"; sendfile*
somebody@example.com/tmp/z000129.zip. The final step is the execution of this script.

10   Conclusions

The principal scientific results described in this paper are as follows. Firstly, a new mathematical model of a linguistic database is developed. This model considers less formal entities in comparison with the models proposed in [10, 11] due to the union of the dictionaries of verbal-prepositional frames and of prepositional frames. Secondly, a new formal way of describing semantic-syntactic structure of NL-texts is proposed in comparison with matrix semantic-syntactic representation introduced in [10, 11]. Besides, a new algorithm of semantic parsing of instructions is presented, it satisfies the following conditions: (a) being able to process complex NL-instructions including several verbs; (b) being easy to implement and to expand; (c) finding semantic-syntactic relationships in the input text without constructing a pure syntactic representation of the input text; (d) being convenient for processing texts not only from English but also from Russian, German, French and many other languages. A program implementation of this algorithm with the help of the language Huskell is developed.

Acknowledgements. We are grateful to four anonymous referees of this paper for precious remarks.

References

1. Artzi, Y., Zettlemoyer, L.: Weakly supervised learning of semantic parsers for mapping instructions to actions. Trans. Assoc. Comput. Linguist. **1**, 49–62 (2013). Action Editor: Jason Eisner. https://aclweb.org/anthology/Q/Q13/Q13-1005.pdf. Accessed 13 Mar 2016

2. Babes-Vroman, M., MacGlashan, J., Gao, R., Winner, K., Adjogah, R., desJardins, M., Littman, M., Muresan, S.: Learning to interpret natural language instructions. In: Proceedings of the Second Workshop on Semantic Interpretation in an Actionable Context, pp. 1–6, Montreal, Canada, 3–8 June 2012. https://aclweb.org/anthology/W/W12/W12-2801.pdf. Accessed 13 Mar 2016

3. Banarescu, L., Bonial, C., Cai, S., Georgescu, M., Griffitt, K., Hermjakob, U., Knight, K., Koehn, P., Palmer, M., Schneider, N.: Abstract meaning representation for sembanking. In: Proceedings of the 7th ACL Linguistic Annotation Workshop and Interoperability with Discourse, Sofia, Bulgaria, 8–9 August 2013 (2013). www.aclweb.org/anthology/W13-2322. Accessed 12 Mar 2016

4. Bastianelli, E., Castellucci, C., Croce, D., Basili, R.: Textual inference and meaning representation in human robot interaction. In: Proceedings of the Joint Symposium on Semantic Processing. Textual Inference and Structures in Corpora, pp. 65–69 (2013). https://aclweb.org/anthology/W/W13/W13-3820v2.pdf. Accessed 13 Mar 2016

5. Benotti, L., Villalba, M., Lau, T., Cerruti, J.: Corpus-based interpretation of instructions in virtual environments. In: Proceedings of the 50th Annual Meeting of the Association for Computational Linguistics (Volume 2: Short Papers), pp. 181–186, Jeju, Republic of Korea, 8–14 July 2012 (2012). https://aclweb.org/anthology/P/P12/P12-2036.pdf. Accessed 13 Mar 2016

6. Carlos, C.S.: Natural language programming using class sequential rules. In: Proceedings of 5th International Joint Conference on Natural Language Processing, pp. 237–245, Chiang Mai, Thailand, 8–13 November 2011. https://aclweb.org/anthology/I/I11/I11-1027.pdf. Accessed 13 Mar 2016

7. Fomichov, V.A.: A mathematical model for describing structured items of conceptual level. Informatica. Int. J. Comput. Inform. (Slovenia) **20**(1), 5–32 (1996)

8. Fomichov, V.A.: Theory of restricted K-calculuses as a comprehensive framework for constructing agent communication languages. Informatica. Int. J. Comput. Inform. (Slovenia) **22**(4), 451–463 (1998). In: Fomichov, V.A., Zeleznikar, A.P. (eds.) Special Issue on NLP and Multi-Agent Systems

9. Fomichov, V.A.: An ontological mathematical framework for electronic commerce and semantically-structured web. Informatica. Int. J. Comput. Inform. (Slovenia) **24**(1), 39–49 (2000). In: Zhang, Y., Fomichov, V.A., Zeleznikar, A.P. (eds.) Special Issue on Database, Web, and Cooperative Systems

10. Fomichov, V.A.: The Formalization of Designing Natural Language Processing Systems. MAX Press, Moscow (2005). (in Russian)

11. Fomichov, V.A.: Semantics-Oriented Natural Language Processing: Mathematical Models and Algorithms. Springer, New York (2010)

12. Fomichov, V.A.: Theory of K-representations as a comprehensive formal framework for developing a multilingual semantic web. Informatica. Int. J. Comput. Inform. **34**(3), 387–396 (2010)

13. Fomichov, V.A.: A broadly applicable and flexible conceptual metagrammar as a basic tool for developing a multilingual semantic web. In: Métais, E., Meziane, F., Saraee, M., Sugumaran, V., Vadera, S. (eds.) NLDB 2013. LNCS, vol. 7934, pp. 249–259. Springer, Heidelberg (2013)

14. Fomichov, V.A.: SK-languages as a comprehensive formal environment for developing a multilingual semantic web. In: Decker, H., Lhotská, L., Link, S., Spies, M., Wagner, R.R. (eds.) DEXA 2014, Part I. LNCS, vol. 8644, pp. 394–401. Springer, Heidelberg (2014)
15. Fomichov, V.A., Razorenov, A.A.: A new method of extracting structured meanings from natural language texts and its application. In: Métais, E., Roche, M., Teisseire, M. (eds.) NLDB 2014. LNCS, vol. 8455, pp. 81–84. Springer, Heidelberg (2014)
16. Lei, T., Long, F., Barzilay, R., Rinard, M.: From natural language specifications to program input parsers. In: Proceedings of the 51st Annual Meeting of the Association for Computational Linguistics (Volume 1: Long Papers), pp. 1294–1303, Sofia, Bulgaria, 4–9 August 2013. https://aclweb.org/anthology/P/P13/P13-1127.pdf. Accessed 13 Mar 2016
17. Marge, M., Rudnicky, A.: Comparing spoken language route instructions for robots across environment representations. In: Proceedings of the SIGDIAL 2010 Conference, pp. 157–164, The University of Tokyo, 24–25 September 2010. https://aclweb.org/anthology/W/W10/W10-4328.pdf. Accessed 13 Mar 2016
18. Misra, D.K., Tao, K., Liang, P., Saxena, A.: Environment-driven lexicon induction for high-level instructions. In: Proceedings of the 53rd Annual Meeting of the Association for Computational Linguistics and the 7th International Joint Conference on Natural Language Processing (Volume 1: Long Papers), pp. 992–1002, Beijing, China, 26–31 July 2015. https://aclweb.org/anthology/P/P15/P15-1096.pdf. Accessed 13 Mar 2016
19. She, L., Yang, S., Cheng, Y., Jia, Y., Cha, J, Xi, N.: Back to the blocks world: learning new actions through situated human-robot dialogue. In: Proceedings of the 15th Annual Meeting of the Special Interest Group on Discourse and Dialogue (SIGDIAL), pp. 89–97, Philadelphia, U.S.A., 18–20 June 2014 (2014). https://aclweb.org/anthology/W/W14/W14-4313.pdf. Accessed 13 Mar 2016
20. Razorenov, A.A., Fomichov, V.A.: The design of a natural language interface for file system operations on the basis of a structured meanings model. Proc. Comput. Sci. **31**, 1005–1011 (2014). Elsevier; Open access. http://authors.elsevier.com/sd/article/S1877050914005304
21. Web-resource: Ninety-Nine Haskell Problems: Construct all spanning trees. https://wiki.haskell.org/99_questions/Solutions/83. Accessed 21 May 2015

Ontology-Based Deep Restricted Boltzmann Machine

Hao Wang$^{(\boxtimes)}$, Dejing Dou, and Daniel Lowd

Computer and Information Science, University of Oregon, Eugene, USA
{csehao,dou,lowd}@cs.uoregon.edu

Abstract. Deep neural networks are known for their capabilities for automatic feature learning from data. For this reason, previous research has tended to interpret deep learning techniques as data-driven methods, while few advances have been made from knowledge-driven perspectives. We propose to design a semantic rich deep learning model from a knowledge driven perspective, by introducing formal semantics into deep learning process. We propose ontology-based deep restricted Boltzmann machine (OB-DRBM), in which we use ontology to guide architecture design of deep restricted Boltzmann machines (DRBM), as well as to assist in their training and validation processes. Our model learns a set of related semantic-rich data representations from both formal semantics and data distribution. Representations in this set correspond to concepts at various semantic levels in a domain ontology. We show that our model leads to an improved performance, when compared with conventional deep learning models in classification tasks.

1 Introduction

Deep learning has achieved state of the art performance on many cutting-edge applications, including computer vision [1], speech and phonetic recognition [2], natural language processing [3], multi-task and multi-modal learning [4], and many others. Deep learning is often called *representation learning* [5], which emphasizes its aspect of automatic representation learning from data. Features in learned representations are formulated in a bottom up way, such that higher-level features are defined recursively from lower-level ones. For this reason, previous research tended to interpret deep learning techniques from data-driven perspectives. Few efforts have been made for semantic-rich deep learning methods, especially, for the ones using formal semantics.

In practice, data-driven approaches often carry various limitations. In deep learning, it is often difficult to interpret representations learned from data with accurate high-level semantics [6]. Over-fitting is a prevalent issue in deep neural networks that have a large number of parameters [7]. While we expect a well-trained deep representation to encode a non-local generalization prior over input space, it has often been proved to be sensitive to training data distribution. Poorly distributed data can result in an inferior or even wrong generalization. For similar reasons, deep representations often fail to generalize to examples that

© Springer International Publishing Switzerland 2016
S. Hartmann and H. Ma (Eds.): DEXA 2016, Part I, LNCS 9827, pp. 431–445, 2016.
DOI: 10.1007/978-3-319-44403-1_27

fall outside original training sample domain. For instance, deep neural networks can mis-classify images, when imperceptible perturbation is applied [8]. Or they can interpret images that are completely unrecognizable to humans, with almost full confidence [9].

One prevalent way to solve the afore-mentioned issues in data-driven approaches is to augment machine learning tasks with domain knowledge. Previously, domain knowledge has been applied on a wide range of applications in various forms. However, for those methods with task-dependent domain knowledge, making generalizations to new applications are usually difficult due to their labor-intensive knowledge crafting process. On the other hand, formal semantics, the formal encoding of domain knowledge, has provided a way to systematically encode, share, and reuse knowledge across applications and domains. In practice, formal semantics can support a wide range of key aspects in machine learning, data mining, and artificial intelligence techniques. For instance, formal semantics can help filter out redundant or inconsistent data, and can generate semantic rich results [10]. It can work as a set of prior knowledge or constraints, to help reduce search space and to guide search path [11].

It turns out to be an intriguing question to wonder what roles formal semantics can play in the recent trend of machine learning research, deep learning. Based on previous research, we expect formal semantics to assist in deep learning process from the following perspectives:

- Directing deep learning architecture design, resulting in learning models that better fit with current application domains.
- Assisting in representation learning processes, leading to data representations that encode critical factors from both data and formal semantics.
- Guiding training processes that capture critical semantics of data, with a representation that well generalizes a non-local prior over input space.
- Assisting in the resulting generation processes with expressive representation interpretations for high level semantics.

In this paper, we address the above goals with a semantic-rich deep learning framework that learns representations from both data distribution and formal semantics. Specifically, we propose an ontology-based deep restricted Boltzmann machine (OB-DRBM) model, in which formal ontology is used to guide architecture design of deep restricted Boltzmann machines (DRBM) [12], as well as to assist in their training and validation processes.

An ontology provides a formal representation of domain knowledge, through concepts, relationships, axiomatic constraints, and individuals. Figure 1 shows a sample ontology for news reports, recreational sports domain, used in one of our experiments. It contains a set of concepts for recreational sports in news reports, and relations between the concepts. Using a domain ontology, we can design an OB-DRBM model to learn a set of representations, each of which corresponds to a concept in the ontology. This set of representations learns to encode regularities from data with various semantic granularities for the current domain. For instance, using the news report ontology, we can learn representations that correspond to

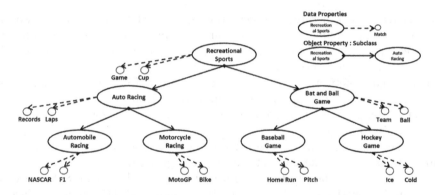

Fig. 1. A sample ontology for news reports, recreational sports domain. Each concept represents a type or category of recreational sport.

concepts, *"recreational sports," "auto racing,"* and *"automobile racing."* Furthermore, our model provides a solution to semantic rich representation learning, in that representations learned for higher level semantics can support representation learning processes for their lower level subclass semantics. For instance, as shown in Fig. 2, in our model, representation learned for *"recreational sports"* can serve as a priori for the representation learning of *"auto racing"* and *"bat and ball games."* The inspiration for our OB-DRBM design primarily comes from the robustness theory of cognitive development process in biological neural networks.

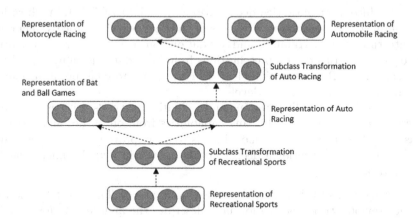

Fig. 2. Representation learning for news report ontology. For each concepts in this ontology, we use RBM layers to learn its data representation, which are further feed into RBM layers for related concepts, as auxiliary information.

In biological neural networks, many activities related to cognitive development process, such recognition and categorization, are often learned as a representation of a shared ontology [13]. Humans learn to categorize objects starting from early age by a hierarchal representation of object taxonomy in the world. The cognitive development process, for human being, usually begins with learning basic categories, such as ball, then progressively evolves into categories with more details, such as basketball [14]. Based on past knowledge and experience, a biological neural network learns by taking advantage of knowledge coming from previously learned categories, rather than learning from scratch. It leads to an efficient learning system that requires fewer samples to develop new generalization or ability promotion.

On the contrary, current deep neural networks not only require a large amount of data to make efficient learning and generalizations, but they also generalize poorly to data instances in a new but related domain [8,9]. Current deep neural networks not only require a large amount of data to make efficient learning and generalizations, but they also generalize poorly to data instances in a new but related domain [8,9]. Following the inspirations from robustness of human cognitive development process [14], we model representation learning in our model with a shared higher-level representation. We model representation learning in our model with a shared higher-level representation. We expect that representations learned for concepts at a higher semantic level, such as car and computer, can well assist in the process of learning representations, such as sedan and laptop, at more detailed semantic levels. It also renders our model the potential to explore the semantic relations between data instances, as well as the capability to learn a set of semantic rich representations with various semantic granularities.

Our contributions of this paper are as follows:

- We introduce a semantic-rich deep learning model, OB-DRBM, in which formal ontology has assisted in all stages of the deep learning process, including architecture design, training, and validation. Such architecture can learn a set of semantic-rich data representations from both data distribution and formal semantics. Representations learned correspond to concepts in a domain ontology, at various semantic levels.
- We propose corresponding training and validation methods, with assistance of inference and consistency-checking capabilities from ontologies and semantic reasoners. We show that our model leads to an improved performance, when compared with conventional deep learning models in text document classification tasks.

The remainder of this paper paper is structured as follows: Sect. 2 describes relevant previous works; Sect. 3 formally describes the architecture formulation of our model; in Sect. 4, we present our experiment result, when we apply OB-DRBM model to problems in various domains; in Sect. 5, we conclude our work by discussing potential future directions and their applications.

2   Related Work

In this paper, we propose to use formal ontology to assist in the deep learning process. Our OB-DRBM model learns a set of semantically related representations for each concept in a domain ontology. This set of representations also constitutes a formal semantics embedding based on both formal semantics and data distribution. Fields closely related to our model include, but are not restricted to, deep learning, knowledge engineering, and knowledge base embedding.

2.1 Deep Learning

In recent years, the rich set of deep neural network variations has lead to successes in numerous applications. Popular architectures of deep neural networks include, restricted Boltzmann machine (RBM) [12], convolutional neural networks (CNN) [1], and recurrent neural networks (RNN) [15]. RBM models have demonstrated exceptional performances for tasks with both labeled and unlabeled data [12]. CNN can effectively train data with topological structures and strong local correlations, such as image and speech [1]. RNN has been successfully applied on time series data and natural languages as a memory and latency model [15].

2.2 Knowledge Engineering

Knowledge engineering (KE) [16] is a research field that dedicates to develop techniques to build and reuse formal knowledge in a systematic way. In the past few decades, the proliferation of knowledge engineering has remarkably enriched the family of formal semantic representations. Ontology is one of the successful knowledge engineering advances. The encoded formal semantics in ontologies is primarily used for effectively sharing and reusing of knowledge and data. Prominent examples of domain ontologies include the Gene Ontology (GO [17]), Unified Medical Language System (UMLS [18]), and more than 300 ontologies in the National Center for Biomedical Ontology (NCBO [19]).

2.3 Knowledge Base Embedding

Recent research has developed methods to learn embeddings of knowledge base (KB) systems, such as WordNet, FreeBase, and DBPedia [20,21]. Entities in knowledge bases are embedded as low-dimensional vector representation. Syntactics, operations, and relations between entities are embedded as linear and bi-linear translations, matrix and matrix factorizations, and tensors. Bordes et al. [20] propose to learn vector-matrix embedding of knowledge base, in which knowledge bases are considered as graph models. Socher et al. [21] developed knowledge base embedding systems based on neural tensor networks for knowledge base completion. The key difference between our OB-DRBM model and previous knowledge base embedding model is, our model can learn embeddings from both data distributions and formal semantics, while previous methods learn only from a knowledge base.

3 Ontology-Based Deep Restricted Boltzmann Machine

In this section, we introduce our method to build an OB-DRBM model. We begin with a review of related techniques, including ontology in Sect. 3.1, semantic reasoner in Sect. 3.2, and restricted Boltzmann machine (RBM) in Sect. 3.3. We discuss the architecture design of our OB-DRBM model in Sect. 3.4 and corresponding training and validation methods in Sect. 3.5.

3.1 Ontology

Ontology [22] is an explicit specification of a shared conceptualization. The formal specification of an ontology can be defined as a quintuple, $\mathcal{O} = (C, P, I, V, A)$ where C, P, I, V, A are the set of classes, properties, individuals, property values and other axioms respectively [23]. Classes C, also referred to as concepts, describe the collections, concepts, types of objects and entities in a domain discourse. Properties P, also referred to as object properties, define relations between classes. Individuals I, are the instances or ground level objects of classes. Property values V, also referred to as data type properties, define features, attributes, parameter values that classes can have. Axioms A, define the ground truth of the domain discourse. The architecture design of our OB-DRBM model primarily uses the set of classes C and properties P in a domain ontology following the subclass relations in P. For concepts c, $s \in C$, we use $subclass(c, s)$, $superclass(c, s) \in P$ to denote the subclass and superclass relations between c and s. For each $c \in C$, $\pi(c) = \{s \,|superclass(s, c), s, c \in C\}$ and $\rho(c) = \{s \,|subclass(s, c), s, c \in C\}$ are used to denote the set of its subclass and superclass concepts.

3.2 Semantic Reasoner

A semantic reasoner [24] (also referred to as inference engine or reasoning engine) is a piece of software that infers logical consequences from a set of explicitly asserted facts or axioms. Prominent semantic reasoners of ontologies includes Pellet [25] and HerMit [26], and many more. It typically provides automated support for reasoning tasks such as deducting new knowledge, checking consistencies, verifying facts, and answering queries. Specifically, given a domain ontology $\mathcal{O}$ and a semantic reasoner $\mathcal{R}$, semantic reasoner can deduct an answer of query q based on the ontology $\mathcal{O}$ and axiom a, that $q = \mathcal{R}(O, a)$.

In our OB-DRBM model, the semantic reasoner is used in as a component for data semantics promotion and result validation. For a data instance $\{x, y\}$, a semantic reasoner $\mathcal{R}$ can return with promoted data instances with labels at a higher semantic level using $x \to \pi(y) = \mathcal{R}(\mathcal{O}, x \to y)$. For instance, for data instance $\{x, AutomobileRacing\}$, a semantic reasoner $\mathcal{R}$ can deduct with the valid promoted data instance, $\{x, AutoRacing\}$, using:

$$\frac{\forall\, x\; AutomobileRacing(x) \qquad \forall\, x\; AutomobileRacing(x) \to AutoRacing(x)}{\forall\, x\; AutoRacing(x)}$$

By recursively applying $x \rightarrow \pi(y) = \mathcal{R}(\mathcal{O}, x \rightarrow y)$ k times, it can deduct with promoted data at even higher semantic levels, $x \rightarrow \pi^{(k)}(y) = \mathcal{R}(\mathcal{O}, x \rightarrow y)$.

Semantic reasoner can also validate the consistency of a set of axioms. For model with multiple representations and outputs, such as our OB-DRBM model; inconsistency can happen without consistency regulations from formal semantics. For instance, for classification outputs, $o_1 = x \rightarrow MotorcycleRacing$ and $o_2 = x \rightarrow BatAndBallGames$, a semantic reasoner can deduct with inconsistency state, $\perp = \mathcal{R}(\mathcal{O}, \{o_1, o_2\})$ using:

$$
\frac{
\begin{array}{c}
\forall x \; MotorcycleRacing(x) \\
\forall x \; MotorcycleRacing(x) \rightarrow AutoRacing(x)
\end{array}
}{
\begin{array}{c}
\forall x \; AutoRacing(x)
\end{array}
}
$$

$$
\frac{
\forall x \; AutoRacing(x) \rightarrow \neg BatAndBallGames(x) \wedge BatAndBallGames(x)
}{
\perp.
}
$$

3.3 Deep Restricted Boltzmann Machine

A deep restricted Boltzmann machine (DRBM) is a deep neural network model with a stacking of many restricted Boltzmann machines (RBM) layers. RBM is a deep learning structure with bidirectionally connected binary stochastic processing units. Typically, a RBM contains a layer of visible units v and a layer of hidden units u, which are connected as a bipartite graph. RBM is a probabilistic graphic model that is based on an energy function defined on the exponential family. The joint probability that RBM assigned to visible units v and hidden units u are:

$$
p(v, h) = \frac{exp(-E(v, h))}{Z}, \tag{1}
$$

where $E(v, h)$ is a energy function defined on all RBM units, which indicates the degree of harmony of the network, Z is the partition function,

$$
Z = \sum_{u,v} exp(-E(v, h)). \tag{2}
$$

For RBM with binary visible units, $E(v, h)$ is defined as:

$$
E(v, h) = -\sum_i a_i v_i - \sum_j b_j h_j - \sum_{i,\,j} v_i h_j w_{ij}. \tag{3}
$$

For RBM with Gaussian visible units, $E(v, h)$ is defined as:

$$
E(v,\ h) = -\sum_i \frac{(v_i - a_i)^2}{2\sigma_i^2} - \sum_j b_j h_j - \sum_{i,\,j} \frac{v_i}{\sigma_i} h_j w_{ij}. \tag{4}
$$

where σ_i is the standard deviation for the Gaussian noise for visible unit i, a_i, b_i are the bias parameters for visible and hidden units and w_{ij} is the weight parameter of a RBM respectively.

Algorithm 1. OB-DRBM Architecture Design

Input: Ontology $\mathcal{O} = \{C, P, I, V, A\}$, Semantic Reasoner $\mathcal{R}$
Output: OB-DRBM structure $\mathcal{T}$

1: Let $r \in C$ be root concept of $\mathcal{O}$
2: Let s_c be an empty set
3: Add r into set s_c
4: **while** s_c is not *empty* **do**
5:    **for each** concept c in s_c **do**
6:       Initialize DRBM $\mathcal{D}_c$ for concept c
7:       Let $\rho(c) = \{s \mid s = \mathcal{R}(\mathcal{O}, subclass\ of\ c)\}$
8:       **if** $\rho(c)$ is not *empty* **then**
9:          Initialize $\mathcal{M}_c = mhmv\_layer(c, \rho(c))$
10:          Let $o_c = \{c \mid c \in C, c \notin \rho(c)\}$
11:          Let $t = \rho(c) \cup o_c$
12:          Initialize $\mathcal{S}_c = softmax\_layer(c, t)$
13:          Connect $\mathcal{S}_c, \mathcal{D}_c$ with $\mathcal{M}_c$
14:          Add $\mathcal{M}_c, \mathcal{S}_c, \mathcal{D}_c$ into $\mathcal{T}$
15:       **end if**
16:       Let $\pi(c) = \{s \mid s = \mathcal{R}(\mathcal{O}, superclass\ of\ c)\}$
17:       **if** $\pi(c)$ is not *empty* **then**
18:          Connect $\mathcal{D}_{\pi(c)}$ and $\mathcal{M}_{\pi(c)}$
19:       **end if**
20:       Let $s_c = \rho(c)$
21:    **end for**
22: **end while**
23: **return** $\mathcal{T}$

3.4   OB-DRBM Architecture Design

In this section, we present the architecture design of our OB-DRBM model. In Algorithm 1, we present the method of the model construction. Given an ontology $\mathcal{O}$ and a semantic reasoner $\mathcal{R}$, we compose the OB-DRBM model $\mathcal{T}$ following the subclass relations $\rho(c) \in P$ for each concept $c \in C$, for $C, P \in \mathcal{O}$. In Fig. 3, we show a sample OB-DRBM model following the sample news reports ontology in Fig. 1. The architecture design follows a top down process from higher level concepts to lower level concepts in C. The model construction process starts by adding the top level class $r \in C$ in the subclass hierarchy into the building sequence set s_c. For each concept $c \in s_c$, we first build a DRBM module $\mathcal{D}_c$ for the representation learning of concept c (lines 1–6). For top class r of the ontology, the DRBM module $\mathcal{D}_r$ takes only its own features as input. For other classes $c \in C, c \neq r$, the DRBM module $\mathcal{D}_c$ takes inputs from both its own features and transformed representations from its superclass modules $\mathcal{D}_{\pi(c)}$.

For each concept c and its corresponding DRBM module $\mathcal{D}_c$, we attach a semantic softmax layer $\mathcal{S}_c$, for semantic rich representation learning. The semantic softmax layer $\mathcal{S}_c$ is a layer that contains target output units at the corresponding semantic level. For each concept c, let $\rho(c)$ be the set of subclass concepts

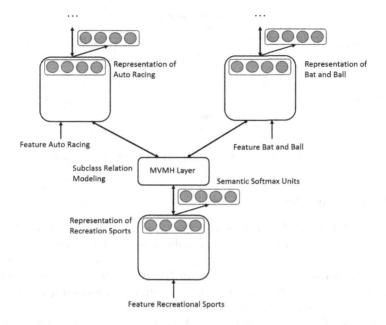

Fig. 3. An OB-DRBM architecture from news report ontology

of c. Each output unit in $\mathcal{S}_c$ corresponds to one concept in the subclass concept set, $\rho(c)$, plus one out of domain unit o_c. The unit o_c is used to model data that falls out of the domain of class c. For example, the semantic softmax layer $\mathcal{S}_{AutoRacing}$ contains three output units, for *AutomobileRacing, MotorcycleRacing, OutofDomain* respectively. For data instance $\{x, \ MotorcycleRacing\}$, the target output for $\mathcal{S}_{AutoRacing}$ is $(0, 1, 0)$. At the training phase, through semantic reasoner query $\mathcal{R}(\mathcal{O}, x \rightarrow y)$, we can convert each labeled data $l = \{x, \ y\}$ to a set of promoted data instances, $l^{(k)} = \{x, \pi^{(k)}(y)\}$, for each semantic softmax layer.

For each concept c and its corresponding DRBM module $\mathcal{D}_c$, we also attach a multiple hidden multiple visible restricted Boltzmann machine (MHMV-RBM) layer $\mathcal{M}_c$ for subclass relation modeling. As shown in Fig. 4, a MVMH-RBM layer is a RBM variation designed to model the subclass transformation from a superclass to its subclasses. In our OB-DRBM model, each DRBM module $\mathcal{D}_c$ for concept c is attached to its own semantic softmax layer $\mathcal{S}_c$. The representation learned in $\mathcal{D}_c$ encodes the high level feature abstractions for concept c. Before feeding such a representation to subclass modules $\mathcal{D}_{\rho(c)}$, the MVMH-RBM layer learns a generative representation for both the input of subclasses features and representation in $\mathcal{D}_c$. The subclass representation and raw input are further feed into subclass modules $\mathcal{D}_{\rho(c)}$ as input.

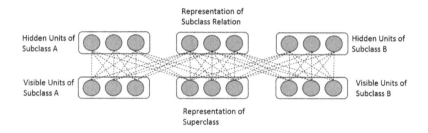

Fig. 4. A MHMV-RBM layer for subclass relation modeling

3.5   Training and Validation

We train our OB-DRBM model using a similar way as the conventional DRBM model. An OB-DRBM model was first trained with greedy module wised and layer wised contrastive divergence (CD) [27]. Then we use stochastic gradient descent across all semantic softmax output to further fine-tune our model with labeled data. In this process, we minimize the sum of cross entropy error for all softmax outputs of each concept in our model.

At the validation phase, the output of our OB-DRBM model contains a set of consistent outputs from all semantic softmax layer units. For example, $\hat{y} = \{Motorcycle\ Racing,\ Auto\ Racing,\ Recreational\ Sports\}$ is a consistent output for input data, $\{x,\ MotorcycleRacing\}$. We enforce this consistency using a logistic regression across all semantic softmax output configurations with consistency validation from a semantic reasoner. Specifically, let $\mathcal{S}$ be the set of all softmax output values, the set of outputs is,

$$\hat{y} = \operatorname*{argmax}_{s \subset \mathcal{S}} \frac{\prod\limits_{c \in s} f_c(x, w)[\mathcal{R}(\mathcal{O}, x \to s)]}{\sum\limits_{s \subset \mathcal{S}} \prod\limits_{c \in s} f_c(x, w)[\mathcal{R}(\mathcal{O}, x \to s)]}, \tag{5}$$

in which $f_c(x, w)$ is the softmax confidence value for unit c, $[\mathcal{R}(\mathcal{O}, x \to s)]$ is the activation function that ensures a valid output configuration.

4   Experiment

We present experiments on two problems related to text documents: topic classification and sentiment analysis. For selected text documents, we adopt a continuous bag of word model [28] in our experiment to convert text documents into continuous vector representations. From the frequent word set, we remove stop words and the most frequent 100 words, then keep the next 5000 most frequent words. In our experiments, we adopt the bag of word model primarily for its simplicity. We understand that the bag of word model might not be the best fit and state of the art approach for the datasets to which we have applied our method on. However, our primary goal is to explore the effect of formal semantics in deep learning process. We verify our theory by comparing our OB-DRBM model with conventional DRBM model under the same context, including data distribution, meta parameters, training time and algorithms, and so on. In all experiments, we divide the dataset into 70 % training, 15 % validation, and 15 % testing. The number of iterations over the training set was determined using early stopping according to the validation set classification error with an additional 100 iterations.

4.1 News Topic Classification

We first evaluated our model on the news topic classification problem on 20 Newsgroups dataset [29]. The data is organized into 20 different newsgroups, each corresponding to a different topic, across four domains of *computer company, recreational sports, science,* and *public talks.* We define domain ontologies for each of those domains, based on the natural taxonomy relations of the topics. We have shown one example domain ontology defined for this dataset in the recreational sports domain in Sect. 1, Fig. 1. Other domain ontologies defined for our experiments can be found in our website [30].

Table 1 gives the classification performances on the four topic domains. Our OB-DRBM model outperforms the conventional DRBM models in 3 out of the 4 domains, including *company, sports,* and *social,* In the *science* domain, DRBM model outperforms our model but only by a less than 1 % margin. This is mostly because the 4 topics in the *science* domain, *sci.electronics, sci.medicine, sci.space,* and *sci.crypt* share very few common characteristics. The best domain ontology that fits with the data is an ontology with a flat structure. In this case, our OB-DRBM model cannot benefit from the shared representation of super-class in this ontology.

Table 1. Classification performance on news topics

Topic domain	OB-DRBM	DRBM
Company	**77.46** %	75.83 %
Sports	**82.11** %	79.57 %
Social	**74.20** %	72.69 %
Science	70.46 %	**71.32** %

4.2 Sentiment Analysis Datasets

We further conduct our experiment upon document datasets on sentiment analysis tasks. We test our OB-DRBM model on the Pang/Lee movie review data [31] and sentiment analysis dataset from sentiment tree bank [32]. In both datasets, movie reviews are labeled as four categories, *positive, neutral positive, neutral negative,* and *negative.* Table 2 gives the classification performances on sentiment analysis tasks. In both datasets, our OB-DRBM model outperforms conventional DRBM model by a large margin.

Table 2. Classification performance on sentiment analysis

Data set	OB-DRBM	DRBM
Pang/Lee	**68.09** %	64.20 %
Sentiment tree bank	**60.19** %	54.45 %

4.3   Data Simulation of Formal Semantics Embedding

One primary motivation of our work is to learn a structured set of representations from both the formal semantics and the data distribution. We expect this set of semantic rich representations can encode regularities of the data at various semantic levels, such that representations of higher-level semantics can encode the common data regularities of their lower-level subclass semantics. We exam our hypothesis through visualization of the representation the learning process. In Fig. 5, we present the low-dimensional principle component analysis (PCA) embedding of the representations learned in our OB-DRBM model at various epochs of the supervised-training process. It shows the representation embeddings of three concepts with subclass relations, "*recreational sports*, "*automobile racing*," and "*bat and ball games*" in our recreational sports ontology. Before the supervised-training process, the OB-DRBM model was first trained with unsupervised-training using contrastive divergence (CD) [27].

In Fig. 5(a), we show the set of representations learned in our OB-DRBM model after the unsupervised-training. At this phase, the model can only learn from the data distribution. There is neither any data semantics, nor any formal knowledge semantics involved during this phase. After the unsupervised-training, representation learned for superclass and subclasses are roughly of the same distribution. Without the direction of formal semantics, each of the three representations plays a similar role in the model. At the 500th epoch, as shown in Fig. 5(b), the distributions of the three data representations are still similar. However, with assistance of formal semantics and labeled data, the representation of superclass, "*recreational sports*," as diverged into a different principle components compared with the representations of its subclasses.

At the 3000th epoch, as shown in Fig. 5(c), principle components of the representations for the two subclass concepts, "*moto racing*" and "*bat and ball*," start to show difference as well. Distinction of distributions has started to emerge between the representation of the superclass "*recreational sports*" and the two subclasses. At the 5000th epoch, as shown in Fig. 5(d), the model learns a set of data representations with three distinct principle components and distributions. At this stage, the representations of the superclass and the subclasses has encoded data representations with different levels of semantics. We can see through this process, how the set of semantic rich data representations influence each other through the assistance of formal semantics. When the superclass representation starts to model the common semantics of "*recreational sports*" gradually, the representations of the two subclasses were set free to learn its local semantics as well.

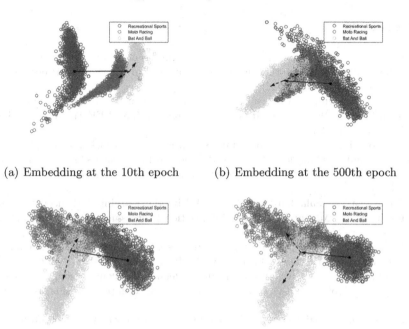

(a) Embedding at the 10th epoch   (b) Embedding at the 500th epoch

(c) Embedding at the 3000th epoch   (d) Embedding at the 5000th epoch

Fig. 5. Visualization of OB-DRBM representations.

5 Conclusion and Future Work

We have evaluated the potential of semantic rich deep learning using our OB-DRBM model. We have demonstrated that, with assistance of formal semantics, deep learning models can learn a set of semantic rich representations from both formal semantics and data. This set of representations constitute a structured embedding of formal knowledge under the data distribution. They also lead to improved performances in document classification tasks.

For future work, we would like to investigate the embedding learning of formal semantics in more forms, such as convolutional neural networks, or matrix vector embeddings. We would like to explore the potential to learn from unsupervised-data with assistance of formal semantics, as well.

Acknowledgment. This work is supported by the NIH grant R01GM103309. We acknowledge Ellen Klowden for her contributions to the manuscript. We also thank anonymous reviewers for their constructive comments, which helped improve the paper.

References

1. Krizhevsky, A., Sutskever, I., Hinton, G.E.: Imagenet classification with deep convolutional neural networks. In: Advances in Neural Information Processing Systems, pp. 1097–1105 (2012)
2. Le, H.S., Oparin, I., Allauzen, A., Gauvain, J., Yvon, F.: Structured output layer neural network language models for speech recognition. IEEE Trans. Audio Speech Lang. Process. **21**(1), 197–206 (2013)
3. Mikolov, T., Deoras, A., Kombrink, S., Burget, L., Cernockỳ, J.: Empirical evaluation and combination of advanced language modeling techniques. In: Annual Conference of the International Speech Communication Association, pp. 605–608 (2011)
4. Srivastava, N., Salakhutdinov, R.R.: Multimodal learning with deep Boltzmann machines. In: Advances in Neural Information Processing Systems, pp. 2222–2230 (2012)
5. Bengio, Y., Courville, A., Vincent, P.: Representation learning: a review and new perspectives. IEEE Trans. Pattern Anal. Mach. Intell. **35**(8), 1798–1828 (2013)
6. Zeiler, M.D., Fergus, R.: Visualizing and understanding convolutional networks. In: Proceedings of the 13th European Conference on Computer Vision, pp. 818–833 (2014)
7. Srivastava, N., Hinton, G., Krizhevsky, A., Sutskever, I., Salakhutdinov, R.: Dropout: a simple way to prevent neural networks from overfitting. J. Mach. Learn. Res. **15**(1), 1929–1958 (2014)
8. Szegedy, C., Zaremba, W., Sutskever, I., Bruna, J., Erhan, D., Goodfellow, I., Fergus, R.: Intriguing properties of neural networks (2013). arXiv preprint arXiv:1312.6199
9. Nguyen, A., Yosinski, J., Clune, J.: Deep neural networks are easily fooled: high confidence predictions for unrecognizable images. In: IEEE Conference on Computer Vision and Pattern Recognition, pp. 427–436 (2015)
10. Dou, D., Wang, H., Liu, H.: Semantic data mining: a survey of ontology-based approaches. In: IEEE International Conference on Semantic Computing, pp. 244–251 (2015)
11. Balcan, N., Blum, A., Mansour, Y.: Exploiting ontology structures and unlabeled data for learning. In: Proceedings of the 30th International Conference on Machine Learning, pp. 1112–1120 (2013)
12. Salakhutdinov, R., Hinton, G.E.: Deep Boltzmann machines. In: International Conference on Artificial Intelligence and Statistics, pp. 448–455 (2009)
13. Kolb, B., Whishaw, I.Q.: Fundamentals of Human Neuropsychology. Macmillan, London (2009)
14. Rosch, E., Mervis, C.B., Gray, W.D., Johnson, D.M., Boyes-Braem, P.: Basic objects in natural categories. Cogn. Psychol. **8**(3), 382–439 (1976)
15. Socher, R., Lin, C.C., Manning, C., Ng, A.Y.: Parsing natural scenes and natural language with recursive neural networks. In: Proceedings of the 28th International Conference on Machine Learning, pp. 129–136 (2011)
16. Russell, S.J., Norvig, P.: Artificial Intelligence: A Modern Approach, 2nd edn. Pearson Education, Upper Saddle River (2003)
17. Consortium, T.G.O.: Creating the gene ontology resource: design and implementation. Genome Res. **11**(8), 1425–1433 (2001)
18. Lindberg, D., Humphries, B., McCray, A.: The unified medical language system. Methods Inf. Med. **32**(4), 281–291 (1993)

19. NCBO: The National Center for Biomedical Ontology. http://www.bioontology.org/
20. Bordes, A., Weston, J., Collobert, R., Bengio, Y.: Learning structured embeddings of knowledge bases. In: Proceedings of the AAAI Conference on Artificial Intelligence, pp. 301–306 (2011)
21. Socher, R., Chen, D., Manning, C.D., Ng, A.: Reasoning with neural tensor networks for knowledge base completion. In: Advances in Neural Information Processing Systems, pp. 926–934 (2013)
22. Gruber, T.R.: A translation approach to portable ontology specifications. Knowl. Acquis. **5**(2), 199–220 (1993)
23. Wimalasuriya, D.C., Dou, D.: Components for information extraction: ontology-based information extractors and generic platforms. In: Proceedings of the 19th ACM Conference on Information and Knowledge Management, pp. 9–18 (2010)
24. Dentler, K., Cornet, R., Ten Teije, A., De Keizer, N.: Comparison of reasoners for large ontologies in the OWL 2 EL profile. Seman. Web **2**(2), 71–87 (2011)
25. Sirin, E., Parsia, B., Grau, B.C., Kalyanpur, A., Katz, Y.: Pellet: a practical OWL-DL reasoner. Web Seman. **5**(2), 51–53 (2007)
26. Motik, B., Shearer, R., Horrocks, I.: Hypertableau reasoning for description logics. J. Artifi. Intell. Res. **36**, 165–228 (2009)
27. Hinton, G.E., Salakhutdinov, R.R.: Reducing the dimensionality of data with neural networks. Science **313**(5786), 504–507 (2006)
28. Mikolov, T., Chen, K., Corrado, G., Dean, J.: Efficient estimation of word representations in vector space (2013). arXiv preprint arXiv:1301.3781
29. Lang, K.: Newsweeder: learning to filter netnews. In: Proceedings of the International Conference on Machine Learning, pp. 331–339 (1995)
30. AIMLAB: Ontologies. http://aimlab-server.cs.uoregon.edu/ontologies
31. Pang, B., Lee, L.: Seeing stars: exploiting class relationships for sentiment categorization with respect to rating scales. In: Proceedings of the Annual Meeting on Association for Computational Linguistics, pp. 115–124 (2005)
32. Socher, R., Perelygin, A., Wu, J.Y., Chuang, J., Manning, C.D., Ng, A.Y., Potts, C.: Recursive deep models for semantic compositionality over a sentiment treebank. In: Proceedings of the Conference on Empirical Methods in Natural Language Processing, p. 1642 (2013)

Author Index

Aldana-Montes, José F. I-405
Alwan, Ali Amer II-377
Amagasa, Toshiyuki I-336
Amagata, Daichi I-37
Amer, Nawal Ould II-235
Andersen, Ove II-437

Baba, Satoshi II-27
Bah, Ashraf II-410
Baioco, Gisele Busichia I-355
Batko, Michal II-185
Beer, Martin II-244
Behzadnia, Peyman II-315
Berdun, Luis II-335
Bezerra, Karen Aline Alves I-185
Bhatnagar, Vasudha I-287
Bischoff, Holger II-137
Boissier, Martin II-137
Bou, Savong I-336
Bouaziz, Rafik II-167
Boukhalfa, Kamel I-223
Boukhelef, Djillali I-223
Boukhobza, Jalil I-223
Brahmia, Safa II-167
Brahmia, Zouhaier II-167
Brzykcy, Grażyna II-275

Campo, Marcelo II-335
Cao, Jialiang II-326
Cao, Zhongsheng II-454
Caroprese, Luciano II-368
Carterette, Ben II-410
Casanova, Marco A. II-68
Chao, Dong I-321
Chao, Han-Chieh I-71
Chelghoum, Kamel I-136
Chen, Guihai II-326
Chen, Hanxiong II-87
Cheng, Kai II-426
Chevalier, Jules II-287
Cuzzocrea, Alfredo I-185

Dash, Debasis I-287
de Almeida, Eduardo Cunha I-207

de La Robertie, Baptiste II-19
de Melo, Gerard I-370
de Oliveira Moraes, Regina Lúcia I-355
Desmontils, Emmanuel II-303
do Nascimento, Ben Hur Bahia I-185
Dong, Yuyang II-87
Dou, Dejing I-431
Dylla, Maximilian I-370

Eisenreich, Katrin II-137

Faghihi, Usef I-88
Färber, Franz II-137
Faron-Zucker, Catherine II-52
Faust, Martin II-137
Fegaras, Leonidas I-240, I-305
Filho, Edson Ramiro Lucas I-207
Fomichov, Vladimir A. I-416
Fournier-Viger, Philippe I-71, I-88
Fugini, Mariagrazia I-121
Furuse, Kazutaka II-87

Gan, Wensheng I-71
Gao, Xiaofeng II-326
Gao, Yunjun I-153
García-Godoy, María Jesús I-405
Georgoulas, Konstantinos I-169
Géry, Mathias II-235
Goda, Kazuo II-389
Golenberg, Konstantin II-399
Grandi, Fabio II-167
Gravier, Christophe II-287
Guedes, Gustavo Bartz I-355
Guesmi, Soumaya II-11

Hara, Takahiro I-37
Haughian, Gerard II-152
Hayamizu, Yuto II-389
He, Guoliang I-272
He, Qinming I-153
Hoang, Bao-Thien I-136
Horie, Shintaro II-102
Hsu, Jing Ouyang II-254

Hsu, Wynne   I-3
Huang, Hao   I-153

Ibrahim, Hamidah   II-377
Ienco, Dino   I-387
Ingalalli, Vijay   I-387

Jaśkowiec, Krzysztof   I-257
Jenkins, Elliot   II-201
Jia, Xianyan   I-3

Kacem, Imed   I-136
Kantere, Verena   II-355
Kato, Chihiro   II-389
Kaur, Sharanjit   I-287
Kayem, Anne V.D.M.   I-105
Keller, Marvin   II-137
Keyaki, Atsushi   II-216
Kießling, Werner   II-3
Kim, Dongsun   I-207
Kiritoshi, Keisuke   II-102
Kitagawa, Hiroyuki   I-321, I-336, II-87
Kitsuregawa, Masaru   I-55, II-389
Kluska-Nawarecka, Stanislawa   I-257
Knottenbelt, William J.   II-152
Komatsuda, Takuya   II-216
Kotidis, Yannis   I-169
Krishna Reddy, P.   I-55

Laforest, Frédérique   II-287
Latiri, Chiraz   II-11
Le Traon, Yves   I-207
Lee, Mong Li   I-3
Legien, Grzegorz   I-257
Leme, Luiz A.P.Paes   II-68
Li, Xuhui   II-263
Li, Yifei   I-272
Li, Yuanxiang   I-272
Liao, Husheng   II-445
Lin, Jerry Chun-Wei   I-71, I-88
Liu, Jia   II-445
Lloret-Gazo, Jorge   II-343
Lopes, Giseli Rabello   II-68
López-Camacho, Esteban   I-405
Lowd, Daniel   I-431
Lu, Wei   I-153

Ma, Qiang   II-27, II-102
Marilli, Guido   I-121

Martinez-Gil, Jorge   II-295
Meinel, Christoph   I-105, II-118
Meira, Jorge Augusto   I-207
Menendez, Elisa S.   II-68
Michel, Franck   II-52
Miyazaki, Jun   II-216
Molli, Pascal   II-303
Montagnat, Johan   II-52
Muhammad Fuad, Muhammad Marwan   II-418
Mulhem, Philippe   II-235

Nakayama, Yuki   I-37
Nalepa, Filip   II-185
Nassopoulos, Georges   II-303
Navas-Delgado, Ismael   I-405
Nawarecki, Edward   I-257
Ngu, Anne H.H.   II-254
Nishimura, Kazuya   I-321
Nunes, Bernardo P.   II-68

Osman, Rasha   II-152

Paik, Hye-young   II-254
Pankowski, Tadeusz   II-275
Paoletti, Alejandra Lorena   II-295
Pitarch, Yoann   II-19
Plattner, Hasso   II-137
Poncelet, Pascal   I-387

Qian, Tieyun   II-263
Qin, Yongrui   II-37

Razorenov, Alexander A.   I-416
Ren, Zhaochun   I-370
Ribeiro, Leonardo Andrade   I-185
Rodríguez, Guillermo   II-335

Saad, Nurul Husna Mohd   II-377
Safiya, Al Sharji   II-244
Sagiv, Yehoshua   II-399
Saxena, Rakhi   I-287
Schewe, Klaus-Dieter   II-295
Schwalb, David   II-137
Serrano-Alvarado, Patricia   II-303
Shaabani, Nuhad   II-118
Shaikh, Salman Ahmed   I-321
Sheng, Quan Z.   II-37
Sidi, Fatimah   II-377

Sniezynski, Bartlomiej   I-257
Soria, Álvaro   II-335
Subercaze, Julien   II-287

Teimourikia, Mahsa   I-121
Teste, Olivier   II-19
Teyseyre, Alfredo   II-335
Theobald, Martin   I-370
Torp, Kristian   II-437
Trabelsi, Chiraz   II-11
Tseng, Vincent S.   I-88
Tu, Yi-Cheng   II-315

Uday Kiran, R.   I-55
Uruchurtu, Elizabeth   II-244

Venkatesh, J.N.   I-55
Vester, C.T.   I-105
Vidal, Vânia M.P.   II-68

Wang, Hao   I-431
Wang, Song   I-153
Wang, Wei   I-20
Wang, Xiaorui   II-315
Wang, Yafang   I-370

Wenzel, Florian   II-3
Wilk-Kołodziejczyk, Dorota   I-257
Wu, Cheng-Wei   I-88
Wu, Shuangke   I-153

Xia, Xuewen   I-272
Xu, Jianliang   I-20

Yaakob, Razali   II-377
Yang, Yanyan   II-201
Yao, Lina   II-37
Ying, Shi   I-153
You, Zhenni   II-263
Yuan, Wei   II-315

Zeng, Bo   II-315
Zezula, Pavel   II-185
Zhan, Liming   II-254
Zhang, Caicai   II-454
Zhang, Yatao   II-326
Zhou, Xiaoling   I-20
Zhu, Hong   II-454
Zhu, Peisong   II-263
Zumpano, Ester   II-368

Printed in the United States
By Bookmasters